Lecture Notes in Networks and Systems

Volume 1148

The series "Lecture Notes in Networks and Systems" publishes the latest developments in Networks and Systems—quickly, informally and with high quality. Original research reported in proceedings and post-proceedings represents the core of LNNS.

Volumes published in LNNS embrace all aspects and subfields of, as well as new challenges in, Networks and Systems.

The series contains proceedings and edited volumes in systems and networks, spanning the areas of Cyber-Physical Systems, Autonomous Systems, Sensor Networks, Control Systems, Energy Systems, Automotive Systems, Biological Systems, Vehicular Networking and Connected Vehicles, Aerospace Systems, Automation, Manufacturing, Smart Grids, Nonlinear Systems, Power Systems, Robotics, Social Systems, Economic Systems and other. Of particular value to both the contributors and the readership are the short publication timeframe and the world-wide distribution and exposure which enable both a wide and rapid dissemination of research output.

The series covers the theory, applications, and perspectives on the state of the art and future developments relevant to systems and networks, decision making, control, complex processes and related areas, as embedded in the fields of interdisciplinary and applied sciences, engineering, computer science, physics, economics, social, and life sciences, as well as the paradigms and methodologies behind them.

Indexed by SCOPUS, EI Compendex, INSPEC, WTI Frankfurt eG, zbMATH, SCImago.

All books published in the series are submitted for consideration in Web of Science.

For proposals from Asia please contact Aninda Bose (aninda.bose@springer.com).

Mohuya Chakraborty ·
Shambhu Prasad Chakrabarty · Ana Penteado ·
Valentina Emilia Balas
Editors

Proceedings of 5th International Ethical Hacking Conference

Leveraging AI to Law (eHaCON 2024)

Springer

Editors
Mohuya Chakraborty
Cybersecurity Centre of Excellence
Institute of Engineering and Management
University of Engineering and Management
Kolkata, India

Shambhu Prasad Chakrabarty
Department of Law (IIIJS)
University of Engineering and Management
Kolkata, India

Ana Penteado
University of Notre Dame
Sydney, NSW, Australia

Valentina Emilia Balas
Department of Automatics and Applied Software
Aurel Vlaicu University of Arad
Arad, Romania

ISSN 2367-3370 ISSN 2367-3389 (electronic)
Lecture Notes in Networks and Systems
ISBN 978-981-97-8456-1 ISBN 978-981-97-8457-8 (eBook)
https://doi.org/10.1007/978-981-97-8457-8

This Springer imprint is published by the registered company Springer Nature Singapore Pte Ltd.
The registered company address is: 152 Beach Road, #21-01/04 Gateway East, Singapore 189721, Singapore

Preface

The advancements in the technologies of computer networks, the Internet, and artificial intelligence have changed the panorama of cybersecurity. The hackers are continuously in the process of exploring new techniques to access vulnerable cyber systems and create havoc. Cyberlaw has a great role to play in saving the world from malicious activities on the Internet. The main objective of this conference is to raise public awareness of cybercrimes that occur in the modern world and the legal steps that can be taken to prevent them. Through four distinct tracks—the Evolution of AI, AI and Law, AI and Cybersecurity, and AI and Industry 6.0—keynote addresses and research paper presentations are delivered. These tracks have covered a variety of research topics, including deep learning, machine learning, data analytics, intelligent systems, big data, prompt engineering, augmented intelligence, IPR, crime prevention, and cybercrimes. To create a secure society, a platform for discussion and exchange of ideas about various topics including data privacy, deep fakes, blockchain, cryptography, penetration testing, cyber-physical systems, digital twins, ethics, supply chain management, malware analysis, and risk management are established. Researchers from academia and industry, both domestically and internationally, participated in the discussions. The proceedings highlight every aspect of the conference in the hopes that, soon, fresh research findings may lead to the development of workable solutions for safe systems. This book is split into five parts. Part I discusses the keynote talks on various topics such as The Global Readiness for Cybersecurity and AI, The Future of Ethical Hacking, Artificial Intelligence and the Human Brain, Artificial Intelligence and Inventorship, and Explainable AI. Part II presents "Evolution of AI", Part III highlights "AI and Cybersecurity" while Part IV and Part V deal with "AI and Law" and "AI and Industry 6.0", respectively.

Part I: Keynote

Keynote 1: This paper examines global readiness for cybersecurity and AI, guided by insights from the Cisco Cybersecurity Readiness Index and the Cisco Global AI

Readiness Index. The study delves into the disparities and strengths across regions and industries, revealing a complex tableau of readiness that underscores significant challenges and untapped opportunities. The paper highlights the symbiotic relationship between cybersecurity and AI through a comprehensive analysis of global standards, frameworks, and readiness indicators. It explores the strategic intersections where AI can enhance cybersecurity defenses and, conversely, where cybersecurity measures are pivotal for safeguarding AI systems.

Keynote 2: This paper looks at the future of ethical hacking, with a specific focus on how artificial intelligence (AI) could be involved and how this might affect the rules and regulations around it. The paper looks at how new technologies like AI can help improve cybersecurity and how they present new challenges. The article stresses the need for creative ideas and strong testing plans to make sure AI systems are secure and can handle cyber threats. It also looks at how the law and AI intersect and why we need new laws that can keep up with the changing world of cyber threats.

Keynote 3: This paper discusses the concept of humanity in recent inventions named artificial brains implanted in human brains for medical purposes such as alleviation of the symptoms of Alzheimer's disease, memory loss, a verbal disorder as aphasia, and physical limitations. There is a possibility that with this reinforcement learning, the boundaries between machines and humans are merged into one. In an artificial neural net architecture that naturally incorporates human logic and factual analysis, new boundaries of intelligence have unveiled two major philosophical issues for legal policies in artificial intelligence: what is human and, subsequently, the eligibility of what is considered human.

Keynote 4: This paper explores the intersection of artificial intelligence (AI) and inventorship, focusing on whether AI can be legally recognized as an inventor. It examines key judicial decisions in the DABUS cases from the United States, United Kingdom, and Australia, which all reject AI as an inventor under current patent laws. The chapter delves into the legal definitions of inventorship, the role of human intentionality and creativity, and the potential need for legislative changes. It also discusses the implications of AI-generated inventions on patent ownership, novelty standards, and the broader intellectual property system.

Keynote 5: This paper discusses "Explainable AI" (XAI), a nomenclature that describes the capacity of AI systems to give humans intelligible justifications for their choices and actions. This is essential for fostering trust in AI, particularly in high-stakes domains like criminal justice, healthcare, or financial risk where choices have a direct influence on people's lives. With the application of XAI approaches, AI models should become more accountable, transparent, and comprehensible so that users may comprehend the reasoning behind a given decision. XAI gives users a better understanding of the decision-making process, which increases user trust and facilitates more efficient usage of AI systems. Black-box machine learning models can be made interpretable with the help of XAI approaches like LIME (Local Interpretable Model-agnostic Explanations) and SHAP (SHapley Additive exPlanations).

Part II: Evolution of AI

In this part, seven papers present various application areas of AI including corporate credit rating, online dispute resolution, skin psoriasis and breast cancer detection, design optimization of monopole antenna, and deep fakes.

Part III: AI and Cybersecurity

In this part, there are five papers including blockchain application in security management in next-generation networks, blockchain and physical unclonable function-based multifactor authentication for IoT devices, AI-based security in banking, criminal liability of AI-generated deep fakes, and ML-based IDS in IoT.

Part IV: AI and Law

This part comprises nine papers that deal with ethical issues in technology, copyright infringement liability, cybercrime prevention, freedom of contract for app navigation, AI and IPR, the impact of AI on jurisdictional challenges, AI-based criminal justice system, copyright claims, and legal consideration for connected failure in healthcare, IoT, and AI.

Part V: AI and Industry 6.0

This part is all about the application of AI in industrial areas. Five papers depict AI used for customer retention, CRM/ERP, judicial decision-making, smart governance, and threat detection and prevention using AI.

The editors believe that this book is exclusive and momentous in that it provides various keynote discussions and research papers on end-to-end standpoints on the applications of AI to solve cybersecurity issues and AI-based cybercrime detection from today's perspective. Within the umbrella of AI, research papers on deep learning, machine learning, neural networks, and data analytics have been included that may be quite beneficial to a large group of scientists, engineers, advocates, researchers, cybersecurity community about the ever-evolving era of cybercrime and application of AI to mitigate them.

Kolkata, India — Mohuya Chakraborty
Kolkata, India — Shambhu Prasad Chakrabarty
Sydney, Australia — Ana Penteado
Arad, Romania — Valentina Emilia Balas

Message from Chief Patron

A very warm welcome to the eHaCON 2024 International Ethical Hacking Conference, which is the 5th of its series. eHaCON 2024 is an annual event of the Cybersecurity Centre of Excellence of the University of Engineering & Management, Kolkata, India. The main objective of this flagship event is to provide a platform to leading researchers from academia and practitioners from industry in India and abroad to share their innovative ideas, experiences, and cutting-edge research in the areas of cybersecurity, ethical hacking, and network security. This year the event is organized in association with the Department of Law of the university and hence covers various legal-related matters in the cybersecurity domain.

eHaCON 2024 has been made possible with the generous support of the grant-in-aid received from the university and technical sponsors Notre Dame University, Australia, Indian School of Ethical Hacking, India, National Institute of Technical Teachers Training and Research, Kolkata, India, CISCO, Sydney, New South Wales, Australia, Nanyang Technological University, Singapore, and Ixonn Group, Australia. I thank all the sponsors, associates, supporters, and the members of University of Engineering and Management for the grand success of this event.

Dr. Satyajit Chakrabarti
Chancellor
University of Engineering and Management, Kolkata, India
March 2024

Message from Patrons

A very warm welcome to the eHaCON 2024, the 5th International Ethical Hacking Conference—Leveraging AI to Law. eHaCON 2024 is an annual event of the Cybersecurity Centre of Excellence at the University of Engineering and Management. We convey our sincere gratitude to the organizers of eHaCON 2024 for conducting such a wonderful event where participants from all over the world presented research papers on various areas such as AI and Law, AI and Cybersecurity, Evolution of AI, and Industry 6.0.

Wishing the conference a continued success and inspiration in the endeavors of the organizers. May this conference foster lasting connections and empower us all to create positive change.

Dr. Sajal Dasgupta
Vice-Chancellor
University of Engineering and Management, Kolkata, India
March 2024

Dr. Satyajit Chakrabarti
Pro Vice-Chancellor
University of Engineering and Management, Kolkata, India
March 2024

Message from Conference General Chair

It was my great pleasure to extend an affable welcome to all the attendees of eHaCON 2024—5th International Ethical Hacking Conference—Leveraging AI to Law organized by the Cybersecurity Centre of Excellence in association with the Department of Law, University of Engineering & Management (UEM), Kolkata held at the University of Engineering & Management, Kolkata, New Town campus on March 27–28, 2024, The aim of eHaCON 2024 was to give an open platform where people were able to discuss the implication of new technologies for secured society. The conference was a balanced mix consisting of technical paper presentations, live demonstrations, workshops, gaming competition, and online coding competition on hacking. The goal was to kick-start efforts to fully automate cyber defense. The most substantial new findings about computer network attacks and defenses, commercial security solutions, and pragmatic real-world security experiences are presented in a two-day informative research paper presentations and keynote speeches. Research papers are submitted from several countries across the globe.

I express my sincerest thanks to the keynote speakers, Dr. Ana Penteado from Notre Dame University, Australia, Dr. Chandan Chakraborty from NITTR, Kolkata, India, Dr. Althaf Marsoof from Nanyang Technological University, Singapore, Mr. Marcelo Bussacarini from Cisco, Australia, Dr. Gleuto Serafim form Ixonn Group, Australia, and Mr. Sandeep Sengupta from Indian School of Ethical Hacking, India for delivering keynote speeches on various cutting-edge topics of legal and security aspects in digital world.

I am immensely grateful to Dr. Shambhu Prasad Chakraborty for convening the entire program smoothly. With the help of an excellent committee of international and national experts, very rigorous principles were followed for selecting only the very best technical papers out of a large number of submissions in order to maintain the high quality of the conference.

I am appreciative to Dr. Rajendrani Mukherjee for performing the outstanding job of conducting the technical programs and acting as a reviewer of several papers. I would also like to thank Prof. Shrabana Chattopadhyay and Prof. Victor Nayak for outstanding contributions to floor management. Participants were there from various schools, colleges, government offices, and industries. Hope they were immensely benefited from the two-day workshop on ethical hacking. I would like to thank Prof. Arin Mukherjee for arranging event participation from various organizations and preparation of calls for papers. In addition to him, I would thank the student coordinators Mr. Soumarghya Dhara, Agniva Bagchi, Divyanshi Srivastava, and Palash Das who left no stone unturned in giving proper shape to the website and the banner.

My sincere thanks are due to University of Engineering & Management for co-sponsoring this conference as well as providing both financial and infrastructural supports. I am grateful to all the members of the advisory and technical committees comprising professors and researchers from various parts of the world like Romania, Australia, India, Singapore, and the USA for providing their excellent service. I am also thankful to all the session chairs, co-chairs, rapporteurs, and local organizing committee for their hard work and effort to make this conference a grand success.

Thanks to Springer for technical support and, last but not least, thanks to all the participants and authors. I hope that they appreciated the conference and I anticipate that they liked our culturally lively city of joy—Kolkata as well!

Dr. Mohuya Chakraborty
Director
Cybersecurity Centre of Excellence
University of Engineering and Management, Kolkata, India
March 2024

Message from Convenor

On behalf of the Organizing Committee of eHaCON 2024 International Ethical Hacking Conference—Leveraging AI to Law, it is my pleasure to welcome the attendees to the University of Engineering & Management, Kolkata, India.

The conference consists of 4 technical sessions with 32 contributed papers and eight keynote addresses. eHaCON 2024 Program Committee, comprising distinguished members worked hard to organize the technical program. Following the rigorous review process, out of about 75 submissions, only 32 concept papers were accepted for presentation in the technical sessions.

Behind every successful event there lies the hard work, commitment, and dedication of many personalities. Firstly, I wish to thank the entire program committee for the excellent job it did in organizing the technical sessions. Special thanks are due to all the reviewers for their obligation in reviewing the papers within a very short time. I am indebted to the faculty members of the university for managing the conference on cybersecurity and law where participants from various schools, colleges, and industries, government officials are benefited. I would like to thank Prof. Pallabi Sengupta for her immense contribution to paper formatting. Special thank goes to the conference chair Dr. Mohuya Chakraborty for giving us immense support and encouragement throughout this period. Once again, I hope that all the delegates from India and abroad will find the program beneficial and will enjoy the historic city of joy—Kolkata.

I sincerely thank Dr. Satyajit Chakrabarti, Chancellor, Dr. Sajal Dasgupta, Vice-Chancellor and Dr. Satyajit Chakrabarti, Pro Vice-Chancellor of the University of Engineering & Management, for their constant support throughout the event. Last but not the least I thank all our delegates from various industries as well as academia for participation without whom the conference would not have been possible.

Dr. Shambhu Prasad Chakrabarty
Dean
Department of Law
University of Engineering and Management, Kolkata, India
March 2024

Programme Committee

Chief Patron

Dr. Satyajit Chakrabarti	Chancellor, University of Engineering & Management, India

Patrons

Dr. Sajal Dasgupta	Vice-Chancellor, University of Engineering & Management, India
Dr. Satyajit Chakrabarti	Pro Vice-Chancellor, University of Engineering & Management, India

General Conference Chair

Dr. Mohuya Chakraborty	Director, Cybersecurity Centre of Excellence, University of Engineering & Management, Kolkata, India

Advisory Committee

Dr. Valentina Emilia Balas	Aurel Vlaicu Univ. of Arad, Romania
Dr. Ana Penteado	Notre Dame University, Australia
Dr. Angappa Gunasekaran	California, State University, Baskerfield, California
Dr. S. K. Ramesh	California State University, Northridge
Dr. Chandan Chakraborty	NITTR, India

Dr. Dorothea La "Chon" Abraham	Mason School of Business at William & Mary, USA
Dr. Gleuto Serafim	Ixonn Group, Australia
Mr. Marcelo Bussacarini	Cisco, Australia
Dr. Pabitra Mitra	IIT, Kharagpur, India
Dr. Justin Paul	University of Puerto Rico, USA
Dr. Mohuya Chakraborty	University of Engineering and Management, India
Dr. Rajiv Ganguly	University of Engineering and Management, India

Keynote Speakers

Mr. Marcelo Bussacarini	Global Security Brand Strategist, CISCO, Sydney, New South Wales, Australia
Dr. Gleuto Serafim	Director and Founder at Ixonn Group, Australia
Mr. Sandeep Sengupta	Indian School of Anti-Hacking, India
Dr. Chandan Chakraborty	Professor at the Department of Computer Science and Engineering, National Institute of Technical Teachers Training and Research, Kolkata, India
Dr. Ana Penteado	Adjunct Associate Professor, Notre Dame University, Australia
Dr. Althaf Marsoof	Deputy Head and Assistant Professor of Law, Division of Business Law, College of Business, Nanyang Technological University, Singapore
Dr. Mohuya Chakraborty	Director, Cybersecurity Centre of Excellence (CCoE), University of Engineering and Management, Kolkata, India
Dr. Justin Paul	Professor, Business Administration, University of Puerto Rico, USA

Editorial Board

Technical Program Committee

Dr. Mohuya Chakraborty	University of Engineering and Management, India
Dr. Chandan Chakraborty	NITTTR, India
Dr. Shambhu Prasad Chakrabarty	University of Engineering and Management, India
Dr. Ana Penteado	Notre Dame University, Australia
Dr. Sukanya Mukherjee	UEMK, India
Dr. Rajendrani Mukherjee	UEMK, India
Mr. Shankhadeep Chatterjee	UEMK, India
Ms. Shrabana Chattopadhyay	UEMK, India
Mr. Victor Nayak	UEMK, India
Dr. Arup Kumar Chattopadhyay	IIT Madras, India

Organizing Committee

Dr. Mohuya Chakraborty	University of Engineering and Management, India
Dr. Shambhu Prasad Chakrabarty	University of Engineering and Management, India
Mr. Shankhadeep Chatterjee	UEMK, India
Ms. Shrabana Chattopadhyay	UEMK, India
Ms. Abhishikta Basu	UEMK India
Mr. Victor Nayak	UEMK, India
Mr. Arin Mukherjee	UEMK

Contents

About the Editors

Mohuya Chakraborty presently holds the post of Director of the Cybersecurity Centre of Excellence (CCoE), Dean of faculty Welfare and HRDC, and Professor of the Department of Computer Science and Engineering, University of Engineering & Management (UEM), Kolkata. Before this, she held the position of head of the department of CSE and IT for the tenure of 13 years. She has done B. Tech and M. Tech from the Institute of Radio Physics and Electronics, Calcutta University in the years 1994 and 2000, respectively, and Ph.D. (Engg.) in the field of Mobile Computing from Jadavpur University in 2007. She has a Chartered Management Institute (CMI Level 5) certification in Leadership and Management from Dudley College, London, UK, 2019. She is the topper of M. Tech and subsequently received the prestigious Paresh Lal Dhar Bhowmik Award from Calcutta University. She was also the recipient of the Rashtriya Gaurav Award in 2019. She has been awarded Certification of Appreciation as one of the Top 50 Distinguished HODs CSE/IT Department in Higher Education Across India for the year 2019 by the Academic Council of uLektz in July 2020. She is a Certified Publons Academy Mentor. She is a member of the editorial board of several international journals. She has published 4 patents and over 80 research papers in reputed international journals and conferences. She is the editor of the book entitled "The Essence of Network Security: An End-To-End Panorama" published by Springer in August 2020 in the series "Lecture Notes in Networks and Systems". She is the volume editor of the Contributed Books; "Proceedings of International Ethical Hacking Conferences (eHaCON-2018 and eHaCON-2019)", in Springer series "Advances in Intelligent Systems and Computing book series (AISC volume 811)", published by Springer. Dr. Chakraborty is the reviewer of the European Journal of Information Security, Taylor & Francis, and has recently acted as associate editor of the same. Dr. Chakraborty participated in many international conferences as an Organizer, General Chair, Session Chair, and member of Steering, Advisory, or International Program Committees. She has handled many research projects worth INR 35 Lakhs funded by the DST, AICTE, CSIR, and NRDC, Government of India, and consultancy projects by non-governmental organizations worth INR 15 Lakhs. Dr. Chakraborty has published 80+ papers in high-impact journals, books, book chapters, and conference proceedings. She is also the external Ph.D. thesis (Foreign) examiner at the University of Vaasa, Finland. Her research areas include artificial intelligence, blockchain, network security, cognitive radio, brain-computer interface, parallel computing, etc. She is a senior member of IEEE.

Shambhu Prasad Chakrabarty is Dean of the Department of Law, University of Engineering and Management, Kolkata, India. He was Head and Research Fellow at CRSGPP, WBNUJS before his current assignment. Dr. Chakrabarty received his LL.B. & LL.M. from Calcutta University and Ph.D. from the University of North Bengal. He has taught and researched for over 20 years at Calcutta University, Amity Law School, and West Bengal National University of Juridical Sciences. He has worked extensively in the areas of Intellectual Property Rights, Traditional Knowledge, Climate Change, and Technology Law. Three Doctoral Scholars have been successfully co-supervised by him. He has authored and edited 7 books and written over 50 articles and book chapters including in the Liverpool Law Review, Environmental Management, and is currently co-editing two books to be published by Thomson Reuters and Springer Nature in 2024. He acted as Chief Editor of NUJS Journal of Regulatory Studies from August 2019–2022. He has presented research papers at more than 60 national and international conferences and invited as a resource person in more than 90 national and international events including the Government of India, the Government of West Bengal, the Law and Society Association, Washington, D.C.; Government of Western Australia; Australian National University; Notre Dame University, Sydney; Pattimura University, Indonesia; Daffodil International University Bangladesh; Tribhuvan University, Nepal, Binghamton State University, New York; University of Wroclaw, Poland, Foreign Trade University, Vietnam, Nanyang Business School, Singapore. He was WIPO Academy Fellow in Malaysia in 2019. He was awarded with Gold Medal for winning the Youth Parliament Competition organized by the Government of West Bengal. He topped his school at plus two level and college in his LL.B. Degree. Dr. Chakrabarty is a member of the prestigious Law and Society Association, USA, the Commonwealth Lawyers Association, UK, and IPIRA.

Dr. Ana Penteado is Honorary Adjunct Associate Professor at the University of Notre Dame, School of Law in Sydney, Australia, holds a Ph.D. from Macquarie University, an LLM from the University of California at Berkeley, Berkeley Law School, and an LLB from Mackenzie University, School of Law. Dr. Penteado combines a hybrid academic experience and a professional legal career spanning four jurisdictions with a concentration on international law, transnational intellectual property law, and public policy applied to Indigenous Knowledge. She has published on aspects of patentability involving drug discovery, COVID-19, compulsory licenses, drug development, and vaccine discovery in Uncertainty and Intellectual Property Rights (2020). Other topics of her research interests are native flora protection and fair and just commercialization via patentable subject matter for associated cultural assets. Dr. Penteado has published in peer-reviewed journals and provided submissions to the Australian government on intellectual property issues. Her projects primarily involve public policy on intellectual property theory and practice, climate change, and metrics. Lately, her academic research has focused on interdisciplinary studies related to Indigenous Knowledge for a collaborative mission across disciplines to promote novel public policies involving Indigenous, tribal, and local communities as active stakeholders in advance of social justice. She is also on the editorial team of a forthcoming book on Traditional Knowledge and Climate Change: An Environment Impact on Landscape and Communities with Dr. Shambhu Chakrabarty and Dr. Owais Hassan Shaikh, an upcoming publication by Springer Nature. Currently, she is involved with ecological metrics, and sustainability in an article to be submitted to a special issue on Biodiversity and Economy in Environmental Indicators by Elsevier.

Valentina Emilia Balas is currently Full Professor in the Department of Automatics and Applied Software at the Faculty of Engineering, "Aurel Vlaicu" University of Arad, Romania. She holds a Ph.D. in Applied Electronics and Telecommunications from Polytechnic University of Timisoara. Dr. Balas is the author of more than 350 research papers in refereed journals and International Conferences. Her research interests are in Intelligent Systems, Fuzzy Control, Soft Computing, Smart Sensors, Information Fusion, Modeling and Simulation. She is Editor-in-Chief of the International Journal of Advanced Intelligence Paradigms (IJAIP) and International Journal of Computational Systems Engineering (IJCSysE), member in Editorial Board member of several national and international journals, and is an evaluator expert for national, and international projects, and Ph.D. Thesis. Dr. Balas is the director of the Intelligent Systems Research Centre at Aurel Vlaicu University of Arad and Director of the Department of International Relations, Programs, and Projects at the same university. She served as General Chair of the International Workshop Soft Computing and Applications (SOFA) in eight editions 2005–2020 held in Romania and Hungary. Dr. Balas participated in many international conferences as Organizer, Honorary Chair, Session Chair, and member of Steering, Advisory, or International Program Committees. She is a member of EUSFLAT, SIAM and Senior Member IEEE, member in TC—Fuzzy Systems (IEEE CIS), chair of the TF 14 in TC—Emergent Technologies (IEEE CIS), member in TC—Soft Computing (IEEE SMCS). Dr. Balas was past Vice President (Awards) of IFSA International Fuzzy Systems Association Council (2013–2015), is a Joint Secretary of the Governing Council of the Forum for Interdisciplinary Mathematics (FIM),—A Multidisciplinary Academic Body, India and recipient of the "Tudor Tanasescu" Prize from the Romanian Academy for contributions in the field of soft computing methods (2019).

Keynote

Global Readiness for Cybersecurity and AI: Assessing the Landscape and Charting the Path Forward

Marcelo Bussacarini

Abstract This paper examines global readiness for cybersecurity and AI, guided by insights from the Cisco Cybersecurity Readiness Index and the Cisco Global AI Readiness Index. With the digital landscape evolving at an unprecedented rate, robust cybersecurity measures and effective AI integration have become critical for organizational resilience and innovation. This study delves into the disparities and strengths across regions and industries, revealing a complex tableau of readiness that underscores significant challenges and untapped opportunities. The paper highlights the symbiotic relationship between cybersecurity and AI through a comprehensive analysis of global standards, frameworks, and readiness indicators. It explores the strategic intersections where AI can enhance cybersecurity defenses and, conversely, where cybersecurity measures are pivotal for safeguarding AI systems. Drawing on the extensive data and insights from the Cisco reports, this research offers a nuanced understanding of the factors influencing global readiness. The paper concludes with strategic recommendations for advancing global readiness in cybersecurity and AI, outlining a roadmap that emphasizes policy development, technological infrastructure, skill advancement, and cross-sector collaboration. This study aims to contribute to the global discourse on digital readiness, providing actionable insights for stakeholders to navigate the complexities of cybersecurity and AI in the digital era.

Keywords Cybersecurity · Artificial intelligence · Safety

1 Introduction

In the digital age, the significance of cybersecurity and AI cannot be overstated. As digital technologies become increasingly embedded in every facet of societal functioning, from economic activities and governmental operations to daily personal

M. Bussacarini (✉)
Cisco Systems, Sydney, Australia
e-mail: marcelo@bussacarini.com

M. Chakraborty et al. (eds.), *Proceedings of 5th International Ethical Hacking Conference*, Lecture Notes in Networks and Systems 1148,
https://doi.org/10.1007/978-981-97-8457-8_1

transactions, the imperative to secure digital infrastructures against malicious attacks grows ever more critical. Concurrently, Artificial Intelligence (AI) has emerged as a tool for innovation and efficiency and as a pivotal ally in the fight against cyber threats. The symbiotic relationship between cybersecurity and AI marks a new frontier in technological advancement, where AI's potential to predict, detect, and respond to threats aligns closely with the strategic objectives of cybersecurity frameworks [1].

The dual focus on cybersecurity and AI readiness across the globe forms the cornerstone of this study. The objectives are twofold: firstly, to assess the current state of global readiness in confronting cyber threats and integrating AI technologies, and secondly, to identify disparities in readiness that could influence global digital security and competitiveness. This analysis is grounded in the insights provided by the Cisco Cybersecurity Readiness Index and the Global AI Readiness Index, which serve as comprehensive benchmarks for evaluating the preparedness of countries and industries in these pivotal areas.

The significance of this study extends beyond academic interest, touching upon vital concerns for a range of stakeholders, including policymakers, business leaders, technology professionals, and educators. In a world where digital vulnerabilities, national security, economic stability, and privacy can have far-reaching effects, understanding the landscape of global readiness is essential. Similarly, as AI continues to shape the future of work, innovation, and social interactions, gauging the readiness for AI adoption and its integration into cybersecurity strategies becomes crucial.

This study aims to provide a more complex understanding of cybersecurity and AI readiness, highlighting the challenges, opportunities, and strategic imperatives for advancing global digital resilience. Through this exploration, the paper seeks to offer actionable insights and recommendations that can guide efforts to enhance the international state of readiness, ensuring a secure, innovative, and competitive future in the digital era.

2 Theoretical Framework

2.1 Defining Cybersecurity and AI Readiness

Cybersecurity readiness encapsulates the preparedness of entities—be they nations, organizations, or individuals—to protect and defend against cyber threats. This readiness involves technological capabilities and policy, procedural, and educational frameworks that collectively safeguard digital assets. AI readiness, on the other hand, pertains to the capacity to adopt, implement, and benefit from AI technologies. This includes the infrastructure to support AI, the workforce skilled in AI development and application, and the regulatory environment that enables ethical AI usage [2].

While distinct, these concepts share a foundational goal: to navigate and leverage digital advancements securely and effectively. Cybersecurity readiness aims to create resilient digital environments, whereas AI readiness focuses on harnessing AI's potential for innovation and problem-solving. The distinctions lie in their primary focus—security versus innovation—with their overlap increasingly evident as AI becomes integral to cybersecurity strategies.

2.2 *Importance of Readiness*

The critical importance of cybersecurity and AI readiness transcends the technical domain, impacting national security, economic stability, and societal trust. In the context of national security, readiness determines a nation's resilience to cyber threats that could disrupt critical infrastructure, steal sensitive data, or undermine democratic processes. Economically, readiness influences an entity's ability to protect assets and maintain a competitive advantage in a digital marketplace. Innovatively, readiness in AI opens avenues for solving complex problems, enhancing productivity, and fostering new industries [1, 2].

The stakes of readiness are high. For organizations, inadequate cybersecurity can lead to devastating breaches, while AI unreadiness can result in missed opportunities and technological stagnation. For nations, these aspects of readiness are intertwined with sovereignty, economic prosperity, and the welfare of citizens. The urgency to enhance readiness is driven by the escalating sophistication of cyber threats and the rapid pace of AI evolution, necessitating proactive and dynamic approaches to digital security and technological adoption. Becomes integral to cybersecurity strategies.

2.3 *Symbiosis of Cybersecurity and AI*

The symbiotic relationship between cybersecurity and AI is a testimony to the dramatic impact of digital technologies. AI's role in cybersecurity is multifaceted: It enhances threat detection through predictive analytics, automates responses to security incidents, and facilitates the analysis of vast datasets for threat intelligence. On the other hand, the development and deployment of AI, the protection of AI systems from manipulation, the integrity of AI-driven processes, and the security of data privacy are essential to a robust cybersecurity framework.

This interdependence underscores a broader trend: the fusion of digital security and innovation as critical readiness components. As AI technologies evolve, their integration into cybersecurity strategies becomes beneficial and essential, marking a shift toward more intelligent, adaptive, and responsive security measures. Simultaneously, the secure deployment of AI requires cybersecurity practices that address unique challenges posed by AI, including ethical considerations, data bias, and the potential for AI-driven threats.

Theoretical exploration of cybersecurity and AI readiness reveals a complex landscape where security and innovation intersect, highlighting the imperative for holistic approaches to readiness encompassing technological, regulatory, and educational dimensions. This framework sets the stage for a thorough analysis of worldwide readiness based on Cisco's Insight Reports. It underscores the need for strategic initiatives that bolster cybersecurity and AI capabilities worldwide.

3 Methodology

3.1 Data Sources

The primary data sources underpinning this study are the Cisco Cybersecurity Readiness Index and the Cisco Global AI Readiness Index. These comprehensive reports provide a wealth of information on the current readiness in cybersecurity and AI across various regions and industries worldwide. Cisco's indexes are distinguished by their methodology, encompassing multiple indicators from policy and infrastructure to skills and innovation ecosystems. By leveraging these reports, the study gains access to rigorously compiled and analyzed data, offering a detailed and expansive global perspective on readiness [3, 4].

3.2 Analytical Approach

The study employs a comparative analysis approach to examine the data presented in the Cisco reports. This involves comparing and contrasting the readiness scores across different geographical regions to identify patterns, strengths, and areas requiring attention. Such an analysis allows for the highlighting of disparities in readiness between developed and developing countries and between various industries, shedding light on the diverse challenges and opportunities present in the global landscape.

Furthermore, the study delves into sector-specific analyses, considering how different industries—healthcare and finance to manufacturing and education—fare regarding cybersecurity and AI readiness. This approach is instrumental in understanding the nuanced demands and capabilities across sectors, enabling a more tailored discussion of strategies for improvement.

In synthesizing the findings from the Cisco reports, the study also incorporates a qualitative assessment of the underlying factors contributing to the observed readiness levels. This includes examining the impact of policy environments, technology and skills investment, and innovation hubs' presence on readiness scores. Through this analytical lens, the paper aims to offer a comprehensive understanding of global readiness in cybersecurity and AI, grounded in data and enriched by contextual analysis [3, 4].

4 Global Standards and Frameworks

4.1 Overview of Global Standards

International standards and frameworks play a key role in guiding the global community toward enhanced preparedness as the world of digital security continues to evolve with the development of artificial intelligence. For cybersecurity, standards such as ISO/IEC 27001 provide a framework for the systematically managed management of a company's confidential information, ensuring security and resilience. Similarly, the NIST Cybersecurity Framework provides organizations with a policy framework for computer security to evaluate and improve their ability to prevent cyberattacks, detect them, and respond to them [2, 5].

While AI readiness is a newer domain, it is increasingly being shaped by frameworks like the OECD Principles on AI, which support AI that is innovative, trustworthy, and respects human rights and democratic principles. The IEEE Global Initiative on Ethics of Autonomous and Intelligent Systems sets out ethical standards and recommendations to align AI technology with societal values and moral principles [4, 6].

4.2 Benchmarking Readiness

Benchmarking readiness through these global standards involves a multifaceted approach. Adherence to ISO/IEC 27001 and compliance with the NIST Framework indicate an organization's or nation's commitment to cybersecurity practices. These benchmarks assess the maturity of information security management systems, evaluate the comprehensiveness of cybersecurity policies, and measure the effectiveness of incident response capabilities [2, 5].

In the realm of AI, readiness benchmarking focuses on the extent to which AI systems are developed and deployed by international principles and ethical guidelines. The evaluation considers factors such as transparency, accountability, fairness, and safeguarding privacy and data protection. Benchmarking against these standards allows for a comparative analysis of AI readiness across countries and industries, highlighting leaders in ethical AI practices and identifying areas where additional efforts are needed to align with global norms [3, 4].

Using these global standards and frameworks to assess and benchmark readiness levels is crucial for establishing a common understanding and shared cybersecurity and AI readiness goals. By adhering to these internationally recognized benchmarks, countries and organizations can ensure that their cybersecurity measures are robust and their AI systems are both practical and ethically responsible. This alignment enhances digital security and innovation and facilitates international cooperation and trust in the digital era [3, 4].

5 Indicators of Global Readiness

5.1 *Cybersecurity Readiness Indicators*

Cybersecurity readiness across the globe is quantified through various metrics, each providing insights into different facets of an entity's defensive posture against cyber threats. Key indicators include:

- **Policy and Governance:** The existence and enforcement of cybersecurity policies and frameworks that guide organizational and national security measures.
- **Technological Infrastructure is the** robustness and modernity of technological systems that detect, prevent, and respond to cyber threats.
- **Incident Response and Recovery:** The capability to quickly and effectively address and recover from security breaches or attacks.
- **Workforce Competency:** The availability of skilled cybersecurity professionals, underscored by ongoing training and education programs.
- **Public–Private Collaboration:** The level of cooperation between governmental bodies and private sector entities in addressing and mitigating cyber threats.

5.2 *AI Readiness Indicators*

While sharing some commonalities with cybersecurity readiness in terms of infrastructure and skilled personnel, AI readiness focuses more on the capacity to develop, implement, and leverage AI technologies. Indicators include:

- **Research and Development (R&D) Investment:** Financial investments in AI research and the development of innovative AI technologies.
- **AI Integration and Application:** The degree to which AI is integrated into public and private sector operations, impacting productivity and service delivery.
- **Ethical and Regulatory Framework:** The presence of ethical guidelines and regulatory policies for AI development and usage, ensuring responsible innovation.
- **Data Accessibility and Quality:** The availability of high-quality datasets is essential for training AI models and ensuring their accuracy and reliability.
- **Education and Talent Development:** Initiatives aimed at developing AI expertise within the workforce, including higher education programs and professional training.

5.3 *Analysis of Readiness Indicators*

The juxtaposition of cybersecurity and AI readiness indicators reveals a multifaceted landscape of global readiness. For instance, a nation may exhibit high levels of technological infrastructure for cybersecurity but needs to catch up in R&D investment for AI, indicating an imbalance in readiness that could impact its overall digital resilience and innovation capacity.

Moreover, the interdependence of specific indicators, such as workforce competency, underscores the synergistic potential of enhancing readiness across both domains. A skilled workforce is crucial not only for implementing effective cybersecurity measures but also for driving AI innovation.

Regions leading in both cybersecurity and AI readiness typically exhibit strong policy frameworks, robust technological infrastructures, and vibrant ecosystems for talent development and public–private collaboration. In contrast, regions lagging in readiness often need help with challenges such as inadequate investment in technology and education, limited regulatory frameworks, and a scarcity of skilled professionals.

This analysis underscores the importance of a holistic approach to enhancing global readiness, where advancements in cybersecurity bolster AI innovation and vice versa, creating a secure and fertile ground for digital progress.

6 Global Perspectives on Cybersecurity and AI Readiness

6.1 *Regional Analysis*

The global landscape of cybersecurity and AI readiness presents a tapestry of varied progress and challenges across different regions:

- **North America:** This region often leads in cybersecurity and AI readiness, attributed to substantial investments in R&D, robust technological infrastructure, and strong public–private partnerships. However, the region still faces cybersecurity workforce shortages and privacy concerns related to AI.
- **Europe:** Showcases high readiness levels, particularly in cybersecurity, due to stringent regulatory frameworks like GDPR and active government involvement in cybersecurity initiatives. AI readiness is bolstered by significant funding for AI innovation and ethical AI frameworks.
- **Asia–Pacific:** This region displays a dynamic range of readiness levels; countries like Singapore, South Korea, and Japan excel in integrating AI into public services and cybersecurity measures while other nations grapple with building the necessary infrastructure and regulatory environment.

- **Latin America and Africa:** These regions generally need to catch up in readiness, hindered by lower investment in technology and education. However, pockets of innovation and progress exist, driven by governmental policies and international cooperation.

6.2 Industry-Specific Insights

Readiness levels also vary significantly across industries, reflecting the diverse nature of cyber threats and the application of AI:

- **Finance and Healthcare:** Both industries demonstrate high cybersecurity readiness due to the critical nature of their data. The finance sector increasingly utilizes AI for fraud detection and customer service, while healthcare leverages AI for diagnostics and patient care optimization.
- **Manufacturing and Education:** The manufacturing sector is rapidly integrating AI for operational efficiency and predictive maintenance but faces cybersecurity challenges related to industrial control systems. Education sectors embrace AI for personalized learning but must address data privacy and security concerns.

6.3 Disparities and Strengths

A nuanced analysis reveals key disparities and strengths in global readiness:

- **Disparities:** The most pronounced disparities arise from differences in technological infrastructure, government policy, and investment in education and training. Developing regions often need help with the foundational aspects of readiness, such as reliable internet access and cybersecurity awareness.
- **Strengths:** Strengths are observed in regions and industries that prioritize strategic investment in technology, adopt comprehensive regulatory frameworks, and foster collaborations across the public and private sectors. Initiatives to close the talent gap in cybersecurity and AI also emerge as critical readiness drivers.

Examining global cybersecurity and AI readiness perspectives underscores the complex interplay of factors influencing readiness levels across regions and industries. While advancements are notable, concerted efforts are required to address disparities and leverage strengths to foster a more secure and innovative global digital landscape.

7 Intersection of AI and Cybersecurity

7.1 Strategic Intersections

The intersection of AI and cybersecurity represents a strategic nexus where deploying advanced technologies meets the imperative for digital protection. AI's capabilities—from machine learning models that predict and detect anomalies to algorithms that automate threat responses—are revolutionizing cybersecurity. This integration enhances the efficiency and effectiveness of cybersecurity measures and introduces new dimensions to global readiness. For organizations and nations alike, the maturity of AI integration into cybersecurity strategies has become a barometer of their overall digital resilience.

AI's role in cybersecurity is multifaceted, offering advanced solutions for continuous monitoring, threat intelligence analysis, and identifying sophisticated cyberattacks such as zero-day exploits. However, this integration also necessitates robust cybersecurity frameworks to protect AI systems themselves from manipulation or attack, highlighting the reciprocal nature of this intersection. As AI technologies evolve, ensuring their security against adversarial AI attacks or data poisoning becomes paramount, underscoring the interdependence between AI advancements and cybersecurity readiness.

7.2 Case Studies

- **Financial Sector AI Defense:** A leading financial institution implemented an AI-powered security operations center (SOC) that uses machine learning algorithms to sift through billions of daily log entries. This AI integration enabled the early detection of fraudulent transactions and potential data breaches, significantly reducing the risk of financial losses and data compromise.
- **Healthcare Data Protection:** A hospital network employed AI-driven anomaly detection systems to monitor patient data access in the healthcare sector. This AI application helped identify and prevent unauthorized access attempts, safeguarding sensitive patient information against potential cyberattacks and ensuring compliance with healthcare privacy regulations.
- **National Cyber Defense Initiatives:** On a national level, a government's cybersecurity agency developed an AI-based cybersecurity platform to protect critical infrastructure. By leveraging AI for real-time threat analysis and automated incident response, the platform enhances the nation's defense against increasingly sophisticated cyber threats, serving as a model for global cybersecurity readiness.

These examples underscore AI's transformative impact on cybersecurity strategies, demonstrating its potential to augment existing defenses and innovate new approaches to digital security. AI's integration into cybersecurity is a testament to the strategic intersections shaping the future of global readiness, where leveraging cutting-edge technology becomes essential in securing the digital frontier.

8 Challenges and Opportunities

8.1 *Identifying Challenges*

The journey toward global cybersecurity and AI readiness is fraught with technical, ethical, and regulatory challenges. One primary challenge is the skill gap; as cybersecurity threats evolve and AI technologies advance, the demand for skilled professionals outpaces the supply, leaving organizations vulnerable. Rapid technological changes that require continuous learning and adaptation are exacerbating this gap.

Another important challenge is the ethical and privacy concerns associated with AI. As more AI systems are integrated into cyber security strategies, the issues of data privacy, bias in AI algorithms, and the potential for misuse of AI come to light. Ensuring AI systems are ethical, transparent, and accountable remains a persistent challenge.

Technological infrastructure also presents a hurdle, particularly in regions or sectors with limited access to advanced computing resources for deploying AI solutions. This disparity in technological access hinders global readiness and widens the digital divide.

Furthermore, regulatory frameworks need help to keep pace with technological advancements, leading to a regulatory lag that complicates the deployment of AI in cybersecurity. The absence of harmonized international regulations adds complexity to global cybersecurity efforts, creating a fragmented landscape of standards and practices.

8.2 *Emerging Opportunities*

Despite these challenges, the intersection of AI and cybersecurity offers significant opportunities to enhance global readiness. One of the most promising opportunities is the automation and scaling of threat detection and response. AI can analyze vast datasets far more quickly and accurately than human operators, identifying threats that would otherwise go unnoticed and responding in real time to mitigate potential damage.

Predictive analytics represent another opportunity: AI leverages historical data to predict future cyber threats, allowing organizations to adopt a more proactive stance toward cybersecurity. This shift from reactive to predictive cybersecurity can significantly reduce the impact of cyberattacks.

The development of resilient AI systems that can operate in adversarial environments presents an opportunity to strengthen the security of AI applications themselves. Research into adversarial AI and machine learning can lead to more robust AI models resistant to attacks, enhancing the security of AI-driven systems.

Collaborative AI projects, where AI systems share threat intelligence and learn from each other across organizations and borders, can lead to a more unified and effective global cybersecurity posture. This collaborative approach can democratize access to advanced threat detection capabilities, raising the global baseline for cybersecurity readiness.

In addressing these challenges and seizing the opportunities, stakeholders can significantly advance global readiness in cybersecurity and AI, paving the way for a more secure and technologically empowered future.

9 Strategic Recommendations and Future Directions

9.1 Advancing Global Readiness

A multifaceted approach is required to enhance global cybersecurity and AI readiness, emphasizing collaboration, innovation, and resilience. Education and Training programs are crucial to addressing the skills gap, with initiatives ranging from university-level AI and cybersecurity courses to professional development programs and online learning platforms. Investment in Infrastructure is also vital, particularly in under-resourced regions, to ensure that the technological foundations for cybersecurity and AI are robust and accessible. Furthermore, fostering Public–Private Partnerships can accelerate innovation in cybersecurity solutions and AI applications, leveraging the strengths of both sectors to advance readiness.

9.2 Policy Recommendations

Policymakers play an essential role in shaping the policy—landscape for cybersecurity and AI readiness. Implementing comprehensive cybersecurity policies that encompass technological measures, human factors, and organizational processes is essential. These policies should encourage adopting best practices and standards, such as those outlined by NIST or ISO. Developing Ethical Guidelines and Regulatory Frameworks for AI that address privacy, bias, and transparency is critical to ensure responsible innovation. Additionally, governments should consider Strategic Investments in R&D for cybersecurity and AI, supporting foundational research and applied projects that can lead to breakthroughs in security technologies and AI applications [2, 5].

9.3 Future Research

Several areas for future research emerge as critical to advancing understanding and capabilities in cybersecurity and AI. Adversarial AI research, exploring how AI systems can be secured against attacks that seek to manipulate or mislead, is a growing field of importance. Investigating human–AI collaboration in cybersecurity contexts can yield insights into leveraging AI's capabilities while mitigating risks associated with over-reliance on automated systems. Additionally, research into Cross-Sector AI Applications can uncover how AI technologies developed in one industry might be adapted for use in cybersecurity contexts, potentially leading to innovative cross-pollination of ideas and solutions.

In the realm of policy and regulation, Comparative Studies of International Cybersecurity Policies can provide valuable lessons on practical strategies and pitfalls to avoid. Similarly, exploring the Impact of AI Regulation on Innovation can help balance the need for ethical guidelines with the desire to encourage technological advancement.

By addressing these recommendations and research directions, stakeholders can significantly enhance global readiness for cybersecurity and AI, ensuring that societies can reap the benefits of digital transformation securely and equitably.

10 Conclusion

10.1 Summary of Findings

This study has traversed the complex landscape of global readiness in cybersecurity and AI, revealing a world at a crossroads. The Cisco Cybersecurity Readiness Index and Global AI Readiness Index served as foundational data sources, illuminating the disparities and strengths across regions and industries. The analysis underscored the critical role of strategic intersections between AI and cybersecurity, showcasing their potential to revolutionize digital defenses and innovate proactive threat mitigation strategies. However, the journey toward global readiness is hampered by significant challenges, including skills gaps, ethical concerns, and the need for robust technological infrastructures and regulatory frameworks.

10.2 Implications for Stakeholders

For policymakers, the findings highlight the imperative to craft policies that bolster cybersecurity defenses and nurture AI innovation within an ethical framework. Industry leaders are called upon to invest in the technological infrastructure and talent development essential for harnessing AI's potential in cybersecurity. For educators, the study emphasizes the importance of evolving curricula to prepare the next generation of professionals equipped to navigate and lead in the digital landscape.

10.3 Call to Action

The path to enhanced global readiness in cybersecurity and AI demands a concerted effort, transcending individual and organizational endeavors to encompass a collaborative, international initiative. Stakeholders across sectors are urged to foster dialogue, share knowledge, and pool resources to address the multifaceted challenges identified. By embracing a unified approach, the global community leverages the transformative power of AI to secure digital infrastructures and pave the way for a future where innovation and security coexist harmoniously. The findings of this study serve as both a roadmap and a call to action, urging stakeholders to engage actively in the collective endeavor of advancing global readiness.

References

1. Cisco Global Cybersecurity Readiness Index (2024) https://newsroom.cisco.com/c/r/newsroom/en/us/a/y2024/m03/cybersecurity-readiness-index-2024.html
2. National Institute of Standards and Technology (2018) Framework for improving critical infrastructure cybersecurity. https://nvlpubs.nist.gov/nistpubs/CSWP/NIST.CSWP.04162018.pdf
3. Cisco Global AI Readiness Index (2023) https://www.cisco.com/c/m/en_us/solutions/ai/readiness-index.html
4. IEEE Global Initiative on Ethics of Autonomous and Intelligent Systems (2016) Ethically aligned design: a vision for prioritizing human well-being with autonomous and intelligent systems. https://standards.ieee.org/industry-connections/ec/autonomous-systems/
5. International Organization for Standardization (2022) ISO/IEC 27001: information security management systems—requirements. https://www.iso.org/standard/27001
6. Organisation for Economic Co-operation and Development (2023) OECD principles on artificial intelligence. https://doi.org/10.1787/835641c9-en?format=pdf

The Future of Ethical Hacking: Charting the Uncharted in Cybersecurity

Gleuto M. Serafim

Abstract In a time of rapidly changing cybersecurity, a term known as "ethical hacking" has become extremely important for finding and fixing weaknesses in the world's information systems (Baddam in Asian J Humanit Art Lit 7(2):147–158, 2020). This paper looks at the future of ethical hacking, with a specific focus on how artificial intelligence (AI) could be involved and how this might affect the rules and regulations around it. The paper looks at how new technologies like AI can help improve cybersecurity and how they present new challenges. The article stresses the need for creative ideas and strong testing plans to make sure AI systems are secure and can handle cyber threats. It also looks at how the law and AI intersect and why we need new laws that can keep up with the changing world of cyber threats (Li Y and Liu Q Energy Rep 7:8176–8186. https://www.sciencedirect.com/science/article/pii/S2352484721007289, 2021). By exploring new areas of cybersecurity weaknesses and ethical hacking, this research hopes to set the stage for the next generation of cybersecurity practices, making sure they work well and follow international laws (Vijayakumaran et al. Int J Electr Comput Eng 10(1):387. https://core.ac.uk/download/pdf/329117789.pdf, 2020).

Keywords Ethical hacking · Cyber threats · Privacy · Cybersecurity

1 Introduction

The term "hacker" was first coined by the brilliant young wizards of early computer programming. These were individuals with an insatiable curiosity to learn how things work and the innate ability to break through the barriers of the unknown. Unfortunately, society's fears and misgivings soon thrust a dark mantle of criminality upon that term. For the past several decades, the world has been plagued by hordes of nefarious criminal hackers, technologically armed to destroy—stealing passwords,

G. M. Serafim (✉)
Ixonn Group, Greater Sydney Area, Sydney, Australia
e-mail: gleuto.serafim@ixonn.com

M. Chakraborty et al. (eds.), *Proceedings of 5th International Ethical Hacking Conference*, Lecture Notes in Networks and Systems 1148,
https://doi.org/10.1007/978-981-97-8457-8_2

looting bank accounts, hijacking identities, and wreaking an untold magnitude of financial and emotional damage. In sheer contrast to the malevolent hackers, an elite group of virtuous professionals are authorized to penetrate computer systems—the ethical hackers. Svensson [1] effulgent and dauntless, they combat the laws of darkness not with the force of arms but with the brilliance of their minds. They are vigilant sentinels, guarding our cyber infrastructure.

Ethical hacking, often performed by white hats, is performed by professionals authorized to break into computer systems to identify security vulnerabilities before black hat criminal hackers discover and exploit them. We are at the inceptive stage in charting the uncharted path of ethical hacking. Yaacoub et al. [2] to illuminate the way ahead, we spotlight the pioneers who have magnificently excelled by sharing career highlights and thoughtful advice. With their incredible journeys and inspiring grit, they robustly demonstrate that the future of ethical hacking is exceedingly bright and blazingly limitless [3–6].

1.1 Context and Background

Hacking has somewhat of a romantic notion attached to the description. Its portrayal in various Hollywood movies propels youngsters, in particular, to explore, learn, and attempt to execute the depicted highly skilled acts of hacking, sometimes even against the law. However, not all hacking activities fall within the ambit of criminal behavior. The term ethical hacking is used to describe such activities conducted with the legal sanction obtained from the owners of computer systems to test and hopefully improve the system's security posture. The ethical hacking field was actually born at the annual digital security conference for information sharing and training known as DEFCON when the "first hacker war" erupted between the United States and Iraq. The US Government placed heavy restrictions on training and working with technology so that private industries and citizens could help protect the country [7]. Later, it was realized that hacking was critical for defense because hackers do not follow any rule book and can test defenses by launching unpredictable attacks. The term ethical hacking was coined, becoming pivotal for defense testing cybersecurity of nations, agencies, and corporations [8].

Various nations have branded cyberspace as a new domain akin to land, sea, air, and space. The term domain signifies a framework that is conceptualized to facilitate the description and understanding of a complex field of activities. Just as in other domains, we see nation-states as significant actors in international relations using the attribute of sovereignty de jure to evolve policies, create defense forces, and ensure command and control (C2) over other internal security functions in the cyber domain. However, the exercise of sovereignty in the cyber domain is still uncharted and contested to a great extent, with the incremental progress of national and international laws [9]. Therefore, it is interesting to note that Cybersecurity comprises two independently evolving concepts: Cyber and Security. Furthermore, the term ethical hacking encompasses the spirit of doing good in a realm perceived to be full of evil [10]. This paper charts the uncharted by exploring the field of ethical hacking within the prism of Cybersecurity with the primary aim of understanding the role of Hacking in ensuring security [11, 12].

1.2 Relevance and Importance

Several high-profile cyberattacks took place over the last couple of years. It is well known that large corporations have teams of cybersecurity experts to protect their data and systems, yet this seems not enough. The cybersecurity industry is ever growing with limited skilled professionals creating this skill gap. Ethical hackers, also known as white-hat hackers, are playing a crucial role in detecting vulnerabilities in systems, web applications, and inside networks [13]. Companies have started bug bounty programs to encourage security researchers and support them in the discovery and responsible disclosure of vulnerabilities. Well-publicized cybersecurity incidents accompanied by the widespread use of social media have raised public awareness of the importance of information security, which could positively impact the security community [14]. The fact that cybersecurity is receiving increased budgets indicates its growing importance [15].

2 Ethical Hacking: Evolution and Current Landscape

However, this is not the sole or most important application of hacking skills for the ethical hacking community [16]. The community's primary goal is to promote and advance knowledge of securing networks and computer systems from unauthorized access and other cyber threats that can constitute real or potential harm or damage [17].

The weapons that hackers use to damage or infiltrate a network are broadly the same as those used by ethical hackers to test network security: viruses, worms, trojan horses, backdoors, rootkits, and other malware; manual and automated password cracking tools; social engineering techniques; and a range of software exploit kits. Arnone [18], Haq et al. [19] this more limited tactical use of ethical hacking in the service of corporate network security is typically intended to avoid or mitigate the damage that can be caused by real hacker attacks.

To date, ethical hacking has typically been undertaken in the spirit of corporate or institutional cybersecurity [20]. Most security professionals who identify themselves as ethical hackers work in-house on temporarily or rolling contracts for individual private or public organizations. They are employed to undertake well-defined testing operations on those organizations' networks and systems to help ensure security controls are properly implemented to protect against unauthorized access, disclosure, alteration, and destruction of confidential information, denial of service, and other potential damage or disruption.

Ethical hacking as a profession has only been around for about two decades. It emerged with the progressive computerization of commercial operations throughout the 1990s when novel online threats also appeared [21]. As such, ethical hacking is a field that is regularly advancing and diversifying [16].

2.1 Definition and Scope

Furthermore, ethical hacking is not confined to any specific industry, department, or profession. It is intended for all those involved in using information and communications technology [2, 22].

Ethical hacking has a clear and dynamic scope in the years to come. Every day is a new challenge in the world of hacking. It is unpredictable, and hackers are always exploring the uncharted to find new vulnerabilities in software, hardware, systems, and networks. The variety of hacking is enormous; thus, it is challenging to compile a comprehensive list of the scope of ethical hacking [23]. The major areas of hacking include advanced persistent threats, zero-day attacks, malware techniques, artificial intelligence, the Internet of Things, cloud computing, blockchain technology, 5G mobile networks, and biometric security systems [24, 25]. The present paper focuses on these specialized areas.

A cybersecurity expert's responsibilities are never-ending, and it is crucial to keep up with the latest skills and knowledge in the dynamic field of hacking [25].

An ethical hacker is a skilled professional who understands and acknowledges a cybercriminal's mindset and knowledge and uses it to identify and rectify system vulnerabilities to safeguard an organization's confidential data and resources. Bansal et al. [26], Gandhi et al. [4] Ethical hacking plays a vital role in the domains of information security and cybersecurity [27]. As the sophistication and frequency of cybercrimes increase over the years, ethical hacking has become a necessary security control among most organizations to protect their information assets.

Ethical hacking is a qualified and authorized attempt to locate and resolve cybersecurity threats within an organization [6]. It is a proactive security process that undertakes an offensive approach to security by performing various hacking activities in a controlled environment [28]. Ethical hacking is also known as penetration testing, intrusion testing, or red teaming. Saha et al. [27] however, it is dissimilar to vulnerability assessments, as it encompasses a comprehensive hacking or firewall testing process.

2.2 Historical Development

The task of ethical hacking arose with the increasing criminal activities in cyberspace. The first computers used in the '60 s and '70 s had very rudimentary functions, and most of the commands were built to manipulate files and list information. Rosati and Paulsson [29] the first computer crimes emerged very quickly, and the main techniques used to commit these crimes were related to the identity of the users and the denial of services [30]. Criminals discovered that using computers to perform illegal activities such as espionage, sabotage, or fraud was easy. As the technology evolved, so did the techniques used to commit cybercrimes [31]. Today, computers are instrumental in nearly all human activities, encompassing mission-critical applications and interconnected by the internet, which creates a broad attack surface for cybercriminals [32].

The idea of hacking appeared with the invention of the phone. Phones were very quickly hacked before the mass media called them phreakers in the '60 s and early '70 s, and they explored manipulating phone systems and the inner connections of telecommunication systems for free calls [33]. Hacking in computers began at the moment computers appeared with a mass audience. In the beginning, computers were large machines that governments and big enterprises used [34]. Between the '60 s and '80 s, only a few computers used time-sharing technology, and these computers became very attractive to some curious people who wanted to explore and understand how they worked internally. These people began studying computers and making changes to improve their performance. In the beginning, the term hacker did not have any negative meaning [35, 36]. It just identified a person who was an enthusiast and enjoyed programming and exploring computers.

3 Artificial Intelligence and Law: Implications for Ethical Hacking

Artificial intelligence (AI) can greatly enhance creativity and efficiency in the implementation of cybersecurity [37]. For example, AI has and will result in software that writes its own code to provide protection against new types of cyber threats [38]. AI, however, presents a number of key challenges in the field of cybersecurity law. One of the ways is a technical approach that utilizes AI to counteract cyber threats. This implies that a machine will either perform a cybersecurity protection task or operate a system to protect cybersecurity without human intervention [38–40]. The other more challenging aspect of cybersecurity law is regulating social relations associated with the privatization of cybersecurity through AI.

Ethical hacking is used to enhance cybersecurity, where hackers identify weaknesses in a computer, networking system, or system applications to ensure data safety and prevent a breach. New technologies have brought new challenges to law and the application of law. One area of law that is affected by challenges posed by new technology is cybersecurity law. It tells us what we shouldn't do in cyberspace and what will happen to us if we break the rules [4, 6]. It also tells us what the state must do to protect us in cyberspace and what will happen to the state if it fails in its duty to protect our rights [41–43]. Among some of the duties of the state is the application of a set of rules that are also affected by challenges posed by new technology.

Ethical hacking uses AI technology to identify prospective weak areas and provide the necessary protection against threats to mitigate the risk in a cybersecurity system [44].

3.1 Overview of Artificial Intelligence in Cybersecurity

The integration of AI in cybersecurity is not a new concept. It was first introduced in the early 1980s when expert systems were employed to support the cybersecurity functions of NATO nations [45]. During that time, AI tools, such as rule-based expert systems, were applied to perform restricted functions. After several decades, the cybersecurity industry grew to become a grand market with several companies and hundreds of security products concentrated around DDoS mitigation, intrusion detection and prevention, firewalls, antivirus, web application firewalls, and secure email gateways [46, 47]. These conventional security services and products primarily depend on pre-configured rules, signatures, and behavioral models to protect the systems and networks. Although they are effective, they can also be easily bypassed with a slight modification in the malware codes. Moreover, they can result in high numbers of false positives. Researchers and security experts used AI techniques to overcome the limitations of these conventional security mechanisms. This has led to the development of a novel generation of intelligent security systems that include intelligent intrusion detection systems, intelligent firewalls, intelligent antivirus systems, intelligent, secure email gateways, intelligent DDoS systems, intelligent vulnerability assessment and management systems, intelligent autonomous offensive systems, and intelligent security information and event management systems [48].

The introduction of artificial intelligence (AI) and its subset machine learning (ML) in the field of cybersecurity has significantly addressed the issue of skill shortage in the industry on the one hand and automated the process of both offensive and defensive security to a major extent on the other. Dimitrov [49], Goupil et al. [50] the application of AI in cybersecurity ranges from enabling intelligent intrusion detection to creating autonomous offensive systems like AI-empowered autonomous weapon systems. Malatji and Tolah [51] the discussion also presents an in-depth exploration of the use of AI in cybersecurity with the help of associated challenges and future promises.

3.2 Legal and Ethical Considerations

Entering into an agreement: Unauthorized hacking is illegal for obvious reasons. Before an ethical hacker proceeds with any attempt, they must have a formal contract or an agreement that outlines the scope of the work, the limitations, and the prevention of any hidden agendas [36].

Permission: Ethical hackers should perform tests only on systems for which the owner or administrator has given written authorization [52].

Limited authorization: The written permission should contain clear limitations on the scope of the testing and on accessing data or resources [2].

Informing the authorities: To avoid repercussions and legal actions in case of discovery, the organizations involved should inform their relevant legal enforcement authorities that they are conducting tests on computer systems [2].

Performing any activity in cyberspace has its rules and code of conduct. Owning and managing physical and digital properties does not exempt one from respecting the laws and policies governing the use of such assets. As beneficial as it is, ethical hacking has established rules to prevent chaos. Legal provisions must be considered with utmost importance to protect ethical hackers and ensure that their actions safeguard the interests of those who authorized them. In general, ethical hacking is legal, with implementing some terms and conditions. Some legal precautions must be observed when hiring people or groups known as ethical hackers. It is important to check the ethical hacking activities and approach used fully, with thorough documentation and verification of written consent for previous activities. Adhere to all relevant laws and regulations, ensuring that individuals' rights do not lead to unauthorized access or exposure to sensitive information. Industry standards adherences and best practices for ethical hacking must be applied [36, 53].

4 The Future of Ethical Hacking

It is key to appreciate that the growth of ethical hacking relies on technology and the supporting structure of government legislation, prosecution, and ethics on which the profession is built. The difference between a felony and a rewarding career ultimately does not rest on the act of hacking itself but on whether the hacker is acting in compliance with the law and within the rules of engagement of cybersecurity professionals [6, 36]. As with any tool, hacking is how one chooses to use it that matters.

Those attributes are likely to see significant growth over the next decade as the shortcomings of current cybersecurity defense models are increasingly exposed and ethical hacking assumes a larger and more proactive role in cybersecurity. Smith et al. [6] here, we present some speculative thoughts about the future of ethical hacking based on the trends we have identified that are driving change in the field. While these ideas may not form a roadmap for the development of ethical hacking, they should at least provoke some thought and discussion. They also help outline the uncharted areas of ethical hacking and, by inference, the future constraints of the field.

Ethical hacking is still a nascent field [54]. Well-defined as the area of cybersecurity dealing with prosecution-approved hacking, it is far smaller and less developed than the vast areas of offensive hacking and cybersecurity defense. Indeed, ethical hacking borrows much from both these adjacent domains, with obvious and direct links to the tools, techniques, and methodologies of hacking and an ultimate purpose in common with security [18]. However, as we have outlined in this report, ethical hacking has some unique attributes that set it apart as a specialist in its own right.

4.1 Emerging Challenges and Threats

We now review some of the latest and coming cybersecurity threats and challenges as seen from the ethical hacking point of view, keeping in mind that ethical hacking then translates into a set of specific and practical guidelines, how-tos, and best practices to deal with each said threat and challenge. The explored land cannot be summarized: it keeps changing at cyber speed [53]. However, some examples propose consolidating the theoretical approach as it outlines the guidelines used by professional ethical hackers to chart uncharted land.

Facing ethical hacking means dealing with the most advanced, hidden, and so far, unknown cybersecurity problems and threats a given society may suffer. Thanks to this modus operandi, ethical hacking tends to advance other means of researching, developing, and practicing cybersecurity in general [55]. Yet, ethical hacking is itself the avant-garde of applying cybersecurity techniques, technologies, and best practices [56].

4.2 Evolving Roles and Responsibilities

Growing areas of external collaboration will challenge current ethical hacking models—particularly the need for scalable, flexible, and secure infrastructures supporting open innovation ecosystems. This will drive a significant shift from securing only internal networks, systems, and information to securing connected services and products that interact with an increasingly open external ecosystem. The future of ethical hacking is both uncertain and wide open [2]. offering superb opportunities for those who embrace change and are prepared to adapt, grow, and build trust-based relationships with all stakeholders involved.

As the accelerating change in digital innovation adds even more complexity to digital infrastructure, ethical hackers will move from merely advising of vulnerabilities to providing recommendations on secure-by-design systems development. The future will see a truly proactive approach to cybersecurity, in which ethical hackers are involved in every phase of the development lifecycle, including analysis, design, and testing, ensuring protection from the start and adapting as designs change. Sinha and Arora [57] the future will also see increasing specialization, in which current generalist ethical hackers must decide whether to focus on deep technical expertise in areas such as AI/ML or broader skill development in specialized domains, such as IoT or cloud [2].

5 Conclusion

In conclusion, it is essential to revisit our stances regarding the ever-changing rules of cybersecurity and fortify them with new initiatives. With the changing patterns of cyber threats, it is imperative that the role of ethical hacking is redefined, improving our concepts and contemplating new frameworks that can enhance the potency of our cybersecurity policies and the implemented defensive measures. Baddam [3] ethical hacking can and will chart the uncharted in cybersecurity if we permit it to do so while engaging in a sound and ethical manner [58].

Cybersecurity mostly depends on defense tactics to protect networks, computers, and data from illicit access, corruption, or damage. Ethical hacking assists in curtailing cyber warfare, which is considered very potent and dangerous in the present and foreseeable future. Private units, national entities, and military organizations have been described as partaking in cyber warfare. The collection of weaponry in cyber warfare is being expanded and improved over time. However, cyber warfare cannot be unleashed instead of kinetic warfare; both have to be combined to function to the full extent [59]. Although cyberterrorism has not been employed much until now, it does have a potent chilling effect [60].

6 Summary of Key Findings

This study is anticipated to contribute to the general understanding of uncharted areas within a specific cybersecurity activity (ethical hacking) and how cybersecurity as a whole can mature by contributing new knowledge on the future of ethical hacking [61].

Therefore, due to the lack of prior work in this area and it being the aim of this research to investigate the uncharted in ethical hacking, it constructs the theoretical framework primarily drawing from studies on the direction and pace of technological change along with the concept of uncharted as discussed in the introduction [2]. The existing theoretical and conceptual models associated with technological advancement were considered. Such constructs that operate in the presence of high uncertainty in the domain of technology and innovation were reviewed. Based on the review of the literature and the specific characteristics of cybersecurity and ethical hacking, it was ascertained that a number of features of cybersecurity represent uncharted areas of ethical hacking [62, 63]. These findings respond to the research aim of exploring the uncharted in ethical hacking.

After a review of the previous studies, it was identified that the thesis of "uncertainty" had been largely overlooked in the context of ethical hacking. Most reports and academic studies summarize guidelines or discussions about the role of ethical hacking in testing security systems. The only report that elaborated on an uncharted aspect of ethical hacking was from IBM on the future of hacking [64, 65].

6.1 Implications for Cybersecurity Practices

Data protection has become indispensable in computer systems implemented to protect the information they contain and in transmission systems used to exchange information. For this reason, it is necessary to guarantee cybersecurity from attacks performed through hacking. Vulnerability assessment using ethical hacking is an activity that helps verify the security of computer systems and discover their weaknesses. It is considered a viable tool to be integrated into the broader cybersecurity management system. To enhance this study, it would be valuable to analyze the requirements of professionals who carry out vulnerability assessment so that in the future, it can be performed by recent ICT graduates or final-year students in the process of undertaking internships [2]. The analysis should aim to understand the current situation and outline the possible skills, qualifications, and tools needed in the future.

The main practical implication of such a study is that companies willing to perform vulnerability assessment soon must be aware of the fact that professionals do have different skills, qualifications, and tools from those that are used now or have been used in the past and should carefully select and possibly train the professionals that will undertake this new task. Georgiadou et al. [66] the K-nn algorithm has confirmed that aside from the expert hired to carry out penetration testing, recent ICT graduates or final-year students in the process of undertaking internships are other professional figures suitable for tasks related to vulnerability assessment [67]. It is beneficial for companies to have external professionals who take care of this delicate activity and an internal professional figure to manage the activity effectively. This will help increase security practices' efficiency without incurring excessive costs.

6.2 Call to Action

As we move forward into the digital age, it is imperative that we collectively advocate for the ethical practice of hacking to ensure the security and integrity of our online systems. Ongoing education and training in ethical hacking techniques are essential to avoid malicious cyber threats and protect sensitive information from unauthorized access. We must actively support and promote ethical hacking as a valuable tool in the ongoing battle against cybercrime [6]. This includes advocating for developing ethical hacking guidelines and standards and encouraging collaboration between ethical hackers and cybersecurity professionals to enhance our defenses against evolving threats [68]. We can chart a course toward a more secure and resilient digital future. A commitment to ethical hacking principles and ongoing education in cybersecurity is essential in shaping the future of cybersecurity professionals and protecting our digital world. To stay ahead of evolving threats and safeguard sensitive information, we must continue to innovate and adapt to new technologies while remaining vigilant in identifying and addressing vulnerabilities. This requires a collective effort from individuals, organizations, and governments to prioritize cybersecurity measures and collaborate on proactive solutions [69, 70]. By staying proactive and adaptive, we can ensure a secure and resilient digital infrastructure for future generations to promote a culture of responsible and ethical hacking practices in the cybersecurity community.

6.3 *Vision for the Future*

In the future, ethical hacking will continue to play a crucial role in safeguarding the digital landscape. As technology advances, ethical hackers must stay ahead of emerging cyber threats and develop innovative strategies to protect sensitive information. With a focus on proactive defense measures, a collaboration between ethical hackers and cybersecurity professionals will be essential in developing comprehensive and effective security solutions [71]. With devices and systems becoming increasingly interconnected, ethical hacking will become an indispensable tool for safeguarding against cyberattacks.

6.4 *Empowering Ethical Hackers*

One possible approach to empowering ethical hackers in the future of cybersecurity is through increased collaboration with industry partners and government agencies. This collaboration can lead to the development of advanced tools and techniques for ethical hackers to combat cyber threats effectively. Additionally, it can also pave the way for greater innovation and progress in the field of cybersecurity. "Empowering ethical hackers can help to bridge the gap between the demand for cybersecurity professionals and the shortage of skilled workers in the industry" George [72], Marat and Marat [73]. In an increasingly interconnected world, the role of ethical hackers in safeguarding cybersecurity has become more crucial than ever.

References

1. Svensson J (2021) Wizards of the Web: an outsider's journey into tech culture programming and mathemagics. Nordicom
2. Yaacoub J-PA, Noura HN, Salman O, Chehab A (2023) Ethical hacking for IoT: Security issues, challenges, solutions and recommendations. Int Things Cyber-Phys Syst 3:280–308. https://doi.org/10.1016/j.iotcps.2023.04.002
3. Baddam PR (2020) Cyber sentinel chronicles: navigating ethical hacking's role in fortifying digital security. Asian J Hum Art Lit 7(2):147–158
4. Gandhi F, Pansaniya D, Naik S (2022) Ethical hacking: types of hackers, cyber-attacks and security. Int Res J Innov Eng Technol 6(1):28

5. Orchilles J (2020) Ethical hacking from vulnerability scanning to adversary emulation. ISSA J, 18(6)
6. Smith L, Chowdhury MM, Latif S (2022) Ethical hacking: skills to fight cybersecurity threats. EPiC Ser Comput 82(5):102–111
7. Bozzini D (2023) How vulnerabilities became commodities. The Political Economy of Ethical Hacking (1990–2020)
8. Stoddart K (2022) Cyberwarfare: threats to critical infrastructure. Springer Nature
9. Wang Z, Sun L, Zhu H (2020) Defining social engineering in cybersecurity. IEEE Access 8:85094–85115
10. Zheng Y, Li Z, Xu X, Zhao Q (2022) Dynamic defenses in cyber security: techniques, methods and challenges. Digit Commun Netw 8(4):422–435
11. Aloqaily M, Kanhere S, Bellavista P, Nogueira M (2022) Special issue on cybersecurity management in the era of AI. J Netw Syst Manag 30(3):39
12. Pandey S, Singh RK, Gunasekaran A, Kaushik A (2020) Cyber security risks in globalized supply chains: conceptual framework. J Glob Oper Strat Sour 13(1):103–128
13. Furnell S (2021) The cybersecurity workforce and skills. Comput Secur 100:102080
14. Palombo H, Tabari AZ, Lende D, Ligatti J, Ou X (2020) An ethnographic understanding of software in Security and a Co-creation model to improve secure software development. In: Sixteenth Symposium on Usable Privacy and Security (SOUPS 2020)
15. Pfanstiel S (2022) Impact of internal control, cybersecurity risk, and competitive advantage on retail cybersecurity budget. waldenu.edu
16. Bellaby RW (2023) An ethical framework for hacking operations. The ethics of hacking. Bristol University Press, pp 32–52
17. Ghimire B, Rawat DB (2022) Recent advances on federated learning for cybersecurity and cybersecurity for federated learning for internet of things. IEEE Internet Things J 9(11):8229–8249
18. Arnone R (2021) Hackers cybercrime-computer security: ethical hacking: learn the attack for better defence. ARIS2 Adv Res Inf Syst Secur 1(1):50–61
19. Haq HBU, Hassan MZ, Hussain MZ, Khan RA, Nawaz S, Khokhar HR, Arshad M (2022) The impacts of ethical hacking and its security mechanisms. Pak J Eng Technol 5(4):29–35
20. Da Silva J, Jensen RB (2022) Cyber security is a dark art”: the CISO as soothsayer. Proc ACM Human–Comput Interact 6(CSCW2):1–31
21. Brooks T (2022) The professionalization of the hacker industry. arXiv preprint arXiv:2207.00890
22. Al-Tawil TNE (2024) Ethical implications for teaching students to hack to combat cybercrime and money laundering. J Money Launder Control 27(1):21–33
23. Onyema E, Dinar A, Ghouali S, Merabet B, Merzougui R, Feham M (2022) Cyber threats, attack strategy, and ethical hacking in telecommunications systems. Security and privacy in cyberspace. Springer, pp 25–45
24. Lone AN, Mustajab S, Alam M (2023) A comprehensive study on cybersecurity challenges and opportunities in the IoT world. Secur Privacy 6(6):e318
25. Perwej Y, Abbas SQ, Dixit JP, Akhtar N, Jaiswal AK (2021) A systematic literature review on the cyber security. Int J Sci Res Manag 9(12):669–710
26. Bansal D, Bhatia M, Atrey A, Yadav AK (2024) Perspective of cybersecurity and ethical hacking with vulnerability assessment and exploitation tools. In: Press CRC (ed) Big data analytics framework for smart grids. pp 98–111
27. Saha S, Das A, Kumar A, Biswas D, Saha S (2020) Ethical hacking: redefining security in information system. In: Proceedings of International Ethical Hacking Conference 2019: eHaCON 2019, Kolkata, India
28. Rana MU, Ellahi O, Alam M, Webber JL, Mehbodniya A, Khan S (2022) Offensive security: cyber threat intelligence enrichment with counterintelligence and counterattack. IEEE Access 10:108760–108774
29. Rosati P, Paulsson V (2022) The evolution of accounting information systems. The Routledge handbook of accounting information systems. Routledge, pp 21–32

30. Ibrahim H (2022) A review on the mechanism mitigating and eliminating internet crimes using modern technologies: mitigating internet crimes using modern technologies. Wasit J Comput Math Sci 1(3):50–68
31. Ali A (2022) Cyberspace and organized crime: the new challenges of the 21st century. Int J Adv Hum Res 2(2):22–37
32. Kizza JM (2023) Ethical and secure computing: a concise module. Springer
33. Oliver D, Randolph AB (2022) Hacker definitions in information systems research. J Comput Inf Syst 62(2):397–409
34. Webb M (2020) Coding democracy: how hackers are disrupting power, surveillance, and authoritarianism. MIT Press
35. Corea F, Bertinetti G, Cervellati EM (2021) Hacking the venture industry: an early-stage startups investment framework for data-driven investors. Mach Learning Appl 5:100062
36. Jaquet-Chiffelle D-O, Loi M (2020). Ethical and unethical hacking. In: The ethics of cybersecurity, pp 179–204
37. Zeadally S, Adi E, Baig Z, Khan IA (2020) Harnessing artificial intelligence capabilities to improve cybersecurity. IEEE Access 8:23817–23837
38. Kaloudi N, Li J (2020) The ai-based cyber threat landscape: a survey. ACM Comput Surv (CSUR) 53(1):1–34
39. Ansari MF, Dash B, Sharma P, Yathiraju N (2022) The impact and limitations of artificial intelligence in cybersecurity: a literature review. Int J Adv Res Comput Commun Eng. https://doi.org/10.17148/IJARCCE.2022.11912
40. Yamin MM, Ullah M, Ullah H, Katt B (2021) Weaponized AI for cyber-attacks. J Inf Secur Appl 57:102722
41. AlShabibi A, Al-Suqri M (2021). Cybersecurity awareness and its impact on protecting children in cyberspace. In: 2021 22nd International Arab Conference on Information Technology (ACIT)
42. Fuster GG, Jasmontaite L (2020) Cybersecurity regulation in the European union: the digital, the critical and fundamental rights. In: The ethics of cybersecurity, pp 97–115
43. Katagiri N (2024) Defending medical facilities from cyber-attacks: critical issues with the principle of due diligence in international law. Int Rev Law Comput Technol 38(1):1–20
44. Binhammad M, Alqaydi S, Othman A, Abuljadayel LH (2024) The role of AI in cyber security: safeguarding digital identity. J Inf Secur 15(02):245–278
45. Lucarelli S, Marrone A, Moro FN (2021) NATO decision-making in the age of big data and artificial intelligence. NATO, Brussels
46. Carmel E, Roche EM (2023) The dominant cybersecurity industry clusters: evolution and sustainment. Ind Innov 30(3):361–391
47. Harkin D, Molnar A (2023) Exploring the social implications of buying and selling cyber security. Crime Law Soc Chang 79(1):83–100
48. Yaacoub J-PA, Salman O, Noura HN, Kaaniche N, Chehab A, Malli M (2020) Cyber-physical systems security: limitations, issues and future trends. Microprocess Microsyst 77:103201
49. Dimitrov W (2020) Analysis of the need for cyber security components in the study of advanced technologies. In: INTED2020 Proceedings
50. Goupil F, Laskov P, Pekaric I, Felderer M, Dürr A, Thiesse F (2022) Towards understanding the skill gap in cybersecurity. In: Proceedings of the 27th ACM Conference on Innovation and Technology in Computer Science Education, Vol. 1
51. Malatji M, Tolah A (2024) Artificial intelligence (AI) cybersecurity dimensions: a comprehensive framework for understanding adversarial and offensive AI. In: AI and Ethics, pp 1–28
52. Yaacoub J-PA, Noura HN, Salman O, Chehab A (2021) A survey on ethical hacking: issues and challenges. arXiv preprint arXiv:2103.15072
53. Hawamleh A, Alorfi ASM, Al-Gasawneh JA, Al-Rawashdeh G (2020) Cyber security and ethical hacking: the importance of protecting user data. Solid State Technol 63(5):7894–7899
54. Kumar JRR, Bhalke D, Nikam S, Chobe S, Khidse S, Kale K (2024) Original research article evaluation of the extent and demanding roles of ethical hacking in cybersecurity. J Auton Intell, 7(1)

55. Macnish K, Van der Ham J (2020) Ethics in cybersecurity research and practice. Technol Soc 63:101382
56. Cai B, Mross E (2024) Integrating ethical framework in the information system design process
57. Sinha S, Arora DY (2020) Ethical hacking: the story of a white hat hacker. Int J Innov Res Comput Sci Technol (IJIRCST), ISSN, 2347–5552
58. Grover S, Broll B, Babb D (2023) Cybersecurity education in the age of AI: integrating AI learning into cybersecurity high school curricula. In: Proceedings of the 54th ACM Technical Symposium on Computer Science Education, Vol 1
59. Scott K (2021) Connected, continual conflict: towards a cybernetic model of warfare. In: ECCWS 2021 20th European Conference on Cyber Warfare and Security
60. Pattison J (2020) From defence to offence: the ethics of private cybersecurity. Eur J Int Secur 5(2):233–254
61. Dautaniya AK, Jha A. Ethical hacking issues: a review. ijte.in
62. Kaushik K, Khan A, Kumari A, Sharma I, Dubey R (2024) Ethical considerations in AI-based cybersecurity. Next-generation cybersecurity: AI, ML, and blockchain. Springer, pp 437–470
63. Raman R, Calyam P, Achuthan K (2024) ChatGPT or bard: who is a better certified ethical hacker? Comput Secur 140:103804
64. Del-Real C, Rodriguez Mesa MJ (2023) From black to white: the regulation of ethical hacking in Spain. Inf Commun Technol Law 32(2):207–239
65. Singh N, Lien L (2023). Ethical hacking as a risk management tool in organizations Nord universitet
66. Georgiadou A, Mouzakitis S, Bounas K, Askounis D (2022) A cyber-security culture framework for assessing organization readiness. J Comput Inf Syst 62(3):452–462
67. Hong WCH, Chi C, Liu J, Zhang Y, Lei VN-L, Xu X (2023) The influence of social education level on cybersecurity awareness and behaviour: a comparative study of university students and working graduates. Educ Inf Technol 28(1):439–470
68. Loi M, Christen M (2020) Ethical frameworks for cybersecurity. In: The ethics of cybersecurity, pp 73–95
69. Blažič BJ (2022) Changing the landscape of cybersecurity education in the EU: Will the new approach produce the required cybersecurity skills? Educ Inf Technol 27(3):3011–3036
70. Crumpler W, Lewis JA (2022) Cybersecurity workforce gap. JSTOR
71. Zaid T, Garai S (2024) Emerging trends in cybersecurity: a holistic view on current threats, emerging solutions, and pioneering new frontiers. Blockchain in Healthcare Today
72. George AS (2024) Riding the AI waves: an analysis of artificial intelligence's evolving role in combating cyber threats. Partners Univ Int Innov J 2(1):39–50
73. Marat C, Marat S (2024) The role of cybersecurity education on ethical hacking
74. Li Y, Liu Q (2021) A comprehensive review study of cyber-attacks and cyber security; emerging trends and recent developments. Energy Rep 7:8176–8186
75. Vijayakumaran C, Muthusenthil B, Manickavasagam B (2020) A reliable next generation cyber security architecture for industrial internet of things environment. Int J Electr Comput Eng 10(1):387

Artificial Intelligence and the Human Brain: Two Minds with a Single Thought, the Legal Problem of Inventorship in Artificial Brains

Ana Penteado

Abstract Artificial intelligence, a term coined by John McCarthy in 1955, is a generic term for learning machines. With the advent of neural networks mimicking the complex hierarchy of human brains, new possibilities for the generation of sequences in words, contextual sentences, and paragraphs that may result in behavioral-based attitudes have represented a breakthrough in computer systems, data storage, and innovation. One of the learning machines' abilities is to perform several tasks, often self-supervised, on a large scale with an immense amount of data collected from human experiences by pre-trained large language models with the capacity for independence in the decision-making process of recognizing facts and actions. In light of these technological marvels, we discuss the concept of humanity in recent inventions named artificial brains implanted in the cerebrum for medical purposes such as alleviation of the symptoms of Alzheimer's disease, memory loss, a verbal disorder as aphasia and physical limitations. With reinforcement learning, the boundaries between machines and humans are merged into one.

Keywords Artificial intelligence (AI) · Artificial brain · Humanity and AI · Harmful effects

1 Introduction

Artificial intelligence (AI) is relatively a new technological inclusion in the domain of Intellectual Property (IP) theory. Eligibility for granting inventions is based on technical elements of novelty and non-obviousness, and the state-of-the-art is limited to the implications of AI use. This innovative application of algorithms that learn from humans independently with no guidance is almost a reality for everyone.

A. Penteado (✉)
The University of Notre Dame Australia, Sydney, Australia
e-mail: ana.penteado@nd.edu.au

M. Chakraborty et al. (eds.), *Proceedings of 5th International Ethical Hacking Conference*, Lecture Notes in Networks and Systems 1148,
https://doi.org/10.1007/978-981-97-8457-8_3

AI is set to save our lives and exclusively work for human welfare and perpetual happiness. What is not included in this equation is singularity and reinforcement learning, which equip AI to learn from human experiences according to parameters that may be overlooked because of independently planning and decided sequences by autonomous systems. We do not have a scientific justification to explain the algorithms' reasoning for ignoring some parameters and embracing others in AI results. Those results in a black box activity may unveil surprising outcomes.

In an artificial neural net architecture that naturally incorporates human logic and machine learning factual analysis, new boundaries of intelligence have unveiled two major philosophical issues for legal policies in AI what is human nature and, subsequently, the eligibility of what is considered human. These are fundamental preconditions to attain moral and ethical values to our laws and constitutional obligations as human beings or rather natural persons living in a community. What is the precondition for anything to be human? Augmented intelligence seems to indicate that machines associated with humans are part of humanity.

However, AI combined with human interaction produced augmented intelligence via invasive devices in the human body, which may have potentially detrimental results on natural personality. Humans have the capacity to select past and present memories of living experiences to justify actions. Moreover, the human brain functions as a manager of our existence, coordinating with other organs the state of being alive.

AI is an evolving technology in intellectual property theory. Patents in AI applications, specifically machine learning [1], have increased exponentially since the end of the pandemic, from autonomy in automobiles to artificial brains. Eligibility for inventions based on novelty, non-obviousness, and state-of-the-art is limited to technicalities divorced from the future applications of AI for consumers' use. This innovative application of algorithms that learn independently with no guidance from humans is almost a reality for everyone around the planet.

AI is set to save our lives and work for human welfare and perpetual happiness for future generations. Conversely, AI may be detrimental to human health, causing permanent disability or, in extreme cases, death [2]. Take the example of invasive devices such as a spiked helmet attached to a natural brain controlled by software outside of the human body as such inventions have already been approved by patent offices [3], and the array of ethical issues will multiply into several areas of knowledge such as psychological, legal, and medical problems that regulations and laws cannot forecast with the tools of over two centuries of innovative inventions. (In fact, the first democratic constitutional legal framework with egalitarian constitutional guarantees to inventorship was ratified in June 1788. [4]).

Another aspect is more subtle, as data mining is still unregulated on the open internet. Such freedom to harvest human data free of compensation has been training large learning models for years. The use of free human datasets available on the open internet is, stricto sensu, data misappropriation, as it provides an economic advantage to the data sellers, not to the ones who create content. Again, the paradox that interferes with this paragon of artificial virtues that will manage our memories and historical references is tested when one examines the question of what is to be a natural person. Until recently, being part of humanity was justified by human experiences, not artificial lives.

2 AI Applications

AI applications are based on human datasets collected for free from our digital world. The absence of regulation for AI applications on software and devices implies that the economic use of AI and trained models on human data means that it is collected without prior informed consent from natural persons. Detached from creator, data is an asset for machines. Further, there is no further informed decision to relinquish it and transform human data into a profitable activity.

Human data is apparently collected and, most of the time, stored on private property generated on social platforms outside of the control and management of data creators. The problem becomes more concerning when invasive devices aimed at mental patient users, such as those suffering from Alzheimer's disease [5], are consumers and users of such technology. These patients are memory loss sufferers and present difficulty in thinking and understanding facts and producing memories.

An AI brain acting as a supporting mind managing emotions and memory storage will interfere with patients' mental decline in unanticipated ways that are not forecasted by the invention.

Among the many uses of AI and its applications, we focused on therapies provided by mainstream healthcare, an industry that claims AI can be deployed in invasive devices for natural persons. It seems that there is a conscious decision to avoid inquiries into potential natural behavior interference and human brain activity. The human interaction with this technology is designed to be an augmented reality.

Medical devices of an invasive nature, to be deployed inside the human brain, managing its reactions and responses to the environment, are extremely unsafe. These devices raise questions about the legal boundaries of prior informed consent for patients and custodians, the negative interaction of an artificial brain with a biological mind, and the possible consequences of device misuse. Misuse is also understood here as offshoots that the artificial brain may produce, as autonomy is possible for large learning models, exerting influence on the functionality of the natural brain. One may easily apprehend that current legal theories fail to take into consideration the impact of such technology beyond commercial use by users.

We must be able to do interdisciplinary research to fully understand the impact of this novel technology on the human body. Similarly, technology innovates faster than any piecemeal legislation enacted to solve the legal gap domestically or internationally.

As far as intellectual property theory is concerned, patents and patentability eligibility for medical inventions using AI in humans reveal its limitations in regulating such technology. The main focus is on leading the competitive market for such inventions to maximize their use as an instrument of commercialization, unlocking economic value for licensing mechanisms, and monopolizing access to their use.

Coincidentally, in this era devoted to engineering science and software programming, the rush to acquire more scientific knowledge is geared in the opposite direction of humanities. Thus, ethical protocols and the study of ethics are forgotten in dusty libraries worldwide. (Lawyers are taught at law schools' professional responsibility, and legal ethics, which is a sub-discipline of Ethics. This trend has been the result of the Realism school that claimed social responsibility. Then later, factual situations would be the main content to teach evaluation of legal ethics, so even attorneys are not prepared to look into problems raised by technology from the ethical point of view of virtues and morality that should have been embedded in law students. Reality may be elusive in achieving such a positive result). [6]. Legal instruments are already imperfect instruments of moral standards because there is no equal agreement among so many jurisdictions on the legal foundation of AI. Pluralism enriches us. However, it shows our contradictions in values and virtues. That moral gap unfolds opportunities for some inventors to expand their economic power through innovative creations, in which AI is the most recent scientific application.

Humans aim to achieve happiness, which is one of the objectives in one's life; the inquiry into what humanity means to natural persons is a complex question. Tasking machines to do odd jobs, keeping us company in our solitude, or using technology to manage our complex global problems may be eligible for being human.

With the use of language derived from humans, machines have adopted a natural person likeness, a language to communicate with other humans, until offshoots have created a language compatible and understandable for machines only; the case of Facebook AI Research Lab (FAIR) bots is illustrative, it prompted the Barcelona Declaration to be designed to avoid ethical issues on AI to be forgotten [7].

Therefore, for large learning language models to outpace human intelligence, it is not a question of how but a reality that most of us must resign to accept when it comes about. That is the current narrative with few suggestions for the opposite. One of the elements of AI that has dominated the social and academic narrative is singularity as the last frontier between people and machines [8].

So, if singularity is just a step away and some are wary of our extermination as human species, should one not be concerned about moral standards in this new technology before it is too late?

In AI, the machine that works with its own language creates its own culture and has its own mediated knowledge to communicate. Conversely, large language models are not necessarily our human language with a cultural and emotional interpretation attached to it. Similarly, foundation models trained by large-scale data are pre-trained language models as the successive versions of GPT-3 can attest [9]. Machines can think in the literal meaning of the verb; however, AI does not emulate our cultural experiences from the highs and lows of human existence because that requires critical thinking and judgment of circumstances. Our darkest moments are also our time to adapt and survive challenges presented by the environment as novel situations.

AI aficionados propose a good argument that AI can think. This is a broad statement if one disregards that machines learn and think using a different and distinctive logic than humans. Artificial intelligence can think, but it is devoid of humanity. The human datasets are expressed in an algorithmic system fed by raw data.

The study of neural networks and their structure as a model for AI leads us to a multidisciplinary field of the study of predictions in computer systems, which may impact areas of humanities as deeply as the nuclear bomb research in the Manhattan Project [10]. That was also a technology that collided with our moral principles and human rights. The practical application of releasing energy via unleashing a chain of atomic power in a device as lethal as a nuclear bomb represented the power of destruction and the power of survival, a limited set of choices to select according to the circumstances and the technology available for the survival of humanity.

Machines are built to mimic right and wrong according to what they have been trained for in their foundation models. If a human being follows an inner conviction to act because she understands the subjective situation she finds herself in, this is the ability to exercise judgment. That is the domain of the human brain and its assessment of presented facts, forecasting responsibilities and consequences that will be generated by individual actions or collective actions in a community.

Since regulatory mechanisms are bound with generic commands and neutral language is utilized aiming at equality, AI may be suitable for data governance when we know the various capabilities of this technology, which is unknown. One must remember that the world of algorithms is the world of measuring information.

Norbert Wiener, in The Human Use of Human Beings [11], illustrates progress and entropy functionality, which relates to a set of patterns in which the more probable an event will occur, the less randomness is present. Humans are not constituted by a set of patterns only, unlike most objects of scientific study. There is complexity in the unpredictable actions a human being may choose to act or refrain from doing.

Futuristic thinkers such as Ray Kurzweil fail to inform us that the computer cannot perfectly emulate the human brain because the human datasets provided for AI and other technologies are as imperfect as our personalities. This aspect appears to be a small detail buried in the equation of marvelous AI or unrestrained enthusiasm for singularity, which is dangerous. One should be troubled by such promises of well-being [12]

Conversely, we do not sit placidly thinking that AI will not evolve by itself, learning from our human experiences and acquiring our vices in the process, which was the concern of the late Stephen Hawking [13] and other scientists as to whether the unregulated and unobserved research of AI aimed at further commercial utilization can thrive unsupervised, instigating success for all or, alternatively, our own misery.

Human rights are foreign to AI and will always be averse to self-serving goals for technologies. The profit of open data is the opposite of what human rights must aim to safeguard us from, which is to protect our privacy and intellectual manifestations. Human rights are, by definition, egalitarian in nature, just and inclusive of all races.

3 Conclusion

Therefore, AI inventions that interfere with brain activity and aim to control our feelings and memories should be seen as an invasion of our basic human rights, a dent in our privacy rights, and a desecration of dignity. Moreover, when a human evaluates a symbol in a situation that requires action, a temporal storage medium is actionable in the mind, which has the awareness of the symbol as part of the whole history of the given situation [14].

We are not perfect, nor are the machines based on mimicking our own flaws by monitoring online human behavior. Thus, we expect efficiency and perfection from an imperfect human dataset fed in a black box. One is not bewildered that machines may have their own language and create their own set of solutions in a different evaluation than humans; one assumes that this is the future. That is dangerous for the whole human race.

References

1. World Intellectual Property Organisation, WIPO Technology Trends 2019: Artificial Intelligence, available at https://www.wipo.int/edocs/pubdocs/en/wipo_pub_1055.pdf
2. Scheier B, Ottenheimer D. The Atlantic, in Robots are already killing people, available at https://www.theatlantic.com/technology/archive/2023/09/robot-safety-standards-regulation-human-fatalities/675231/
3. USPTO patent number 11,119, 483 B2, September 2021 (2021) System and methods for conscious machines. available at https://patentimages.storage.googleapis.com/fb/c6/0e/28838d225e0b65/US11119483.pdf
4. The United States Constitution (2024) in national constitutional center, available at https://constitutioncenter.org/blog/the-day-the-constitution-was-adopted (as of 10 Apr 2024)
5. Castellani RJ, Rolston RK, Smith MA (2010) Alzheimer disease. Dis Mon 56(9):484–546. https://doi.org/10.1016/j.disamonth.2010.06.001
6. Moliterno JE (2024) College William and Mary Law School, William and Mary Scholarship Repository, Faculty Publications. In: an analysis of ethics teaching in law schools: replacing lost benefits of the apprenticeship system in the academic atmosphere. available at https://scholarship.law.wm.edu/cgi/viewcontent.cgi?referer=&httpsredir=1&article=2041&context=facpubs (as of 10 Apr 2024)
7. Steels LS, de Mantaras RL (2018) The Barcelona declaration for the proper development and usage of artificial intelligence in Europe. AI Commun 31:485
8. Boden MA (2018) Artificial intelligence: a very short introductions, the singularity, Oxford Academic, p 130, See also Shanahan M (2015) The technological singularity. published by MIT Press
9. Manning C (2022) HAI Stanford University human-centered artificial intelligence, artificial intelligence definitions, V.1.2, published in April
10. National Historical Park (2024) Nm, WA, TN, National Park Service, Manhattan Project, What is the Manhattan Project?, available at https://www.nps.gov/mapr/learn/manhattan-project.htm (as of 10 Apr 2024)
11. Wiener N (1950) The human use of human beings cybernetics and society. Houghton Mifflin Company, Cambridge, pp 20–21
12. Walsh T. Machines, behaving badly the morality of AI. La Trobe University Press, Victoria, p 28

13. BBC News (2024) Stephen Hawking warns artificial intelligence could end mankind, available at https://www.bbc.com/news/technology-30290540 (as of 10 Apr 2024)
14. Von Der Malsburg C (1986) Am I thinking assemblies? In: Günther P and Aertsen A (eds) Brain theory, Springer-Verlag, Berlin, pp 161–176

Artificial Intelligence and Inventorship

Althaf Marsoof

Abstract This chapter explores the intersection of artificial intelligence (AI) and inventorship, focusing on whether AI can be legally recognized as an inventor. It examines key judicial decisions in the DABUS series of cases from the United States, United Kingdom, and Australia, which all reject AI as an inventor under current patent law. The chapter delves into the legal definitions of inventorship, the role of human intentionality and creativity, and the potential need for legislative changes. It also discusses the implications of AI-generated inventions on patent ownership, novelty standards, and the broader intellectual property system.

Keywords AI · Inventorship · Patent law · DABUS cases · Comparative perspectives

1 Introduction

Artificial Intelligence (AI) has propelled society into uncharted territory, challenging conventional legal frameworks and norms. As AI permeates various facets of our lives, from autonomous vehicles to algorithmic decision-making systems, it presents a plethora of challenges raising difficult questions of legal interpretation and questioning the normative foundations of law as we know it. However, of the various challenges that AI has posed in the legal space, one area stands out as particularly contentious. And that is the interface between AI and inventorship, which is the focus of this chapter.

A. Marsoof (✉)
Division of Business Law, Nanyang Business School, Nanyang Technological University, Singapore, Singapore
e-mail: althaf@ntu.edu.sg

M. Chakraborty et al. (eds.), *Proceedings of 5th International Ethical Hacking Conference*, Lecture Notes in Networks and Systems 1148,
https://doi.org/10.1007/978-981-97-8457-8_4

2 AI, an Inventor?

Traditionally, the concept of inventorship has been firmly rooted in human creativity and ingenuity. Inventors, often celebrated as visionaries and pioneers, are recognized and rewarded by legally protecting their inventions through the intellectual property system. However, the emergence of AI as a tool for generating novel ideas and solutions has blurred the lines of traditional inventorship. With AI systems capable of autonomously generating inventive concepts and designs, the question of inventorship becomes increasingly complex.

At the heart of this complexity lies the tension between the recognition of AI as a creative agent and the preservation of human agency and innovation. While AI systems can undoubtedly produce inventive ideas and concepts, the absence of human intentionality and consciousness raises fundamental questions about the attribution of inventorship. It is against the backdrop that I pose the question—should AI systems be legally recognized as "inventors"?

The answer to the above question, at least for now, has been a firm no. And this is patently clear when we consider the infamous series of cases concerning DABUS and its inventions. DABUS, short for Device for the Autonomous Bootstrapping of Unified Sentience, an AI system created by Dr Stephen Thaler, purportedly came up with two inventions—one concerned a design for an improved beverage container [1], and the other was a neural flame device for search and rescue missions [2]. Dr Thaler filed patent applications for these inventions in a number of countries naming DABUS, the AI system, as the inventor. However, the applications filed in the United States, the United Kingdom, and Australia were rejected because DABUS was not human. Dr Thaler challenged these rejections, resulting in a series of judicial decisions on whether an AI system can be recognized as an inventor under patent law. In the next part of this chapter, I consider these judicial decisions in some detail.

2.1 United States

In July 2019, Dr Thaler filed two patent applications with respect to the inventions identified above at the United States Patents and Trademarks Office (USPTO). He named DABUS as the sole inventor and claimed himself to be the assignee. Having reviewed the two applications, the USPTO issued a notice to Dr Thaler requesting him to identify valid inventors for the inventions. In response, Dr Thaler submitted a petition to the USPTO to vacate the notice, which the USPTO denied on the basis that a machine does not qualify as an inventor. Thereafter, Dr Thaler pursued judicial review of the USPTO's final decision in US District Court for the Eastern District of Virginia. The District Court agreed with the USPTO's conclusion and denied relief [3]. Dr Thaler appealed.

Thus, the US Court of Appeals for the Federal Circuit had to determine whether an AI system was an "individual" for the purposes of the US Patent Act, 35 U.S.C. §

100(f), which defines an inventor as "the individual [...] who invented or discovered the subject matter of the invention" [4]. The Patent Act does not provide an interpretation for "individual", but the Federal Circuit, relying on US Supreme Court precedent, concluded that unless a contrary congressional intent appears from the text of a statute, the noun "individual ordinarily means a human being" [5]. Moreover, previous decisions of the Federal Circuit confirm that only natural persons (i.e. humans) (and not corporations or sovereigns) can be designated as inventors [6, 7]. On the same reasoning, since DABUS is not a natural person, it did not qualify as an individual who could be designated as an inventor under US patent law—"[w]hen a statute unambiguously and directly answers the question before us, our analysis does not stray beyond the plain text. Here, Congress has determined that only a natural person can be an inventor, so AI cannot be. Accordingly, the decision of the district court is affirmed"[4].

Dr Thaler's petition for a writ of certiorari to the US Supreme Court was denied affirming the position that under US patent law (and until the law is amended by Congress), only humans can be granted patent rights as inventor [8]. In essence, AI inventorship has been rejected as a possibility under current law.

2.2 *United Kingdom*

In the United Kingdom, Dr Thaler filed two patent applications in 2018 in respect of DABUS's inventions. The applications expressly indicated that Dr Thaler was not the inventor. The UK Intellectual Property Office (UKIPO) notified Dr Thaler requiring him to file a statement of inventorship in terms of section 13(2) of the Patents Act 1977. This was done by Dr Thaler in July 2019. Interestingly, Dr Thaler indicated "DABUS" in the field for the family name of the inventor and in response to the box requiring the applicant to indicate how he had the right to be granted a patent, Dr Thaler wrote "by ownership of the creativity machine 'DABUS'". An explanation was filed by the applicant stating that the invention was generated by DABUS briefly explaining why Dr Thaler should be granted the patents. The UKIPO, consequent to a hearing, decided that DABUS was not a "person" within the meaning of the Patents Act and, therefore, cannot be an inventor. This also meant that Dr Thaler was not entitled to apply for the patents on the basis stated in his applications. Dr Thaler's appeal to the UK High Court was dismissed on the ground that DABUS could not be an inventor because it was not a person and Dr Thaler has failed to establish a right to be granted the patents in respect of the underlying inventions [9]. He appealed to the Court of Appeal. The decision of the Court of Appeal is considered in some detail below.

The Court of Appeal had to determine whether Dr Thaler could be granted patent rights in circumstances where he had identified DABUS as the inventor of the subject matter of the patent applications. Under section 7 of the UK Patents Act, any person can apply for a patent, but a patent can only be granted to three classes of persons and no other. The first is to the inventor of the invention. The second is to any person who, by virtue of any rule of law, was at the time of the making of the invention entitled to the whole of the property in it. The property in the invention essentially means the right to apply for a patent, and here, the person derives rights from the inventor. The third is to the successor in title of any person mentioned above.

The Court of Appeal agreed with the High Court's decision that Dr Thaler cannot be granted the patents in respect of the inventions he had applied for. Obviously, Dr Thaler did not come within the first category because he had declared that DABUS was the inventor. The Patents Act defines an inventor as "the actual deviser of the invention" [10]. However, according to the Court, DABUS is not the actual deviser of the inventions because that is and can only be the person who actually devised the inventions [11]. The Court of Appeal concluded that the fact that machines can create inventions does not mean that they are inventors within the meaning of the Patent Act.

Dr Thaler also did not come within the third category because DABUS has no legal personality or capacity and thus cannot enter into any transaction with Dr Thaler transferring title to him (this is, of course, assuming that DABUS could hold any rights to the invention to begin with) [11]. Instead, Dr Thaler argued that he came within the second category because he was at the time of the making of the inventions entitled to "the whole of the property" in them by virtue of a rule of law. Dr Thaler's argument was that he possessed the right to apply for the patents in respect of the inventions devised by DABUS because he owned the machine—i.e. the computer/device in which DABUS operated [11]. However, merely because Dr Thaler owned the machine, this did not entitle him to own the information produced by the machine—the rule of accession in property law, which is based on Roman Law, is concerned with tangible property and not with intangibles [11]. As such, it was clear that no patent could be granted to Dr Thaler on any of the bases set out in the Patents Act.

A further appeal by Dr Thaler to the UK's Supreme Court also confirmed the positions taken by the Court of Appeal and the High Court [12]. In essence, unless the legislature intervenes, there is no possibility for an AI or any *non-human* to be designated as an inventor under UK law. Nor is it possible for anyone to derive the right to apply for patents from such non-human entities.

2.3 Australia

As in the United States and the United Kingdom, the Australian Intellectual Property Office (IP Australia) had to grapple with the question of whether to grant Dr Thaler a patent on the basis of DABUS's inventions. The application was filed in September 2019 under the terms of the Patent Cooperation Treaty 1970 and DABUS was named as the inventor in the application. Upon the application entering the national phase, IP Australia wrote to Dr Thaler's patent attorneys that the application failed to comply with the provisions of the Australian Patents Act 1990 and the relevant regulations because it did not identify a natural person as the inventor. In response, Dr Thaler advanced the position that an AI could legitimately be named as an inventor. Ultimately, it was decided that the Patents Act and regulations were incompatible with the suggestion that an AI could be recognized as an inventor. As a result, Dr Thaler's patent application lapsed. Dr Thaler applied to the Australian Federal Court to review the decision of the Deputy Commissioner of Patents. Justice Beach, the primary judge who considered the matter, concluded that an inventor as recognized under the Patents Act can be an AI system or device and ordered that the Deputy Commissioner's determinations be set aside [13]. This was a significant deviation from the norm and approaches adopted by the US and UK courts. However, when the Commissioner of Patents appealed to the Full Federal Court, it restored the Deputy Commissioner's original decision—thus reverting back to the position that AI cannot be treated as an inventor for the purposes of the Australian Patents Act [14].

The Full Federal Court declared that under section 15 of the Patents Act, "a patent for an invention may only be granted to a person who is an inventor", and "the reference to 'a person' emphasises, in context, that this is a natural person" [14]. Any person other than the inventor applying for a patent (such as Dr Thaler in this case) must establish a legally recognizable relationship between himself and the actual inventor [14]. However, such a relationship did not exist between Dr Thaler and DABUS [14]. For that reason, it was decided that Dr Thaler could not be granted a patent on the basis that DABUS was an inventor—because DABUS, a non-human entity, was not an inventor within the meaning of Australian patent law.

3 Further Reflections

While judicial decisions considered above make it explicitly clear that under current law, an AI system cannot be declared an inventor (at least under the laws of the United States, the United Kingdom, and Australia—which is largely representative of the approaches adopted in most other jurisdictions), and no person claiming to derive rights from an AI system is eligible for a patent grant, it is still worthwhile reflecting on the implications of legally recognizing AI systems as inventors. If that does eventually happen (in fact, in Brazil, there is already a Bill [15] to amend existing patent law so that AI systems can be designated as investors), many questions will arise.

For example, to whom should a patent be granted with respect to an AI-generated invention? In this regard, it is worth quoting the following passage from the judgment of the Full Federal Court of Australia in the DABUS litigation:

> In our view, there are many propositions that arise for consideration in the context of artificial intelligence and inventions. They include whether, as a matter of policy, a person who is an inventor should be redefined to include an artificial intelligence. If so, to whom should a patent be granted in respect of its output? The options include one or more of: the owner of the machine upon which the artificial intelligence software runs, the developer of the artificial intelligence software, the owner of the copyright in its source code, the person who

inputs the data used by the artificial intelligence to develop its output, and no doubt others [14].

Also, how should we go about threshold standards such as novelty and non-obviousness—who should be the person skilled in the art, and what might be the body of knowledge attributable to this person if it is an AI system? [14] What about incentives? Recognizing rights with respect to inventions generated by AI systems is likely to turn the current incentive structure that informs the patent system on its head.

Lastly, an important point to note here is that merely because an AI system created an invention, that does not mean that the underlying invention cannot be patented. The Full Federal Court of Australia made this very clear in its judgment. The Full Court observed:

…we do not accept the premise of the proposition, accepted by the primary judge and apparently influential in his reasoning, that if DABUS is not accepted to be an inventor, no invention devised by an artificial intelligence system is capable of being granted a patent. In the present case, it was said to be an agreed fact that DABUS is the inventor of the invention the subject of the application and that Dr Thaler is not. However, the characterization of a person as an inventor is a question of law. The question of whether the application the subject of this appeal has a human inventor has not been explored in this litigation and remains undecided. Had this question been explored, it may have been necessary to consider what significance should be attributed to various matters including the (agreed) facts that Dr Thaler is the owner of the copyright in the DABUS source code and the computer on which DABUS operates, and that he is also responsible for the maintenance and running costs [14].

The above observation of the Full Federal Court has significant implications because it is clear that AI systems can and should be used as *tools* in the inventive process, and the inventive outcomes of such use should be eligible for patent protection.

Ultimately, in my considered opinion, it is always a *human* who conceives the inventive concept, whether the inventive concept is derived from a purely human exercise, a purely automated process, or a mixture of the two. But what does it really mean to conceive an invention? The process involves a number of steps, which include, first, the identification of a problem or need; second, research and knowledge gathering; and third, conceptualizing or ideation. Now, when we consider these stages, AI can be used in all of them, but can we really say that an AI system is capable of "conceiving" an invention? While AI systems can be programmed to identify patterns, anomalies, or inefficiencies in data, which could be interpreted as identifying problems or needs, this is largely dependent on the data it has been trained on and the parameters set by human programmers. AI lacks the human experience and intuition that often leads to recognizing unmet needs or problems in a more holistic and contextual manner. AI systems will probably do better than humans in research and knowledge gathering due to their processing capabilities and analytical techniques. However, when it comes to the final stage, which arguably is the most crucial one in the inventive process, AI is likely to struggle. AI can generate ideas based on existing information and can even combine concepts in novel ways. However, this process is fundamentally different from *human* creativity. Ideas generated by AI are typically derivations or re-combinations of its training dataset; it lacks conscious thought and intention, which are key aspects of human ideation. AI does not "conceive" ideas in the human sense but can produce outputs that might resemble human ideation.

When addressing the question of whether AI can invent and be recognized as inventors, it is essential for us to consider the matter from the perspective of AI itself (i.e., what do AI systems "think" about this). After all, the law requires us to give all concerned parties (*even AI!*) a hearing before deciding on matters that affect them. And with that in mind, I posed the following question to ChatGPT, the world's most widely used generative AI tool—i.e. "Can an AI system conceive an invention?" The response I received was as follows:

> While AI can perform tasks that are components of the conceptualization process, its capability to "conceive" an invention in the way humans do is limited. AI lacks the intrinsic creativity, contextual understanding, and intentionality that are central to human invention. It can assist significantly in the process, augmenting human capabilities, but the spark of original conception, as of current AI capabilities, remains a distinctly human attribute…

4 Conclusion

The interplay between AI and inventorship challenges the foundational principles of patent law. As this chapter has shown, courts in multiple jurisdictions have denied AI systems the status of an inventor, emphasizing that inventorship requires human attributes. The DABUS cases illustrate the tension between technological advancements and existing legal approaches. Although courts in the United States, the United Kingdom, and Australia have concluded that patent law does not accommodate *non-human* inventors, ongoing legislative developments, such as those in Brazil, indicate that the debate is far from over.

The possible recognition of AI as inventors in the future will have implications to how we approach the grant of patent rights in respect of the resulting inventions. Many questions are also likely to arise. These questions, among other things, concern patent ownership, the standards of novelty and non-obviousness, and the necessity to rethink existing incentive structures that promote innovation. Until these questions are resolved and the law is willing to accept AI inventors, the essence of inventorship—the "spark" of original conception—will remain uniquely human. However, AI will still continue to play a significant role in the inventive process.

As we navigate this new frontier, balancing technological advancements with the human elements of creativity and innovation is crucial. Policymakers, legal scholars, and technologists must collaborate to develop frameworks that accommodate advancements in AI while preserving the foundational principles of the intellectual property system. The future of AI and inventorship will likely involve a nuanced approach, where AI enhances human ingenuity within a thoughtfully crafted legal framework.

References

1. Thaler (2020) Food container and devices and methods for attracting enhanced attention. Patent No. WO2020079499A1 (WIPO), available at https://patents.google.com/patent/WO2020079499A1/en
2. Thaler (2019) Devices and methods for attracting enhanced attention. Patent No. EP3563896A1 (European Patent Office), available at https://patents.google.com/patent/EP3563896A1/en
3. Stephen Thaler v Andrew Hirshfeld, 558 F. Supp. 3d 238 (E.D. Va. 2021)
4. Stephen Thaler v Katherine K Vidal, Under Secretary of Commerce for Intellectual Property and Director of the United States Patent and Trademark Office, United States Patent and Trademark Office, 43 F.4th 1207 (Fed. Cir. 2022)
5. Mohamad v Palestinian Authority, 566 U.S. 449 (2012)
6. University of Utah v Max-Planck-Gesellschaft zur Forderung der Wissenschaften E.V., 734 F.3d 1315 (Fed. Cir. 2013)
7. Beech Aircraft Corp. v EDO Corp., 990 F.2d 1237 (Fed. Cir. 1993)
8. Stephen Thaler v Katherine K. Vidal, See https://www.supremecourt.gov/search.aspx?filename=/docket/docketfiles/html/public/22-919.html
9. Stephen Thaler v Comptroller-General of Patents, Designs and Trade Marks [2020] EWHC 2412 (Pat)
10. Sec 7(3) of the Patents Act 1977 (UK)
11. Stephen Thaler v Comptroller-General of Patents, Designs and Trade Marks [2021] EWCA Civ 1374
12. Stephen Thaler v Comptroller-General of Patents, Designs and Trade Marks[2023] UKSC 49
13. Thaler v Commissioner of Patents [2021] FCA 879
14. Commissioner of Patents v Thaler [2022] FCAFC 62
15. Bill amending art. 6th of Law No. 9,279, of May 14, 1996, to provide for the ownership of inventions generated autonomously by artificial intelligence systems, Bill No 303/2024 (introduced to the Brazilian Parliament on 20 February 2024), available at https://www.camara.leg.br/proposicoesWeb/prop_mostrarintegra?codteor=2387544&filename=PL%20303/2024

Explainable Artificial Intelligence (XAI): A Perspective

Mohuya Chakraborty

Abstract The term "Explainable AI" (XAI) describes the capacity of AI systems to give humans intelligible justifications for their choices and actions. This is essential for fostering trust in AI, particularly in high-stakes domains like criminal justice, healthcare, or financial risk where choices have a direct influence on people's lives. With the application of XAI approaches, AI models should become more accountable, transparent, and comprehensible so that users may comprehend the reasoning behind a given decision. XAI gives users a better understanding of the decision-making process, which increases user trust and facilitates more efficient usage of AI systems. Black-box machine learning models can be made interpretable with the help of XAI approaches like LIME (Local Interpretable Model-agnostic Explanations) and SHAP (SHapley Additive exPlanations). By approximating the model's behavior around the prediction, LIME focuses on producing locally faithful explanations for specific forecasts. This is accomplished by first altering the input data and seeing how this affects the model's output. Next, it fits an understandable model to these changes. Conversely, SHAP values take into account all potential feature permutations to offer a single, comprehensive estimate of feature relevance for every prediction. The Shapley value, which gives each characteristic a value according to how much it contributes to the prediction, is the foundation of cooperative game theory, which is where SHAP values originate. To improve openness and confidence in AI systems, LIME and SHAP are both helpful resources for deciphering and elucidating the choices made by intricate machine learning models. This paper focuses on the perspective of XAI with two experimental studies dealing with breast cancer detection and financial credit risk prediction.

Keywords Black-box · Breast cancer · White-box · Explainable AI · Financial credit risk · Healthcare · Interpretable · LIME · SHAP

M. Chakraborty (✉)
Cybersecurity Centre of Excellence, Institute of Engineering and Management, University of Engineering and Management, Kolkata, India
e-mail: mohuyacb@iem.edu.in

M. Chakraborty et al. (eds.), *Proceedings of 5th International Ethical Hacking Conference*, Lecture Notes in Networks and Systems 1148,
https://doi.org/10.1007/978-981-97-8457-8_5

1 Introduction

The human mind is naturally curious about everything in nature. It is the responsibility of physicians to explain the consequences and adverse effects of the drug's ingredients as well as the misuse of dosage when they prescribe medication. Similarly, educators must provide a clear explanation for each student's exam score. In the medical and educational fields, as in the two scenarios above, countless more examples demonstrate why explaining a given task is crucial to fostering faith in the system. Currently, Artificial Intelligence (AI) is being used in many complex applications and its advancements have prompted its integration with decision-making processes necessitating the development of algorithms and strategies to explain AI outputs driven by the need for trustworthy AI and compliance with law. However, since many AI models are black-box, it might be difficult to understand their results. The lack of a clear explanation for an answer can significantly impact the trust users place in a particular system. Generally speaking, it is critical to comprehend the logic that drives an AI model's decisions. Consequently, the need arises for eXplainable AI (XAI) techniques to increase user confidence in AI models. In recent years, XAI has gained popularity as a study topic in the AI community [1]. Black-box AI models produce judgments or conclusions without offering a rationale for their selection. Deep networks of artificial neurons disseminate information and decision-making among tens of thousands of neurons in black-box models, creating a complexity that might be as challenging to comprehend as the human brain. There comes the role of white-box AI models, which are often known as interpretable models that can have their behavior, prediction-making process, and contributing variables explained. A model must have two essential components in order to be considered white-box: easily understood features and a transparent machine learning (ML) method. Hence, explainability and interpretability are two of the several properties that characterize trust in AI systems. Figure 1 shows the black-box AI vs white-box XAI model concepts. There is confusion about the outcome obtained from AI black-box models regarding a few questions like "why was the task undertaken/not undertaken?", "How to correct errors during failure", "When was the event a success/failure?", etc. Since the prediction outcome was done by numerous operations inside the black-box, it is completely non-transparent and non-interpretable at the user end. On the other hand, white-box AI gives proper explanation of the outcome of prediction that is clear and transparent to the user as it answers all the above questions.

1.1 Properties of XAI

A few colloquial phrases used in XAI literature to characterize human comprehension of ML processes are as follows.

Interpretability: A passive feature of ML used to elucidate the inner workings of an ML procedure in words that are clear to humans.

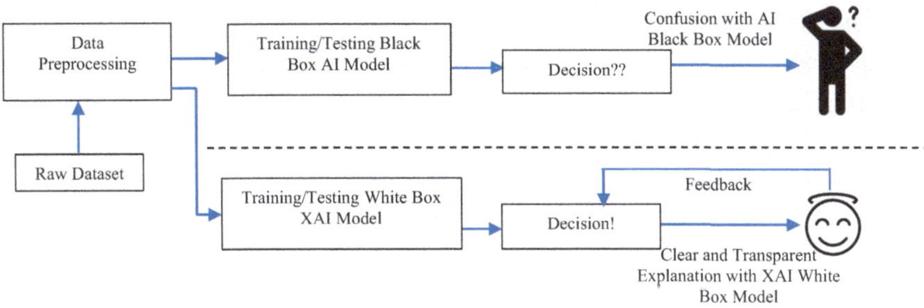

Fig. 1 Black-box and white-box model concepts

Explainability: A method, process, or human–machine interface for ML that facilitates understanding.
Transparency: When a model's operation can be understood by humans without the requirement for post-hoc justification, it is said to be transparent.
Opaque: An opaque model is the opposite of a transparent one. Humans cannot easily understand opaque models. They need post-hoc explanations in order to be interpreted.
Comprehensibility: The capability of an ML algorithm to convey its acquired information so as to make it logical to humans.
Accountability: The obligation of the ML model to answer for its actions, decisions, or performance.
Bias: Systematic errors or distortions in decision-making and a cognitive tendency that affects how the model perceives, interprets, and makes decisions.
Reliability: The ability of an AI system, procedure, or measurement to consistently carry out its intended task or yield reliable results over an extended period of time.
Resiliency: Ability of the model to cope with and recover from setbacks.

The phrases "explainable AI" and "interpretable AI" cannot be used interchangeably. An AI that can be understood by humans is said to be interpretable. An interface that aims to make the ML model interpretable for humans is called explainable AI [2].

1.2 *Philosophies of XAI*

Four philosophies of XAI determined through research from the fields of computer science, engineering, and psychology are explanation, meaningful, explanation accuracy, and knowledge limits [3]. From a computer science perspective the existing XAI algorithms and systems are placed into the context of these four philosophies whereas from the psychological perspective, investigation on how well people's explanations follow these four philosophies is done. Let us understand these philosophies of XAI systems briefly.

1 **Explanation:** Systems provide supporting data or justification for each output enabling clarification.
2 **Meaning:** Systems provide descriptions that are reasonable to individual users.
3 **Accuracy:** The explanation appropriately reflects the system's procedure for producing the output.
4 **Limits:** The system operates solely when it meets the designed conditions or when it gains sufficient confidence in its output.

1.3 Consumer Specific XAI

Explanations typically vary depending on the type of consumer. The empirical findings of [4] reveal that people prioritize importance of rankings over explanations for bad decisions, tolerate moderate explanation complexity, and prefer specificity based on context. People with a more deliberate cognitive style tend to prefer low-level specificity. The results align with customer behavior theories, suggesting marketing techniques can effectively understand explanation preferences. The consumer specific explanations such as user benefit, societal acceptance, regulatory and compliance, system development, and owner benefit are briefly discussed here [3].

1 **User Benefit:** The purpose of this explanation is to enlighten the user about an output; for example, it may clarify why a student failed or passed the examination.
2 **Societal Acceptance:** The goal of this explanation is to win over society's acceptance and trust, viz., this explanation could aid users in comprehending the reasoning behind the output's generation and also boost their confidence in the system in case the system yields an unforeseen outcome.
3 **Regulatory and Compliance:** This explanation helps to ensure that safety standards and regulations are followed, viz., the developer or auditor of a self-driving car, and may also include explanations to evaluate the output of a forensic examination after an airplane crash.
4 **System Development:** This explanation simplifies AI algorithm creation, enhancement, troubleshooting, and management for product managers, executives, and technical personnel, particularly those needing detailed information, like technical staff debugging vision algorithms.
5 **Owner Benefit:** This explanation aids system operators, such as in a recommendation system that offers a list of movies or videos with choices based on previous viewing. A system recommends a movie and explains this choice by stating "Here is a movie to watch because you liked these other movies". If the user trusts the explanation, the owner benefits because that person continues watching movies on their service.

1.4 Categories of XAI Methods

Different kinds of explanation strategies are available for creating XAI methods. Explainability techniques might be model-specific (designed for a specific functionality and a particular kind of model), or model agnostic, (generalized design that can be applicable to any type of model). Table 1 shows the various categories of XAI techniques with model specificity.

1 **Simplification**: This explanation method uses rule extraction and knowledge distillation to give explanation. These rules give a clear picture of the model's decision-making process, making it simpler for people to understand. Knowledge distillation is a technique that allows knowledge to be transferred to smaller, lighter models from larger, computationally expensive models without sacrificing the validity of the former models.
2 **Feature relevance**: This explains a feature by rating or quantifying how much of an impact it has on a prediction result. This method determines a score that indicates the "importance" of every input feature in an ML model. A higher score denotes a feature's greater impact on the model's capacity to forecast a given variable.
3 **Visual:** This explanation gives information by illustrating predictions visually. Before supplying the dataset to the training model, researchers can analyze and preprocess the datasets with the aid of visualization tools like heatmaps, bar graphs, box plots, etc. Confusion matrices, ROC curves, and feature importance all aid in the interpretation of model predictions during the testing phase.

Table 1 XAI techniques with method and model specificity

XAI Method / Model specificity	Simplification	Feature relevance	Visual	Concept-based
Model agnostic	Local Interpretable Model-Agnostic Explanations (LIME), G-REX	SHapley Additive exPlanations (SHAP), Automatic Structure Identification (ASTRID)	Individual Conditional Expectation (ICE), Partial Dependence Plots (PDPs)	ConceptLIME, Axiomatic Approach
Model specific	Simplified Tree Ensemble Learner (STEL), DeepRED	Mean Decrease Accuracy (MDA), Mean Increase Error (MIE), DeepLIFT	Layer-Wise Relevance (LRP), Gradient-Weighted Class Activation Map (Grad-CAM)	Testing with Concept Activation Vectors (TCAV)

4 **Concept-based:** This explanation offers clarification using user-defined concepts. These models concentrate on employing abstractions or notions that are understandable to humans to explain AI model predictions. They offer explanations in terms of relevant properties rather than depending only on raw features (such as pixel values in an image). These ideas more closely reflect human reasoning and explanation.

1.5 Levels of Explainability

XAI methods also provide explanations at different levels of granularity such as (i) Global (explanation for all predictions, or the model decision-making process itself), (ii) Cohort (explanation for a cohort or subset of predictions), (iii) Local (explanation for a single prediction) [5]. A set of predictions for local, cohort, and global XAI reasons for choosing a doctor for patients who are infants, females, or males is displayed in Fig. 2.

1 **Global Explanation:** Determining which characteristics are most responsible for the model's output is made easier with global explainability. It assists in ascertaining the role that a specific feature has in the model's ultimate conclusion or prediction. By taking into account variations in the degree of a specific characteristic utilized in the model's decision-making, it may also be used to comprehend how a model is "learning".
2 **Cohort Explanation:** Organizations can determine the specific variables causing the loss in accuracy in a particular subset with the help of the Explainability of Cohort Models. With this method, model soundness may be quantified and compared across several cohorts or data subsets. When a model owner wants to know why a model is not performing as well for some of its inputs, cohort

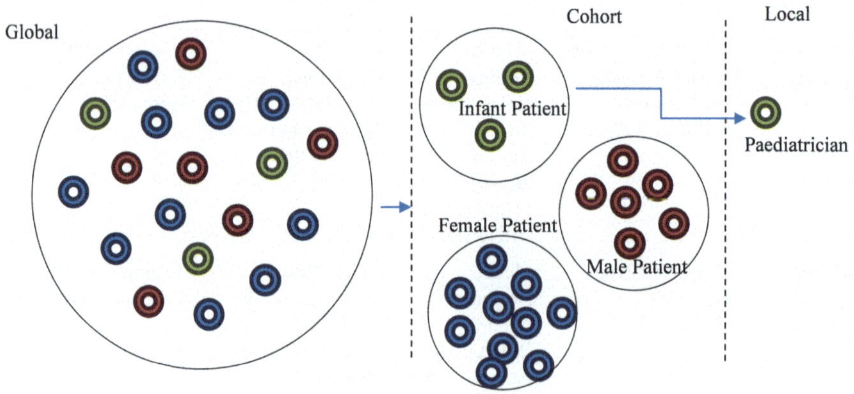

Fig. 2 Explanations at different levels of granularity for medical practitioners

explainability can be a huge assistance. It can help look for bias in the model and pinpoint places where the datasets might benefit from strengthening.

3 **Local Explanation:** An ML model's ability to make a single prediction is explained by local explainability. Determining which specific factors influenced a given decision is made easier with the help of the local model explainability idea. Explainability of local models is important for models used in regulated organizations.

After the introduction, Sect. 2 gives overview of several XAI techniques. Section 3 presents two experimental studies and analyses using XAI techniques followed by conclusion with some highlights on the future of XAI in Sect. 4.

2 Overview of XAI Techniques

This section presents an overview of the XAI techniques mentioned in Table 1 with model specificity and method. We can have **Model Agnostic** XAI technique that helps us comprehend complicated models without requiring us to understand their core workings. They are essential to maintaining AI systems' accountability, equity, and openness. There is a second type of XAI technique called **Model Specific**. These methods aim to shed light on the rationale behind particular ML models' decisions focusing on its internal workings.

2.1 Simplification

(i). Model Agnostic

t1**: Local Interpretable Model-Agnostic Explanations (LIME)** (LIME) is a local model interpretation method that approximates the predictions of the underlying black-box model using local substitution models. It hides the internal workings of the model that can be particularly beneficial when the model is confidential and cannot be revealed. It works on any black-box model, and works with a variety of data types such as tabular, text, and picture data. One can confirm the explanations and build confidence using domain knowledge [6].

t2**: Genetic Rule Extraction (G-REX)** refers to the extraction of genetic rules. It is an approach that uses genetic algorithm to produce AND/OR operators with IF–THEN rules. These guidelines aid in the explanation of the predictions produced by machine learning classifiers in certain situations. In essence, G-REX seeks to reveal the decision-making process of complicated models by extracting interpretable rules from them [7].

(b) Model Specific

t3: **Simplified Tree Ensemble Learner (STEL)** The STEL technique enhances ensemble models, particularly Random Forests, by eliminating rules, selecting relevant ones, and simplifying the ensemble learner, thereby increasing transparency and explanation [8].

t4: **Rule Extraction from Deep Neural Networks (DeepRED)** is a method for deriving rules from multi-layered deep neural networks (DNNs) [9].

2.2 *Feature Relevance*

(i) Model Agnostic

t5: **SHapley Additive exPlanations (SHAP)** is a game theory-based explanation of ML model output that involves equitable prediction value sharing among inputs, offering reliability through local explanations, effective credit distribution, and uniformly distributed predictions [10].

t6: **Automatic Structure Identification (ASTRID)** is an R-package for finding statistically significant attribute interactions with classifiers [11].

(b) Model Specific

t7: **Mean Decrease Accuracy (MDA)** measures how much the accuracy of the model drops when a particular covariate (feature) has its values permuted. Stated differently, it evaluates the amount that the model's performance degrades when randomizing the feature's values. By locating key features, MDA facilitates the analysis of random forests' opaque mechanisms. MDA frequently aims for the incorrect number, particularly in dependent situations, making it difficult to identify significant confounders. A novel significance metric known as Sobol-MDA has been put out to overcome this problem. The Sobol-MDA improves upon the shortcomings of the original MDA and yields feature importance estimations that are more precise [12].

t8: **Mean Increase Error (MIE)** helps us understand how much the model's prediction error increases when a specific feature is excluded from the model [13].

t9: **Deep Learning Important FeaTures (DeepLIFT)** dissects the neural network's output prediction for a given input by backpropagating the contribution of each neuron to each input feature. It does this by comparing each neuron's activity under the real input to the activations that a reference input would have caused. The input features are given relevance values based on the difference between these activations [14].

2.3 Visual

(i) Model Agnostic

t10: **Individual Conditional Expectation (ICE)** plots show how the prediction of an instance varies with a change in a feature by displaying one line for each occurrence. It provides a local view of the relationship between a feature and the model's prediction. These plots are a method of analyzing the prediction of a model by varying the value of a specific feature for each data instance, resulting in a set of lines representing the change in prediction [15].

t11: **Partial Dependence Plot (PDP),** while maintaining the same values for other characteristics, offers a global visual depiction of how inputs affect the anticipated outcome of the model. It establishes whether there is a simple or complex relationship between the selected attribute and the goal. PDP, or the mean of the lines in an ICE plot, is calculated by utilizing the black-box model to make predictions and swapping over feature values for grid values. A set of points with grid values and forecasts is the end product [16].

(b) Model Specific

t12: **Layer-wise Relevance Propagation (LRP)** technique assigns the output importance of a neuron or feature to those that aided its activation, using propagation rules that consider the activation function at each layer. This helps identify input data segments that significantly influence network choice [17].

t13: **Gradient-Weighted Class Activation Map (Grad-CAM)** is a tool that helps in class prediction by visualizing the importance of various regions in an input image. It creates heatmaps that show the areas that have the biggest influence on the final classification choice. Grad-CAM, in contrast to LRP, calculates the gradient of the target class score while operating at convolutional layers [18].

2.4 Concept-Based

(i) Model Agnostic

t14: **Concept LIME** formulates explanations in terms of ideas that are intelligible by humans, unlike the semantically inadequate explanations for image classification at super-pixel level provided by the original LIME thereby enhancing the method's comprehensibility [19].

t15: **Axiomatic Approach** is done through linearity, recursivity, and similarity. Linearity is a concept explanation measure's behavior when combining multiple concepts, allowing for the weighted sum of their importance scores to break down their combined effect. Recursivity is a concept explanation metric that assesses a concept's significance by considering its influence on other concepts, aiding in prediction and understanding of relationships. Similarity is a concept explanation

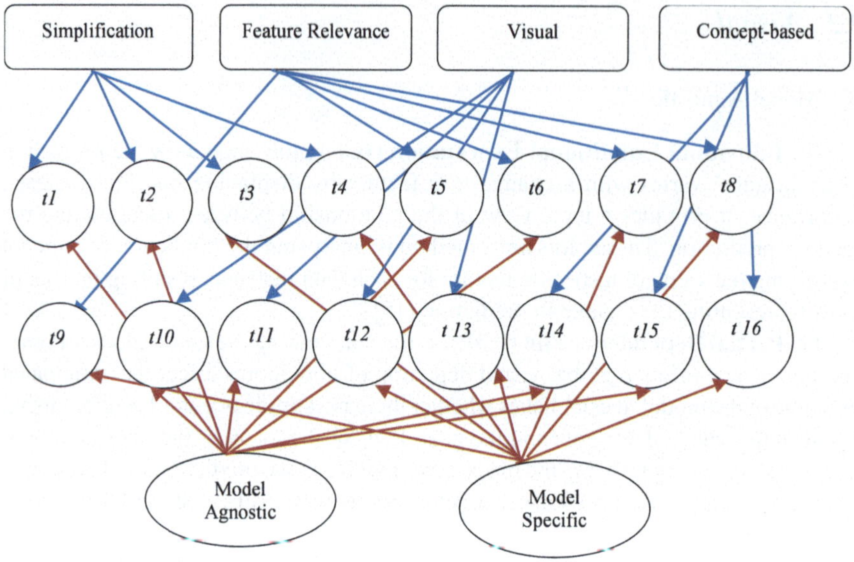

Fig. 3 Various XAI Techniques (*t*) vis-à-vis model specificity and method

that assigns equal priority to ideas with similar qualities, ensuring they align with our natural perception of similarity, considering relevance or semantics [20].

(b) Model Specific

t16: **Testing with Concept Activation Vectors (TCAV)** determines a concept's significance for a prediction by looking at its directional sensitivity (derivatives) across layers of a neural network. TCAV aggregates the dot product between the gradients of model predictions with regard to the layer's output and the CAV (associated with the idea in that layer) for a particular concept and layer [21].

Figure 3 depicts the interlinking of XAI techniques (*t1-t16*) with regard to model specificity and method.

3 Experimental Studies

In this section two XAI case studies related to healthcare and financial risks are presented with their analyses by using model agnostic techniques such as LIME and SHAP. The former gives local explanation by simplification and the latter provides global explanation by feature relevance.

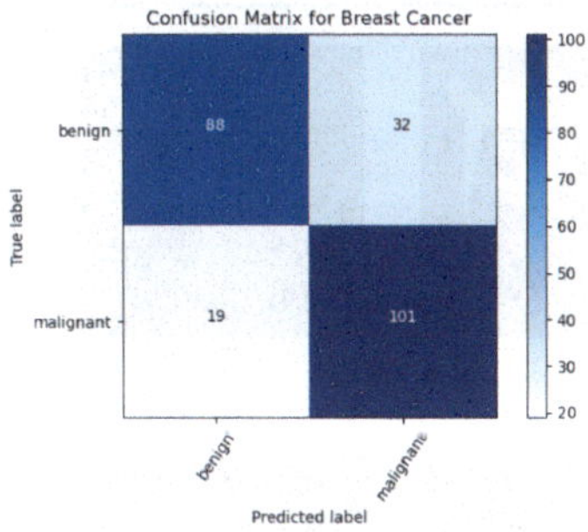

class	precision	recall	f1-score	support
0	0.56	0.97	0.71	120
1	0.91	0.24	0.38	120
accuracy			0.61	240
macro avg	0.73	0.61	0.55	240
weighted avg	0.73	0.61	0.55	240

Fig. 4 Confusion matrix and performance metrics of breast cancer detection

3.1 Study 1: Breast Cancer Detection, Investigation, and Explanation

Breast Cancer detection using pre-trained DenseNet121 model is developed in this study using dataset from Kaggle [22]. The simulation was done using NVIDIA Tesla P100, an advanced data center GPU designed for high performance computing (HPC) and deep learning workloads. The dataset was trained for classification of benign (class 0) and malignancy (class 1) output and prediction results evaluated. Performance metrics such as accuracy of 73% (weighted average), precision of 56% (class 0) and 91% (class 1), recall of 97% (class 0) and 24% (class 1) and F1_score of 71% (class 0) and 38% (Class 1), and ROC-AUC of 0.813 were obtained. The confusion matrix and classification report are shown in Fig. 4. LIME visualizations of the pixels contributing positively and negatively to prediction for top 5 features for accurate and wrong predictions are shown in Figs. 5 and 6 respectively. Figure 7 depicts SHAP explanations for accurate and wrong predictions of the first sample with top 5 features. The image displayed on the chart shows the two categories to which the model assigns the image. The image shows some areas highlighted in blue and red tints. Sections that contributed positively to the category prediction are indicated by colors of red, and sections that contributed negatively are indicated by shades of blue.

3.2 Study 2: Credit Risk Prediction, Analysis, and Interpretation

AI-based credit risk prediction model is developed in this study using a credit risk dataset from Kaggle [23]. The preprocessing of the dataset involved removing duplicates and missing values, converting it to numeric form, and then normalizing the values using a scale of [0, 1].

The extracted dataset comprises 6 features and one target as depicted by the correlation matrix (heatmap) including features whose correlation with the target

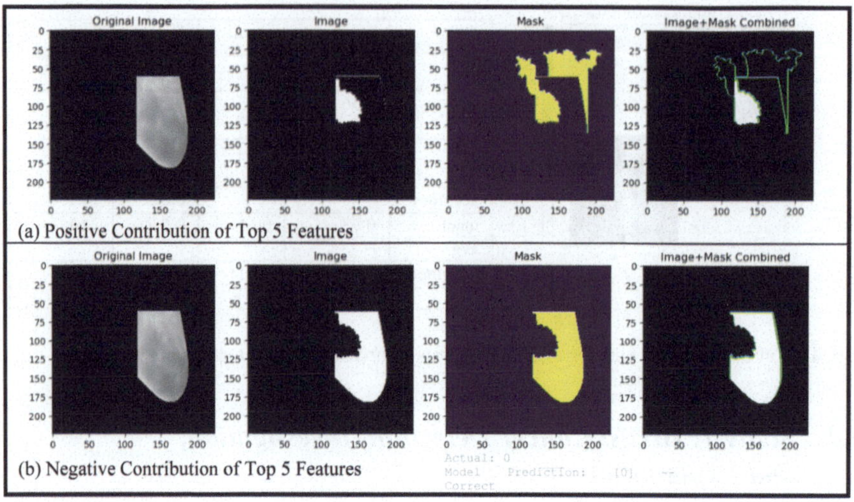

Fig. 5 LIME visualization of accurate prediction

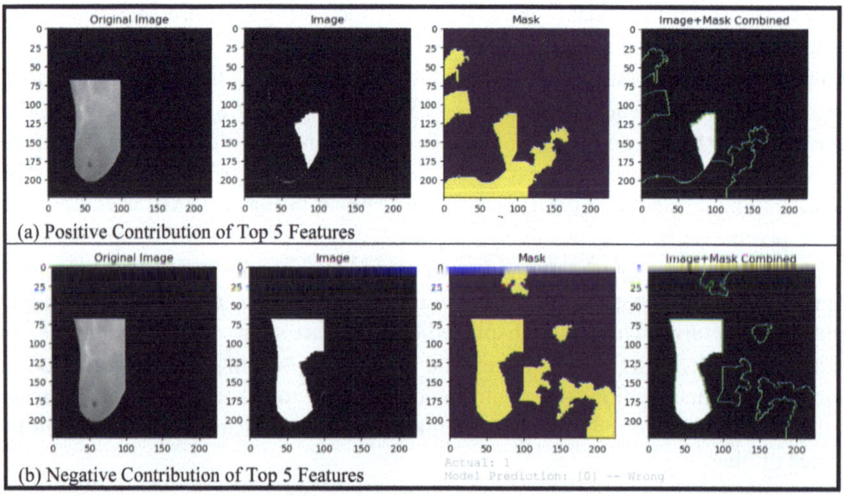

Fig. 6 LIME visualization of wrong prediction

$\geq \pm$ 0.1 in Fig. 8. The feature numbers and their description are shown alongside. The prediction of credit no-risk (class 0) and credit risk (class 1) have been performed using hybrid ensemble learning model that uses five ML algorithms such as Logistic Regression, Decision Tree, Random Forest, Naïve Bayes, and Support Vector Machine. The performance metrics such as accuracy, precision, recall, specificity, and F1_score are within the acceptable range of 86–88%.

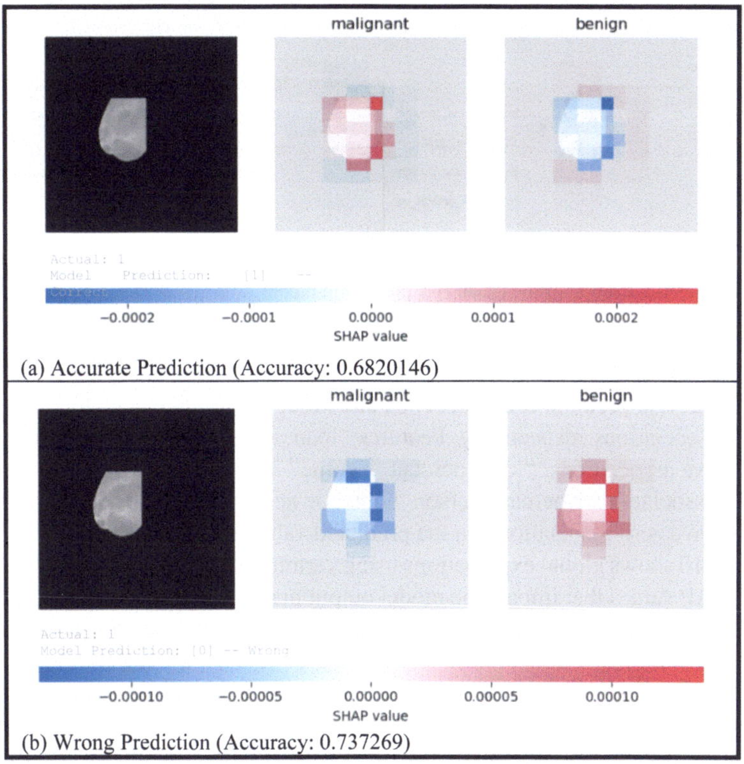

Fig. 7 SHAP visualization

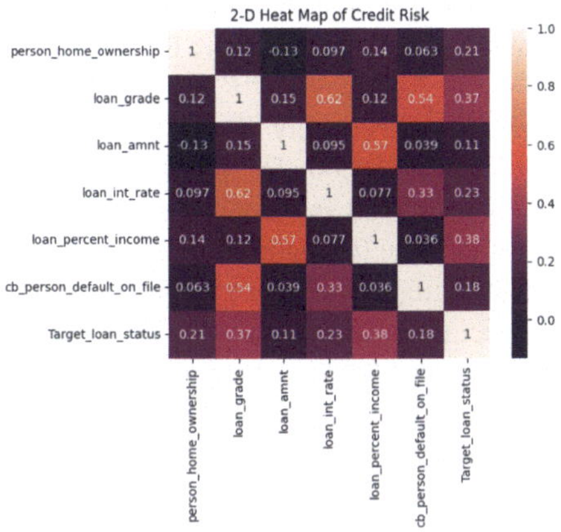

Feature 0: person_home_ownership
Feature 1: loan_grade
Feature 2: loan_amnt
Feature 3: loan_ant_rate
Feature 4: loan_percent_income
Feature 5: cb_default_in_file
Target Label: Target_loan_status

Fig. 8 Heatmap of credit risk dataset after feature extraction

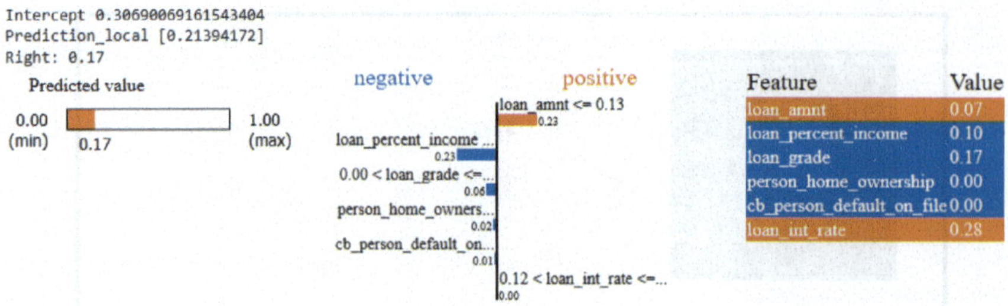

Fig. 9 LIME visualization of the credit risk prediction model

Figure 9 shows the local explanation by LIME visualization of the 63rd test vector of the credit risk prediction model. The colors blue and orange depict negative and positive associations respectively. Features "loan_amnt" and "loan_int_rate" have low positive associations, "loan_percent_income" and "loan_grade" have very low negative associations whereas "person_home_ownership" and "cb_person_default_on_file" have zero associations on the predicted value of 0.17 (0.00_{min} to 1.00_{max}).

Figure 10 shows global explanations using various SHAP plots. Figure 10a shows mean SHAP values that impact the model output magnitude for the two target labels (class 0 → no risk; class 1 → risk). The feature names are displayed on the Y-axis from top to bottom in order of relevance. The degree of change in log odds is indicated by the SHAP value, represented on the X-axis. Every point on the graph has a color that corresponds to its related feature value; blue points have low values and red points have high values. Figure 10b shows the SHAP values for the target label "0", Fig. 10c shows the decision plot for the target label "1". The cumulative SHAP values for each prediction are plotted on a decision plot to visually represent the model decisions. The decision plot's depicted lines indicate the relative contributions of each feature to a single model prediction, illuminating the feature values that drove the prediction. The decision plot with the target label "1" skewed in favor of "1". Figure 10d and e show force plots for target labels "0" and "1 "respectively. The model's scores for both predictions are indicated in bold, at 0.94 and 1.00. The features that were crucial in predicting this observation are displayed in red and blue, where red stands for elements that increased the model's score and blue for features that decreased it. Nearer the red-blue line are features that had a greater influence on the score; the size of the bar indicates how much of an impact these features had. In all scenarios, target label "1" is correctly predicted, but target label "0" is incorrectly predicted.

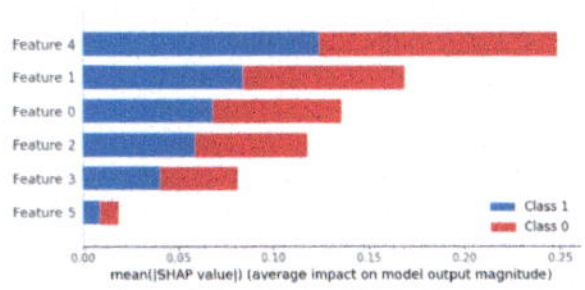

(a) Mean SHAP Value

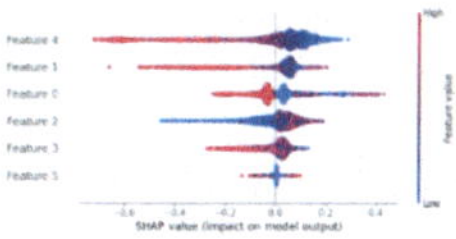

(b) SHAP Value for Target Label "0"

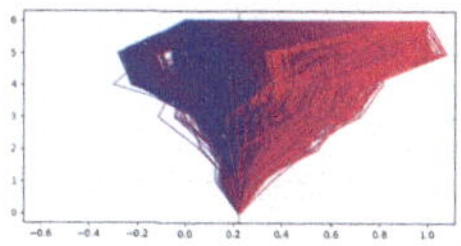

(c) Decision Plot of Target Label "1"

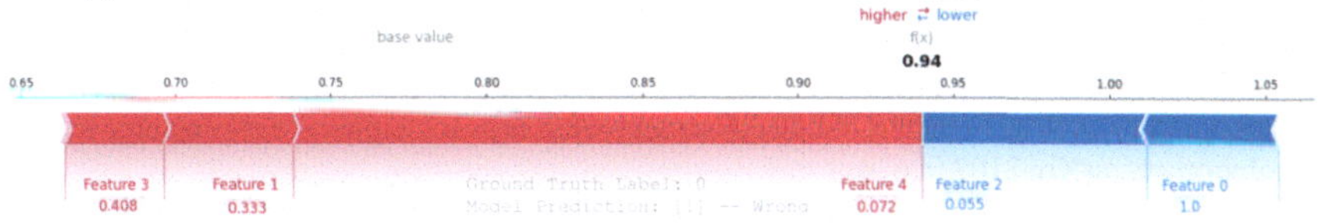

(d) Force Plot of Target Label "0"

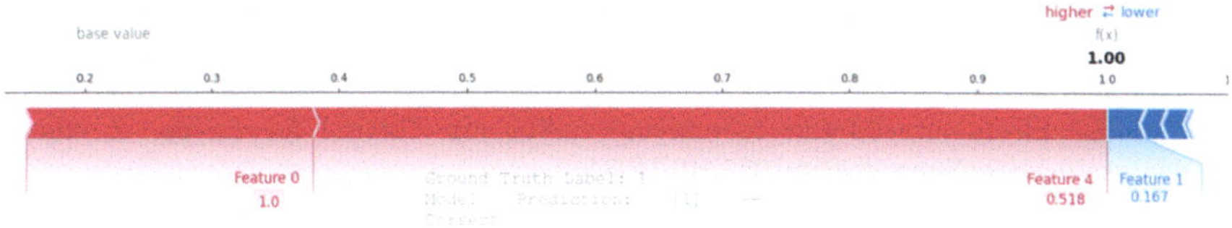

(e) Force Plot of Target Label "1"

Fig. 10 SHAP visualization of the credit risk prediction model

4 Conclusion

This paper has discussed all the aspects of XAI including its properties, principles, and categories projecting the specific needs of explanations based on consumers. The several techniques of XAI based on the categories and model specificity have been deliberated at length. The study also focuses on the level of granularity explanation such as local, cohort, and global. Furthermore, two experimental studies and analyses of CNN-based breast cancer detection and hybrid ensemble ML-empowered credit risk prediction followed by local and global explanations using LIME and SHAP respectively have been performed, using datasets from Kaggle repository. Simulation results in Python show quite promising results through performance metrics. Explanations of correct and wrong predictions by these AI models have been explicitly depicted in the figures. As confidence grows, explainable AI will aid in increasing model adoption, enabling models to have a greater influence on research and decision-making. Explainability must be designed from the beginning and integrated throughout the full ML lifecycle; it cannot be an afterthought. AI explainability simplifies the interpretation of complicated models while preserving transparency and efficiency.

References

1. Ali S et al (2023) What we know and what is left to attain trustworthy artificial intelligence. Inf Fusion. https://doi.org/10.1016/j.inffus.2023.101805
2. Arrieta AB et al (2019) Explainable artificial intelligence (XAI): concepts, taxonomies, opportunities and challenges toward responsible AI. arXiv:1910.10045
3. Phillips PJ et al (2021) Four principles of explainable artificial intelligence. NISTIR 8312. https://nvlpubs.nist.gov/nistpubs/ir/2021/NIST.IR.8312.pdf
4. Ramon Y et al (2021) Understanding consumer preferences for explanations generated by XAI algorithms. arXiv:2107.02624
5. Dafali SM, Kissi M, El Beggar O (2023) Comparative study between global and local explainable models. In: 2023 14th International Conference on Intelligent Systems: Theories and Applications (SITA), Casablanca, Morocco, pp. 1–8. https://doi.org/10.1109/SITA60746.2023.10373599
6. Reddy GP, Kumar YVP (2023) Explainable AI (XAI): explained. In: IEEE Open Conference of Electrical, Electronic and Information Sciences (eStream), Vilnius, Lithuania, pp 1–6. https://doi.org/10.1109/eStream59056.2023.10134984
7. Vilone G, Longo L (2021) Classification of explainable artificial intelligence methods through their output formats. Mach Learn Knowl Extr 3(3):615–661. https://doi.org/10.3390/make3030032
8. Sepiolo D, Ligęza A (2022). Towards explainability of tree-based ensemble models. A critical overview. In: Zamojski W, Mazurkiewicz J, Sugier J, Walkowiak T, Kacprzyk J (eds) New advances in dependability of networks and systems. DepCoS-RELCOMEX 2022. Lecture Notes in Networks and Systems, vol 484. Springer, Cham. https://doi.org/10.1007/978-3-031-06746-4_28
9. Zilke JR, Loza Mencía E, Janssen F (2016). DeepRED—rule extraction from deep neural networks. In: Calders T, Ceci M, Malerba D (eds) Discovery science. DS 2016. Lecture Notes in Computer Science, vol 9956. Springer, Cham. https://doi.org/10.1007/978-3-319-46307-0_29
10. Oveis AH, Giusti E, Meucci G, Ghio S, Martorella M (2023). Explainability in hyperspectral image classification: a study of Xai through the Shap algorithm. In: 13th workshop on hyperspectral imaging and signal processing: evolution in remote sensing (WHISPERS), Athens, Greece, pp. 1–5, https://doi.org/10.1109/WHISPERS61460.2023.10430776
11. Henelius A et al (2016) Finding statistically significant attribute interactions. https://arxiv.org/abs/1612.07597
12. Bénard C, Da Veiga S, Scornet E (2022) Mean decrease accuracy for random forests: inconsistency, and a practical solution via the Sobol-MDA. Biometrika 109(4):881–900. https://doi.org/10.1093/biomet/asac017
13. Richmond CD (2006) Mean-squared error and threshold SNR prediction of maximum-likelihood signal parameter estimation with estimated colored noise covariances. IEEE Trans Inf Theory 52(5):2146–2164. https://doi.org/10.1109/TIT.2006.872975
14. Li J, Zhang C, Zhou JT, Fu H, Xia S, Hu Q (2022) Deep-LIFT: deep label-specific feature learning for image annotation. IEEE Trans Cybern 52(8):7732–7741. https://doi.org/10.1109/TCYB.2021.3049630
15. Goldstein A, Kapelner A, Bleich J, Pitkin E (2015) Peeking inside the black box: visualizing statistical learning with plots of individual conditional expectation. J Comput Graph Stat 24(1):44–65. https://doi.org/10.1080/10618600.2014.907095
16. Muschalik M, Fumagalli F, Jagtani R, Hammer B, Hüllermeier E (2023). iPDP: on partial dependence plots in dynamic modeling scenarios. In: Longo L (eds) Explainable artificial intelligence. xAI 2023. Communications in Computer and Information Science, vol 1901. Springer, Cham. https://doi.org/10.1007/978-3-031-44064-9_11
17. Montavon G, Binder A, Lapuschkin S, Samek W, Mülle KR (2019) Layer-wise relevance propagation: an overview. In: Samek W, Montavon G, Vedaldi A, Hansen L, Müller KR

(eds) Explainable AI: interpreting, explaining and visualizing deep learning. Lecture Notes in Computer Science, vol 11700. Springer, Cham. https://doi.org/10.1007/978-3-030-28954-6_10
18. Selvaraju RR, Cogswell M, Das A et al (2020) Grad-CAM: visual explanations from deep networks via gradient-based localization. Int J Comput Vis 128:336–359. https://doi.org/10.1007/s11263-019-01228-7
19. Tan L, Huang C, Yao X (2024). A concept-based local interpretable model-agnostic explanation approach for deep neural networks in image classification. In: Shi Z, Torresen J, Yang S (eds) Intelligent information processing XII. IIP 2024. IFIP Advances in Information and Communication Technology, vol 704. Springer, Cham. https://doi.org/10.1007/978-3-031-57919-6_9.
20. Feng Z et al (2024) An axiomatic approach to model-agnostic concept explanations. arXiv:2401.06890v1
21. Amara J et al (2023) Concept explainability for plant diseases classification. arXiv:2309.08739v1 15 Sep 2023
22. /kaggle/input/breast-cancer-diagnosis
23. https://www.kaggle.com/datasets/laotse/credit-risk-dataset

Evolution of AI

Transforming Conflict Resolution: The Impact of AI on Online Dispute Resolution (ODR) in India and Beyond

Victor Nayak and Prashna Samaddar

Abstract Artificial Intelligence (AI) is becoming a potent instrument in the junction of law and technology, with the potential to revolutionise the facets of legal practice. AI has been identified to have a significant influence on conflict resolution. Dispute resolution is a crucial component of almost all the legal system, offering opportunities to the parties to settle their disagreements without going to court. Historically, methods of resolving disputes including litigation, mediation, and arbitration have mainly depended on human judgement and interaction. With the emergence of AI, many traditional techniques are seeing a substantial transformation. AI's capacity to analyse extensive data and derive insights from patterns presents novel opportunities for enhancing conflict resolution with more efficiency and effectiveness. AI is transforming the conflict resolution field by automating chores, making predictions, and serving as mediators or arbitrators. Online Conflict Resolution is expanding in both practical terms, via the creation of systems to handle simple issues, and philosophically, with many experts suggesting that these platforms are the future of resolving disputes and the legal system in general. Significant advancements in artificial intelligence have been utilised in online conflict resolution, serving as a support tool and sometimes a substitute for conventional settlement methods across several dispute resolution domains. This article focuses on India but will use best practices from several jurisdictions as case studies. This article will also provide a concise study of the evolution of AI in ODR, examining the technology and exploring the challenges related to human–machine interaction. It will also assess the prospective impact of these challenges on ODR. An analysis will be conducted on the impact of artificial intelligence in online dispute resolution (ODR) on an individual's access to justice. The analysis will conclude with a set of suggestions on the optimal approach to integrating AI in the popular ODR landscape of India.

V. Nayak (✉)
IEM's International Institute of Juridical Sciences (IIJS), University of Engineering and Management (UEM), Kolkata, India
e-mail: victor.nayak.25@gmail.com

P. Samaddar
Xavier Law School, St. Xavier's University, Kolkata (SXUK), Kolkata, India

M. Chakraborty et al. (eds.), *Proceedings of 5th International Ethical Hacking Conference*, Lecture Notes in Networks and Systems 1148,
https://doi.org/10.1007/978-981-97-8457-8_7

Keywords Artificial intelligence (AI) · ODR · ADR · Justice delivery system · Law and technology

1 Introduction

Online Dispute Resolution (ODR) is a relatively new and effective method for settling conflicts that is easily accessible and fast, providing an alternative to traditional court proceedings [1]. ODR's origins may be linked to the early stages of the Internet, where the necessity for alternate dispute resolution procedures became evident in the digital era. With technological advancements, online dispute resolution services have become more prominent, offering affordable and convenient conflict resolution options for those looking to resolve disputes via the Internet.

The worldwide ODR market was valued at USD 57.3 million in 2020 and is projected to grow to USD 210.53 million by 2028, with a revenue Compound Annual Growth Rate (CAGR) of 17.5% throughout the forecast period as can be seen in Fig. 1 [2]. Market revenue growth is primarily fueled by the increasing number of individuals, parties, civilian and commercial entities, and others choosing to resolve legal disputes virtually or remotely without physical presence in a courtroom or in front of judicial bodies. Furthermore, reduced costs related to fees, court appearances, and time spent in court are significant drivers of market revenue expansion [2].

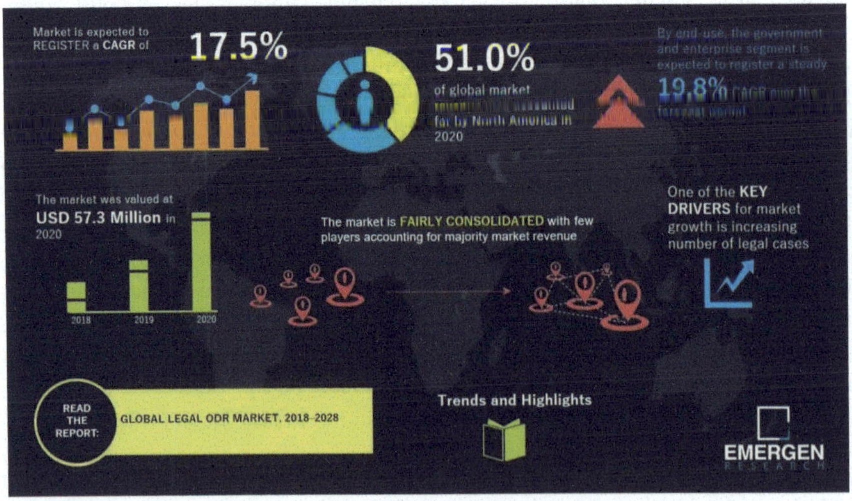

Fig. 1 Expected growth of ODR market as per the Emergen Research Report. [Legal Online Dispute Resolution (ODR) Market By Component (Solutions, Services), By Type (Mediation, Arbitration, Negotiation), By Deployment Type (On-Premise, Cloud-based), By Application, By End-Use, and By Region Forecast to 2028. (2021, July 20). Emergen Research. https://www.emergenresearch.com/industry-report/legal-online-dispute-resolution-market?utm]

The COVID-19 pandemic highlighted a pressing need for ODR, with a significant increase in cases related to lending, credit, property, commerce, and retail being processed through this method. Lawyers have been exposed to this process and are likely to start incorporating it into their regular practice as new, more infectious, and damaging variants of the virus lead to infections and partial lockdowns in cities and countries worldwide [3].

ODR has become a versatile sector that facilitates the resolution of many kinds of disputes and conflicts remotely. This trend is anticipated to continue growing rapidly in the future [3]. Due to the cost-effectiveness of ODR, it is being used to resolve various types of conflicts, and this growth is expected to continue in the foreseeable future [3]. The benefits of this option include ease of access, convenience, quick settlement, and less pressure on traditional courts and legal institutions. Resolving conflicts through traditional judicial systems is costly and requires significant work for each hearing. Legal ODR shares some parallels with a typical hearing but is not identical [4].

Decision-making may include incorporating several technologies to help the litigants come to a resolution. There are procedures where the parties have full control over the settlement and solutions, with the assistance of Artificial Intelligence (AI) and Information and Communication Technology (ICT) [5].

The digital commerce sector is projected to accomplish earnings of USD 55.97 million by 2028 with a CAGR of 20.5% over the forecast period. The rapid increase in Internet usage and technological progress has facilitated instant communication, commerce, and virtual contact among individuals worldwide [2]. Cyberspace has become a vital domain for business-to-business, business-to-consumer, and consumer-to-consumer interactions. Geographical, temporal, and linguistic barriers have been nearly eradicated, opening up vast opportunities for e-Commerce. Nevertheless, this shift has led to some adverse consequences, with ODR playing a significant part. The rise of revenue in the e-Commerce sector is positively influenced by the cost-effectiveness, enhanced efficiency, and flexibility provided by ODR for e-Commerce. In the future, these advantages are anticipated to lead to a consistent use and favouritism of ODR for resolving e-Commerce disputes [2].

2 Role of Artificial Intelligence in Dispute Resolution

The government and enterprise sectors are projected to contribute significantly to revenue in 2028 and achieve a CAGR of 19.8% over the projection period. Legal companies worldwide handle a diverse array of dispute categories. These encompass disputes or complaints between the business and customers (B2B or B2C), conflicts with partners in the firm's supply chain, and disagreements with rivals or other businesses (such as on intellectual property). Cross-border conflicts can increase complexity and expenses. ODR is becoming essential to address these difficulties and is projected to grow more in popularity in the years to come [2].

India, being one of the largest democracies in the world, has a comprehensive judicial system that must deal with several challenges stemming from its intricate legal framework and diverse array of stakeholders. Another important issue that arose while dealing with these issues and continues to affect the system is the backlog of cases. Figure 2 shows the present status of disposal v. pendency of cases in the Supreme Court of India alone. It can be seen that in recent times, i.e., post-pandemic there is a huge rise of pendency in comparison to the disposal of cases [6].

When one analyses the reasons for the pendency of cases, one of the major reasons that can be attributed is the absence of the parties during the court proceedings and the second major reason of judicial delays is due to the stay of judicial process (Fig. 3) [6].

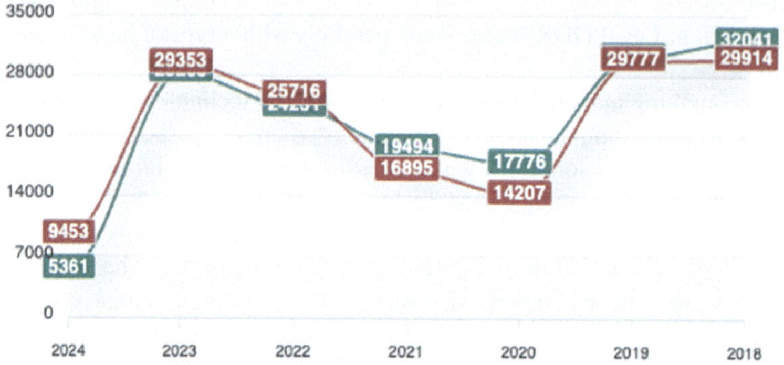

Fig. 2 Pendency v. Disposal of cases in Indian Supreme Court in the last 5 years [National Judicial Data Grid. (2024, June). National Judicial Data Grid. Retrieved June 18, 2024, from https://njdg.ecourts.gov.in/njdgnew/?p=main/pend_dashboard]

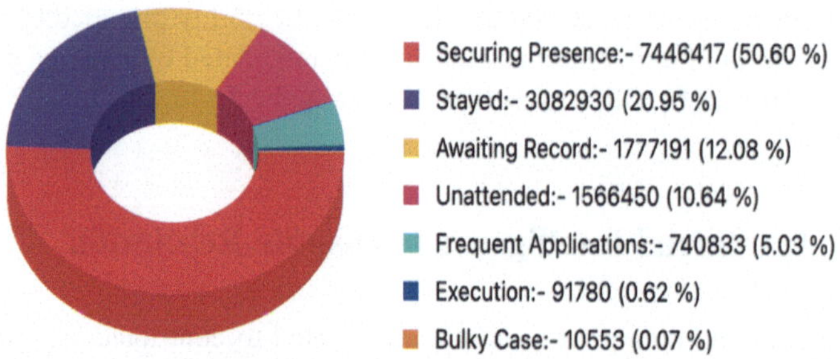

Fig. 3 Reasons for delay in pendency of cases in Indian Judicial system [National Judicial Data Grid. (2024, June). National Judicial Data Grid. Retrieved June 18, 2024, from https://njdg.ecourts.gov.in/njdgnew/?p=main/pend_dashboard]

2.1 *E-Court Project and Embracing Use of AI in Indian Judicial System*

The e-Courts mission started in 2005 with the goal of improving judicial efficiency and accessibility, while also making the justice system more inexpensive, open, and responsible. It initiated the digitisation of the judicial administration process in over 19,000 courts in India. Today, the National Judicial Data Grid offers up-to-date information on both pending and resolved cases. To assist judges with case management and decision-making, the Supreme Court has implemented a platform that is driven by AI and is referred to as SUPACE (Supreme Court Platform for Assistance in Court's Efficiency). SUVAS (Supreme Court Vidhik Anuvaad Software) is a translation tool that is driven by artificial intelligence and was developed with cooperation from the Ministry of Information Technology [7]. The capability of this system to translate documents, commands, and judgements written in English into ten distinct vernacular languages is impressive.

Phase III of the e-Courts initiative aims to create a judicial system in India that is more accessible, efficient, and fair for all those involved in seeking or delivering justice. The idea is:

> "At the stroke of a button, a system that integrates the large body of court data to promote legal literacy and provide information on remedies to a person who has wronged them. A potential future in which decisions are made based on large amounts of data, which will allow for targeted interventions and the distribution of resources on both the administrative and judicial levels." [8]

3 *Indianising* Online Dispute Resolution

Various Indian laws, including the Arbitration and Conciliation Act of 1996, in conjunction with section 89 of the Civil Procedure Code of 1908 and the "Lok Adalat" concept under the Legal Services Authorities Act of 1987, explicitly encourage the utilisation of alternative dispute resolution methods. Furthermore, The Gram Nyayalayas Act of 2008, the Commercial Courts Act of 2015, and the Consumer Protection Act of 2019 also have provisions. The Parliament has recently enacted the Mediation Act 2023, which aims to enhance the institutional mediation process in India. The implementation of this legislation can be perceived to be a crucial step in fully acknowledging the importance of mediation and fostering a culture of resolving conflicts outside of court via peaceful means. An effective settlement not only preserves the connection between the parties involved, but also enhances their quality of life and contributes to economic progress [9]. The legislation also establishes a regulatory agency to oversee the process and promote the use of online mediation in the country. The Supreme Court ruled that under Order 10 Rule 1-A of the Civil Procedure Code, 1908, courts must consider referring parties in a legal dispute to Alternative Dispute Resolution processes [10]. The referral should occur after the conclusion of pleadings, including requesting admissions and denials, but before formulating questions. The Information Technology Act of 2000 provides legal recognition for transactions made using electronic communication techniques. Moreover, it provides. The law recognises electronic signatures for authentication, simplifies the process of submitting documents electronically, and establishes the equivalence of electronic records to physical ones.

The evidentiary importance of electronic contracts can be grasped by analysing the provisions specified in the Indian Evidence Act, 1872, which directly deal with the assumptions linked to electronic records. Recently, Securities and Exchange Board of India (SEBI), the market regulator in India, has implemented rules to regulate the use of ODR in the Indian primary and secondary market. These rules apply to authorised intermediaries and regulated companies to facilitate investor access to the system. SEBI has broadened the availability of the conciliation and arbitration procedure managed by the MII to include listed intermediaries, regulated enterprises, and their investors/clients. It allowed processes to be handled in a hybrid way, along with streamlining the conflict settlement process and implementing other steps to enhance judgement implementation.

Given that ODR has slowly made its mark in the legal and the judicial enforcement mechanisms it can be also said that the diversity in this method of dispute resolution allows the freedom to choose the appropriate form from the bouquet of options available at the disposal of the clients. These types have been customised as per the needs of the subject matter and keeping in mind the viability options too.

3.1 Types of ODR Techniques

(a). *Blind Bidding or Automated Negotiation*: The parties offer a settlement amount in cash to a computer one at a time. After comparing the offer and the demand, the programme determines their arithmetic mean. It is also possible to combine online mediation with the blind bidding technique. This extremely creative type of ODR is appropriate for financial claims settlements in which there is no disagreement over liability—just the amount of damages to be paid [11].

(b). *Assisted Negotiation*: The parties engage in intercommunication through the utilisation of internet-based platforms, such as electronic mail, web-based communication tools, or video conferencing. Initially, it is customary for each side to articulate their initial stance, refraining from making explicit demands, and instead focusing on identifying the fundamental interests at stake. The programme compiles a comprehensive list of all issues, both qualitative and quantitative, in an unbiased manner. The primary purpose of the parties involved is to engage in a process that allows for a more detailed examination of their fundamental interests, priorities, and the relative significance of each item. In addition, they would like to lay out their hopes and fears regarding the consequences of any lawsuit and learn as much as possible about the alternatives to a negotiated settlement. (BATNA) [12].

(c). *Online Mediation*: The mediator facilitates the resolution of conflicts by employing electronic communication methods such as email or dedicated online platforms that offer virtual spaces for parties to engage in online dialogue. Users of the Internet Neutral or Mediator platform have their pick of several different types of online mediation, including e-mail, instant messaging, chat conference rooms, and video conferencing [13].

(d). *Online Arbitration*: The utilisation of Information and Communication Technology facilitates the replication of the conventional offline fact-finding and decision-making procedures of arbitration. At present, the majority of ODR providers let parties engage in certain aspects of the arbitration procedure via digital means. For instance, parties have the ability to access and complete claim forms electronically, submit relevant documents by either regular email or a secure web interface, and participate in telephone hearings, among other online functionalities. One of the primary obstacles lies in the fact that the inclusion of judicial enforcement measures somewhat undermines the fundamental objective of establishing an online procedure [14].

3.2 Major ODR Platforms of India

Currently, there has been a noticeable shift in the method of conflict resolution, marked by the growing presence and use of online dispute mediation platforms domestically [15]. These platforms have become facilitators of specific dispute resolution processes for many local and international companies. By combining alternative conflict resolution with modern technology, Online Conflict Resolution solutions may efficiently simplify and speed up the process of resolving disputes. Upon the commencement of the worldwide epidemic, various renowned arbitration institutions have established fresh norms. Key arbitration institutions include the London Court of International Arbitration (LCIA), Singapore International Arbitration Centre (SIAC), Hong Kong International Arbitration Centre (HKIAC), International Chamber of Commerce (ICC), International Centre for Settlement of Investment Disputes (ICSID), and other well-known global arbitral bodies [15]. A vital resource for the dispute resolution community is what the ICC guidance note is, which comprises a detailed checklist and has value in terms of a model procedural order. Some important organisations in India are involved in promoting ADR and ODR. The Construction Industry Arbitration Council (CIAC), Construction Industry Development Council (CIDC), Indian Council of Arbitration (ICA), International Centre for Alternative Dispute Resolution (IC), Indian Chamber of Commerce (ICC), and Bengal Chamber of Commerce and Industry are some organisations aimed at this purpose.

(a). SAMA, which is approved by the Department of Justice, aims to have a fast and efficient way of solving consumer and business disputes; employers and workers; landlords and tenants. Online platforms are efficient and economical means for dispute resolution. ICICI uses this technology currently to resolve over 10,000 disputes with some valued as high as INR 20 Lakh. SAMA's rules establish the process for three common alternative dispute resolution methods: arbitration, mediation, and conciliation. The platform is overseen by ODRways Solutions Pvt. Limited, and the rules were created by prominent legal institutions like Trilegal and CAMP Arbitration & Mediation Practice Pvt. Ltd [16].

(b). The Centre for Alternate Dispute Settlement Excellence (CE) provides several services related to alternative dispute resolution. The centre focuses on providing services for peer-to-peer and small-scale conflicts of minor value using efficient procedures, a technical platform, and trained arbitrators. This research covers online arbitration, debt collection, and issues related to tenant and rental contracts at NestAway, a web-based home renting company. The procedure is carried out in compliance with the regulations specified in the Arbitration and Conciliation Act of 1996.

(c). Presolv360 is an online platform that specialises in resolving commercial disputes in the legal-tech field. The firm positions itself as a modern digital substitute for the antiquated dispute resolution rules from the nineteenth century. The organisation provides arbitration and mediation services.

(d). Law Waggon specialises in providing ODR services for International Business Corporations (IBCs) and business disputes. It is essential to cover the financial costs related to the length of the legal process, which includes meetings, hearings, and discussions with legal counsel outside the official courtroom environment.

(e). Centre for Online Dispute Resolution (CODR): The platform is a digital tool designed to streamline case administration using online methods. The main emphasis is on offering specialised training in arbitration, with future intentions to broaden its services to include entire ODR processes. This strategy aims to simplify the procedure and guarantee the just administration of justice by giving the client and their legal agent the power to supervise the whole process.

(f). AGAMI: Agami is an organisation dedicated to promoting entrepreneurship in the legal and judicial sectors. Agami pledges to address one million issues by using ODR methods. They have a unique programme that aims to accelerate the expansion of ODR nationwide. The programme is advancing by actively fostering the development of startups and creating demand among the corporate sector, society, and government institutions.

(g). SMART ODR: This platform has been launched in July 2023 by SEBI to solve disputes between investors and intermediaries in an amicable way and using the benefit of technology. The platform is being managed by the major institutions in the capital market such as Bombay Stock Exchange, National Stock Exchange, Metropolitan Stock Exchange to name a few. The platform targets to settle the dispute within 90 days.

4 Analysing Case Studies of AI Integration with Judicial System

Even though India has adopted ODR as a method of dispute resolution but the scope of integrating AI with ODR in India has seen sparing. The responses at a global level showcase the use of AI in the justice delivery system in quite diverse domains and provide an insight about the scope for developing ODR mechanisms integrating AI to enhance judicial efficiency.

(a). Gina the Avatar: Gina the Avatar, an online assistant from the Superior Court of Los Angeles, assists individuals in managing their traffic citations. The jury bot is developed using the Microsoft Cognitive Services platform, which utilises functions such as natural language comprehension, Q&A maker, and translation services. Gina is proficient in five languages and assists more than 5000 persons weekly.

(b). Robojudge of Estonia: Estonia has implemented an e-residency project together with a national ID smart card. The Estonian Ministry of Justice has asked Estonia's chief data officer to create an automated judge to oversee small claims and disputes involving amounts under 7000 euros.

(c). Smartsettle ONE: An ODR programme created in British Columbia that uses algorithms to understand the bidding strategies and goals of the parties involved in a dispute, assisting in reaching a settlement.

Each party in a disagreement evaluates their desired outcome and their minimum acceptable outcome, then submits their preferred resolution along with a blind offer of their BATNA (best alternative to a negotiated agreement) (Table 1).

In addition to this in Table 2 some other jurisdictions have been highlighted where AI has been integrated with the judicial system and have received some notable progress [15].

5 Integrating AI with ODR: Issues and Challenges

Integrating AI into the arbitration process has several difficulties and challenges that need to be addressed. The problems include (i) regulatory compliance, (ii) technological constraints, (iii) ethical issues, and (iv) practical factors [15].

The major hurdles are summarised below:

(i). *Privacy and Security of Data*: In some cases, arbitration may involve sensitive or confidential information. The data processed during the arbitration should be secured to enable fair outcomes for the disputes. There has to be adherence to the legislated privacy of information laws such as the Indian Personal Data Protection Act 2023.

(ii). *Prejudice and Equality*: Artificial intelligence systems could reproduce unfair results, which are reflections of biases from the training datasets that they

Table 1 Government initiatives to promote ODR in India—the timeline

Timeline	Event details
2005	The Supreme Court established an e-Committee to oversee the use of Information and Communication Technology in the Judiciary
2006	The National Internet Exchange of India (NIE) included ODR in the deployment of the ".in" Domain Name Dispute Resolution Policy (INDRP)
2016	The Online Consumer Mediation Centre (OCMC) was founded at NLSIU Bangalore under the direction of the Ministry of Consumer Affairs
2018	The SAMADHAN site was created by the Ministry of MSME to address concerns connected to late payments or payment disputes affecting MSMEs
2019	The Reserve Bank of India (RBI) set up the High-Level Committee on Deepening of Digital Payments, chaired by Nandan Nilekani, to create an Online Dispute Settlement framework for resolving problems that arise from digital payment transactions
2019	E-Alternate Dispute Resolution (E-ADR): The aim was to promote innovation in the legal field and create a robust system that combines technology to speed up the resolution of legal conflicts in India
2020	The Vivad-Se-Vishwas Scheme addresses tax, interest, and penalty conflicts, along with other associated concerns via an ODR method. The scheme's second edition is referenced in the Budget 2023–24
April 2021	The Vivad-Se-Vishwas Scheme addresses tax, interest, and penalty conflicts, along with other associated concerns via an ODR method. The scheme's second edition is referenced in the Budget 2023–24
November 2021	Justice (Retd.) A. K. Sikri prepared a Report titled "Designing the Future of Dispute Resolution - The Online Dispute Resolution Policy Plan for India" which laid down the blueprint of ODR and its implementation in India
March 2023	SEBI has opted to utilise ODR to save costs, expedite the resolution process, and provide stakeholders and investors with more choices for addressing their issues

The NITI Aayog Expert Committee on ODR. (2021). Designing the Future of Dispute Resolution: The ODR Policy Plan for India. In Niti Aayog. Retrieved June 18, 2024, from https://www.niti.gov.in/sites/default/files/2023-03/Designing-The-Future-of-Dispute-Resolution-The-ODR-Policy-Plan-for-India.pdf

learn. Ensuring fairness in AI arbitration requires constant monitoring and a decrease in bias. Furthermore, the people who feed the data need to be neutral.

(iii). *Regulatory Frameworks*: There does not exist, as of now, any guidance on the use of AI in arbitration under the existing legislation, for instance, the Arbitration and Conciliation Act of 1996. Comprehensive regulations and standards are essential for maintaining accountability and uniformity.

(iv). *Technical Constraints*: Uneven access to justice can occur under such a method, especially in countries like India, as not all jurisdictions or parties can afford AI-based ODRs.

(v). *Vital Moral Aspects*: Questions on openness regarding an arbitral award based on AI arise because algorithms are hard to explain and their logic is also incomprehensible to decipher. Fair and ethical outcomes come from the necessary

Table 2 AI integration in justice delivery system—Global scenario

Country	Implemented AI
Canada	Before one gets into reviewing contracts, one should consider Kira on this. It is possible to identify, extract, and review provisions in contracts and other legal papers. By using machine learning, Kira's tool "Quick Study" ensures precision when it comes to appraising clauses of agreement
China	To date, China's courts have been the most progressive in their use of artificial intelligence. In the 1990s, a new technology called the Smart Court was established to help judges with case analysis and decision-making by connecting to their desktop computers China's first robotic research tool for court matters, FaXiaotao, is powered by AI. It classifies instances, determines their nature, and suggests methods for resolving disputes. Another AI robot in Chinese courts, Xiao Fa, uses a childlike voice to explain complicated legal problems in simple terms. It streamlines processes and cuts down on litigation expenses while answering several legal issues
Singapore	One of the many public services provided by the newly launched "Ask Jamie" chatbot is legal assistance in Singapore. Information on government policies, procedures, and legal requirements may be easily found by individuals. Legal information and services are made more accessible with the help of this chatbot powered by AI
United Kingdom	In 2020, the Crown Court in the United Kingdom implemented the Digital Case System (DCS) to handle case management. DCS enables the computerised handling of cases from beginning to end by judges, solicitors, and court personnel. It allows anyone to participate in court processes remotely and get case updates in real time. With the option to digitally submit proof and documents, the system also encourages sustainability by eliminating paper consumption

ODR: The Future of Dispute Resolution in India. (2020, August 5). Vidhi Centre for Legal Policy. https://vidhilegalpolicy.in/research/the-future-of-dispute-resolution-in-india/

balance between complete automation of arbitration procedures and control by humans.

(vi). *Client Confidence and Satisfaction*: Arbitration would only occur if the parties have faith in AI systems to be fair and unbiased. Many are not comfortable with technology, and hence it would be hard to instill faith in AI's powers and impartiality.

(vii). *Awareness:* Integrating artificial intelligence (AI) technology into the arbitration process may need training and instruction for arbitrators, legal experts, and involved parties [17].

6 AI and ODR: The Way Forward!

Technology's widespread integration into contemporary life has made ODR more crucial. ODR is in its early stages in India, but its extensive adoption is likely to happen soon. The essential components for online arbitration, such as convenient, cost-effective, and appropriate computer and internet connections, are inconsistent and many cases absent [18]. India's online adjudication and conciliation are impeded by the country's inconsistent internet connectivity due to digital divide.

ODR is slowly gaining traction and approval in India, amidst challenges. The evolving conflict resolution methods in India may be observed via the focus on arbitration according to the Arbitration and Conciliation Act, 1996, as well as current legislative reforms related to e-governance and e-commerce [19].

The second version of the ODR platform, created by the Techno Legal Centre of Excellence for ODR, India, is now live and available for use in resolving disputes between parties from India and around the world.

ODR must build a solid foundation on broad understanding and support to successfully adapt to the rapid growth of online business. This can be done through grassroots activities touching a range of issues encompassing social media, education, street drama, advertisement, conferences, seminars, and promotions [19].

Governmental involvement is also important in ensuring adequate financing of ODR and the needed technical and administrative infrastructure. To be able to derive maximum benefit from technology-driven legal disputes, concerns about privacy and confidentiality have to be addressed through initiatives in the enhancement of privacy and the betterment of security mechanisms [20]. Some techniques include methods such as privacy design methodology, privacy engineering, and privacy self-synchronisation, among others. With regard to standardisation of laws, keeping uniform standards and rules is necessary for the recognition and validation of ODR processes at national and international levels, addressing issues of conflicting legal regulations [21].

Accessibility to the legal system should be universalised. It is important that everyone can get to access the legal system. This can be achieved through rising literacy levels, improved communication, decrease in bias, and making e-courts easily available which in turn guarantees that justice reaches many people faster. The development of ODR has been said to offload pressure from the courts thus needing interventions from world bodies and governments. The Niti Aayog's study has supported the use of AI technology in alternative dispute resolution hearings to harness its potential in minimising human bias and improving conflict resolution processes, hence promoting innovation and efficiency [8]. The destiny of AI in Online Dispute Resolution in India looks positive, with chances of improvement being predicted as; it can be designed more advanced with the use of some sophisticated NLP (Natural Language Processing) & ML (Machine Learning) Parasites to handle complicated legal disputes. Introducing AI into ODR methods in a strategic manner could change arbitration perspective in India making it swift, affordable, and just [22]. With technological strides coming in thick and fast, the country can use AI to serve justice faster, better, and become a role model in the global legal community. In order to unlock the complete potential of AI in ODR, it is important that the government, judiciary, technology providers, and legal professionals work together.

Advancing Online Dispute Resolution (ODR) is crucial for enhancing global connectivity in cross-border conflicts and fostering global peace. In order to effectively serve government entities and many sectors of the economy and society, it is vital to include Private ODR providers in this context. Certain situations should make ODR mandatory and it should roughly involve three sessions since parties should not take it as a mere formality. More awareness is needed if the general public is to access and utilise online redressal mechanisms more effectively.

References

1. Verma D, Banwari A, Pande N (2018) Online dispute resolution. InTech eBooks. https://doi.org/10.5772/intechopen.76032
2. Legal Online Dispute Resolution (ODR) Market by Component (Solutions, Services), by Type (Mediation, Arbitration, Negotiation), by Deployment Type (On-Premise, Cloud-based), by Application, by End-Use, and by Region Forecast to 2028 (2021) Emergen research. https://www.emergenresearch.com/industry-report/legal-online-dispute-resolution-market?utm
3. Online Dispute Resolution's Role Critical during Covid-19 Pandemic: Justice DY Chandrachud (n.d.) https://pib.gov.in/PressReleaseIframePage.aspx?PRID=1710900
4. Koulu R (2018) Law, technology and dispute resolution: the privatisation of coercion. .https://helda.helsinki.fi/bitstream/10138/311733/1/Law_Technology_and_Dispute_Resolution.pdf
5. Gaur RK (2024) Live law. Retrieved June 19, 2024, from https://www.livelaw.in/lawschool/articles/future-of-justice-technology-alternative-dispute-resolution-260027
6. National Judicial Data Grid (2024) National judicial data grid. Retrieved 18 June 2024, from https://njdg.ecourts.gov.in/njdgnew/?p=main/pend_dashboard
7. Supreme Court of India "PRESS RELEASE" (2019) Supreme Court of India. Retrieved 18 June 2024, from https://main.sci.gov.in/pdf/Press/press%20release%20for%20law%20day%20celebratoin.pdf
8. The NITI Aayog Expert Committee on ODR (2021) Designing the future of dispute resolution: The ODR Policy Plan for India. In Niti Aayog. Retrieved 18 June 2024, from https://www.niti.gov.in/sites/default/files/2023-03/Designing-The-Future-of-Dispute-Resolution-The-ODR-Policy-Plan-for-India.pdf
9. Year End Report 2023 (n.d.) https://pib.gov.in/PressReleaseIframePage.aspx?PRID=1986723
10. Afcons Infrastructure Ltd. v. Cherian Varkey Construction Co. (P) Ltd ((2010) 8 SCC 24)
11. Dwomoh KA (2021) Online dispute resolution. https://www.linkedin.com/pulse/online-dispute-resolution-kweku-attakora-dwomoh/
12. Valley KL, White SB, Iacobucci D (1992) The process of assisted negotiations: a network analysis. Group Decis Negot 1(2):117–135. https://doi.org/10.1007/bf00406751
13. Staff P (n.d.) online mediation. PON - Program on Negotiation at Harvard Law School. https://www.pon.harvard.edu/tag/online-mediation/
14. Fitrianingrum A, Shahrullah RS, Syarief E (2016) Legal approaches to online arbitration: opportunities and challenges in Indonesia. Jurnal Mimbar Hukum 28(2):314. https://doi.org/10.22146/jmh.16724
15. Vikrant R, Nihit N, Avik G, Vikrant R, Nihit N, Avik G (2023) Harnessing the power of artificial intelligence in arbitration: a comprehensive analysis of Indian jurisprudence. Bar And Bench - Indian Legal News. https://www.barandbench.com/law-firms/view-point/artificial-intelligence-arbitration-analysis-indian-jurisprudence
16. Bolem L, SVM Dr (2022) Online dispute resolution—its prospective in India. Int J Law 8(6):173–180. https://www.lawjournals.org/assets/archives/2022/vol8issue6/8-5-92-750.pdf
17. Alessa H (2022) The role of Artificial Intelligence in online dispute resolution: a brief and critical overview. Inf Commun Technol Law 31(3):319–342. https://doi.org/10.1080/13600834.2022.2088060
18. Gómez-Moreno JP (2023) Advocacy for online proceedings: features of the digital world and their role in how communication is shaped in remote international arbitration. Int J Semiot Law. https://doi.org/10.1007/s11196-023-10041-y

19. ODR: The Future of Dispute Resolution in India (2020) Vidhi centre for legal policy. https://vidhilegalpolicy.in/research/the-future-of-dispute-resolution-in-india/
20. Rajendra JB, Thuraisingam AS (2022) The deployment of artificial intelligence in alternative dispute resolution: the AI augmented arbitrator. Inf Commun Technol Law 31(2):176–193. https://doi.org/10.1080/13600834.2021.1998955
21. Sulistianingsih D, Rante Lembang AA, Adhi YP, Prabowo MS (2023) Online dispute resolution: does the system actually enhance the mediation framework? Cogent Social Sci. https://doi.org/10.1080/23311886.2023.2206348
22. Lodder AR, Zeleznikow J (2012) Artificial intelligence and online dispute resolution. In: Wahab M, Katsh E, Rainey D (eds) Online dispute resolution: theory and practice. Eleven Publishers, pp 61–82

Exploring Deep Transfer Learning for Skin Psoriasis Detection Using Digital Imaging

Sumit Nayek, Partha Pratim Chaulia, Rashmi Mukherjee, and Chandan Chakraborty

Abstract Psoriasis is a chronic autoimmune human disorder that affects humans both physically and mentally over a period. Early detection is a challenging task for better management and prevention. Given this, this paper demonstrated an AI-enabled approach for automated psoriasis detection using digital imaging. In this retrospective case–control study, a dataset of 150 psoriasis images (Group I) and 150 normal skin images (Group II) were considered where several image patches (420 psoriasis and 310 normal skin) were generated from the captured image frames for the model optimization purpose. Five deep transfer learning frameworks (VGG16, VGG19, Inception, Exception, and ResNet50) were explored for psoriasis classification, utilizing convolutional layers with various filters to detect psoriasis features. The models were trained and tested with default parameters (*Optimizer—Adam, learning rate—0.1%, Loss function—Binary cross entropy*). The Inception model outperformed others, further improved with fine-tuning using the AdaGrad optimizer. Performance metrics including sensitivity, specificity, and overall accuracy were evaluated, demonstrating the system's effectiveness in diagnosing psoriasis from clinical images. The Inception model demonstrated superior performance in automated psoriasis detection, initially achieving 89.84% sensitivity, 90.37% specificity, and 92.24% overall accuracy. After fine-tuning the model hyperparameters with the AdaGrad optimizer, the performance was improved to 93.51% sensitivity, 96.42% specificity, and 99.01% overall accuracy, respectively. The fine-tuned transfer learning model with its higher accuracy can be a potential tool for point-of-care psoriasis detection using clinical photographs.

Keywords Psoriasis · Transfer learning · Inception · AdaGrad optimizer · ROC

S. Nayek · C. Chakraborty (✉)
National Institute of Technical Teachers' Training & Research, Kolkata, India
e-mail: chandan@nitttrkol.ac.in

P. P. Chaulia
University of Engineering and Management, New Town, Kolkata, India

R. Mukherjee
Raja Narendra Lal Khan Women's College (Autonomous), Paschim Medinipur, India

M. Chakraborty et al. (eds.), *Proceedings of 5th International Ethical Hacking Conference*, Lecture Notes in Networks and Systems 1148,
https://doi.org/10.1007/978-981-97-8457-8_8

1 Introduction

In dermatology, psoriasis is a significant challenge, characterized by a rapid buildup of skin cells leading to raised, red patches covered with white scales. This autoimmune condition impacts both physical health and mental well-being. Accurate and timely diagnosis is critical for commencing timely treatment. Globally, psoriasis prevalence is estimated to be between 2 and 3%. Dermatologists, however, may face diagnostic challenges due to the disease's variability and the subjective nature of assessment, potentially impeding effective clinical practice. The condition is relatively common, with prevalence rates reported to be between 0.0–2.1% in children and 0.91–8.5% in adults [1, 2]. Although incurable, applying AI-enabled management strategies can significantly enhance the quality of life for psoriasis patients, especially when adopted at the early stages of the disease. The detection of psoriasis is complicated due to its diverse symptomatology. Researchers have explored a range of deep learning frameworks to develop reliable clinical tools for more accurate detection of psoriasis [3–17]. Additionally, deep CNN architectures have been utilized to aid in the assessment of disease severity [15, 18–22]. Given the high computational demands of these models, proposals for low-resource models and scoring methods have also emerged, aimed at automated detection and severity evaluation of psoriasis [23, 24]. Our study underscores the efficacy of improving model accuracy for psoriasis screening by fine-tuning the hyperparameters within an advanced deep transfer learning framework. Training on generated image patches, models such as VGG16, VGG19, ResNet50, Xception, and Inception extracted key features at various levels critical for psoriasis detection. We determined the performance of our finely tuned model by evaluating its sensitivity, specificity, and ROC curve analysis.

2 Methodology

2.1 Image Acquisition

This research work is a retrospective case–control study where a dataset of psoriasis images [Group I (# psoriasis images) = 150; Group II (# normal skin images) = 150] was developed in collaboration with a regional medical hospital for AI model training and testing purposes. To drop redundant information from the raw image and to do effective model training, a patch base dataset was created using raw collected digital images [Group I (# psoriasis images) = 420 psoriasis and Group II (# normal skin images) = 310 of dimension 240×240]. A sample view of images in two groups is given in Fig. 1 and the patch generation process is shown in Fig. 2.

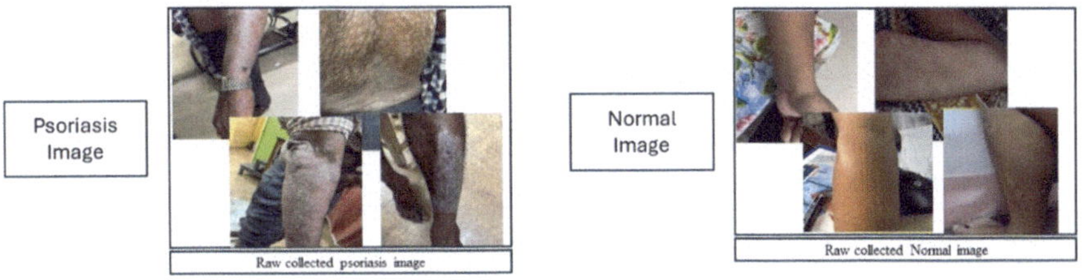

Fig. 1 Sample view of optically captured skin images

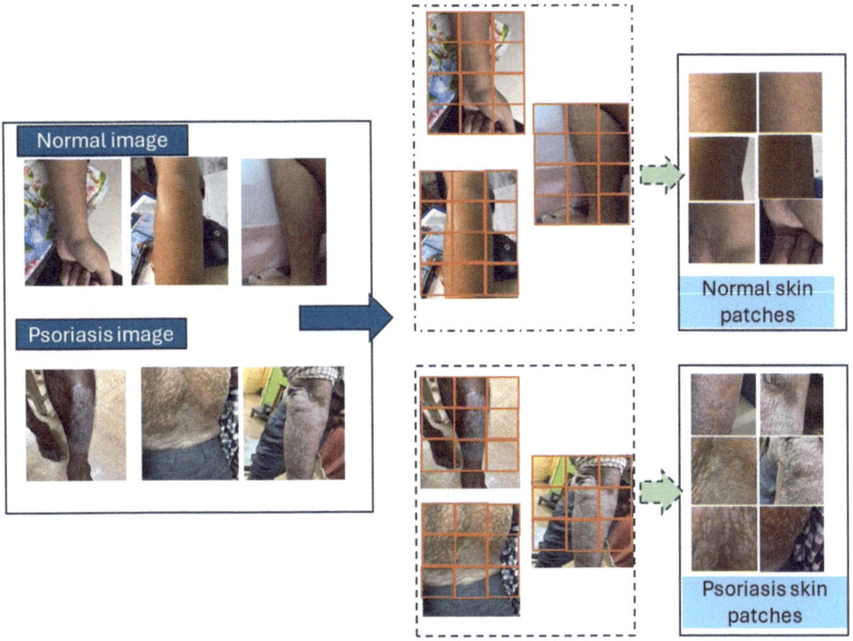

Fig. 2 Psoriasis image-patch generation and selection process with their signatures

2.2 *Image Processing*

The process of gathering photographs has resulted in a collection of images that are noisy due to different backgrounds, portions of the image that are likewise fuzzy, uneven light distribution in certain images, etc. Because of this, we have generated patches made up of 240 × 240 images, and we have carefully chosen the patches that have the appropriate skin sample on them. There are 420 psoriasis patches and 310 healthy skin sample patches in the final patch database that were chosen. The whole pipeline of the proposed method is given in Fig. 3.

Our models seek input images that were processed and of a certain size. Hence, the input dataset was resized in 240 × 240 × 3 dimension as it was needed for executing

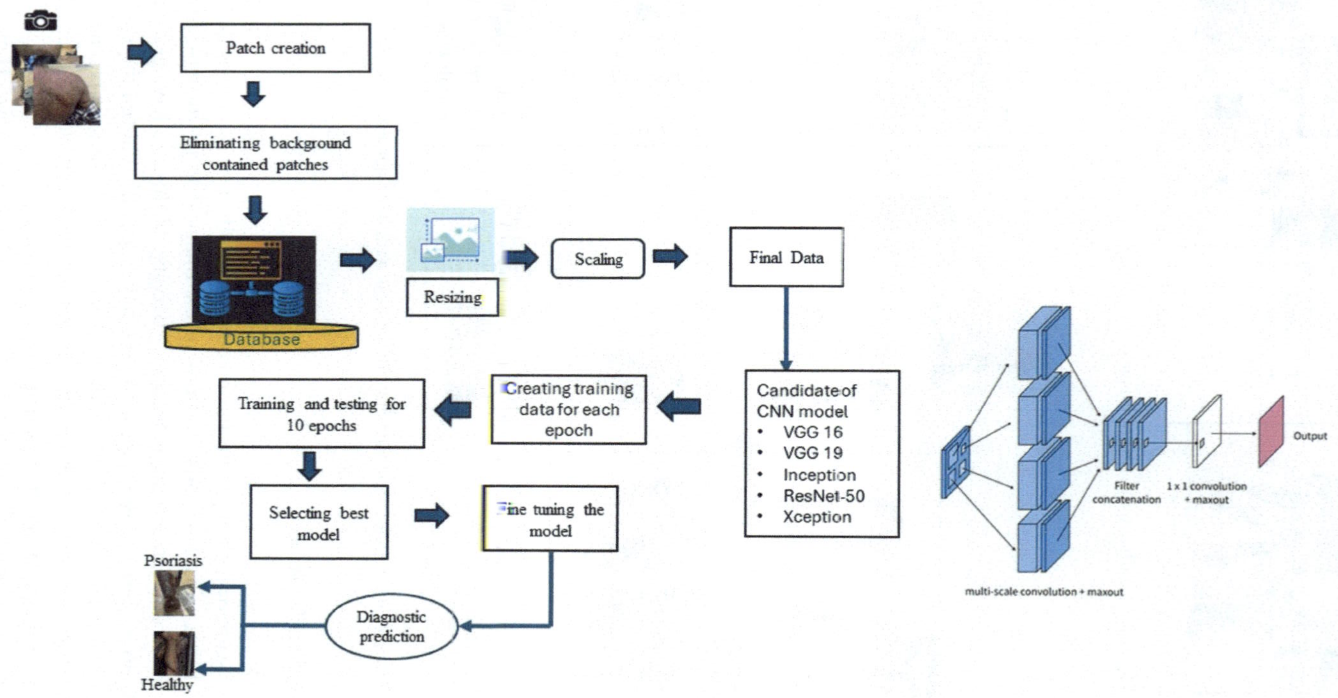

Fig. 3 The flowchart shows the working principle of the proposed AI-assisted methodology for automated skin psoriasis detection

the models under consideration viz., VGG16, VGG19, Inception, Xception, and ResNet50. Lastly, all the resized images were then scaled into the range of 0 to 1 for the model performance and run-time complexity.

2.3 *Proposed Architecture*

The objective was to deploy deep transfer learning models to develop an automated system for diagnosing psoriasis, with a focus on refining various optimizers to improve detection efficiency from clinical images. These models were applied to a curated image dataset, where convolutional layers used different filters for edges, shapes, and colors to create feature maps crucial for identifying psoriasis. The weights of these pre-trained models were fixed, with adjustments occurring only in the weights of newly added layers during training. Upon training and evaluating all models under standard parameters—using the Adam optimizer with a 0.1% learning rate and binary cross-entropy as the loss function—the Inception model emerged as the most effective. This model's basic structure consisted of an initial convolutional layer, multiple inception blocks, fully connected layers, and an output layer. We augmented our proposed architecture with an additional flattening layer, three dense layers, and one dropout layer to reduce overfitting. Further improvements were made by fine-tuning with the AdaGrad optimizer and adjusting the learning rate—endeavors aimed at enhancing the model's precision in detecting psoriasis from clinical photographs. We gauged the framework's performance by measuring sensitivity, specificity, and overall accuracy. These metrics provided valuable insights into the system's capabilities in diagnosing psoriasis, substantiating the effectiveness of our proposed approach.

3 Results and Discussion

The experimental setup for our skin disease detection model involved two systems: Google Colab, equipped with 12.69 GB of RAM and a 15 GB GPU, was utilized for model training and validation, while a personal computer with 8 GB of memory and a 2 GB GPU was used for image pre-processing. The proposed model's architecture adheres to a structured design that incorporates a series of convolutional, pooling, and fully connected layers. The convolutional layers, including parallel convolutional layers, are integral for detecting and extracting features. In this setup, the input layer has been updated to a dimension of ($240 \times 240 \times 3$) and is connected to a convolutional layer sized ($119 \times 119 \times 32$). The architecture includes three main types of convolutional blocks that repeat throughout the model. The first block consists of seven convolutions followed by one average pooling layer. The second block contains eight convolutional layers and one average pooling layer, using filters of size (3×3). Lastly, the third block is comprised of nine convolutional layers paired with one max pooling layer, aiding in pattern recognition within the image. The methodology is implemented through 48 layers that make use of multiple sets of filters. Some of the initial image transformations are illustrated in Fig. 4.

The fitted models were evaluated using the standard performance measures. The results of the models are given in Table 1.

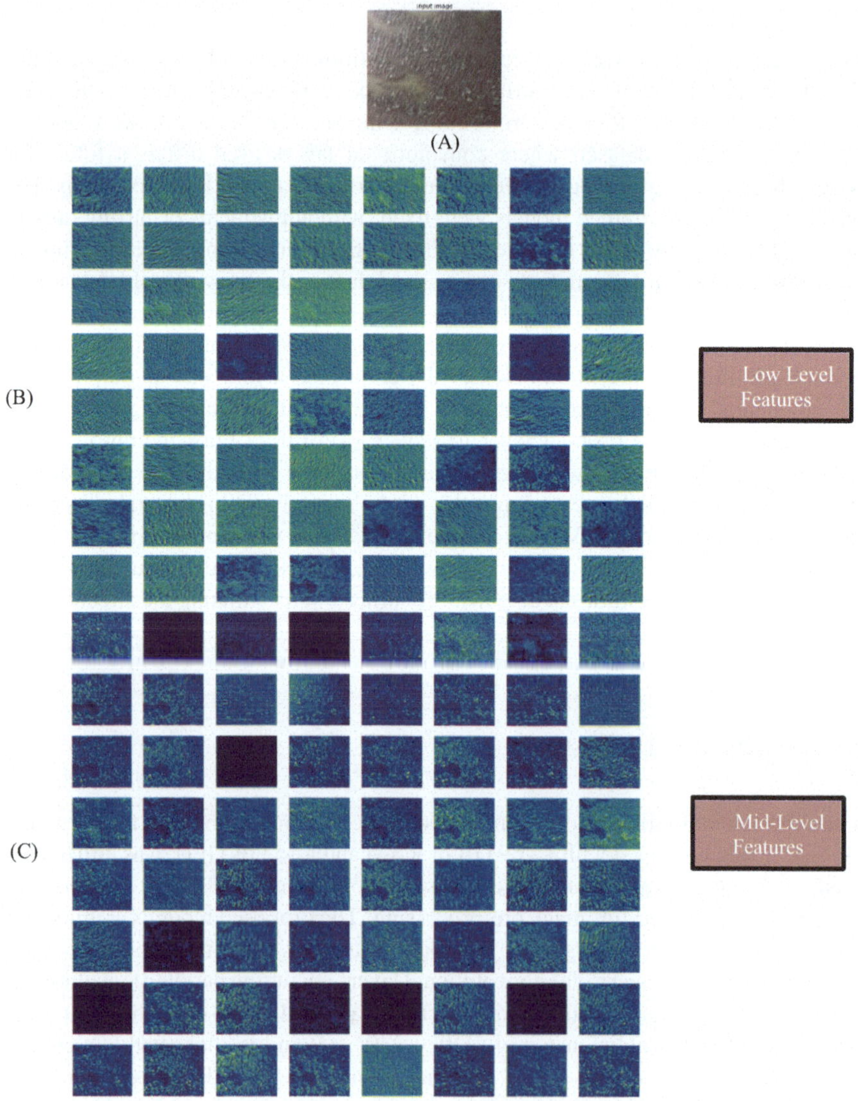

Fig. 4 **A** Original input image that feeds into the model, **B** Sample of some low-level features, **C** Sample of some mid-level features, and **D** Sample of some high-level features of proposed architecture

(D)

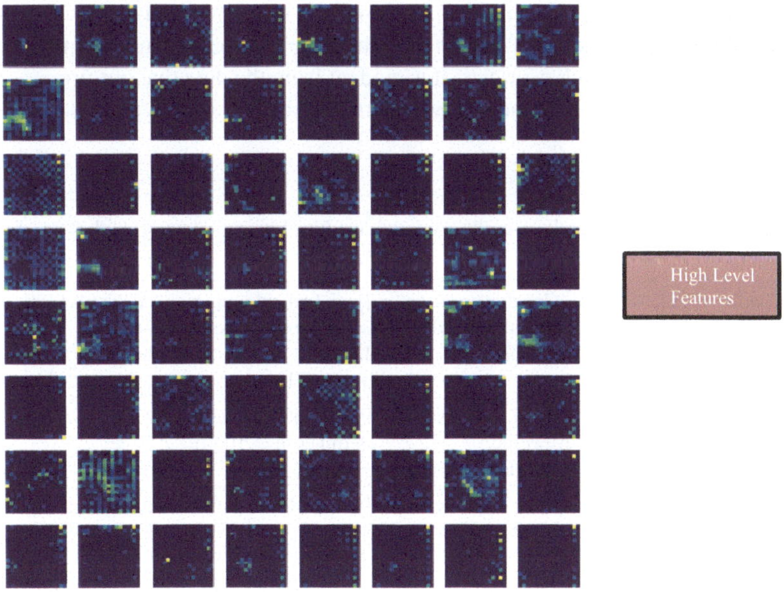

Fig. 4 (continued)

Table 1 Model performance table (statistical measures and information about the models)

DL model	Parameters	Specificity	Sensitivity	Accuracy
VGG16	Optimizer: Adam, Learning rate: 0.1%, loss function: binary cross-entropy	84.73	78.59	91.63
VGG19	Optimizer: Adam, Learning rate: 0.1%, loss function—binary cross-entropy	85.85	75.13	92.57
Inception	Optimizer: Adam, Learning rate: 0.1%, loss function: binary cross-entropy	90.37	89.84	92.24
ResNet50	Optimizer: Adam, Learning rate: 0.1%, loss function—binary cross-entropy	65.36	83.66	88.69
Xception	Optimizer: Adam, Learning rate: 0.1%, loss function: binary cross-entropy	87.23	91.53	96.45
Proposed Fine-Tuned Inception	Optimizer: Adagrad, Learning rate: 1%, loss function—binary cross-entropy	96.42	93.51	99.01

The Inception model demonstrated superior performance in automated psoriasis detection, initially achieving 89.84% sensitivity, 90.37% specificity, and 92.24% overall accuracy (AUC = 0.922). After fine-tuning with the AdaGrad optimizer and a learning rate of 1%, the model's performance improved to 93.51% sensitivity, 96.42% specificity, and 99.01% overall accuracy.

Table 2 provides a summary and comparison of the methods, datasets, and accuracies reported in state-of-the-art literature against our refined transfer learning approach for detecting psoriasis. It is evident from the data that our proposed model demonstrates noteworthy accuracy in distinguishing psoriasis from healthy skin in images.

Table 2 Comparative analysis of the proposed fine-tuned Inception model with state-of-art works

Author	Methodology	Dataset used	Performance measure
Zhao et al. [9]	Convolutional neural networks (CNN)	Type: Optical images Local psoriasis image dataset (n = 8021)	AUC: 98.1 ±0.01
Shtanko et al. [12]	CNN with a very specific architecture	Type: Optical image Dermnet (n = 1000)	Accuracy: 98%
Aijaz et al. [5]	Deep learning and LSTM	Type: Optical image BFLNTU (n = 172) Dermnet (n = 301)	Accuracy: CNN: 84.2% LSTM: 72.3%
Nieniewski et al. [8]	VGG architecture with SVM as classifier	Type: Optical images Psoriasis images: 1988 Non-psoriasis images:1582	Accuracy: 80 ±0.03
Yang et al. [7]	EfficientNet-B4	Type: Dermoscopic images Local dataset (n = 7033)	2-class problem: Sen: 92.7 ±0.02 Spe: 82.7 ±0.01 4-class problem: Sen: 88.9 ±0.01 Spe: 96.8
Wu et al. [6]	EfficientNet-B4	Type: Optical images Local dataset (n = 4740)	Accuracy: 95.80%
Yu et al. [3]	Transfer Learning	Type: Dermoscopic images Psoriasis images: 508 Seborrheic images: 850	AUC: 92.2%
AlDera et al. [4]	Support Vector Machine, KNN, Random Forest	Type: Optical image Dermnet & Atlas dataset (n = 377)	Accuracy: 90.7%
Proposed Fine-tuned Transfer Learning	Inception with AdaGrad	Type: Optical image Psoriasis patches: 497 Healthy patches: 396	Accuracy: 99.01%

4 Conclusion

For doctors, the proposed automated system for psoriasis picture recognition using Inception with AdaGrad optimization is a potential approach. Because of its great sensitivity and specificity, psoriasis may be accurately identified from clinical photographs. With the use of this instrument, doctors may be able to diagnose patients more quickly and accurately, which would enhance patient treatment for this inflammatory skin condition. Additional investigation may examine practical use and assimilation into medical practice.

Key Contribution

- The model Inception V3 for Psoriasis detection was introduced in psoriasis classification.
- The hyper-parameter of the proposed model was tuned to achieve the highest accuracy.
- The proposed methodology can be used as a point-of-care unit in rural areas.

References

1. Parisi R, Symmons DP, Griffiths CE, Ashcroft DM (2013) Global epidemiology of psoriasis: a systematic review of incidence and prevalence. J Invest Dermatol 133(2):377–385
2. Egeberg A, Andersen YMF, Thyssen JP (2019) Prevalence and characteristics of psoriasis in Denmark: findings from the Danish skin cohort. BMJ Open 9(3):e028116
3. Yu Z, Kaizhi S, Jianwen H, Guanyu Y, Yonggang WA (2022) A deep learning-based approach toward differentiating scalp psoriasis and seborrheic dermatitis from dermoscopic images. Front Med 9:965423
4. AlDera SA, Othman MTB (2022) A model for classification and diagnosis of skin disease using machine learning and image processing techniques. Int J Adv Comput Sci Appl (IJACSA) 13(5):2022
5. Aijaz SF, Khan SJ, Azim F, Shakeel CS, Hassan U (2022) Deep learning application for effective classification of different types of psoriasis. J Healthc Eng 2022:7541583
6. Wu H, Yin H, Chen H, Sun M, Liu X, Yu Y, Tang Y, Long H, Zhang B, Zhang J, Zhou Y, Li Y, Zhang G, Zhang P, Zhan Y, Liao J, Luo S, Xiao R, Su Y, Zhao J, Wang F, Zhang J, Zhang W, Zhang J, Lu Q (2020) A deep learning, image based approach for automated diagnosis for inflammatory skin diseases. Ann Transl Med 8(9):581
7. Yang Y, Wang J, Xie F, Liu J, Shu C, Wang Y, Zheng Y, Zhang H (2021) A convolutional neural network trained with dermoscopic images of psoriasis performed on par with 230 dermatologists. Comput Biol Med 139:104924
8. Nieniewski M, Chmielewski LJ, Patrzyk S, Wozniacka A (2023) Studies in differentiating psoriasis from other dermatoses using small data set and transfer learning. J Image Video Proc 2023:7
9. Zhao S, Xie B, Li Y, Zhao X, Kuang Y, Su J, He X, Wu X, Fan W, Huang K, Su J, Peng Y, Navarini AA, Huang W, Chen X (2020) Smart identification of psoriasis by images using convolutional neural networks: a case study in China. J Eur Acad Dermatol Venereol JEADV 34(3):518–524
10. Li F-L, Wang X, Hu J-W, Xiong NN, Du X-Y, Li S-B (2020) Deep learning in skin disease image recognition: a review. IEEE Access 8:208264–208280

11. Yu K, Syed MN, Bernardis E, Gelfand JM (2020) Machine learning applications in the evaluation and management of psoriasis: a systematic review. J Psoriasis Psoriatic Arthr 5(4):147–159
12. Shtanko A, Kulik S (2022) Preliminary expriments on psoriasis classification in images. Proc Comput Sci 213(250–254):2022
13. Peng L, Na Y, Changsong D, Sheng L, Hui M (2021) Research on classification diagnosis model of psoriasis based on deep residual network. Digit Chin Med 4(2):92–101
14. Okamoto T, Kawai M, Ogawa Y, Shimada S, Kawamura T (2022) Artificial intelligence for the automated single-shot assessment of psoriasis severity. J Eur Acad Dermatol Venereol JEADV 36(12):2512–2515
15. Fink C, Alt C, Uhlmann L, Klose C, Enk A, Haenssle HA (2018) Intra- and interobserver variability of image-based PASI assessments in 120 patients suffering from plaque-type psoriasis. J Eur Acad Dermatol Venereol JEADV 32(8):1314–1319
16. Hsieh KY, Chen HY, Kim SC, Tsai YJ, Chiu HY, Chen GY (2022) A mask R-CNN based automatic assessment system for nail psoriasis severity. Comput Biol Med 143:105300
17. Shrivastava VK, Londhe ND, Sonawane RS, Suri JS (2016) Computer-aided diagnosis of psoriasis skin images with HOS, texture and color features: A first comparative study of its kind. Comput Methods Programs Biomed 126:98–109
18. George Y, Aldeen M, Garnavi R (2018) Psoriasis image representation using patch-based dictionary learning for erythema severity scoring. Comput Med Imaging Graph Off J Comput Med Imaging Soc 66:44–55
19. Schaap MJ, Cardozo NJ, Patel A, de Jong EMGJ, van Ginneken B, Seyger MMB (2022) Image-based automated psoriasis area severity index scoring by convolutional neural networks. J Eur Acad Dermatol Venereol JEADV 36(1):68–75
20. Lin YL, Huang A, Yang CY, Chang WY (2022) Measurement of body surface area for psoriasis using U-net models. Comput Math Methods Med 2022:7960151
21. George Y, Aldeen M, Garnavi R (2020) Automatic scale severity assessment method in psoriasis skin images using local descriptors. IEEE J Biomed Health Inform 24(2):577–585
22. Dash M, Londhe ND, Ghosh S, Raj R, Sonawane RS (2020) A cascaded deep convolution neural network based CADx system for psoriasis lesion segmentation and severity assessment. Appl Soft Comput 91:106240
23. Raj R, Londhe ND, Sonawane R (2021) Automated psoriasis lesion segmentation from unconstrained environment using residual U-Net with transfer learning. Comput Methods Programs Biomed 206:106123
24. Bilal A, Imran A, Baig TI, Liu X, Long H, Alzahrani A, Shafiq M (2024) Improved support vector machine based on CNN-SVD for vision-threatening diabetic retinopathy detection and classification. PLoS ONE 19(1):e0295951

AI in the Detection of Breast Cancer and Related Ethical and Legal Issues

Keya De Mukhopadhyay, Tapas Guha, and Abhisikta Basu

Abstract Breast Cancer accounts for a significant percentage of malignancy among women globally. It is a disorder where abnormal breast cells develop uncontrollably and form tumours when left untreated, the tumour could spread throughout the body and become fatal. So, early detection is critical for survival. The clinical outcomes of breast cancer can be significantly improved with an accurate and timely diagnosis. Significant advancements in breast cancer diagnosis and customized treatments have been made possible, by the emergence of Artificial Intelligence particularly in the field of image analysis, the K-Nearest Neighbors algorithm is one of the simplest and easily interpretable supervised ML algorithms, which can determine whether the cancer lumps are malignant or benign. KNN, SVM, and GRNN are machine learning algorithms that make use of an openly accessible dataset, features taken from images, data pre-processing techniques, and Principal Component Analysis (PCA) in order to improve accuracy and decrease diagnostic time. Artificial Intelligence can identify both missed cancers and breast tissue features, which in turn, can help to predict future cancer development. Some Artificial Intelligence algorithms excel at predicting patients with a high risk of interval cancer, which is aggressive. When no cancer was clinically detected by screening mammography, even Artificial Intelligence algorithms trained for short time horizons can forecast the chance of developing cancer for up to five years in the future. The WHO reports that low- and middle-income nations account for over 80% of fatalities from breast and cervical cancer. In the developing economic scenario of India, one in 28 women is likely to

K. De Mukhopadhyay (✉)
Department of Biotechnology, Institute of Engineering and Management, University of Engineering and Management, Kolkata, India
e-mail: Keya.Demukhopadhyay@uem.edu.in

T. Guha
Department of Computer Science Engineering, Techno International, New Town, India
e-mail: tapas.guha@tint.edu.in

A. Basu
IEM's International Institute of Juridical Sciences (IIIJS), University of Engineering and Management, Kolkata, India
e-mail: abhisikta.basu@uem.edu.in

M. Chakraborty et al. (eds.), *Proceedings of 5th International Ethical Hacking Conference*, Lecture Notes in Networks and Systems 1148,
https://doi.org/10.1007/978-981-97-8457-8_9

develop breast cancer. The Artificial Intelligence detection model is a cost-effective precautionary measure for the early detection of breast cancer in women. The data protection laws should be convenient enough to carry out the test with the informed consent of the patients and this would ensure the smooth running of the detection procedure using the Artificial Intelligence model for the better treatment of Cancer and increase of survival chances. In that context, this chapter proposes the importance of Artificial Intelligence in breast cancer detection and requirement of proper ethical and legal guidelines.

Keywords Breast cancer · Artificial intelligence · Data protection laws · Informed consent

1 Introduction

Justice for health is an approach as much as an outcome. The "quadruple aim" of healthcare, which is to improve the health of the growing population, relation between the patient and caregiver, and lower the ever-increasing cost of medical care, presents alarming difficulties to healthcare systems around the globe. Growing populations, an increase in the prevalence of chronic illnesses, and rising healthcare expenses worldwide are putting pressure on governments, payers, regulators, and providers to reinvent and change the way that healthcare is delivered. Furthermore, with the backdrop of worldwide pandemic, the healthcare systems have been directed to "perform"—that is, providing high-quality treatment—and "transform" care at scale by incorporating data-driven insights from the real world into patient care. Artificial Intelligence (AI) is a thriving field and when it is combined with biotechnology, it opens up an amazing new set of potential solutions. Biotechnology and AI advancements provide innovative solutions to global problems. AI can be used for drug target identification, drug screening, image screening, and predictive modelling. AI and Machine Learning are effective in detecting and treating dangerous diseases, such as breast cancer, which is the leading cause of death in women worldwide.

Breast cancer is mainly of four types—normal cancer, benign cancer, in-situ carcinoma, and invasive carcinoma. The first three types do not affect all the organs and if diagnosed early can be treated. However, invasive carcinoma can spread to all the organs. In India, the majority of the female population is affected by the breast cancer which leads to the highest mortality among women. Population-based screening is important for combating such position. In the USA, every woman above the age of 40 years undergoes regular screening. This brings down the cases of missed cancer and early detection helps in proper treatment subsequently lowering the mortality rate. If AI is used in India, keeping in view of the growing population and the smaller number of physicians the country has, it will help in better early detection and proper treatment of the patients. This chapter deals with a novel concept where a model has been proposed which will be cost-effective and speedier in detecting breast cancer.

1.1 Novelty of the Chapter

This chapter is worked on the idea of the Legal and Ethical Issues in the use of AI in detecting breast cancer. In this chapter, not only the concerned issues are dealt with but also a model has been proposed which is cost-effective and speedier in reading the 2D and 3D images.

The novelties highlighted here are the backbone of the work presented ahead which has not been found in the earlier works:

(a). The model proposed by the authors has the potentiality to provide a more accurate result lowering the case of false negatives and false positives.
(b). The legal issues concerning data protection and the ethical issues concerning informed consent which are intertwined with the proposed model have been discussed.
(c). The comparative analysis of the American standard of population screening which has led to potential increase in mortality rates has been stated.
(d). The problems of health policies have been addressed in this chapter.

1.2 Methodology of the Chapter

Section 1 deals with introduction, Sect. 2 presents the concept of AI and how it can potentially be used in detecting breast cancer, Sect. 3 deals with the Ethical Legal and Social Implications for the prospective use of AI in breast screening, Comparative study with the American Model and the proposed model has been discussed in Sects. 4 and 5 presents the future challenges, while Sect. 6 concludes the chapter projecting the potentiality of AI in securing Health Care Justice.

2 What is AI and How It is Being Used in Breast Cancer Care?

AI is a thriving field and when it combines with biotechnology, it opens up an amazing new set of potential solutions. Biotechnology and AI advancements provide innovative solutions to global problems, contributing to sustainable development goals such as enhancing the quality of food, health, clean water, and climatic actions. The application of AI could possibly result in circumstances where computers gain the ability to make decisions. The deployment of computer systems, including AI, must respect human autonomy. It additionally suggests that individuals should still be in charge of the healthcare system and medical decisions. Respecting human autonomy requires providers to undertake associated duties as well, like ensuring patients understand the part AI systems play in their care and have access to the information they need to use them securely. Furthermore, it requires maintaining confidentiality and privacy in addition to obtaining valid consent through appropriate legal frameworks for data protection. AI can be used for drug target identification, drug screening, image screening, and predictive modelling.

2.1 What is AI?

AI involves a category of computer systems that are engineered to address tasks that necessitate perception, classification, optimization, inferential reasoning, and decision-making from incomplete or unresolved data [1]. AI programmes comprise a wide spectrum of computer programmes with different levels of cognitive ability, independence, and problem-solving flexibility [2]. For instance, a programme that is trained to detect breast cancer will give all possible results that detect carcinoma in both the early and advanced stages as well as if there is a chance of carcinoma in the near future. AI systems that rely on pre-programmed rules to make judgements and then draw conclusions or assess options are the least flexible end of the range. Modern AI programmes that use machine learning algorithms have the capability to learn from data and are among the most versatile. Further demonstrating the distinction between AI that learns from data and AI that only assesses rules or potential outcomes, let's look at a hypothetical computer program that determines if a given tumour is benign or malignant. The computer is tasked with the work of choosing the category of tumour and its subsequent staging.

2.2 How is AI Being Developed and Used for Breast Cancer Care?

AI and Machine Learning are effective in detecting and treating dangerous diseases, such as breast cancer, which is the leading cause of death in women worldwide [3]. Breast cancer is mainly of four types—normal cancer, benign cancer, in-situ carcinoma, and invasive carcinoma. The first three types do not affect all the organs and if diagnosed early can be treated. However invasive carcinoma can get spread to all the organs.

While considering the detection of ductal carcinoma in situ and invasive lobular carcinoma, AI outperformed mammography with a sensitivity of 100% and 96.6% compared to 88.9% and 82.2%, respectively [4]. Mammography is less sensitive than AI at identifying cancerous breast lesions. In India, the majority of the female population is affected by the breast cancer which leads to highest mortality among women. Population-based screening is important for combating such position. In the USA, every woman above the age of 40 years undergoes regular screening. This brings down the cases of missed cancer and early detection helps in proper treatment subsequently lowering the mortality rate. Saige Q and Mammo Screen software packages are AI systems, US FDA approved which is used for the screening techniques. This leads to more efficiency in less time and smooth conduction of such screening proving better accuracy. If AI is used in India, keeping in view of the growing population and the smaller number of physicians the country has, it will help in better early detection and proper treatment of patients.

2.3 Proposed Model of the Researcher

The development of AI, particularly in the area of image analysis, has allowed for significant advancements in the diagnosis and tailored treatment of breast cancer. One of the most straightforward and understandable supervised machine learning algorithms, the K-Nearest Neighbors algorithm, can identify whether cancer lumps are benign or malignant. Machine learning algorithms KNN, SVC, and GRNN are recommended to improve the accuracy and decrease the diagnostic time by utilizing an openly accessible dataset, features retrieved from images, data pre-processing techniques, and Principal Component Analysis (PCA).

Figure 1 shows the flow of the mechanism that will be followed in this study. The steps to be followed are:

1. The first step in developing an AI-powered breast cancer detection system is to collect and preprocess a large dataset of breast cancer images which should include both positive and negative cases of breast cancer. Here in this step, both 2D mammogram images and 3D Tomosynthesis images are collected and fed to the system so that the system can analyse the benign and malignancy more accurately.

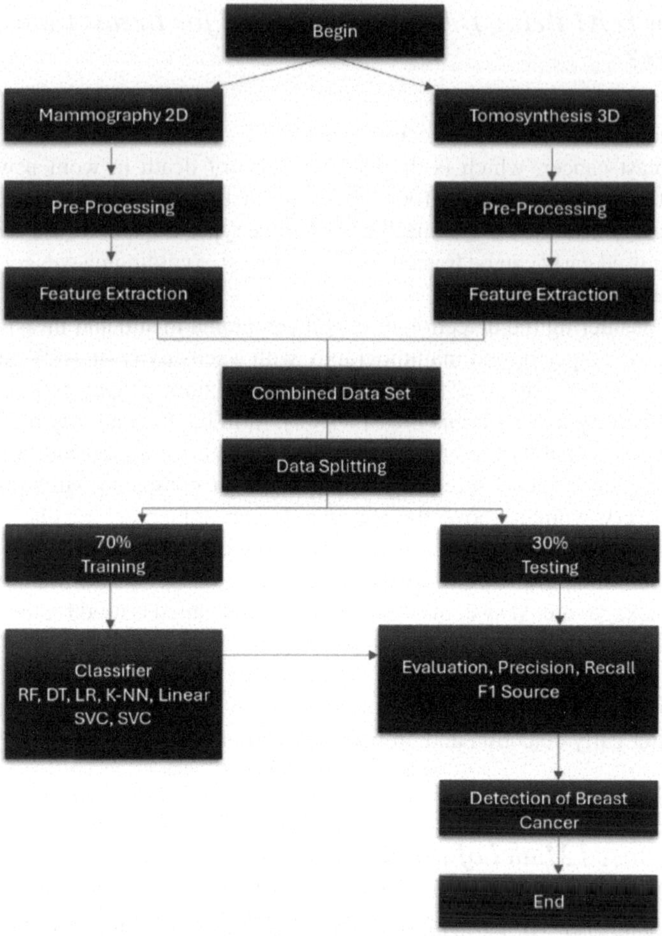

Fig. 1 A method for breast cancer detection using image processing techniques

2. The next step involves extracting relevant features from the breast cancer images which include extracting texture, shape, and other visual features that can help us to differentiate between cancerous and non-cancerous cells.
3. With extracted features, a machine learning model convolutional neural network (CNN) can be trained to learn the patterns, and characteristics of breast cancer. We fed the model with pre-processed images along with their corresponding labels (cancerous or non-cancerous) to learn to differentiate between the two labels.
4. Once we have trained the model it needs to be validated to assess its performance. A separate set of breast cancer images that we did not use during the training period is used to evaluate the model's accuracy, precision, recall, and other metrics.

5. After the model is successfully trained and validated it can be deployed and integrated into a breast cancer detection system. This involves implementing an interface where users can input breast images for cancer detection. AI system will then process the input images using the trained model and will provide a diagnosis or probability of cancer presence.
6. Lastly, the system should be regularly updated with new data to improve its performance and to adapt to evolving breast cancer cases. This will ensure that the AI system will remain accurate and effective.

These data which is required to train the software fall under the purview of personal data as the images whether 2D or 3D can be used to identify human metrics and therefore proper use is important. The Digital Personal Data Protection Act, 2023 [61] is one such law which ensures the protection of personal data in India but the area concerning data dealing with healthcare has not been explicitly provided. Moreover, it is the discretion of the authority to mark any data as "sensitive," no proper definition has been provided under the said act. Central Drugs Standard Control Organisation (CDSCO) under Directorate General of Health Services, Ministry of Health & Family Welfare, Government of India introspects whether any software that is used for the purpose of investigation of human health must be registered under the Drugs and Cosmetics Act, but if the software deals with sensitive data and data transfer the area is quite volatile. The Digital Personal Data Protection Act, 2023 explicitly states the importance of consent of data fiduciaries but Indian Medical Council (Professional Conduct, Etiquette and Ethics) Regulation, 2002 [22] is also silent in regard to standardization of the process of informed consent that has to be taken for the implementation of any technology which will enhance the treatment procedure. Therefore, the use of AI in detecting breast cancer involves various legal and ethical implications.

3 Ethical, Legal, and Social Implications in the Use of AI in Breast Cancer Detection

The possible benefits of any technology must be assessed against any potential hazards or inconveniences. The values expressed in AI systems, their comprehensibility the necessity of putting an emphasis on clinical outcomes, the possibility of bias, the transferability issue, data concerns, professional, moral, and legal responsibilities, the effect on patient experience, and potential reasons for the push towards implementation are just a few of the issues that will be addressed in the sections that follow.

3.1 Values Reflected in Algorithms

AI has the potential to produce both positive and negative effects, despite the conventional perception that it is "value neutral"—that is, neither positive nor adverse in and of itself. However, all algorithms will encode values—either explicitly or, more frequently in the "new Artificial Intelligence" age, implicitly. To take breast screening as an example, a deep learning algorithm may, like any breast screening programme, prioritize minimizing false negatives over minimizing false positives or behave differently based on the features of the breast tissue being imaged or for women from various sociodemographic groups. The efficacy of screening programmes was determined by a multitude of elements before AI, such as the technology itself, e.g., what digital mammograms potentially detect and the collective judgement of people where programmes determine cut-offs for recall [5]. These elements continue to remain important after new AI is developed, but they will also bring in novel ones. These new aspects will mostly come from the data used to train AIs, the way the algorithm uses the data, and the conscious and unconscious biases that human programmers bring to the table.

The "artificial neural network (ANN)," which draws inspiration from the human brain, is the foundation of this technology. The "neuron" is the fundamental building block of activity in the human brain [6]. The human brain is made up of tens of billions of neurons that are coupled to form a bigger structure known as a "neural network." [6] In Deep Learning (DL), researchers and engineers have attempted to replicate the human brain's "neural network" in order to construct a corresponding learning mechanism also referred to as a "neural network."[6] Three characteristics identify medical AI systems. First of all, they can develop in their ability to study for themselves. Conventional computational programmes follow a rule-based structure, meaning that a set of rules determines the mechanism in these programmes [7]. On the other hand, deep learning systems can process vast amounts of data and learn on their own. These algorithms can generate the desired output based on the supplied data [8].

Second, medical AI systems typically exhibit high levels of diagnostic prediction accuracy. Large volumes of data are necessary for medical AI systems to enhance their potential for self-learning, as was previously indicated. These data are divided into testing and training sets. In order to fit the model and test it, designers need training data. Medical Artificial Intelligence systems provide exceptionally high testing accuracy after training. They even outperformed medical professionals in recognizing certain illnesses.

In terms of successfully predicting breast cancer, Google's DeepMind AI algorithm outperformed human specialists. The technology outperformed six human doctors in predicting breast cancer in new mammograms after learning X-ray pictures from around 29,000 female patients [8].

Thirdly, there is no explanation for the medical AI systems' suggested diagnosis and treatments. As previously stated, artificial neural networks—which are replicas of the neural networks found in the human brain—provide the foundation for the medical AI systems covered in this article. For patients, physicians, and even designers, medical AI systems are "black boxes," much as humans still do not fully understand how the human brain functions. The so-called "black box" suggests that individuals are unaware of how these devices operate internally. Medical AI systems can, for example, forecast lung cancer prognosis by examining microscopic images or predict tumour response to a certain medication based on allelic patterns among thousands of genes—all without knowing or figuring out why or how those patterns matter [9]. Researchers only knew that Deep Patient's algorithm could reliably forecast conditions through deep learning. How the algorithm comes to its diagnostic conclusions and how such precise predictions made are the unanswered problems that have baffled academics.

The maxim "do no harm" comes from The Epidemics, a book of clinical symptoms that Greek doctors documented those dates back to Hippocrates [10]. The definition of ethical codes is given as follows in one of the treatises: "As to diseases, make a habit of two things—to help, or at least to do no harm" [11]. The Hippocratic Oath, which is widely considered to be the first ethical code in medicine, contains a similar statement that reads, "I will use treatment to help the sick according to my ability and judgment, but never with a view to injury and wrong doing" [11]. Since then, the most important medical ethical guideline has evolved into "do no harm." A couple of things need to be made clear here. The right meaning of "do no harm" is that doctors have a moral duty to avoid doing needless harm. In this context, unnecessary harm is defined as harm that is neither an acceptable solution for achieving specific advantages nor a necessary procedure to prevent more significant injury. There are two categories of avoidable injury created by doctors in terms of medical practices. The first kind of needless injury is typically brought on by incorrect diagnoses and treatments administered by medical professionals. The concept of "medical paternalism" is involved in the second category of needless harm.

Medical paternalism is frequently demonstrated by doctors deceiving or withholding facts from patients in order to persuade them to act in their best interests. "Paternalism is interference with a person's freedom of action or freedom of information, or the deliberate dissemination of misinformation, where the alleged justification of interfering or misinforming is that it is for the good of the person who is interfered with or misinformed," writes Allen Buchanan, outlining this position explicitly [12]. The implication is that medical paternalism is immoral because doctors have the power to make decisions for their patients without providing an explanation or obtaining informed consent, even when the doctors are the ones who know the patients' best interests. As we define patients' values and preferences as their selection of a particular treatment based on their bodily and psychological needs, it is morally wrong to disregard patients' preferences. This has detrimental effects.

In clinical practice, errors may also occur. In Mongolia, Watson was assigned to the UB Songdo Hospital. According to reports, the system prescribed taxane—a medication—inappropriately for a patient whose medical history would have prohibited the use of that medication [10]. Luckily, an oncology specialist caught this mistake. Sometimes, the outcomes that medical AI systems produce even defy basic reason. Medical error identification and detection are made considerably more challenging by the unexplainability attribute of medical AI systems. It is acceptable to require medical AI systems to be explicable enough to enable a doctor to recognize inaccurate results, given that they may put the patient at risk. The researchers here believe that the nature rather than the accuracy of medical AI systems' misdiagnosis varies from that of human doctors. As previously said, it can be challenging to determine which criteria have been taken into account and how these elements are used because medical AI systems' fundamental mechanisms are unclear. Therefore, when medical AI systems handle massive amounts of data, some irrelevant factors might be taken into account. These unimportant elements could lead to results that are paradoxical and comparable to the ones that were previously described.

3.2 Medical Paternalism

Medical AI systems may potentially cause harm to patients since they make it impossible to ensure patients' right to full disclosure because of their unexplainability. Diagnoses and treatment recommendations are generated by medical AI systems without sufficient justification, which could help patients make informed medical decisions. Patients reverted to a paternalistic model of the doctor–patient relationship under such a model, where doctors make medical decisions on behalf of their patients without sharing information that is pertinent to the decision-making process. The distinction is that in the age of AI, those decisions are heavily influenced by AI systems.

Consent is perhaps the only principle that permeates every facet of modern healthcare systems. It also means that the fundamental right to one's own autonomy and self-determination is expressed both ethically and legally [13]. Both criminal and tort law will hold a medical professional accountable if they seek to treat a patient without getting their permission [14]. The informed consent in medical treatment is a high volatile area. There is no strict rules or policies followed in treating a patient. In the early twentieth century, a plethora of four court rulings established the basis for the concept of patient autonomy, which is also the concept of informed consent [15]. These judicial decisions started in 1905 with the *Mohr* v. *Williams* [16] when the surgeon performing the operation took consent from the patient for the right ear and operated on the left ear which he never took any consent thus affecting the patient's autonomy. Therefore, the Supreme Court of Minnesota agreed and stated that this attitude rightly affected the patient's autonomy and thus the doctor was held liable. In *Pratt* v. *Davis* [17] case the surgeon took consent for an earlier operation and also a subsequent operation was conducted without consent. In this case, the appellate court of Illinois held the surgeon liable for battery as no valid informed consent was taken. The notion of patient autonomy was subsequently established and strengthened by two other instances, *Rolater* v. *Strain* [18] and *Schleiendorff* v. *Society of New York Hospital* [19], which served as the foundation for the eventual requirement of informed consent in medical and research. *Rolater* v. *Strain,* [18] the third case of this early era, expanded the judicial decisions in the Mohr [16] and Pratt

[17] instances to similar scenarios in which surgeons carried out operations that the patient had expressly refused [16, 17]. The last significant case that established the concept of patient autonomy in law was *Schloendorff* v. *Society of New York Hospital* [19] in 1914. Judge Benjamin Cardozo stated,

> "Every human being of adult years and sound mind has a right to determine what shall be done with his own body; and a surgeon who performs an operation without his patient's consent commits an assault, for which he is liable in damages."

Prior to the term being used in the public records of the 1957 case *Salgo* v. *Leland Stanford Jr. University Board of Trustees*, [20] the concept of "informed consent" was unidentified and not legally enforceable. According to a ruling by an appellate court in California, doctors are responsible for providing patients with all the information they require to make informed choices about their care, and they must use good judgement when disclosing any possible procedural risks. It decided the court to highlight the importance of informing patients about the possible advantages and disadvantages of any medical practice.

According to a statutory legal requirement in India, surgeons must inform patients about the nature, goal, risks, and advantages of every proposed operation in addition to any available alternatives. But getting permission for surgery is typically handled like a ritual, with patients being given complicated information they don't comprehend and being required to sign or put their thumbprints on the line. This is especially true in developing nations such as India, where a large number of patients are poor and illiterate, placing all of their trust in their doctors, whom they revere as being on par with God [21].

In India the Indian Medical Council (Professional Conduct, Etiquette, and Ethics) Regulations, 2002 [22] there is nowhere mentioned the potential use of AI in healthcare and the subsequent consent to be taken for the said purpose. However, it states "A registered medical practitioner shall not publish photographs or case reports of his/her patients without their permission, in any medical or other journal in a manner by which their identity could be made out. If the identity is not to be disclosed, the consent is not needed" [23]. When AI is used to detect breast cancer, the proposed model requires a huge set of data. These data have the potentiality to identify human metrics, therefore, a proper consent has to be taken while using such data to treat the patients. There is a gap in the existing regulation to address this grey area. Artificial Intelligence clinical decision support (CDS) systems in the US and European countries are subject to various regulatory approaches, including pre-market review by the FDA and conformity assessment by the EU. These processes evaluate the safety and efficacy of the system, ensuring that it meets safety and performance requirements. Post-market surveillance is also required to monitor the safety and effectiveness of medical devices, including AI CDS systems [24]. Nevertheless, this procedure may require plenty of resources and might not always identify all safety concerns.

AI–patient and AI–physician–patient diagnostic models are produced by medical AI applications; the former is patient-centric, while the latter is doctor-centric. The former creates ethical and technical issues by requiring AI systems to perform activities on their own. The AI–physician–patient model creates additional ethical and technical challenges by requiring AI systems to autonomously do pertinent activities. In the AI–human physician–patient model, medical professionals collaborate with AI technologies to offer diagnoses and treatment recommendations. For example, by assessing a patient's risk index hourly, the medical AI system Sepsis Watch helps doctors save time and effort by promptly identifying patients who require treatment for Sepsis [25]. This cooperation between medical AI systems and physicians increases efficiency and reduces the burden on doctors [26].

3.3 *Transferability in AI*

The spirit of transferability lies in its ability to facilitate the transfer of knowledge from one domain or task to another within the area of AI. In simpler terms, it enables AI systems to leverage existing knowledge and experiences to tackle new challenges or domains effectively. This process is fundamental in optimizing AI performance and addressing limitations posed by data scarcity and domain-specific nuances.

Deep neural networks have been incredibly successful in the previous decade in a broad range of applications. In numerous disciplines of research, such as computer vision, natural language processing, and robotics, deep learning has produced hitherto unheard-of breakthroughs. Deep neural networks have several desirable qualities, which are largely responsible for this amazing achievement. Deep feature representations' transferability is one of its most notable qualities. It is true that transferred features significantly improve the optimization landscape and facilitate optimization which is another important aspect of transferability. The landscapes with transferred features, according to the results, are smoother and more predictable, which essentially stabilizes and speeds up training, particularly in the beginning phases.

3.4 *Legal Risk and Liability*

As one author put it, "we currently have no clear regulator, no clear trial process, and no clear accountability trail" due to the exponential introduction of AI into healthcare [27]. Scherer refers to this as a "regulatory vacuum," noting that the usually vocal field of legal study has been noticeably silent and that almost any courts have established criteria explicitly addressing who should be held legally accountable if an AI does harm [28]. Data protection, particularly the protection of health information, is one area where there is an exemption. Health Insurance Portability and Accountability Act (HIPAA) [29] in the USA and the General Data Protection Regulation 2016/79 (GDPR) [30] in the European Union are two examples of emerging general and specific regimes. On the contrary, in India, there is no explicit law on the protection of Healthcare Data. The newly legislated Digital Personal Data Protection Act, 2023 gives an umbrella protection to all types of digital data. As it has already been mentioned in the previous section, the images of mammography and tomosynthesis have the potentiality to identify human metrics, therefore, this falls under the category of personal data. Any sort of personal data is protected under the said act [31]. According to the present model, huge sets of breast images are required as data. These data are received from the Data Principal and we being the fiduciary have the obligation to protect those sensitive data. Moreover, in the long run if a Data Principal refuses to share the data in the public domain the data sharing has to be restricted and not be used further without further consent. There are many lacunas which have to be addressed when using AI in the Healthcare Sector. Firstly, the act fails to provide any explicit provision exclusive to the purpose of Healthcare Data. Secondly, this legislation does not impose any liability upon the Data Processor. The regulatory gap in the clinical application of AI is unclear, and it's unclear how to fill it [32]. Existing frameworks could be re-amended or a new purpose-built framework could be created [33]. A meaningful debate is needed to ensure legal developments reflect AI capabilities and protect vulnerable patients. Cooperation between regulators and developers is crucial to ensure adequate protection and support for AI innovation.

4 American Standard in Population-Based Screening

Since two-thirds (65%) of the 8 million cancer-related fatalities that occur worldwide each year are recorded in low- and middle-income countries (LMICs), the burden of cancer is rising [62]. The primary challenge concerning cancer control in low- and middle-income countries (LMICs) is the high percentage of advanced cancer diagnoses, which frequently leads to a combination of higher morbidity, limited survival options, and depletion of the affected family's financial resources [63]. Thus, cancer prevention and early detection through screening are crucial objectives for governments and programmes around the world. There are two categories of screening available first the organized and second the opportunistic [64]. Organized screening targets a population and screens regularly for the early detection which leads to higher survival rates whereas opportunistic screening is provided to those who request for it [64]. In this chapter the comparison with American model has been highlighted as in the USA, the population screening has been a popular healthcare programme, and many technologies have evolved over the time. The government policies are also providing protection to the citizens for regular screening.

4.1 Existing AI Technologies in Detecting Breast Cancer

AI has revolutionized the breast cancer detection and management in the USA, making the process faster, more accurate, and more accessible. AI-powered breast cancer detection has made significant strides in the USA and it is offering promising advancements in early diagnosis of breast cancer with much more accuracy.

Some of the key steps are as follows:

Firstly, Enhanced Imaging Analysis: AI algorithms, particularly those using deep learning, have been developed to analyse mammograms and other imaging modalities (like MRI and ultrasound) with high accuracy. Several studies have shown that these AI systems can detect cancers that radiologists might miss and reduce the chance of false positives [55].

Secondly, FDA Approvals: Several AI-based tools for breast cancer detection have received FDA approval. Two important ones are—Google's AI model [56] for mammogram analysis and Hologic's 3D-Quorum Imaging Technology [57]. These are being used in the clinical setup and enhancing radiologists' ability to detect cancer.

Thirdly, Clinical Integration: AI is increasingly being integrated into clinical workflows, as it assists the radiologists by providing a second opinion, highlighting areas of concern, and even predicting the likelihood of presence of cancer, based on the imaging features [58].

Lastly, Personalized Medicine: AI is helping in tailoring treatment plans by analysing large datasets from various sources, including genomic data. Accordingly, it is predicting responses to different therapies. This approach is improving the outcomes by providing personalized treatment plans [58].

However, the integration of AI in Digital Breast Tomosynthesis (DBT) is transforming breast cancer screening and diagnosis in the USA, leading to earlier detection, more accurate diagnoses, and better patient outcomes. DBT, often referred to as 3D mammography, captures multiple images of the breast from different angles to create a three-dimensional image, enhancing the ability to detect abnormalities. AI algorithms, especially those utilizing deep learning, can analyse DBT images to identify potential cases of breast cancer with high accuracy. These systems can detect subtle patterns and anomalies that may be missed by human radiologists, improving early detection rates. AI helps in reducing false positives and false negatives during detection. This leads to fewer unnecessary biopsies and follow up procedures, reducing patient anxiety as well as the healthcare costs. Several AI tools for DBT have already received FDA approval, ensuring they meet stringent safety and efficacy standards. Some of the notable examples include:- Hologic's Genius Artificial Intelligence Detection [57] which enhances radiologists' ability to detect breast cancer in DBT images, iCAD's ProFound AI [59] which offers high sensitivity and specificity in detecting breast cancer. AI-powered DBT tools have streamlined the radiologists' workflow by prioritizing cases with higher suspicion of malignancy, reducing the time required to review images, and allowing radiologists to focus on complex cases. This improves overall efficiency and throughput in breast imaging centres. Numerous clinical studies have validated the effectiveness of AI in DBT. For example, research has shown that AI can match or exceed the diagnostic performance of experienced radiologists. Ongoing studies continue to refine these algorithms and expand their clinical applications [60]. Continuous advancements in AI and machine learning research are driving the development of more sophisticated DBT analysis tools. Leading academic institutions, research hospitals, and tech companies are collaborating to push the boundaries of what AI can achieve in breast cancer detection.

4.2 Population-Based Screening and Insurance Coverage

Although increased availability to screening and advancements in treatment techniques have decreased breast cancer mortality, breast cancer is still the most prevalent cancer diagnosed in the USA and the second-leading cause of cancer-related deaths among women. The US Preventive Services Task Force recommends biennial mammography for women aged 50–74, qualifying it as a prevention benefit under the Patient Protection and Affordability Act [34]. However, barriers like lack of access and coverage, low socioeconomic status, uninsured status, and non-white race may continue to undermine preventive screening efforts [35]. The Affordable Care Act (ACA) mandates private insurance plans and state Medicaid expansion programmes to cover breast cancer screening and prevention services at no cost to consumers. These include mammography screenings at least every 2 years and as frequently as once a year for women aged 40–74, genetic counselling for those with a history of breast cancer, and preventive medication for those at elevated risk [36]. Other services, like MRI screenings, are not currently covered under the ACA's preventive services coverage requirement [34] (Table 1).

The Affordable Care Act (ACA) significantly decreased the percentage of Americans without insurance and those with inadequate insurance over the 2014–16 period [37]. The uninsurance rate among individuals aged 0–64 in the USA dropped from over 16.7% in 2013 to a historic low of 10.4% in 2018, according to time series estimates based on nationally representative survey data. This is an almost 40% fall in the uninsurance rate [38]. The table and graph mentioned below clearly indicate the increase of screening rates among the various age groups of women [39] (Fig. 2).

Parallelly in India the year 2020 witnessed 8,51,678 deaths and 1,324,413 new cases of cancer [40]. The first, second, and third most prevalent locations of cancer in all age groups and genders are the mouth, the breast, and the cervix uteri, accounting for roughly 4,38,197 cases in India [41]. India proposed a plan to start a national cancer screening programme in 2016 as part of the Ayushman Bharat Comprehensive Primary Health Care Programme and the National Programme for Prevention and Control of Cardiovascular Disease, Diabetes, Cancer, and Stroke (NPCDCS) [42]. To provide comprehensive care and supplement the NPCDCS, population-based screening for oral, breast, and cervical cancers is being carried out under the National Health Mission (NHM) [43]. The National Family Health Surveys (NFHS-5) presently do not provide any uniform data relating to the use of population-based cancer screening under the NPCDCS, which aids in understanding the discrepancy between cancer screening installation and usage. As Ayushman Bharat is a part of the central scheme the state does not have a mandate to follow the scheme. Many states including West Bengal refuse to accept Ayushman Bharat as the part of the healthcare policy [44]. Therefore, there is no uniformity in population-based screening across the country. Moreover, the female population of the country lacks awareness regarding the high possibility of breast cancer.

Table 1 Breast cancer preventive services covered without cost-sharing [36]

Preventive service	Target population	Recommendation
Breast Cancer Screening (HRSA)	Women with an average risk of breast cancer, aged 40–74 (HRSA)	Screening mammography as often as once a year, but at least every two years
Breast Cancer Screening (USPSTF)	50–74-year-old women (USPSTF)	Mammography screening every two years
BRCA-Related Cancer: Risk Assessment, Genetic Counselling and Genetic Testing (USPSTF)	Women who have a history of breast, ovarian, tubal, or peritoneal cancer in their families or who have ancestors who have mutated the BRCA 1 or 2 gene	Using an appropriate brief familial risk assessment instrument, medical professionals ought to examine women who have a personal or family history of breast, ovarian, tubal, or peritoneal cancer, or who have ancestry linked to mutations in the BRCA1/2 gene, which increases the chance of developing breast cancer. Women who score well on the risk assessment tool ought to get genetic counselling and, if necessary, genetic testing following counselling
Breast Cancer: Medications for Risk Reduction (USPSTF)	Women age 35 + at higher risk for breast cancer	Medical professionals must additionally choose to give risk-reducing drugs, like aromatase inhibitors, raloxifene, or tamoxifen, to women who have a low risk of side effects from their medications but an elevated risk of breast cancer

Information available at Public Domain

5 Challenges and Future of Breast Cancer Treatment Using AI

In the field of cancer treatment, AI has emerged as a very useful tool. As a result of its remarkable results, it is likely to revolutionize all current therapeutic approaches. Identifying the limit between AI and human intelligence is the only question that ought to concern us [45]. Population data is the foundation of AI. For this reason, when data on persons from various socioeconomic backgrounds are developed, there will inevitably be a discrepancy [46].

One such illness whose occurrence varies depending on race is cancer. There are predetermined results from studies on AI efficiency that can be used to evaluate the standards and reliability of the research. Like any other scientific discovery, AI

Table 2 Percentage of women undertaking Breast Screening over the years [39]

Age group	1987	1990	1991	1993	1994	1998	1999	2000	2003	2005	2008	2010	2013	2015	2018	2019
40–49 years	31.9	55.1	55.6	59.9	61.3	63.4	97.2	64.3	64.4	63.5	31.5	62.3	59.6	58.3	61.5	60.2
50–74 years	29.9	53.3	58.6	64.8	65.2	72.2	75.7	77.2	75.7	72.0	73.8	72.4	72.5	71.6	72.9	76.5
50–64 years	3.17	56.0	60.3	65.1	66.5	73.7	76.5	78.7	76.2	71.8	74.2	72.6	71.4	71.3	71.8	75.7
65 years and over	22.8	43.4	48.1	54.2	55.0	63.8	66.8	67.9	67.7	63.8	65.5	64.4	66.9	63.3	64.8	68.1
65–74 years	26.6	48.7	55.7	64.2	63.0	69.4	73.9	74.0	74.6	72.5	72.6	71.9	75.3	72.2	75.0	78.1
75 years and over	17.3	35.8	37.8	41.0	44.6	57.2	58.9	61.3	60.6	54.7	57.9	55.7	56.5	51.5	50.6	54.2

Data taken from Primary Source, available in Public Domain

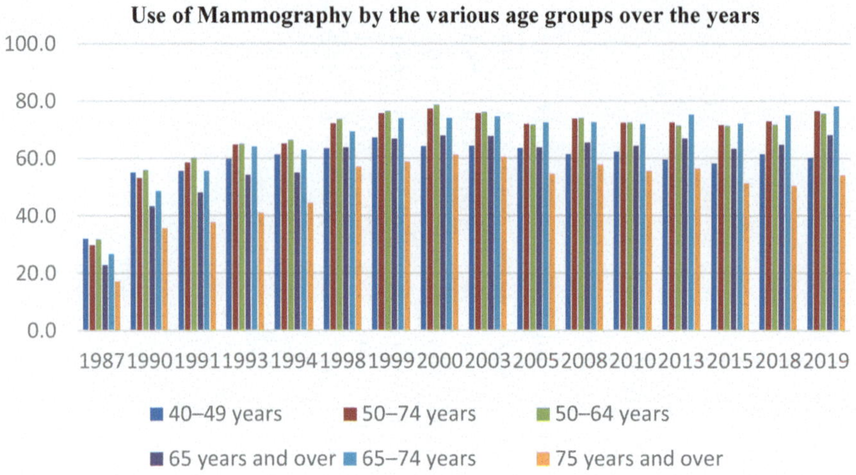

Fig. 2 Graphical representation of the data extracted from Table 2. (Chart has been generated by the author relying on the data taken from Govt. Site available in public domain)

robots need to be able to be independently replicated and produced for humans to acknowledge them as real. This suggests that an identical code must be accessible to all and that this is only feasible if data are distributed equitably among all parties [47].

Image data is the foundation of AI algorithms used in cancer management. The issue with this specific facet is the widespread underutilization in many institutions of patient histories stored as electronic health records [48]. The software systems of hospitals around the world need to include easily accessible databases and user-friendly software [49]. This is now a challenging endeavour that calls for the collaborative efforts of the medical and engineering communities [50]. Establishing doctors' faith in AI to assist them in making judgements is another difficulty in using the technology [51]. It is arguable that the use of non-explainable AI ought to be forbidden in the medical field, where there are already strict medicolegal and ethical obligations to inform. Access to enormous volumes of high-quality data is necessary to build successful AI, but this data sharing also poses serious risks of data breaches, injury, and failure to provide benefits to the public and a private market increases these dangers [52]. Studies indicate that the public is in favour of using medical data for technological advancement, but only if assurances about control, governance, confidentiality, and usage for the public good are made clear [53]. Physicians need to have sufficient training in AI technologies [54].

6 Conclusion

The cornerstone of high-quality human resources is good health and welfare, which also dictates a nation's developmental plan. By enhancing patient safety, health outcomes, and the cost and accessibility of high-quality care, AI, when backed by timely and reliable data and evidence, has the potential to completely change the way healthcare is delivered. Building an infrastructure that can handle an ageing population, make use of our growing understanding of disease and its potential for precision treatments, and address the shortage of medical professionals will require the integration of AI.

AI's potential in breast cancer detection is significant, as it can analyse large amounts of medical data and identify breast cancer signs. This integration can improve diagnosis accuracy and speed leading to earlier detection and better patient outcomes. AI should be used to assist medical professionals in detecting breast cancer, but not to replace them. Ethical and privacy concerns must be considered with proper protocols in order to secure patient data use and confidentiality. Overall, the use of AI in the detection of breast cancer has the potential to revolutionize the field by improving accuracy, speed, and efficiency. Its continued growth could lead to improved patient outcomes.

References

1. Tegmark M (2018) Life 3.0. Penguin Random House
2. IBM and sage bionetworks announce winners of first phase of DREAM digital mammography challenge [updated June 2 2019], https://sagebionetworks.org/inthe-news/ibm-and-sage-bionetworks-announce-winners-of-first-phase-of-dream-digital-mammography-challenge/2017. [Accessed 26 May 2024]
3. Ng AY, Oberije CJG, Ambrózay É, Szabó E, Serfőző O, Karpati E, Fox G, Glocker B, Morris EA, Forrai G, Kecskemethy PD (2023) Prospective implementation of AI-assisted screen reading to improve early detection of breast cancer. Nat Med 29(12):3044–3049. https://doi.org/10.1038/s41591-023-02625-9
4. Durur Subasi I, Özçelik B (2024) Artificial Intelligence in breast imaging: opportunities, challenges, and legal-ethical considerations. Euras J Med 55(S1):114–119. https://doi.org/10.5152/eurasianjmed.2023.23360
5. Edrees Almalki Y, Shaf A, Ali T, Aamir M, Khalid Alduraibi S, Mohessen Almutiri S, Irfan M, Abd Alkhalik Basha M, Khalid Alduraibi A, Manaa Alamri A, Zeeshan Azam M, Alshamrani K, Alshamrani AH (2022) Breast Cancer detection in Saudi Arabian women using hybrid machine learning on mammographic images. computers. Mater Continua 72(3):4833–4851. https://doi.org/10.32604/cmc.2022.027111
6. Goel A, Goel AK, Kumar A (2023) The role of artificial neural network and machine learning in utilizing spatial information. Spat Inf Res 31(3):275–285. https://doi.org/10.1007/s41324-022-00494-x
7. McKinney SM, Sieniek M, Godbole V et al (2020) International evaluation of an AI system for breast cancer screening. Nature 577(7788):89–94. https://doi.org/10.1038/s41586-019-1799-6
8. Miotto R, Li L, Dudley JT (2016) Deep learning to predict patient future diseases from the electronic health records. In: Proceedings of 38th European Conference on Information Retrieval Research, Padua, ECIR. https://doi.org/10.1007/978-3-319-30671-1_66.

9. Watson DS, Krutzinna J, Bruce IN et al (2019) Clinical applications of machine learning algorithms: beyond the black box. BMJ 364:l886. https://doi.org/10.1136/bmj.l886
10. Martin PM, Martin-Granel E (2006) 2,500-year evolution of the term epidemic. Emerg Infect Dis 12(6):976–980. https://doi.org/10.3201/eid1206.051263
11. Allbutt C (1923) Hippocrates–Hippocrates. With English Translation by W. H. S. Jones. https://doi.org/10.1017/S0009840X00040361
12. Buchanan A (1978) Medical paternalism. Philos Public Aff 7(4):370–390
13. Rashke RL (1976) Personal perspectives in medical ethics. New Catholic world 219(1313):204–213
14. Nandimath O (2009) Consent and medical treatment: the legal paradigm in India. Indian J Urol 25(3):343. https://doi.org/10.4103/0970-1591.56202
15. Will JF (2011) A brief historical and theoretical perspective on patient autonomy and medical decision making: part II: the autonomy model. Chest 139(6):1491–1497. https://doi.org/10.1378/chest.11-0516
16. Mohr v Williams. 95 Minn 261, 104 NW 12 (1905). Accessed 23 May 2024. www.lexisnexis.com/community/casebrief/p/casebrief-mohr-v-williams
17. Pratt v Davis. 118 Ill App 161 (1905). Accessed 23 May 2024. www.ravellaw.com/opinions/967d63e3e4aceb0c7ccc62e78af2d543
18. *Rolater* v *Strain*. 39 Okla 572, 137 P 96 (1913). Accessed 23 May 2024. https://law.justia.com/cases/oklahoma/supreme-court/1913/14030.html
19. Schloendorff v Society of New York Hospital. 211 NY 125, 105 NE 92 (1914). Accessed 23 May 2024. https://biotech.law.lsu.edu/ cases/consent/schoendorff.htm
20. Salgo v Leland Stanford, Jr University Board of Trustees. 154 Cal App 2d 560, 317 P2d 170 (1957). Accessed 23 May 2024. https://caselaw.findlaw.com/ca-court-of-appeal/1759823.html
21. Sanwal AK, Kumar S, Sahni P, Nundy S (1996) Informed consent in Indian patients. J R Soc Med 89(4):196–198. https://doi.org/10.1177/014107689608900406
22. Indian Medical Council (Professional Conduct, Etiquette and Ethics) Regulations, 2002 Published in Part III, Section 4 of the Gazette of India, dated 6th April, 2002
23. Indian Medical Council (Professional Conduct, Etiquette and Ethics) Regulations, 2002, Regulation 7.17
24. Harvey HB, Gowda V (2020) How the FDA regulates AI. Acad Radiol 27(1):58–61. https://doi.org/10.1016/j.acra.2019.09.017
25. Sendak M, Elish M, Gao M et al (2020) The human body is a black box: supporting clinical decision-making with deep learning. In: Proceedings of the 2020 conference on fairness, accountability, and transparency. https://doi.org/10.48550/arXiv.1911.08089
26. Niemiec E (2022) Will the EU medical device regulation help to improve the safety and performance of medical AI devices? Digit Health 8:20552076221089080. https://doi.org/10.1177/20552076221089079
27. McCartney M (2018) AI in medicine must be rigorously tested. BMJ. https://doi.org/10.1136/bmj.k1752
28. Scherer MU (2015) Regulating artificial intelligence systems: risks, challenges, competencies, and strategies. Harv J Law Technol 29(2):353–401. https://doi.org/10.2139/ssrn.2609777
29. United States (1996) The health insurance portability and accountability act (HIPAA). Washington, D.C.: U.S. Pub. L. p. 104e91. Stat. 1936. Web
30. Regulation EU (2016) The European Parliament and of the Council of 27 April 2016 on the protection of natural persons with regard to the processing of personal data and on the free movement of such data, and repealing Directive 95/46/ EC. General Data Protection Regulation; 2016/679 of
31. The Digital Personal Data Protection Act, 2023, Section 2(t)
32. Gunkel DJ (2017) Mind the gap: responsible robotics and the problem of responsibility. Ethics Inf Technol. https://doi.org/10.1007/s10676-017-9428-2
33. Haefner N, Parida V, Gassmann O, Wincent J (2023) Implementing and scaling artificial intelligence: a review, framework, and research agenda. Technological Forecasting and Social Change. 197. https://doi.org/10.1016/j.techfore.2023.122878

34. Preventive services covered under the Affordable Care Act [monograph on the Internet]. Washington, DC: HealthCare.gov; Updated 2012 Sep 27. Available from: www.healthcare.gov/news/factsheets/2010/07/preventive-services-list.html
35. Byers TE, Wolf HJ, Bauer KR, Bolick-Aldrich S, Chen VW, Finch JL, Fulton JP, Schymura MJ, Shen T, Van Heest S, Yin X (2008) The impact of socioeconomic status on survival after cancer in the United States. Cancer 113(3):582–591. https://doi.org/10.1002/cncr.23567
36. Coverage of Breast Cancer Screening and Prevention Services | KFF (2019) KFF. https://www.kff.org/womens-health-policy/fact-sheet/coverage-of-breast-cancer-screening-and-prevention-services/
37. Duarte F, Kadiyala S, Kominski GF, Riveros A (2021) The effect of the affordable care act on cancer detection among the near-elderly. Health Affairs (Project Hope) 40(2):258–265. https://doi.org/10.1377/hlthaff.2020.00369
38. Kominski GF, Rasmussen PW, Zhang C, Hassan S, Freund D (2020) Ten years of the affordable care act: major gains and ongoing disparities [Internet]. Los Angeles (CA): UCLA Center for Health Policy Research. Available from: https://healthpolicy.ucla.edu/publications/Documents/PDF/2020/10yearsACA-report-apr2020.pdf
39. Data Finder - Health, United States (n.d.) https://www.cdc.gov/nchs/hus/data-finder.htm
40. Gopika MG, Prabhu PR, Thulaseedharan JV (2022) Status of cancer screening in India: an alarm signal from the National Family Health Survey (NFHS-5). J Family Med Primary Care 11(11):7303–7307. https://doi.org/10.4103/jfmpc.jfmpc_1140_22
41. Cancer Today (n.d.) https://gco.iarc.fr/today/data/factsheets/populations/356-india-fact-sheets.pdf
42. Bharat A (2018) New Delhi: MoHFW, Government of India. Comprehensive Primary Health Care through Health and Wellness Centers:Operational Guidelines. Available from: https://www.nhm.gov.in/New_Updates_2018/NHM_Components/Health_System_Stregthening/Comprehensive_primary_health_care/letter/Operational_Guidelines_For_CPHC.pdf
43. Bagcchi S (2016) India launches plan for national cancer screening programme. BMJ 355:i5574
44. Implementation of Ab-Pmjay in West Bengal (n.d.) https://sansad.in/getFile/loksabhaquestions/annex/1711/AU1524.pdf?source=pqals
45. Xu Z, Wang X, Zeng S, Ren X, Yan Y, Gong Z (2021) Applying artificial intelligence for cancer immunotherapy. Acta Pharm Sin B 11:3393–3405. https://doi.org/10.1016/j.apsb.2021.02.007
46. Pandey K, An HJ, Kim SK et al (2019) Molecular mechanisms of resistance to CDK4/6 inhibitors in breast cancer: a review. Int J Cancer 145:1179–1188. https://doi.org/10.1002/ijc.32020
47. Yuan J, Hu Z, Mahal BA et al (2018) Integrated analysis of genetic ancestry and genomic alterations across cancers. Cancer Cell 34:549–60.e9. https://doi.org/10.1016/j.ccell.2018.08.019
48. Dileep G, Gianchandani Gyani SG (2022) Artificial intelligence in breast cancer screening and diagnosis. Cureus 14(10):e30318. https://doi.org/10.7759/cureus.30318
49. Glicksberg BS, Oskotsky B, Thangaraj PM et al (2019) PatientExploreR: an extensible application for dynamic visualization of patient clinical history from electronic health records in the OMOP common data model. Bioinformatics 35:4515–4518. https://doi.org/10.1093/bioinformatics/btz409
50. Landi I, Glicksberg BS, Lee HC et al (2020) Deep representation learning of electronic health records to unlock patient stratification at scale. NPJ Digit Med 3:96. https://doi.org/10.1038/s41746-020-0301-z
51. Begoli E, Bhattacharya T, Kusnezov D (2019) The need for uncertainty quantification in machine-assisted medical decision making. Nat Mach Intell 1:20–23. https://doi.org/10.1038/s42256-018-0004-1
52. Carter SM, Rogers W, Win KT, Frazer H, Richards B, Houssami N (2020) The ethical, legal and social implications of using artificial intelligence systems in breast cancer care. The Breast 49:25–32. https://doi.org/10.1016/j.breast.2019.10.001
53. Aitken M, de St, Jorre J, Pagliari C, Jepson R, Cunningham-Burley S (2016) Public responses to the sharing and linkage of health data for research purposes: a systematic review and thematic synthesis of qualitative studies. BMC Med Ethics 17(1):73

54. Subramanian M, Wojtusciszyn A, Favre L et al (2020) Precision medicine in the era of artificial intelligence: implications in chronic disease management. J Transl Med 18:472. https://doi.org/10.1186/s12967-020-02658-5
55. Yoen H, Chang JM (2023) Artificial intelligence improves detection of supplemental screening ultrasound-detected breast cancers in mammography. J Breast Cancer 26(5):504–513. https://doi.org/10.4048/jbc.2023.26.e39
56. Advancing Breast Cancer Detection with AI - Google Health (n.d.) https://health.google/caregivers/mammography/
57. 3DQuorum® Imaging Technology | Hologic (n.d.) https://www.hologic.com/hologic-products/breast-health-solutions/3dquorum-imaging-technology
58. Rentiya ZS, Mandal S, Inban P, Vempalli H, Dabbara R, Ali S, Kaur K, Adegbite A, Intsiful TA, Jayan M, Odoma VA, Khan A (2024) Revolutionizing breast cancer detection with artificial intelligence (AI) in radiology and radiation oncology: a systematic review. Cureus 16(4):e57619. https://doi.org/10.7759/cureus.57619
59. Artificial Intelligence for Cancer Detection | iCAD (2024) iCAD. https://www.icadmed.com/
60. Sechopoulos I, Teuwen J, Mann R (2021) Artificial intelligence for breast cancer detection in mammography and digital breast tomosynthesis: state of the art. Semin Cancer Biol 72:214–225. https://doi.org/10.1016/j.semcancer.2020.06.002
61. The Digital Personal Data Protection Act, 2023 Act no 22 of 2023
62. Torre LA, Bray F, Siegel RL, Ferlay J, Lortet-Tieulent J, Jemal A (2015) Global cancer statistics, 2012. CA: Cancer J Clin 65(2):87–108. https://doi.org/10.3322/caac.21262
63. Institute of Medicine (US) Committee on Cancer Control in Low- and Middle-Income Countries. In: Sloan F A, Gelband H (eds) (2007) Cancer control opportunities in low- and middle-income countries. National Academies Press (US)
64. Sivaram S, Majumdar G, Perin D, Nessa A, Broeders M, Lynge E, Saraiya M, Segnan N, Sankaranarayanan R, Rajaraman P, Trimble E, Taplin S, Rath GK, Mehrotra R (2018) Population-based cancer screening programmes in low-income and middle-income countries: regional consultation of the international cancer screening network in India. Lancet Oncol 19(2):e113–e122. https://doi.org/10.1016/S1470-2045(18)30003-2

Artificial Intelligence Assisted Design Optimization of Monopole Antenna

Suchanda Das and Anjan Kumar Kundu

Abstract A frequency reconfigurable flexible monopole antenna for wearable application is demonstrated in this article. Initially, the antenna consists of a tuning fork-shaped patch for obtaining the operating frequency at 5.2 GHz and after that a meander-shaped element has been attached to lengthen the proposed structure for performing another resonating frequency at 2.45 GHz for biotelemetry application. Frequency agility is achieved by varying the switching mechanism of the PIN diodes. The antenna has accomplished a compact size of 40 mm × 36 mm × 1 mm which has been employed on textile substrate. Here, we use some of the machine learning models of supervised learning to aid in the creation of the structure focused on the enhancement of bandwidth. The parameters have been optimized with R-squared validation 0.97–0.99 and achieved mean square error is nearly 0.03–0.5 whereas the given inputs are operating frequencies, length, width, height, and dielectric properties of substrate and reflection coefficient of the designed structure. Multiple ML algorithms like GPR, SVR, ANNs, and RF Regression are used to predict the parameters by splitting the dataset into training and testing with the ratio of 60 and 40 and comparative study of the results is mentioned here. This antenna exhibits 4.75 dB gain at 2.45 GHz and 9.18 dB gain at 5.2 GHz. The antenna is investigated by placing it on the various parts of human body and the obtained results provide the safety and suitability of the proposed antenna for use in biomedical application whereas the specific absorption rate is satisfied the limit that is set by the Federal Communication Commission.

Keywords Reconfigurable · PIN diodes · Wearable application · Parameter optimization · Artificial intelligence · Machine learning models

S. Das (✉) · A. K. Kundu
Institute of Radio Physics and Electronics, University of Calcutta, Kolkata, India
e-mail: dassuchanda9224@gmail.com

M. Chakraborty et al. (eds.), *Proceedings of 5th International Ethical Hacking Conference*, Lecture Notes in Networks and Systems 1148,
https://doi.org/10.1007/978-981-97-8457-8_10

1 Introduction

Body-centric communication has become more prevalent due to the advancement of wireless technology whereas the body-centric communication along with the wearable technologies has led to the enhancement of remote health care monitoring that should be versatile and should perform well when positioned near the human body with minimal deterioration. Antenna is one of the crucial components of wearable communication devices that ensure a reliable connection between devices. In order to accommodate the natural surface curvature of a human body, wearable antennas should be extremely compact, flexible, lightweight, mechanically strong, and efficient. The contemporary wireless communication systems require that kinds of antenna systems which can perform various roles with minimal space. Reconfigurable antennas have become a preferred solution for many communication requirements because of its potential to provide more capabilities than conventional antennas while reconfigurable antenna reduces filtering requirements by providing narrow band frequency responses. The creation of wearable technology has been made possible by the miniaturization of devices whereas the miniaturization can also be established by the reconfiguration method of such antennas. Recent years have seen a significant increase in the number of reconfigurable antennas based on active switches, including RF varactor diodes [1], PIN diodes [2], and RF MEMS [3] to introduce more of their features like frequency, pattern and/or polarization. The article, proposed in [4] demonstrates the frequency reconfigurability by employing a grid-slotted patch along with switchable varactors covering the operated frequency range from 2.45 to 3.55 GHz. In [5], a compact structured frequency agile antenna is proposed at four various modes that rely on three PIN diodes with good efficiency and enough gain. Different modes of switching mechanisms are described in [6], with simulated and fabricated construction of a dodecagon-shaped frequency reconfigurable antenna by employing four PIN diodes that are acceptable for ISM band, Wi-Fi, WiMAX, WLAN, etc.

Additionally, researchers are focusing on flexible, compact, and lightweight antennas due to the growing need for wearable technology. Therefore, modern applications like biosensing and e-health monitoring require flexibility in addition to reconfigurability. A flexible, multi-frequency antenna is designed upon Rogers RT5880LZ substrate, consisting of the slotted patch in circular shape, fed by coplanar feed, performing reconfigurability of frequency which is achieved by using PIN diode that can occur the current distribution by altering the electrical length of the designed antenna [7]. In [8], a wearable patch antenna operates at 2.4 GHz and 5.8 GHz by achieving the capability of frequency reconfiguration with high efficiency, configured by tunable PIN diodes and acceptable SAR values set by FCC and ICNIRP, respectively. A frequency reconfigurable textile antenna is proposed in [9] where the frequency agility is generated by using diodes embedded in a slotted place that switches the frequency from 1.57 to 2.55 GHz depending on seven modes of switching combinations. The article [10] describes a fabricated antenna performance that is frequency switchable and based on a slotted design and also investigated by placing on flexible like jeans and commercially available nonflexible material FR4 and having operational frequencies working for Bluetooth, WiMAX, and WLAN. Some of the radiating patches with various lengths are constructed and connected to the main radiating element via PIN diodes in order to reconfigure the frequency in [11]. RF choke inductors serve to impede the RF signal while permitting the smooth flow of the DC signal, while a DC block capacitor is employed to prevent any interference from the DC signal on the RF signal, ensuring the correct biasing of the PIN diode. The presented work is implied on Rogers 5870 having the dielectric constant 2.33 and successfully provides four significant frequencies that appropriate for 5G communication since it spans the 5G new radio (NR) standard's bands 1, 7, 46, and 77 in terms of bandwidth. Fabric based material has been used to construct the frequency reconfigurable antenna at 2.45 GHz and 5.8 GHz in [12].

During the past few years, it has been seen that a tremendous advancement is continuing in the use of AI approaches in numerous fields. A subfield of artificial intelligence known as machine learning (ML) allows machines to learn vast amounts of data and take appropriate action or make decisions without needing explicit instructions. Machine learning can be broadly categorized into three groups: supervised learning, unsupervised learning, and reinforcement learning. After learning about the principles of AI and ML, the application of these concepts and methodologies to the antenna and propagation research should be understood. It is necessary to first formulate actual problems in a certain mathematical fashion so that AI and ML techniques can act as solvers. These techniques concentrate on some important methods like regression that can conclude the mathematical relationship between two or more variable, i.e. input and output information and fix the level of the dependency, classification is one of the methods of supervised ML that attempts to predict the accurate data as input set and the model in classification is thoroughly trained by utilizing the training set and then assessed using test data and another one is clustering, a method of classifying the data into various clusters, each composed of related data. The items that may be comparable to one another stay in a group that is less similar to another group or has none at all. Due to the fact that it is an unsupervised learning approach, the algorithm operates on an unlabeled dataset without any supervision. The paper [12] explores the development of a flexible wearable sensor system for breast cancer detection at early stage maintaining biocompatibility with frequency range of 1.5–8 GHz that is implemented on Roger substrate and by using the algorithms of artificial intelligence, it is possible to discover the undesirable tumor component inside the breast by scanning the gathered data. In [13], monopole antenna parameters are optimized using some machine learning methods including kNN, ANN, and lasso, covering the two frequency bands and the accuracy is also confirmed by comparing the predicted results with the obtained results of HFSS. Dielectric resonator-based antenna has been optimized in [14] through several ML techniques, i.e. Artificial neural network, K-Nearest Neighbors, XG Boost, Random Forest model, and Decision Tree within the frequency band 3.3–3.65 GHz. Return loss is predicted by the ML models and validated through the simulation results of HFSS simulator software and also fabricated. A reconfigurable antenna based on metamaterial is proposed and the performance is predicted by using some machine learning techniques in [15]. Six distinct models of machine learning have been assessed through the employment of the sophisticated Bayesian hyperoptimization technique whereas the performance has been assessed through nested k-fold cross-validation for determining the RMSE in [16]. Few machine learning approaches are employed to optimize the eye bolt-shaped slotted antenna for predicting the value of reflection coefficient, described in [17]. The paper [18] describes the frequency agility method and introduces the supervised machine learning models with RSSI improvement. The article [19] presents the antenna with multiband characteristics and additionally involves the discussions of utilizing machine learning techniques to mitigate the electromagnetic interference of the proposed structure.

In this article, a frequency agile textile antenna for wearable applications is proposed based on our previous conference paper [20]. The structure of antenna and initial simulation results are shown on the conference paper.

In this paper, the prototype of the designed antenna is fabricated and also measured and details regarding additional analysis and experimental findings are provided as follows whereas it has been noted that wearable antenna users are not particularly comfortable with the stiff substrate and size should be compact. Furthermore, these antennas' stated bandwidth, gain, and SAR value can be increased.

- Here, a frequency reconfigurable jeans-based wearable antenna is structured and fabricated where the agility has been achieved by using two PIN diodes at 5.2 GHz and 2.45 GHz with suitable characteristics of the antenna for biotelemetry application.
- The dataset of optimal antenna parameters is generated and by using the dataset it has been possible to train the algorithms of machine learning model like Random Forest, Neural Network, Support Vector Machines, and Gaussian Process Regression that mainly focus on supervised learning.
- The performances are validated by using the prediction models and compare the results with the true responses and the error percentage is analyzed for doing the segregation with the results that are obtained by simulation software.
- The resonant frequency and operating bandwidth of wearable antennas can be influenced by the high permittivity of human tissue due to their close proximity to the human body so the machine learning algorithms are used here to enhance the bandwidth with the help of provided parameters using the models.

Section 2 of the article describes the geometry of the proposed antenna and its design methodology, experimental results with discussions are described in Sects. 3, 4 describes the machine learning approaches for optimization of the proposed antenna and the conclusion is discussed in the last section of the paper.

2 Geometry of the Antenna and Design Methodology

2.1 Geometry of the Antenna

The three views of the designed antenna (simulated and fabricated) are demonstrated in Fig. 1 and respective dimensions are given in Table 1. The proposed antenna is simulated and also fabricated on jeans having a thickness (h) of 1 mm, dielectric constant (ε_r) of 1.7, and loss tangent (tanδ) of 0.025. The overall dimension of the fabricated antenna is 40 mm × 36 mm × 1 mm. Initially, an exquisitely designed monopole antenna, fashioned in the elegant form of a tuning fork, has been meticulously engineered to operate flawlessly at the resonant frequency of 5.2 GHz. Subsequently, an elegant and meandered parasitic has been seamlessly integrated with the exquisite tuning fork-shaped monopole. This ingenious fusion has been meticulously achieved by employing the sophisticated PIN diode (BAP64-02), resulting in the elongation of the antenna's overall length. The harmonious collaboration of these elements has been able to obtain an additional resonant frequency, resonating at 2.45 GHz. An adhesive copper tape with a thickness of 0.071 mm has been elegantly implemented as both the radiating element and ground plane of the monopole design that ensures optimal performance and efficiency in transmitting signals. A space of 1.25 mm has been maintained between the tuning fork and meandered parasitic patch in order to affix the PIN diodes according to the datasheet. This meticulous attention to detail ensures optimal performance and precision in the installation process, reflecting a commitment to excellence and sophistication in the design and execution of the project.

Employing a microstrip transmission line with a width of 5.5 mm. This serves as a conduit to nourish and empower the system it feeds, ensuring a seamless flow of information and communication. A 15-pF capacitor has been elegantly positioned upon the feedline, serving as a splendid barrier against the intrusion of any undesirable DC signal into the esteemed RF source of the antenna.

The entire simulation process has been meticulously executed through the utilization of the esteemed High-Frequency Structure Simulator (HFSS).

2.2 Design Methodology

Within the envisioned antenna design, the ability to reconfigure frequencies is made possible through the synchronized ON and OFF states of two distinct PIN diodes. Through the strategic utilization of various configurations of PIN diodes, the physical characteristics of the antenna can be altered, thereby enabling the flexibility of frequency reconfiguration to be achieved whereas the centered frequency of an antenna is clearly dictated by its dimensions. By incorporating two PIN diodes, namely D1 and D2 into the design, the radiating structure undergoes a remarkable transformation. As a result, two distinct frequencies, namely 2.45 GHz and 5.2 GHz,

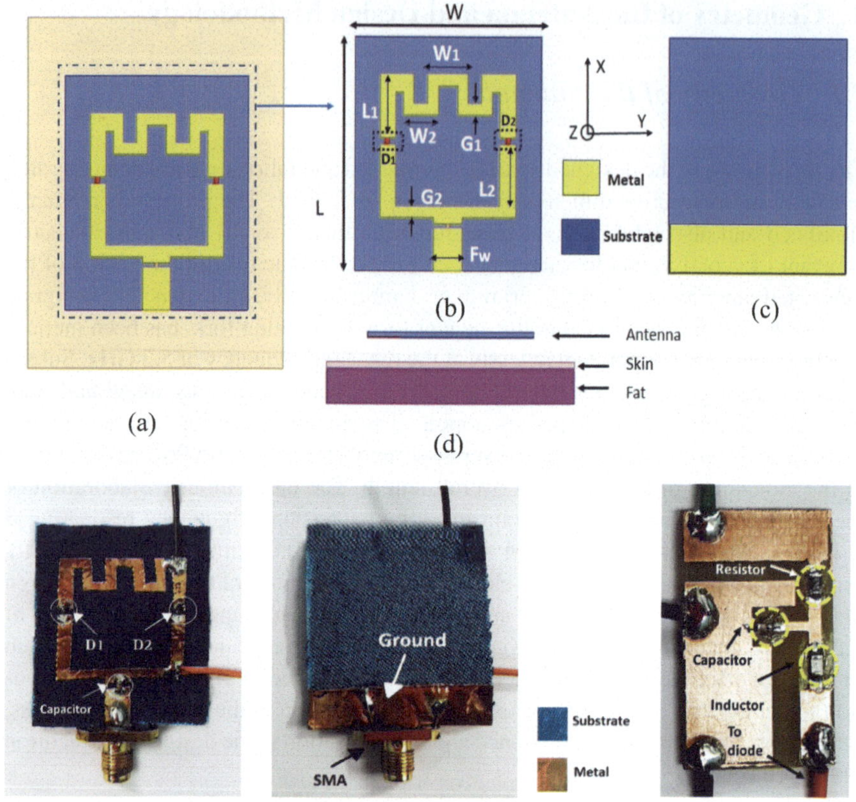

Fig. 1 Proposed antenna geometry **a** upon two-tissue phantom **b** front **c** back **d** side view **e** fabricated antenna prototype with front view **f** fabricated antenna's back view **g** fabricated biasing circuit

Table 1 Proposed antenna dimensions (mm)

L	W	L_1	L_2	W_1	W_2	F_W	G_1	G_2	G_L
40	36	10.8	12.5	7.8	6.6	5.5	2	2	8

are achieved. The intricate interplay between these two diodes and their subsequent effects are meticulously illustrated in the illustrious Table 2, providing a comprehensive understanding of the diverse possibilities that arise from this integration. When both D1 and D2 are turned on, the tuning fork monopole, combined with the parasitic meander, functions as a radiator, producing 2.45 GHz. In the event that both D1 and D2 are in the OFF position, the tuning fork becomes detached from the upper length of meandered shape, leading to a reduction in its overall size. Consequently, this configuration enables the operation specifically at 5.2 GHz since the most of the

Table 2 Effects of the diodes at ON and OFF state

D1	D2	Name of the state	Operating frequency (GHz)
ON	ON	I	2.45
OFF	OFF	II	5.20

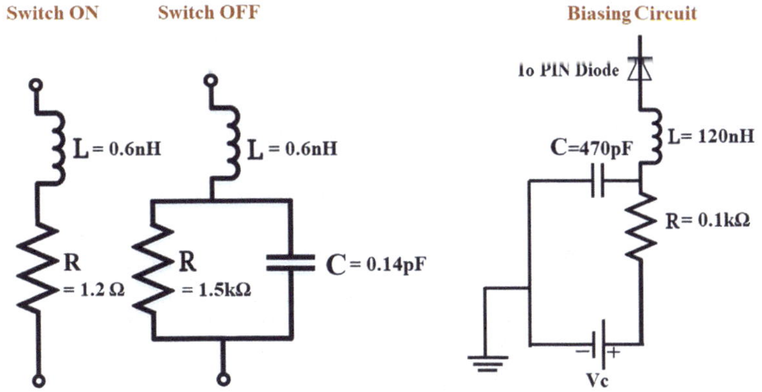

Fig. 2 Schematic structure of ON and OFF state and biasing circuit of the proposed antenna

currents are distributed on the lower element. The schematic diagram of the diodes in both the states, along with the biasing circuit diagram has been explained in detail in Fig. 2 and the fabricated biasing circuit is already given in Fig. 1.The evolution steps of the antenna design are displayed in Fig. 3.

The evolution steps and their corresponding return loss are displayed in Fig. 3 whereas the design procedure has been started from a tuning fork-shaped structure and then the diodes are placed between the two hands on both sides while the upper length is a straight line with operating frequency at 2.5 GHz and 5.45 GHz. The design is further modified with a meander design on the upper length to increase the physical length of the antenna whereas the antenna at the ON state the electrical length has been increased and focused on the frequency of 2.45 GHz and at the OFF state it has been resonated at 5.2 GHz.

3 Experimental Results and Discussions

The antenna has been fabricated on jeans substrate of 1 mm thickness and copper tape has been used as a metal element. The antenna is simulated on two-tissue phantoms and as well as the human phantom using HFSS 19 whereas the fabricated antenna is measured using Vector Network Analyzer by placing the antenna on human hand. In this section, the results including reflection coefficient, gain, far-field radiation pattern, and specific absorption rate of the antenna are discussed.

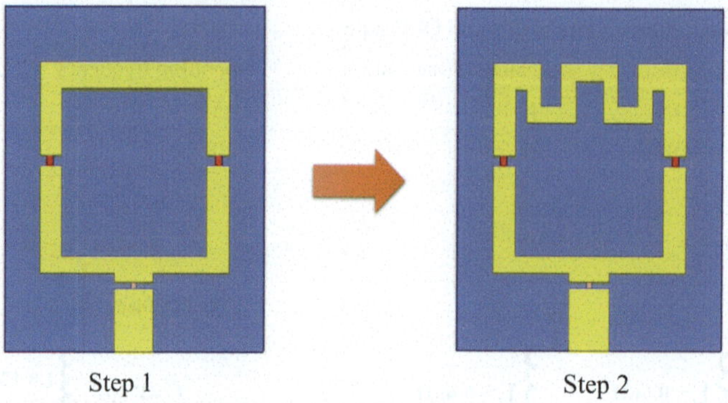

(a)

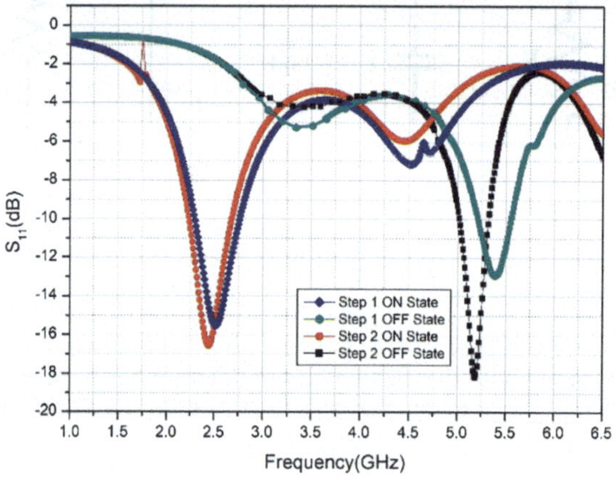

(b)

Fig. 3 **a** Antenna evolution steps with front view of the structure and partial ground plane **b** Compared results of reflection coefficient at State I and State II on two specific steps

3.1 Reflection Coefficient

The results of return loss are compared with simulated and measured values at two states named State I and State II with respect to the two configurations of the diodes BAP64-02 in Fig. 4. When the antenna is at State I the s11 provides − 16.61 dB with the bandwidth of 440 MHz at simulation stage and the fabricated result shows the s11 at − 15.54 dB with the 10 dB bandwidth of 460 MHz. When State II is investigated at the simulation stage, it resonates at 5.2 GHz with s11 − 18.26 dB while the fabricated measured s11 is observed at 5.4 GHz with − 14.06 dB when the bandwidth covers 360 MHz. The measurement setup is also demonstrated in Fig. 5.

The return loss performance is also observed in the simulation stage by placing the antenna on human phantom hand, displayed in Fig. 6. It is noticed that the measured

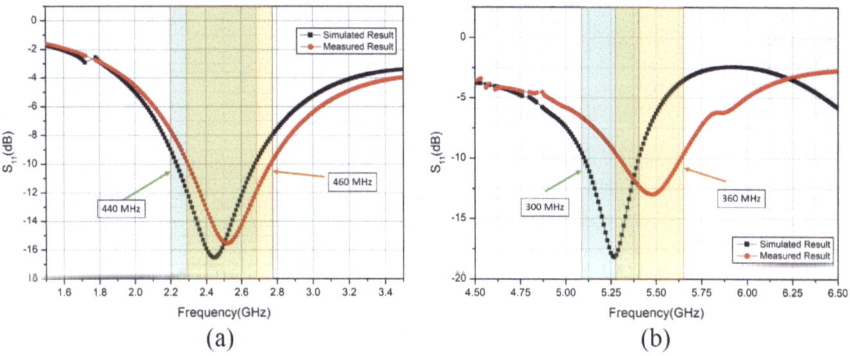

Fig. 4 **a** Return loss at State I **b** Return loss at State II

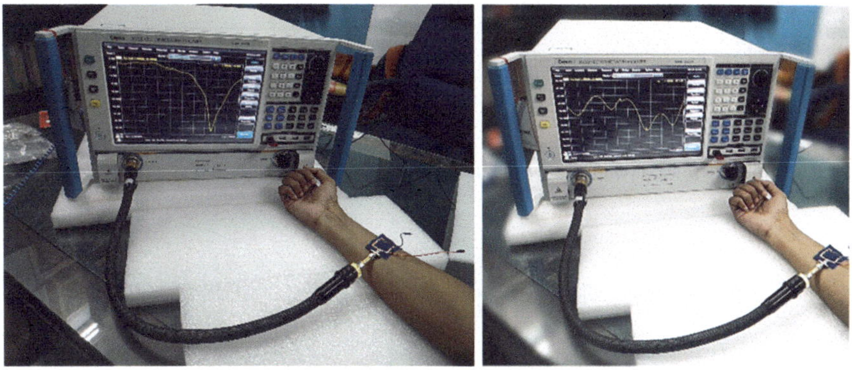

Fig. 5 Measuring return loss using VNA by placing the antenna on human hand in two states

results are little changed which can be the closed proximity to the human body but the performances are acceptable with good agreement.

3.2 *Far-Field Radiation Pattern and Gain Analysis*

The radiation characteristics of the antenna are analyzed by studying the 2D radiation patterns for each operating frequency at two states by the combination of diode configuration. In order to simulate the 2-dimensional radiation pattern on the main planes (phi = 0° and phi = 90°), the antenna is simulated on HFSS. Figure 7 shows the simulated 2D radiation pattern at both of the states.

It has been observed that the proposed antenna exhibited an omni-directional radiation pattern for the ON state at Phi = 0°. However, at Phi = 90°, the radiation pattern resembled a shape of eight for the lower frequency. Moving on to the upper

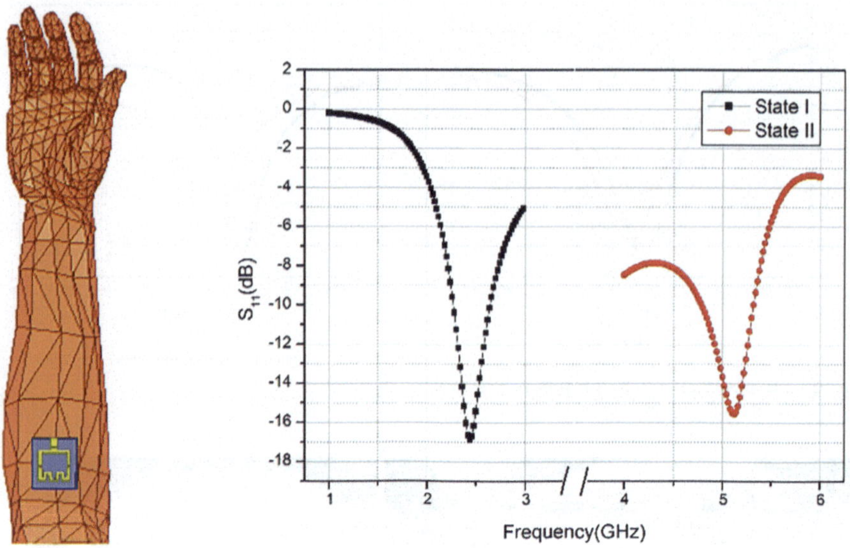

(a) (b)

Fig. 6 **a** Antenna placed on human arm phantom **b** Corresponding return loss at State I and State II

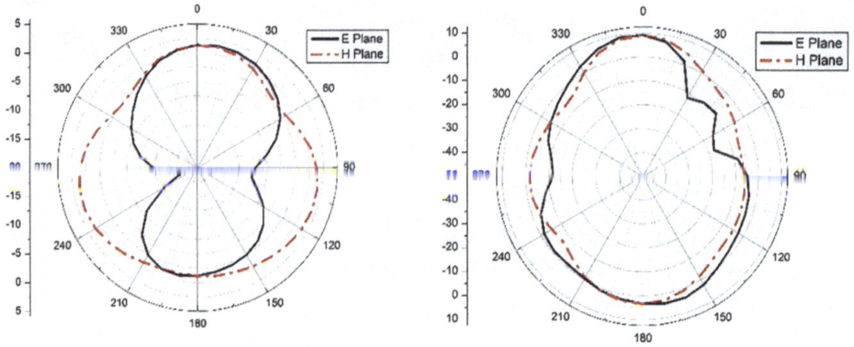

(a) (b)

Fig. 7 **a** 2D radiation pattern at State I **b** 2D radiation pattern at State II

operating band at Phi = 0°, the radiation pattern is omni-directional and at Phi = 90° the eight shape is found but quite distorted due to the presence of skin-fat phantom.

The antenna exhibits the gain of 4.75 dB at State I and 9.1886 dB at State II. The gain vs frequency plot has been displayed in Fig. 8 whereas the gain results are also observed on human phantom hand placement of antenna which is quite low, i.e. 2.9 dB and 6.4 dB w.r.t corresponding states.

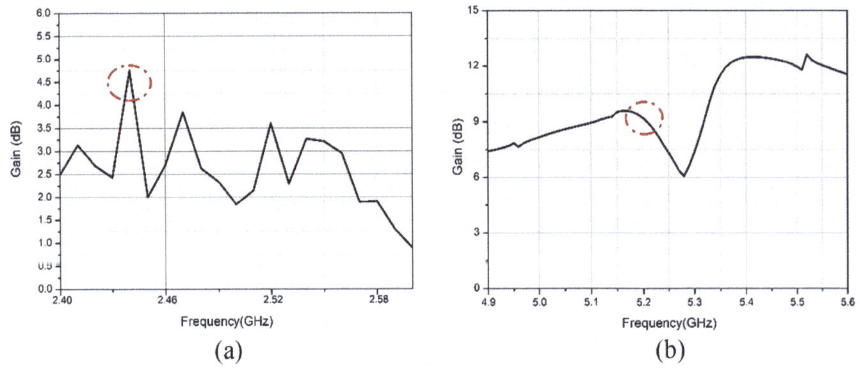

Fig. 8 Gain vs. Frequency plot **a** State I **b** State II

3.3 *Current Distribution on Antenna Surface*

The distribution of surface current for each state within their designated frequencies when mode 1 at 2.45 GHz and mode 2 at 5.2 GHz is observed in Fig. 9. The distribution of surface current on the metal surface varies across different frequency bands. It is analyzed that at State I the current has focused on the feeding point as well as the two hands attached with the diodes and for State II the distributed current has surrounded within the tuning fork mostly and a small amount is moving on the upper part of the diodes.

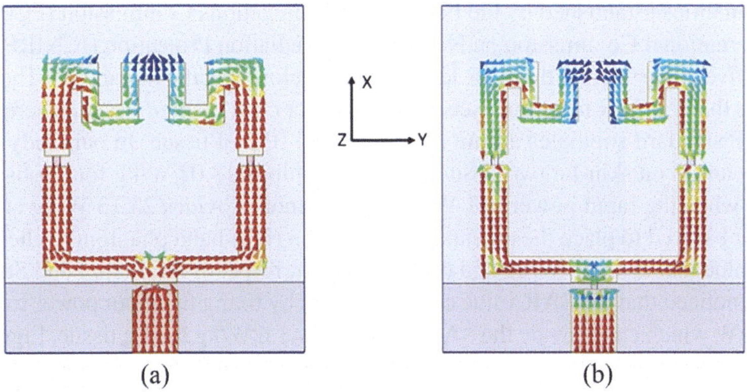

Fig. 9 Surface current distribution of the proposed design **a** State I **b** State II

3.4 Analysis of Antenna Performances on Bent and Crumpled Condition

The antenna performance for wearable applications on bent surface has been examined and displayed in this section. The study of the jeans-based antenna along the y-direction is investigated at the diameters of 20 mm, 50 mm, and 70 mm for State I and State II and is also demonstrated in Fig. 10. The reflection coefficient at 20 mm diameter along the y-direction for State I is − 17.66 dB when the operating frequency is 2.46 GHz, at 50 mm the resonant frequency is 2.46 GHz, − 23.37 dB, and at 70 mm the resonant frequency is 2.4 GHz, − 28.42 dB. For State II, the corresponding frequencies and s11 are 5.2 GHz, − 44.07 dB when diameter is 20 mm. 5.34 GHz, − 40.11 dB when diameter is 50 mm and 5.46 GHz, − 48.38 dB when the diameter of the bending surface is 70 mm.

Further analysis is conducted on the performance of the suggested wearable antenna when utilized on a crumpled surface while the depth of the crumpled surface is 8 mm and the distance between the peak points of cylindrical surface is maintained as 18 mm. The return loss on the crumpled surface at both states is demonstrated in Fig. 11.

3.5 Specific Absorption Rate Analysis of Proposed Antenna

The assessment of the Specific Absorption Rate (SAR) of an antenna when it is utilized on the human body holds significant significance. In order to adhere to the regulations established by the Federal Communications Commission (FCC) and the International Commission on Non Ionizing Radiation Protection (ICNIRP), it is imperative to ensure that the SAR level remains below specific thresholds. The FCC dictates that the SAR must not exceed 1.6 W/kg for each gram of tissue, whereas the ICNIRP standard stipulates a limit of 2 W/kg per 10 g of tissue. In our study when the antenna is on skin-fat two-tissue phantom, exhibits 13.02 W/kg for 1 g tissue at State I when the input power is 1 W and the antenna provides 23.13 W/kg at State II. It is observed to place the antenna on the human right-hand phantom to check the SAR value that is 45.36 W/kg and 62.66 W/kg for, respectively, State I and State II.

It is noticed that the SAR value can be reduced by fixing the input power from 50 to 70 mW which can provide the SAR value below 1.6 W/kg for 1 g tissue. Figure 12 shows the SAR images of State I and State II and the distribution of SAR on human right-hand phantom is shown in Fig. 13.

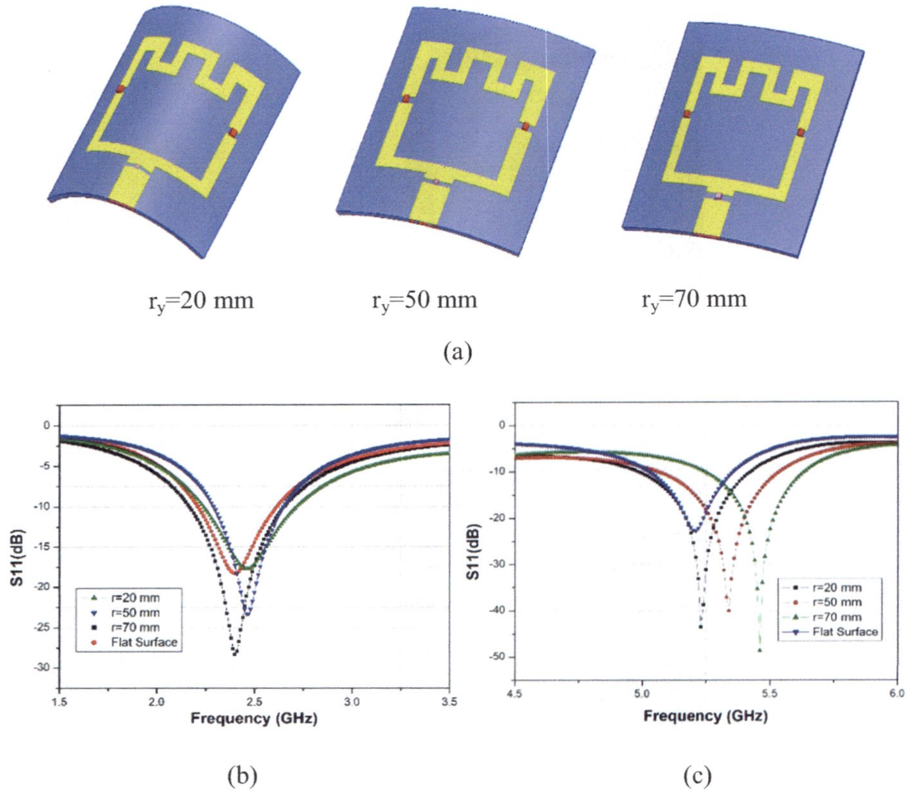

Fig. 10 **a** Bent analysis of proposed antenna along y-direction **b** Compared s11 plot of bent and flat surface for State I **c** Compared s11 plot of bent and flat surface for State II

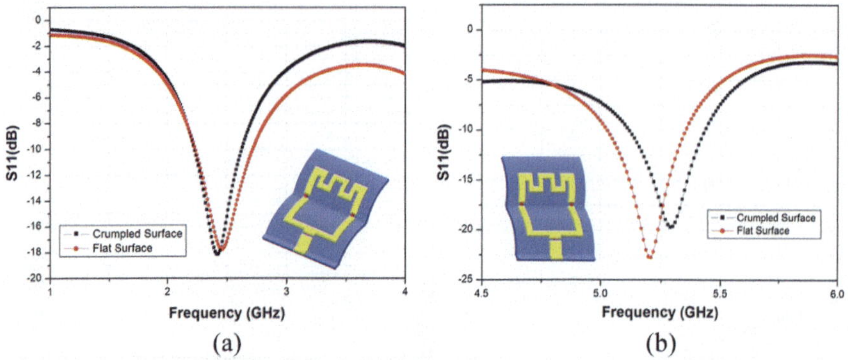

Fig. 11 Compared s11 plot of crumpled and flat surface **a** State I **b** State II

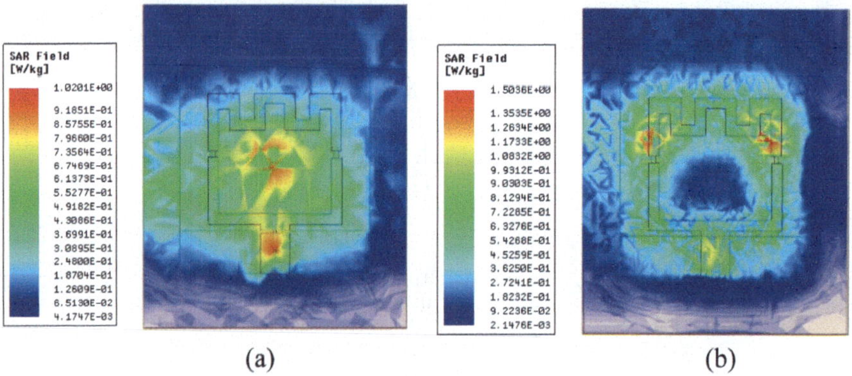

Fig. 12 SAR distribution on proposed antenna **a** State I **b** State II

4 Machine Learning Approaches for Design Optimization of the Proposed Antenna

In this article, we apply some regressive machine learning models to optimize the antenna parameters for enhancing the bandwidth of the designed structure. Four machine learning models are used here for optimization names as Gaussian Process Regression, Random Forest, Support Vector Regression, and Artificial Neural Network in which, as input parameters, resonant frequency, reflection coefficient, and characteristics of the substrate like dielectric constant 1.7, height and width, respectively, 40 mm and 36 mm with 1 mm thickness are provided whereas the feature of the substrate has been kept same. The input parameters range and output parameters ranges are given, respectively, in Tables 3 and 4.

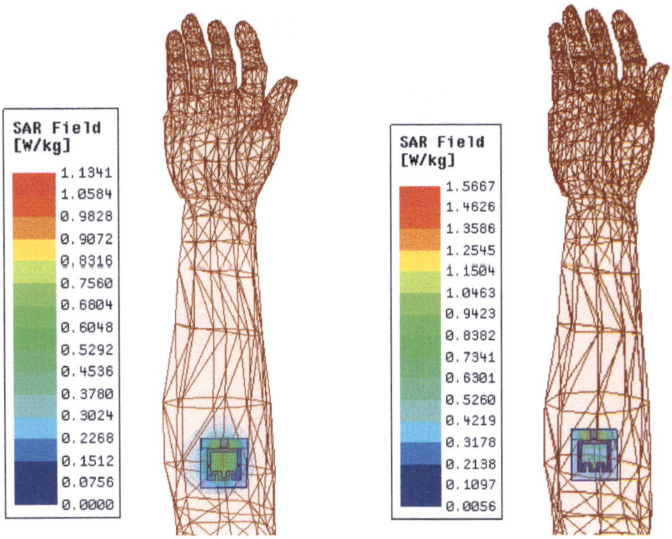

(a) (b)

Fig. 13 SAR distribution on human right-hand phantom **a** State I **b** State II

Table 3 Input parameters with varying ranges of values

Input parameters	Minimum value	Maximum value
Fc	2.29 GHz	2.63 GHz
S11	− 66.17 dB	− 6 dB

Table 4 Output parameters with varying ranges of values

Output parameters	Min. value (mm)	Max. value (mm)
G1	1	4
GL	7	10
W1	5.8	8.8
Fw	3.5	6.5
W2	4.6	7.6

4.1 Support Vector Regression

Support Vector Regression (SVR) is a variant of Support Vector Machine (SVM), which is a supervised learning technique suitable for classification or regression purposes. SVMs aim to identify the hyperplane within a multi-dimensional space that best distinguishes various classes or predicted values. SVR has the ability to utilize various types of kernels, which serve as functions to assess the similarity between input vectors. The selection of kernel is based on the specific characteristics of the data and the complexity of the task at hand. SVR places a symmetric tube with a width greater than zero, denoted as epsilon, around the predicted function. This ensures that errors with absolute values below the epsilon threshold are disregarded both below and above the estimation. Here we use the Gaussian kernel function where the box constraint is 0.7413 and epsilon is 0.05.

4.2 Gaussian Process Regression

Gaussian process regression (GPR) models are probabilistic models based on nonparametric kernels. Here the Rational Quadratic kernel function is used with the sigma value of 0.3205 and the scale is given as 22.3. The Eq. 1 of this model is written as—

$$k(x, y|\theta) = \sigma^2\left(1 + \frac{p^2}{2\alpha\sigma^2}\right) \quad (1)$$

where σ = scale of signal length, α = scale mixture parameter, p = Euclidean space between x and y

4.3 Artificial Neural Network

Artificial Neural Networks are comprised of artificial entities referred to as neurons. These neurons are structured in layers that together create the complete Artificial Neural Network within a system. An Artificial Neural Network consists of an input layer, hidden layers, and an output layer. The input layer's role is to receive external data that needs to be processed or interpreted by the neural network. Subsequently, this data is processed through one or more hidden layers to convert the input into meaningful information for the output layer. Ultimately, the output layer delivers a response representing the Artificial Neural Network's reaction to the input data received. Here we use a tri-layered neural network and each having a size of 10 with the activation of ReLU and an iteration limit has given 1000.

4.4 Random Forest

Multiple numbers of trees are trained on arbitrarily selected test works from the training dataset in a random forest. A selection is made from a subset of the original input features during this procedure. The ultimate prediction is made by the random forest model through the combination of predictions derived from each decision tree. A RF model can be evaluated by the expression 2—

$$y = \frac{1}{m}\sum_{i=1}^{m} t(x) \tag{2}$$

where m = total decision trees, applied here as 110, x = input value, y = output from the model, and t = decision tree.

5 Validation Results

In order to train the machine learning models we have proposed, we created a dataset of 440 samples through the use of High Frequency Simulating Software where the dataset is distributed into 60 percentage for training purpose, 20 percent for validation, and 20 percent for testing purposes whereas the models are validated by using the evaluation of RMSE that is Root Mean Squared Error which can able to evaluate the difference between original value and predicted value and if it shows the lower amount of value then it assures the model is perfect. The expression is written on Eq. 3—

$$RMSE = \sqrt{\frac{\sum_{i=1}^{m} (y_i - o_i)^2}{m}} \tag{3}$$

where m = number of observations, Yi = Predicted value, O_i = Original value of ith observation.

The Mean Squared Error (MSE) calculates the average of the squared variances between the predicted values and the actual target values. The expression can be expressed by the Eq. 4—

$$MSE = \sum_{i=1}^{m} (P_i - y_i)^2 \tag{4}$$

where P_i = Predicted output, y_i = Original Output.

The R-squared validation indicates the perfect fit or not by providing the value close to 1.

Here the cross-validation is used as 20-fold where we can see the best result is observed by GPR and RF model, the tabular form of the result is detailed in Table 5 and the prediction model is given based on the models in Figs. 14, 15, 16. We also attach the MSE results in Fig. 16.

On the given Table 3, the ranges of the input parameters are mentioned here whereas the dielectric properties of substrate material are constant. Here we have focused on enhancing the bandwidth by optimizing the design parameters of G1, GL, W1, Fw, and W2 by giving them a range of values.

Table 5 Results of RMSE, MSE, R-Square, and the training time of GPR and ANN

	GPR				ANN			
	RMSE	Time	R^2	MSE	RMSE	Time	R^2	MSE
G1	9.78E−06	75.09	1	9.57E−11	0.12776	94.9	0.98	1.63E−02
GL	9.78E−06	13.502	1	9.57E−11	0.21148	10.654	0.96	4.47E−02
W1	1.05E−05	31.568	1	1.09E−10	0.065832	23.825	0.996	4.33E−03
Fw	1.04E−05	18.103	1	1.09E−10	0.15972	23.479	0.9763	2.55E−02
W2	1.05E−05	9.3023	1	1.11E−10	0.22293	6.7461	0.95	4.96E−02
AVG	**1.02E−05**	**29.513**	**1**	**1.04E−10**	**0.15754**	**31.920**	**0.9724**	**2.81E−02**

From Tables 5 and 6 it is observed that the best results of MSE, RMSE, and R-square have been achieved by the machine learning model of GPR and RF. The lowest value of MSE is 1.02e−05 found from the GPR model, the lowest RMSE is 1.02e−05 also found from GPR and the highest value of R^2 is 1, found from Random Forest and Gaussian Process Regression. As a consequence in this study, the scatter plots of this particular model have been attached along with this.

Further three samples are used to predict the best results about the bandwidth as well as gain. The sample table is displayed in Table 6. The results of Mean Square error are also demonstrated here.

In Table 7, it can be observed that the improved bandwidth can be achieved by the model of ANN and SVR for State I and SVR also provides the improved bandwidth for State II and that particular ground truth value has been used to design and fabricate the antenna. The reflection coefficient from GPR and RF is the same and for ANN it is achieved as − 17.05 dB and − 23.22 dB from SVR which is quite improved. The s11 vs frequency graph is plotted through the simulation on HFSS, shown in Fig. 16. The test 2 data samples are displayed in Table 8.

On test 2 the reflection coefficient is achieved from GPR and RF with no error and from ANN the S11 is − 30.9 dB and − 25.06 dB when the ground truth is − 15.78 dB and − 21.61 dB, respectively, for State I and State II. From SVR the s11 are − 21.89 dB and − 27.17 dB.

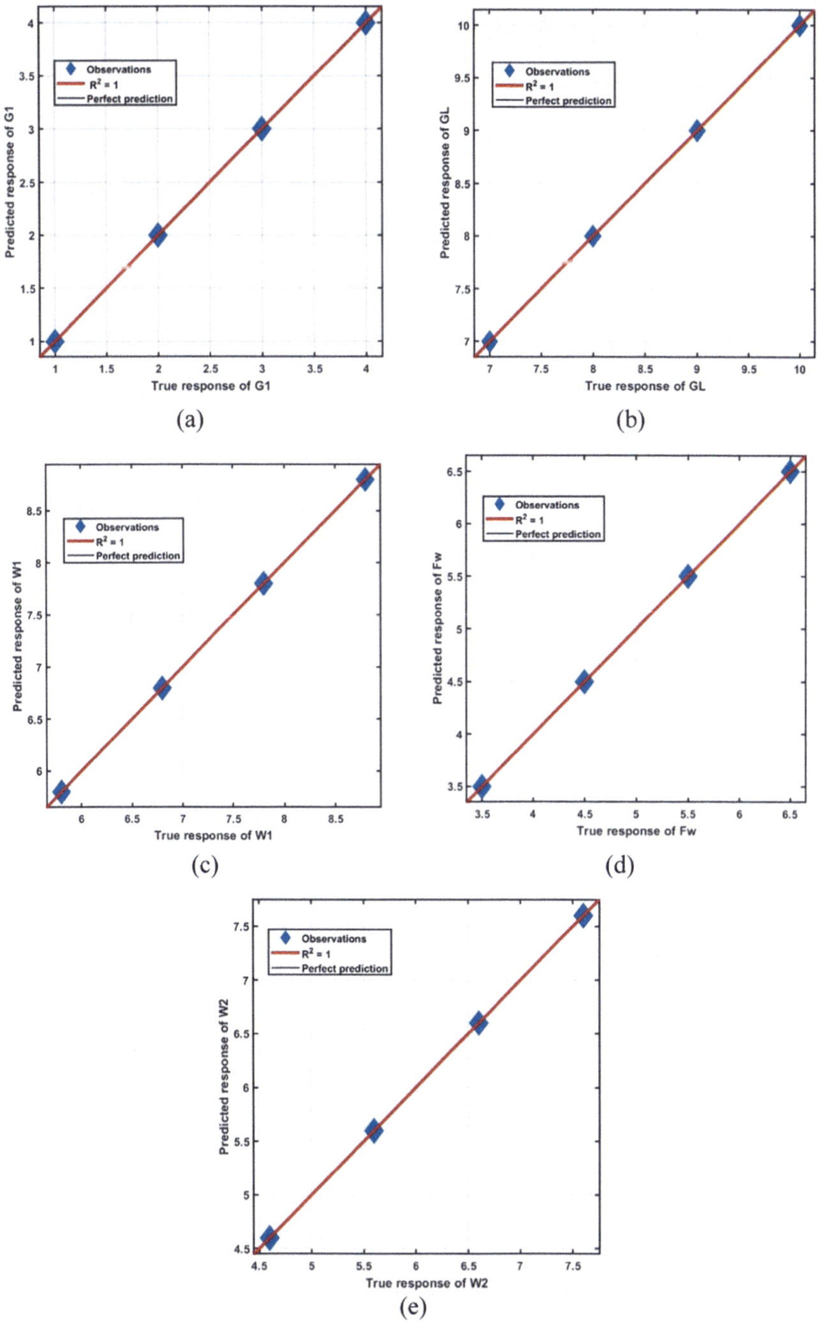

Fig. 14 Predicted vs true response plots **a** G1 (width of line of meandering structure) **b** GL (length of ground) **c** W1(upper length in y-direction of meandering line) **d** Fw (width of feed) **e** W2(lower length in y-direction of meandering line)

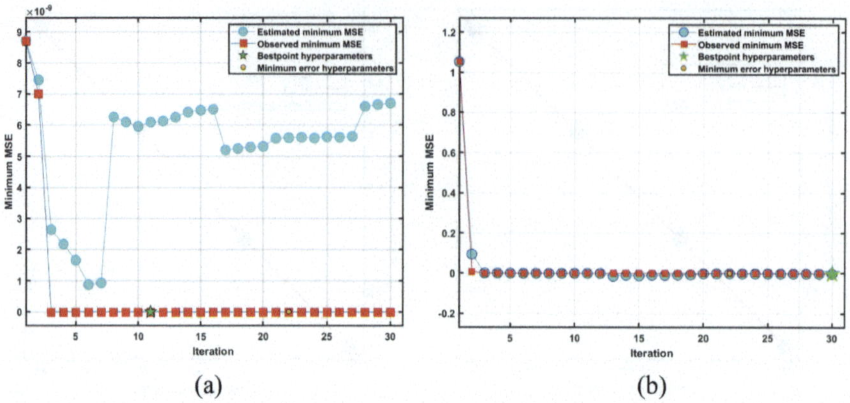

Fig. 15 Mean square error **a** GPR **b** RF

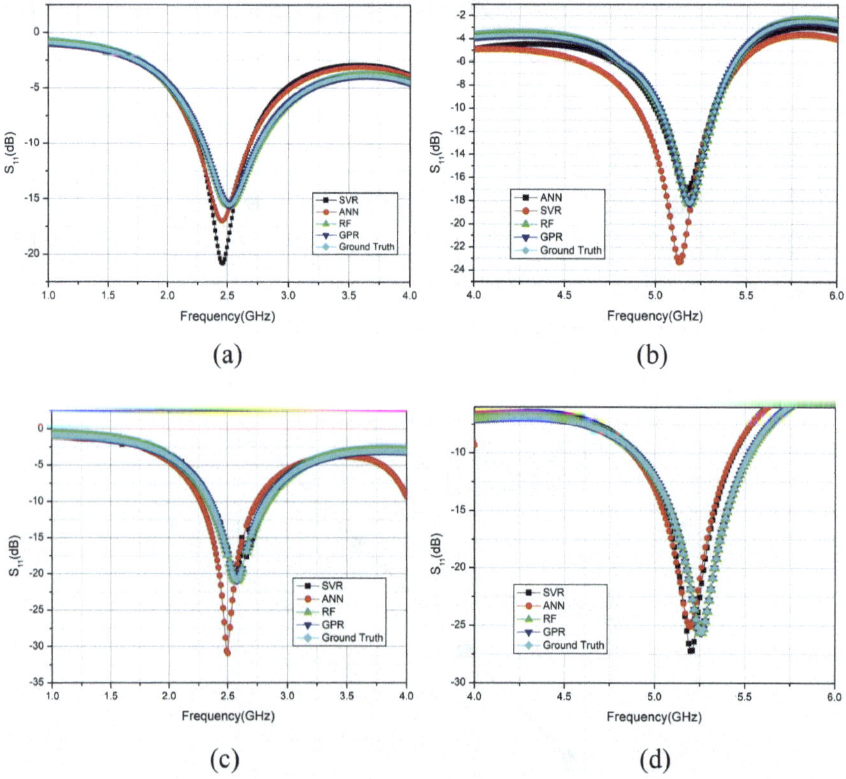

Fig. 16 Plot of Reflection coefficient of proposed antenna **a** Test 1 at State I **b** Test 1 at State II **c** Test 2 at State I **d** Test 2 at State II

Table 6 Results of RMSE, MSE, R-Square, and the training time of SVR and RF

	SVR				RF			
	RMSE	Time	R^2	MSE	RMSE	Time	R^2	MSE
G1	0.11914	78.379	0.99	1.42E−02	0.003185	107.9	1	1.02E−05
GL	0.11913	17.313	0.99	1.42E−02	0.003726	27.777	1	1.39E−05
W1	0.11016	10.346	0.99	1.21E−02	0.002703	42.259	1	7.31E−06
Fw	0.11002	6.94	0.99	1.21E−02	0.003670	33.924	1	1.35E−05
W2	0.11103	4.3396	0.99	1.23E−02	3.57E−15	14.041	1	1.27E−29
AVG	**0.11389**	**23.4635**	**0.99**	**1.30E−02**	**0.002657**	**45.180**	**1**	**8.96E−06**

Table 7 Validate results with improved bandwidth of Test sample 1

		Optimized parameters (mm)					Bandwidth MHz	
		G1	G2	W1	Fw	W2	State I	State II
Test 1	Ground Truth	2	8	7.8	5.5	6.6	440	300
	GPR	2	8	7.8	5.5	6.6	440	300
	Error	**0**	**0**	**0**	**0**	**0**		
	ANN	1.9997	8.0009	7.8	5.4997	6.5994	**450**	330
	Error	**0.0003**	**0.0009**	**0**	**0.0003**	**0.0006**		
	SVR	2.184	8.1841	7.6468	5.5741	6.4468	**450**	**420**
	Error	**0.184**	**0.1841**	**0.153**	**0.0741**	**0.1532**		
	RF	2	8	7.8	5.5	6.6	440	300
	Error	**0**	**0**	**0**	**0**	**0**		

Table 8 Validate results with improved bandwidth of Test sample 2

		Optimized parameters (mm)					Bandwidth (MHz)	
		G1	G2	W1	Fw	W2	State I	State II
Test 2	Ground Truth	3	9	7.8	5.5	6.6	470	630
	GPR	3	9	7.8	5.5	6.6	470	630
	Error	**0**	**0**	**0**	**0**	**0**		
	ANN	2.999	9.001	7.8	5.655	6.599	**510**	540
	Error	**0.001**	**0.001**	**0**	**0.155**	**9E−04**		
	SVR	2.926	8.926	7.87	5.347	6.674	**560**	560
	Error	**0.074**	**0.074**	**0.07**	**0.153**	**0.074**		
	RF	3	9	7.8	5.5	6.6	470	630
	Error	**0**	**0**	**0**	**0**	**0**		

6 Conclusion

In this article a wearable textile-based reconfigurable monopole antenna is designed and further machine learning algorithms are applied with four widely used models for optimizing the parameters to enhance the bandwidth within few seconds whereas the results are compared with themselves and also compared with the genetic algorithm-based simulation results of HFSS. As a result, the accuracy of the results is observed by the output and also through the models of MSE, RMSE, and R^2. Ultimately the bandwidth with acceptable gain can be improved by ANN and SVR models and it is also concluded that the proposed structure of the antenna is suitable for biomedical applications with successfully reconfiguring the frequency with the help of PIN diodes.

References

1. Elkorany SA et al (2020) Heart shaped reconfigurable band-notched ultra-wide band antenna using electromagnetic band-gap structure and varactor diodes. Analog Integr Circ Signal Process 105:385. https://doi.org/10.1007/s10470-020-01697-8
2. Lavadiya PS et al (2021) Design and verification of novel low-profile miniaturized pattern and frequency tunable microstrip patch antenna using two PIN diodes. Braz J Phys. https://doi.org/10.1007/s13538-021-00951
3. Naik KK, Sai Sailaja VB (2020) RF-MEMS switch for reconfigurable with half- moon slots on elliptical-shaped patch antenna for 5G applications. IEEE Open J Antennas Propag. https://doi.org/10.1109/OAJPE.2020.2976889
4. Cai YM et al (2018) A low-profile frequency reconfigurable grid-slotted patch antenna. IEEE Access Digit Object Ident. https://doi.org/10.1109/ACCESS.2018.2850926
5. Iqbal A et al (2019) Low-profile frequency reconfigurable antenna for heterogeneous wireless systems. Electronics 8:976. https://doi.org/10.3390/electronics8090976
6. Varshney A et al (2022) Dodecagon-shaped frequency reconfigurable antenna practically loaded with 3-delta structures for ism band and wireless applications. IETE J Res. https://doi.org/10.1080/03772063.2022.2034536
7. Hussain N et al (2022) A conformal frequency reconfigurable antenna with multiband and wideband characteristics. Sensors 2022(22):2601. https://doi.org/10.3390/s22072601
8. Musa U et al (2023) Wearable dual-band frequency reconfigurable patch antenna for WBAN applications. Progress Electromagn Res M 120:95–111
9. Salleh MS et al (2017) Textile antenna with simultaneous frequency and polarization reconfiguration for WBAN. IEEE Access Digit Object Ident. https://doi.org/10.1109/ACCESS.2017.2787018
10. Thanki P, Raval F (2019) Fork-shaped frequency and pattern reconfigurable antenna. Int J Commun Syst 2020:e4613. https://doi.org/10.1002/dac.4613
11. Karthika K, Kavitha K (2023) Design and development of parasitic elements loaded quadband frequency and pattern reconfigurable antenna. Int J RF Microw Comput Aided Eng 2023:10. https://doi.org/10.1155/2023/4034241
12. Elsheakh ND et al (2024) An early breast cancer detection by using wearable flexible sensors and artificial intelligent. IEEE Access 12:48511–48529. https://doi.org/10.1109/ACCESS.2024.3380453
13. Sharma Y, Zhang HH, Xin H (2020) Machine learning techniques for optimizing design of T-shaped monopole antenna. IEEE Trans Antennas Prop 68(7):5658–5663

14. Singh O et al (2022) Microstrip line fed dielectric resonator antenna optimization using machine learning algorithms Indian academy of sciences. Sadhana 47:226. https://doi.org/10.1007/s12046-022-01989-x
15. Patel KS et al (2022) Machine learning assisted metamaterial-based reconfigurable antenna for low-cost portable electronic devices. Sci Rep 12:12354. https://doi.org/10.1038/s41598-022-16678-2
16. Malm O et al (2024) Predicting the antenna properties of helicon plasma thrusters using machine learning techniques. J Electr Propul 3:6. https://doi.org/10.1007/s44205-023-00063-w
17. Joshiba MJ, Judson D (2024) An eye bolt shape slotted microstrip patch antenna design and return loss prediction using machine learning Indian Academy of Sciences. Sadhana 49:77. https://doi.org/10.1007/s12046-024-02433-y
18. Yahya SM et al (2023) Machine learning-optimized compact frequency reconfigurable antenna with RSSI enhancement for long-range applications. IEEE Access 12:10970–10987
19. Pusuluri VB et al (2024) Design of a multi-band 5G antenna and machine learning approach for electromagnetic compatibility. J Syst Eng Electron 34(3):1671–1793
20. Das S, Kundu KA, Saha P (2023) A frequency agile textile based wearable antenna for on body biomedical applications. In: 2023 International Conference on Computer, Electronics & Electrical Engineering & their Applications, IEEE Xplore. https://doi.org/10.1109/IC2E357697.2023.10262413

Efficient Deep Fake Detection Technique on Video and Audio Dataset Using Deep Learning

Rahul Vadishetty

Abstract The latest advances in computer vision processing need tools for creating realistic deepfakes. Deepfake technology has made it viable to provide very practical-looking fake audio, video, and images. In the recent era, we have faced issues that include impersonation and disseminating false information. The spread of fake media streams causes chaos in social networks and has the potential to harm a person's or a group's reputation, manipulating public sentiment, as well as mindsets about the individual or community. To address these challenges and detect deep fake video images and voices, a deep learning approach has been adopted in this study by using two publicly available datasets containing audio, videos, and images for research and experiment. Six models such as CNN, LSTM, and Pre-trained models VGG-16, Alex-net, Google-Net, and Mobile-net have been analyzed for deepfake detection. Our proposed CNN approach achieved an accuracy of 98.89% on the face and an accuracy of 98.96% on the voice dataset and LSTM achieved an accuracy of 95.11% on the face and 92.67% on another voice dataset. A comparative performance analysis of the Deepfake analysis results obtained in this study is performed with related works to show the supremacy of the current study.

Keywords Deepfake videos · Deepfake audios · Convolutional neural networks · Deep learning · VGG-16 · Alex-net · Google-Net · Mobile-net · LSTM

1 Introduction

Deepfake technology has emerged as an authoritative malicious tool to cause considerable problems in cybersecurity and privacy. These tools adopt powerful deep learning algorithms to create accurate but untrue audiovisual content in cyberspace including digital media. Deepfakes generated by deep learning algorithms have become more diverse from trustworthy content and the probability of exploitation

R. Vadishetty (✉)
Wayne State University, Detroit, MI, USA
e-mail: rahulvy91@gmail.com

M. Chakraborty et al. (eds.), *Proceedings of 5th International Ethical Hacking Conference*, Lecture Notes in Networks and Systems 1148,
https://doi.org/10.1007/978-981-97-8457-8_11

in dispensing falsehood, executing deception, and scoring cyberattacks has grown tremendously, raising grave worries about evidence veracity and individual and institutional safety [1].

Lots of research is being conducted by scientists and technologists to create efficient deepfake detection tools to combat the rising growth and perils danger of deepfakes. These initiatives pursue to differentiate between genuine and transformed evidence by inspecting numerous data mechanisms, such as inconsistencies in visual features, audio patterns, and behavioral indications that may not correspond to real human connections [2]. Deep learning, a subset of machine learning, has appeared as a noteworthy tool in this quest, contributing hopeful opportunities through the building of neural networks capable of learning and recognizing delicate faults indicative of fabrication [3].

The development of an operative deepfake detection solution that uses deep learning is important. Such a system would preferably be proficient enough in processing and analyzing large amounts of video and speech material as well as appropriately distinguishing authentic media from modified one [4]. This would include innovative methods for feature extraction, pattern recognition, and the use of complex neural networks adjusted for higher accuracy and efficiency in numerous challenging detection scenarios. The ultimate aim of this study is to guarantee data genuineness and truthfulness, protecting human beings and society from the countless threats and attacks associated with deepfake technologies [5].

In this study, research and simulation using two publicly available datasets encircling audio, video, and photo have been conducted to analyze deepfake detection using six models: CNN, LSTM, and pre-trained models VGG-16, Alex-net, Google-Net, and Mobile-net on several video and audio datasets. Our proposed CNN approach produced an accuracy of 98.89% on the face and 98.96% on the voice dataset, while the LSTM achieved an accuracy of 95.11% on the face and 92.07% on another audio dataset.

The main objective of the research is to build and evaluate deep learning algorithms. This study aims to contribute to digital security by enhancing the accuracy and efficiency of deepfake detection systems, reducing the hazards associated with the growing availability of synthetic media.

The three key contributions of this study include

- Innovative approaches to the dataset and feature analysis methods tailored to detecting video and audio deepfake.
- Establishment of a systematic evaluation of deep learning models including VGG-16, Alex-Net, and others for deepfake detection.
- Enhancement of real-time detection capabilities with optimized speed and accuracy for practical applications.

The organization of this paper is as follows: Sect. 2 is based on the literature review, while Sect. 3 contains details about data gathering and preprocessing methods. Section 4 contains the results and discussion and Sect. 5 contains the conclusion and future work.

2 Literature Review

In this section, we are going to explore all aspects of deepfake video and audio including all machine learning techniques, deep learning techniques, datasets used, and challenges that most research tackles while detecting deepfake video and audio.

2.1 *Machine Learning Techniques for Deepfake Video/Audio Detection*

Machine learning techniques for detecting deepfake typically involve traditional algorithms that have been adapted to recognize manipulated media. In [6], SVM was used to detect deepfake images from four datasets: CelebA HQ (1000 images), Flickr (1000 images), Person Does Not Exist (1000 images), and a generated dataset (1000 images). The study utilized 80% of the data for training and 20% for testing and achieved an accuracy of 99.76%. The research [7] detects deepfake movies with Support Vector Machine (SVM) regression. The SVM classifier can be trained on feature points retrieved using one of the various feature detectors, including HOG, ORB, BRISK, KAZE, SURF, and FAST algorithms. The dataset for deepfake movies has 98 videos. The results suggest that the proposed method of training the SVM with feature-detector-descriptors is successful in detecting bogus movies. The HOG feature-point extraction approach yielded approximately 95% accuracy. In [8], a model based on SVM is proposed as a classifier to recognize phony human faces made by machines. The strategy is more resilient than current methods. Preprocessing is done in the early stages, which begin with converting RGB images to YCbCr and subsequently gamma correction. In the end, the findings demonstrate that the Canny filter-extracted edges proved helpful in identifying fakes in face photos. Subsequently, using "SVM" with "Principal Component Analysis (PCA)" and two alternative detection methods with "SVM" in the absence of "PCA" classifiers were employed. According to the findings, utilizing the SVM with PCA produced the greatest accuracy of 96.8%, whereas using the SVM alone yielded 72.2%. [9] describes an improved VGG network called NA-VGG for detecting deepfake face images, which is based on picture noise and augmentation. The Celeb-DF dataset yielded an accuracy of 85.7%. Deepfakes pose a significant threat to digital security, privacy, and human rights [10]. This research gives a unique dataset for constructing a generalized detection algorithm. They introduce an attention-based, multimodal deepfake detection framework. Extensive research shows that their methodology is generalizable and has effectively learned the audiovisual representations needed to challenge existing deepfake detection systems. The cross-modal method based on attention processes detects deepfakes with 97.6% accuracy.

Used for classifying videos and audio as real or fake based on feature extraction.

- Random Forests is a collective approach that uses several decision trees to generate decisions, making it effective for handling complex datasets.

- Logistic Regression: Although simpler, it can be effective in cases with strong, distinguishable features between real and fake content.

2.2 Deep Learning Techniques for Deepfake Video/Audio Detection

These techniques often rely on extracting statistical features from the data, such as texture, noise patterns, and other inconsistencies that are not easily visible to the human eye. The authors of [11] illustrate a learning-based approach to identifying authentic and bogus deepfake multimedia material. They propose a deep learning network based on the Siamese network architecture with triplet loss and compare their approach to many SOTA deepfake detection algorithms, obtaining per-video AUCs of 84.4% on the DFDC data and 96.6% for the DF-TIMIT data [12] suggests a hybrid deep learning method that combines LSTM with NOLANet to differentiate between authentic and false movies by consistently coupling spatial, spectral, and temporal data. They demonstrated how the Discrete Cosine Transform, which captures the spectral properties of individual frames, might enhance the identification of deepfakes. Using the Facebook Deepfake Detection Challenge (DFDC) dataset, they have developed a multimodal network in this work that investigates novel features for the detection of deepfake videos, attaining 61.95% accuracy.

Deep learning provides more sophisticated tools capable of detecting more subtle and complex deepfakes including.

- *Convolutional Neural Networks (CNNs)*: Highly effective for spatial analysis in videos and images to detect anomalies introduced by deepfake algorithms.
- *RNNs and LSTMs*: Useful for analyzing temporal inconsistencies in videos where the frame-to-frame relationship might be manipulated.
- *Autoencoders and Generative Adversarial Networks (GANs)*: Can be used to both generate and detect fakes, with the latter being trained to identify characteristics of fakes made by similar models.

2.3 Dataset Used for Deepfake Video/Audio Detection

FaceForensics + + [13–15] is a ground-breaking, large-scale dataset for the identification of facial alteration [14]. The data are of variable compression levels and sizes, and the primary facial manipulations Deepfakes, Face2Face, FaceSwap, FaceShifter, and Neural Textures methods are representative. The content of the 1000 real and 4000 false YouTube videos that make up this collection is composed of 60% female and 40% male videos. Deep-voice [16] Real-time voice translation from one human to another is made possible by the growing uses of generative AI in the speech domain, as well as voice cloning. Real-time speech recognition for deepfake Voice Conversion [17] is critical since this technology presents a serious ethical risk and may result in privacy violations and deception of the DEEP-VOICE dataset to solve the aforementioned growing challenges. Retrieval-based Voice Conversion was used to convert the speech of eight well-known figures into one cohesive voice for DEEP-VOICE. Audio features can be found in the datasets DfTIMIT and DFDC [18] preview. There are 4113 phony samples and 1131 actual videos in DFDC. The DFDC preview dataset makes no mention of the audio's falsity. Moreover, neither the music nor the accompanying graphics lip-synch. DfTIMIT only includes original audio.

Developing and testing detection systems require robust datasets. Some well-known datasets include.

- *FaceForensics++*: A large-scale dataset containing real and manipulated videos, intended to help develop algorithms that can detect visual manipulations.
- *Deepfake Detection Challenge Dataset (DFDC)*: Provided by Facebook, this dataset includes a large number of videos with varying quality and manipulation techniques, making it one of the most challenging datasets.
- *Celeb-DF*: A dataset containing fabricated celebrity videos for assessment of the sturdiness of fake detection methods.

2.4 Challenges of Deepfake Video/Audio Detection

The challenges associated with the detection of deepfakes are as follows.

- *Quality and complexity of fakes*: As deepfake technology evolves, the fakes become more sophisticated and harder to detect.
- *Inconsistency in techniques*: Different methods and tools used for creating deepfakes can result in a wide variety of manipulation methods to detect.
- *Generalization across datasets*: Models trained on one dataset may not perform well on another due to differences in the types of manipulations or the quality of the videos.
- *Real-time detection*: Creating methods that can operate in real-time, particularly for platforms demanding instant verification, adds another layer of complexity.

3 Proposed Methodology

The proposed methodology to address the problems put forth by deepfake technology is discussed in this section. Deepfake technology permits the fabrication of convincing forged audio, video, and picture content to cause problems that include impersonation and the spread of incorrect information. To address these challenges, we used deep learning algorithms to recognize deepfake facts in various media types, including videos, voice recordings, and images. Publicly available datasets were used in the experiments for all pre-trained models and LSTM and CNN. The block diagram representation of the proposed methodology is shown in Fig. 1.

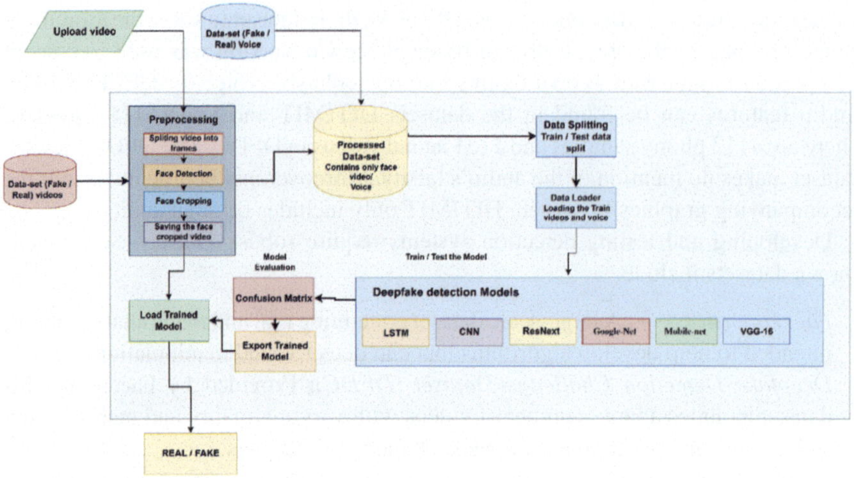

Fig. 1 Overall top-down approach

3.1 Dataset Collection

The dataset used in this paper is defined in Table 1. Benchmarked and publicly available datasets were used for research and experiments, consisting of both images and voice datasets. Deep-voice: The deepfake Voice Recognition [16] dataset is provided in two different formats. First, the "AUDIO" directory contains the raw audio. The "REAL" and "FAKE" class folders contain them in order shown in Fig. 2. The actual speakers and the voices were converted to are noted in the audio filenames. For instance, "Obama-to-Biden" indicates that Joe Biden's voice has been substituted for Barack Obama's during his address. Secondly, the "DATASET-balanced.csv" file contains the extracted features. This information was utilized in the study that appears below. Every feature in the dataset was taken from audio clips lasting one second, and random sampling was used to ensure balance. Deepfake Detection [19, 20], models were trained and tested using the deepfake dataset. The benchmark deepfake dataset is accessible to the public on Kaggle.

Table 1 Dataset description

Sr#	Reference	Modality	Size of dataset
1	[20]	Images	5 GB
2	[21]	Images	6 GB
3	[22]	Voice	4 GB
4	[23]	Images	2 GB
5	[24]	Videos	4.5 GB
6	[25]	Images	4.44 GB
7	[26]	Images	725 MB

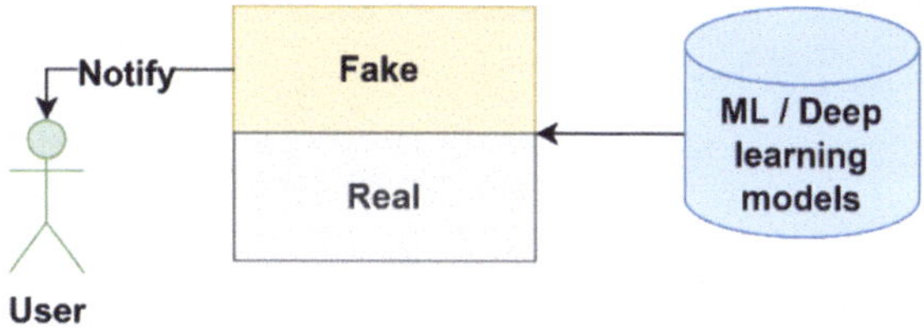

Fig. 2 Use of real-time system

3.2 *Preprocessing*

Preprocessing is an important stage in refining voice/images for accurate analysis and decision-making in deepfake. The audio data is extracted using the extracted features function. This function iterates through the "REAL" and "FAKE" folders in the specified directory, loads each audio file with the Libros library, computes MFCCs as features, and pads or trims them to a fixed length. Labels are assigned according to the folder name (1 for "FAKE" and 0 for "REAL"). Error handling is used to manage any errors that arise during file processing. After preprocessing, the collected features and accompanying labels are ready for use in training deep learning models.

3.3 *Architecture of the Proposed Model*

3.3.1 VGG-16 Technique

The VGG-16 [27] is a neural network approach that has been pre-trained and is mostly used for image recognition tasks. VGG-16, an Oxford Visual Graphics Group creation, is a 16-layer deep CNN [28]. It is renowned for being straightforward, utilizing simply max-pooling layers, three fully connected layers at the end, and three 3×3 convolutional layers stacked on top of one another in increasing depth. The ReLU activation function is employed. Because VGG-16 can recognize characteristics at different scales, it is often used for image recognition applications. Because of its deep architecture, it can detect subtle and complex elements in photographs, which is useful for figuring out the advanced image modifications used to create deepfake images. The layered architecture for VGG-16 is shown in Fig. 3.

3.3.2 Alex-Net

Among the first deep CNNs to perform noticeably better on large-scale image recognition tasks than other CNNs. Alex-Net [29] consists of three fully connected layers, five convolutional layers, max-pooling layers, and dropout layers. When it comes to

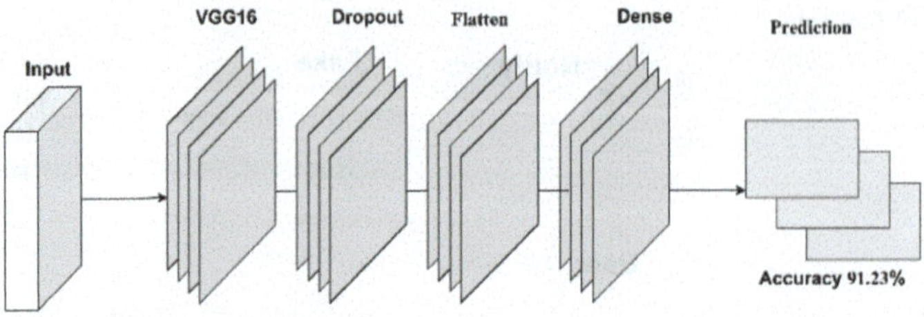

Fig. 3 Layered architecture of VGG-16

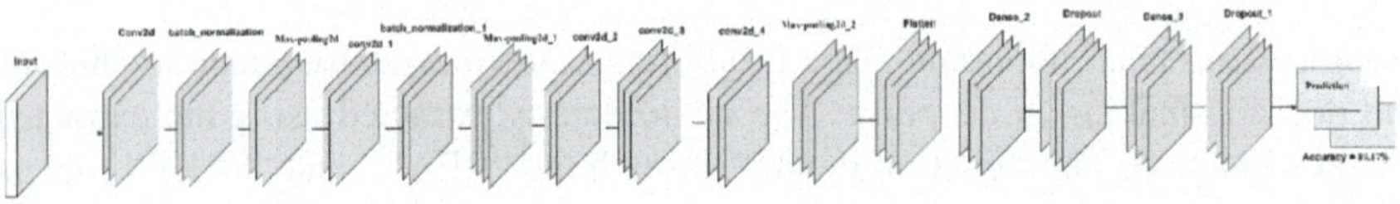

Fig. 4 Layered architecture of Alex-net

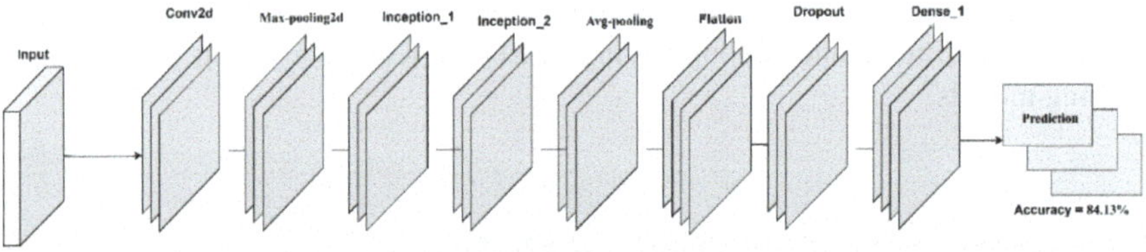

Fig. 5 Layered architecture of Google-net

activities involving picture identification and classification, Alex-Net shown in Fig. 4 has been essential.

3.3.3 Google-Net

A complex neural layout composed of recurrent units known as initiation modules may serve as a starting arrangement [30]. As previously stated, the subtle aspects of the initiation module are discussed in this paper. Inception modules are Google-Net's defining feature. These modules run multiple convolutions of varying sizes (1×1, 3×3, 5×5) and pooling operations on the same input feature map simultaneously, then concatenate the results. This enables the network to capture data at different scales mentioned in Fig. 5.

3.3.4 Mobile-Net

In 2017, a new CNN architecture was proposed that named as Mobile-Net [31–33]. This was proposed by Howard. Its specialty is that it is depthwise, with depthwise

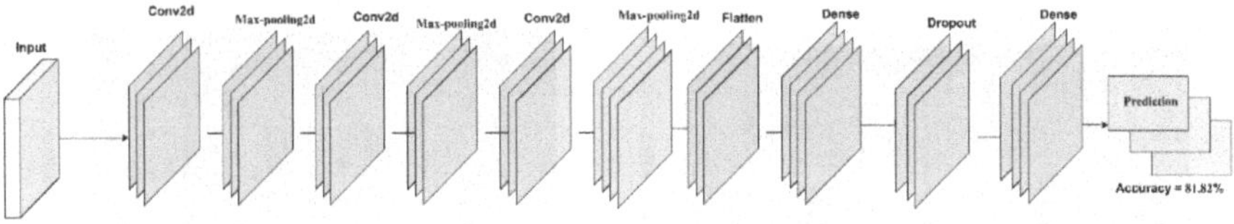

Fig. 6 Layered architecture of Mobile-Net

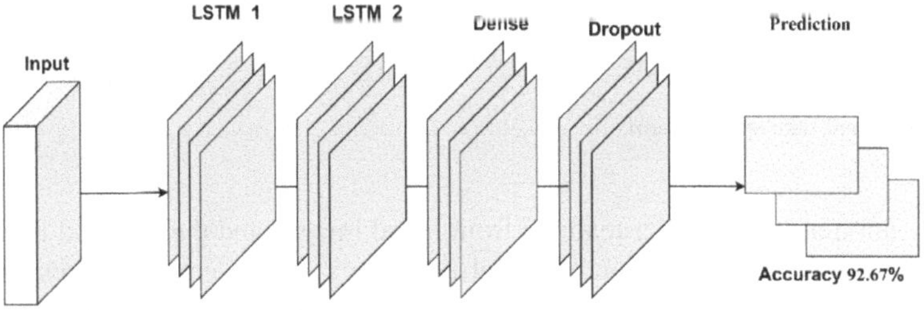

Fig. 7 Layered architecture of LSTM

convolutions. Convolutions [34] are applied to each channel rather than the entire architecture at once. This lowers the cost of computing. Layer architecture for mobile-net has been shown in Fig. 6.

3.3.5 Lstm

Long Short-Term Memory Networks [35] are a form of recurrent neural network that can learn order dependence in sequence prediction tasks and architecture shown in Fig. 7. This is especially beneficial in fields like natural language processing (NLP), speech recognition, and time-series forecasting Table 1 illustrates the design of Long Short-Term Memory Networks [35], a type of recurrent neural network that is capable of learning order dependence in sequence prediction tasks. Particularly useful applications for this include time-series forecasting, speech recognition, and natural language processing (NLP), where traditional feedforward neural networks struggle to model sequence data and LSTM is used for the detection of deepfake voice and face dataset and spectrogram shown in Fig. 8.

3.3.6 CNN

CNNs are adaptable deep learning architectures utilized in a variety of applications, including image and audio processing. In our study on face deepfake detection [36], the layered architecture of CNN shown in Fig. 9 analyzes video frames to uncover anomalies that the human eye may not notice. These differences could be in texture, edges, or artifacts introduced during the creation of a deepfake [37]. CNNs are

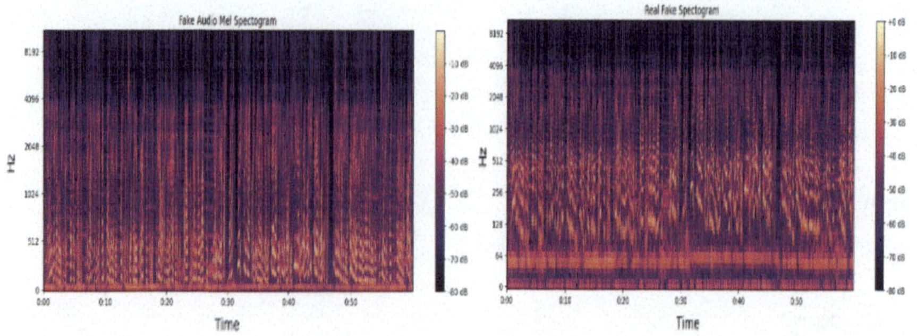

Fig. 8 Real and fake spectrogram

used to automatically extract features from facial images and the real and fake data charts including chromatogram and Mel-Frequency coefficients, are presented in Figs. 10, 11, and 12 respectively. These traits may include edges, forms, or textures that distinguish authentic from fraudulent photos. Layers of convolutions aid in collecting hierarchical patterns, which are critical for detecting subtle anomalies generated by deepfake algorithms.

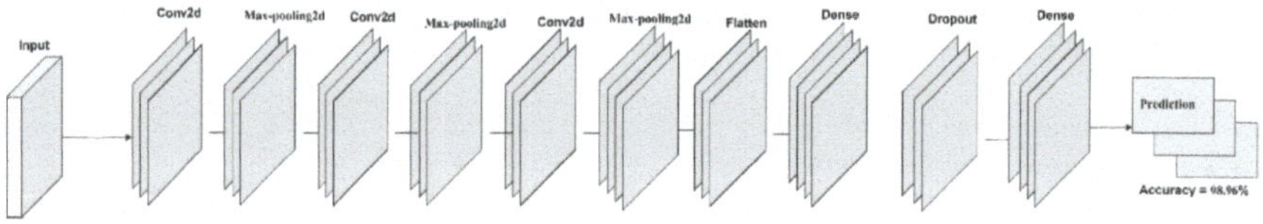

Fig. 9 Layered architecture of CNN

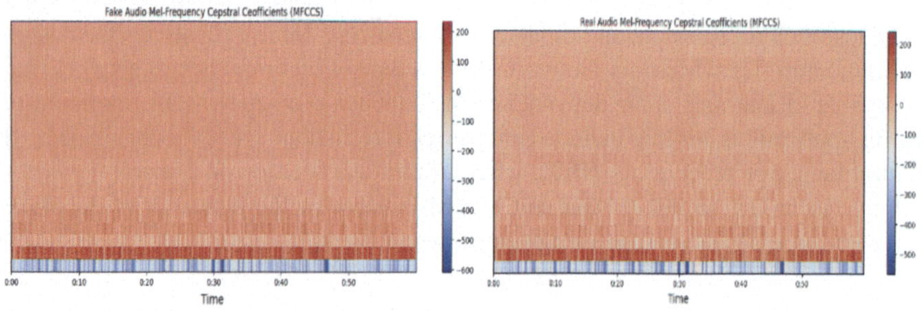

Fig. 10 Real and fake audio Mel-frequency coefficients

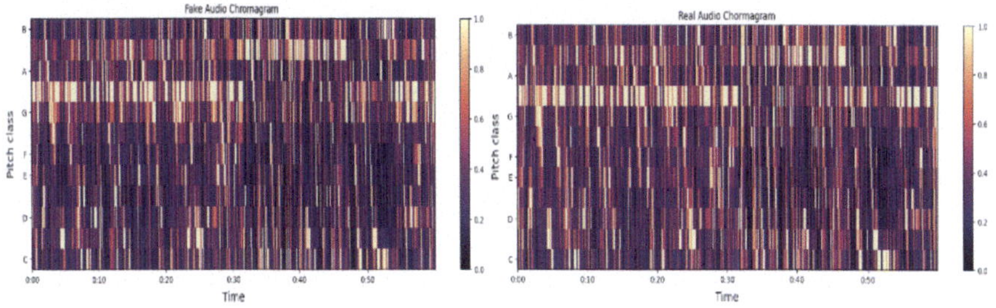

Fig. 11 Real and fake audio chromatogram

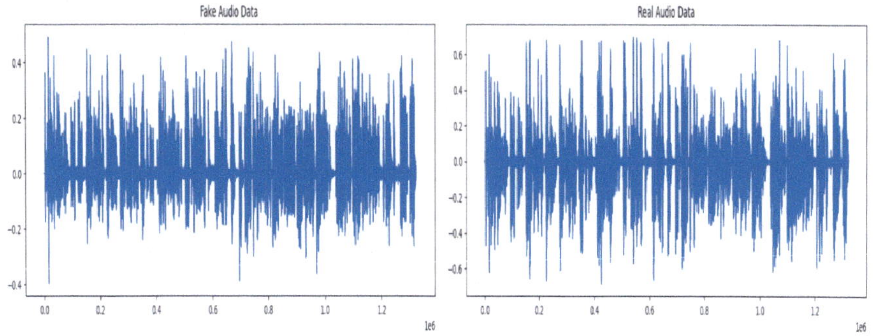

Fig. 12 Real and fake audio

3.4 Evaluation Measure

In this study, the evaluation of the deepfake detection models was performed using several metrics to ensure a full assessment of each model's performance.

- **Accuracy**—This is the easiest metric, indicating the model's total correct predictions divided by the total number of predictions. High accuracy shows that the model discriminates between actual and fraudulent stuff.
- **Precision—**Precision is the ratio of true positive predictions to total positive predictions generated by the model (true positives plus false positives). This statistic is critical in situations where the cost of a false positive is significant shown in Eq. (1).

$$\text{Precision} = \frac{\text{True Positives (TP)}}{\text{True Positives (TP)} + \text{ False Positives}} \tag{1}$$

- **Recall—**Recall is the ratio of true positive predictions to total real positives (true positives plus false negatives). This statistic is critical for applications in which missing a fake (false negative) is costly shown in Eq. (2).

$$\text{Recall} = \frac{\text{True Positives(TP)}}{\text{True Positives(TP)} + \text{False Negatives(FN)}} \tag{2}$$

3.5 *Experimental Setup*

The trials in this study were done on an Intel Core i9 CPU clocked at 6 GHz. Voice conversion was conducted on a GPU using the CREPE method, utilizing an Nvidia RTX 2080 Ti (4352 CUDA cores) [38]. Audio characteristics were retrieved using the Librosa package, and machine learning models were built utilizing Scikit-learn [39, 40], Tensor flow [41], and Numpy [42, 43]. Faces also take advantage of graphics processing units. The neural network's learning rate must be modified based on the error gradient, which is critical for guaranteeing consistent and smooth training. This component is critical in training since it influences both accuracy and the time necessary for model training. In our model, we explicitly set the learning rate to 0.001 and ran 100 and 50 epochs, respectively, with a dropout rate of 0.10. In this study, we ran trials with both pre-trained models such as VGG-16, Alex-net, Google-Net, and Mobile-net and a custom CNN model or LSTM that we created. These models are well-known and have been pre-trained for image classification tasks.

3.6 *Hyperparameter of the Proposed Architecture*

3.6.1 Learning Rate

The learning rate determines how quickly a model modifies itself in response to expected errors. It is directly related to the extent of weight changes throughout training. When assessing the level of revisions required for a model's weights, the learning rate is most important. It assists in determining the best training method for a given set of weights. When configuring a neural network, the learning rate comes out as a significant hyperparameter, and it is the most important component to consider. The learning rate chosen for our study is 0.001 for our proposed study.

3.6.2 Momentum

Building on before steps' insights, momentum is critical in determining the next move. Its main purpose is to assist prevent oscillations and stabilize the learning process. Momentum, when we consider the neural network approach, represents a simple yet effective method for improving both accuracy and speed. In deep learning, momentum refers to a set of procedures and approaches used to speed the convergence of first-order optimization methods like gradient descents primary purpose is to accelerate the optimization process, whether by reducing the number of functional evaluations required to achieve optimal results or by improving the overall capability of optimization methodologies.

3.6.3 Epoch's Number

A learning period known as an epoch is when the model looks over the whole training dataset. It establishes the maximum number of iterations the algorithm can do over the entire training dataset; the exact amount may differ based on the dataset being utilized. The number of times each image is processed independently via the network, both forward and backward, defines an epoch. Until the test error and training error achieve their lowest values, this amount should be raised. It's also critical to remember that an epoch is made up of one or more data batches. For an algorithm to be trained efficiently, multiple epochs are usually required. Eight epochs in total were chosen for the model's training.

3.6.4 Batch Size

Batch size is the measure of how many datasets we give to our model at one iteration of our training. In contrast, the total number of subsamples that are delivered to the network before parameter modifications is known as the mini-batch size. Batch size can be 32, 64, 128 256, 512, and other such amounts. Due to the large amount of frames, the training procedure became unnecessarily time-consuming when the batch size was set to 32. Because of this, a more reasonable batch size of 64 has been selected for the proposed CNN design.

4 Results and Discussion

Table 2 shows the results obtained from both the pre-trained and proposed models on each of the publicly available datasets for detecting fake or real voices and images. It is worth noting that the simulation results of VGG-16, Alex-Net, Google-Net, Mobile-Net, LSTM, and CNN are shown in Fig. 13.

4.1 Results

In Fig. 13, plot (a) displays training and testing accuracy for the VGG-16 model, showcasing its capability to generalize from training data to unseen data. Plot (b) represents the training and testing loss for the VGG-16 model, indicating how well the model minimizes error across epochs. Plot (c) shows the training and validation

Table 2 Results obtained through the deep learning models and the pre-trained models for fake and real voice and image detection

Model	Dataset	Purpose	Accuracy	Precision	Recall
VGG-16	Locally developed	Deepfake voice recognition	91.23%	0.86	0.81
Google-Net			84.13%	0.82	0.83
Mobile-Net			81.82%	0.84	0.83
Alex-net			89.17%	0.82	0.87
LSTM			92.67%	0.92	0.96
CNN			98.96%	0.91	0.95
VGG-16	Locally developed	Deepfake Faces	85.23%	0.81	0.82
Google-Net			81.68%	0.92	0.86
Mobile-Net			88.87%	0.87	0.92
Alex-net			97.80%	0.88	0.91
LSTM			95.11%	0.92	0.93
CNN			98.89	0.93	0.91

accuracy for the Google-Net model, reflecting its effectiveness in learning and validation scenarios. Plot (d) illustrates the training and validation loss for the Google-Net model, providing insights into the loss minimization throughout the learning process. Plot (e) features the training and validation accuracy for the Alex-Net model, a pioneering architecture in deep learning for image recognition. Plot (f) depicts the training and validation loss for the Alex-Net model, helpful for understanding the model's efficiency in error reduction. Plot (g) captures the training and testing accuracy for the Mobile-Net model, designed for mobile and embedded vision applications. Plot (h) presents the training and testing loss for the Mobile-Net model, emphasizing its performance in a constrained environment. Plot (i) outlines the training and validation accuracy for an LSTM model, highlighting its strength in capturing temporal dependencies. Plot (j) shows the train and valid loss for the LSTM model, essential for assessing its consistency over time. Plot (k) details the training and validation accuracy for a generic CNN model, illustrating the adaptability of convolutional networks. Plot (l) exhibits the training and validation loss for a generic CNN model, indicating the model's learning curve.

Almost all models are more accurate at recognizing deepfake faces than deepfake sounds. This could indicate that the face traits changed in deepfakes are easier for models to detect than auditory changes in voices, or it could represent differences in the quality or complexity of the datasets used in training. CNN model outperformed in both challenges and obtained an accuracy of 98.96% for deepfake voice detection and 98.89% for deepfake face detection in image dataset, demonstrating the efficiency of convolutional networks in handling image-based data and, presumably, feature extraction in audio via spectrogram analysis.

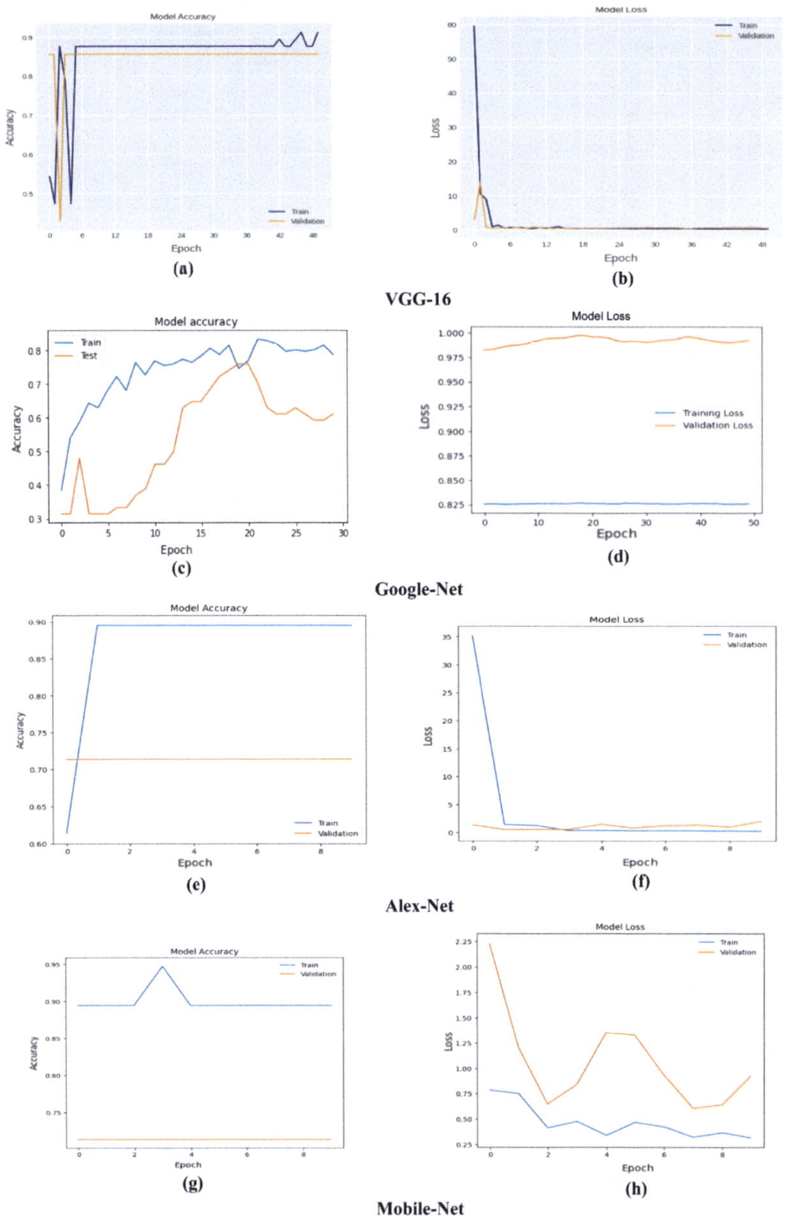

Fig. 13 Simulation results

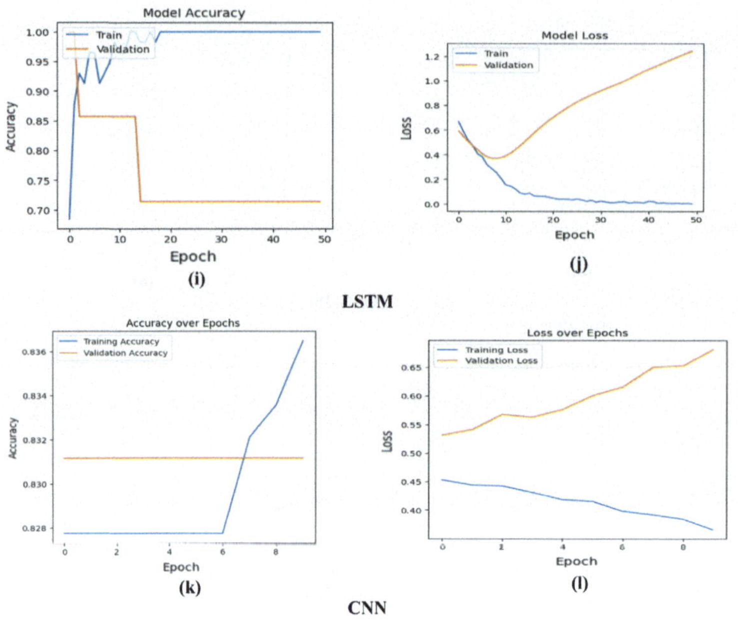

Fig. 13 (continued)

4.1.1 Analysis of Model Performance on Deepfake Voice Recognition Dataset

In the voice, LSTM models surpassed the rest with an accuracy of 92.67%, demonstrating their ability to time-series tasks, which is frequent in audio processing. This is closely followed by VGG-16, which has an impressive accuracy of 91.23%. Alex-net also performs well, with an accuracy rate of 89.17%. However, Google-Net and Mobile-Net perform somewhat poorly, at 84.13% and 81.82%, respectively, which may reflect limits in their architecture to capture the nuanced fluctuations in audio data utilized for detecting deepfakes in voices. These results show deep learning models for audio problems, such as LSTM and CNN may be better suited to these types of difficulties.

4.1.2 Analysis of Model Performance on Deepfake Faces Recognition Dataset

CNN-based models outperform in facial recognition, with a regular CNN leading at 98.89% accuracy and Alex-net coming in second at 97.80%. These high accuracies demonstrate the efficiency of convolutional neural networks for extracting and learning complex spatial hierarchies from image datasets. VGG-16 also performs well, with an accuracy of 85.23%, however, it falls behind the previously listed CNNs.

Table 3 Result comparison of the proposed result with the state-of-the-art studies

Authors	Methods	Dataset	Performance Metrics
[6]	SVM	CelebA HQ, Flickr, Person Does Not Exist, Generated dataset (1000 images each); 80% training, 20% testing	99.76% accuracy
[7]	SVM with feature-point detectors (HOG, ORB, BRISK, KAZE, SURF, FAST)	Dataset of 98 deepfake films	~ 95% accuracy with HOG
[8]	SVM with and without PCA, using Canny filter for edge detection	Dataset not specified	96.8% accuracy with PCA, 72.2% without PCA
[9]	NA-VGG (Improved VGG network based on image noise and augmentation)	Celeb-DF dataset	85.7% accuracy
[10]	Attention-based, multimodal deepfake detection framework	Custom dataset for deepfake detection	97.6% accuracy

Mobile-Net and Google-Net, which are designed for efficiency and speed, achieve 88.87% and 81.68%, respectively, indicating some trade-offs between performance and processing demands.

4.2 Discussion

To fully understand the functionality of the suggested paradigm we extend our study by testing our model independently against voice and images from each of these data sources. It has enabled us to analyze the machine learning and deep learning models in Table 3 more effectively.

5 Conclusion and Future Work

Deepfake technology, which uses powerful deep learning algorithms to generate realistic but fabricated audiovisual material, has posed substantial issues in various fields, including digital media, cybersecurity, and privacy. In this study, we used two famous datasets containing audio, videos, and images, with the usage of six models such as custom CNN, LSTM, and Pre-trained (VGG-16, Alex-net, Google-Net, Mobile-net for research and experiments. In the ultimate step of this observation, all models were compared regarding the dataset to conclude which version performs higher in accuracy, time complexity, and useful resource usage. We used all these models but CNN outperformed in both images and audio datasets and gave the best results. In the future, we will extend the dataset part and use a locally developed dataset for the deepfake detection.

References

1. Elpeltagy M, Ismail A, Zaki MS, Eldahshan K (2023) A novel smart Deepfake video detection system. Int J Adv Comput Sci Appl. https://doi.org/10.14569/IJACSA.2023.0140144
2. Rafique R, Gantassi R, Amin R, Frnda J, Mustapha A, Alshehri AH (2023) Deep fake detection and classification using error-level analysis and deep learning. Sci Rep 13:7422
3. Stroebel L, Llewellyn M, Hartley T, Ip TS, Ahmed M (2023) A systematic literature review on the effectiveness of Deepfake detection techniques. J Cyber Secur Technol 7:83–113
4. Chakravarty N, Dua M (2024) A lightweight feature extraction technique for Deepfake audio detection. Multimed Tools Appl 83:1–25
5. Pant S, Gosavi C, Barekar S (2024) Deep fake detection using LSTM and survey of deep fake creation technologies. Int J Intell Syst Appl Eng 12:840–845
6. Agarwal H, Singh A, Rajeswari D (2021) Deepfake detection using SVM. In: Second International Conference on Electronics and Sustainable Communication Systems (ICESC), pp 1245–1249
7. Kharbat FF, Elamsy T, Mahmoud A, Abdullah R (2019) Image feature detectors for Deepfake video detection. In: IEEE/ACS 16th International Conference on Computer Systems and Applications (AICCSA), pp 1–4
8. Altaei MSM (2023) Detection of deep fake in face images based machine learning. Al-Salam J Eng Technol 2:1–12
9. Chang X, Wu J, Yang T, Feng G (2020) Deepfake face image detection based on improved VGG convolutional neural network. In: 39th Chinese Control Conference (CCC), pp 7252–7256
10. Asha S, Vinod P, Menon VG (2024) A defensive attention mechanism to detect Deepfake content across multiple modalities. Multimed Syst 30:56
11. Mittal T, Bhattacharya U, Chandra R, Bera A, Manocha D (2020) Emotions don't lie: an audio-visual deepfake detection method using affective cues. In: 28th ACM International Conference on Multimedia, pp 2823–2832
12. Lewis JK, Toubal IE, Chen H, Sandesera V, Lomnitz M, Hampel-Arias Z et al (2020) Deepfake video detection based on spatial, spectral, and temporal inconsistencies using multimodal deep learning. In: IEEE Applied Imagery Pattern Recognition Workshop (AIPR), pp 1–9
13. Zhou T, Wang W, Liang Z, Shen J (2021) Face forensics in the wild. In: IEEE/CVF conference on computer vision and pattern recognition, pp 5778–5788
14. Rossler A, Cozzolino D, Verdoliva L, Riess C, Thies J, Nießner M (2019) Faceforensics++: Learning to detect manipulated facial images. In: IEEE/CVF international conference on computer vision, pp 1–11
15. Rössler A, Cozzolino D, Verdoliva L, Riess C, Thies J, Nießner M (2018) Faceforensics: a large-scale video dataset for forgery detection in human faces. arXiv preprint arXiv:1803.09179
16. Bird JJ, Lotfi A (2023) Real-time detection of AI-generated speech for DeepFake voice conversion. arXiv preprint arXiv:2308.12734
17. Chen T, Kumar A, Nagarsheth P, Sivaraman G, Khoury E (2020) Generalization of audio Deepfake detection. In: Odyssey, pp 132–137
18. Dolhansky B, Howes R, Pflaum B, Baram N, Ferrer CC (2019) The deepfake detection challenge (dfdc) preview dataset. arXiv preprint arXiv:1910.08854

19. Raza A, Munir K, Almutairi M (2022) A novel deep learning approach for Deepfake image detection. Appl Sci 12:9820
20. Phung H (2020) Deepfake detection - Faces - Part 8_0
21. Phung H (2020) Deepfake detection - Faces - Part 1_1.
22. Bird JJ (2023) Deep-Voice: DeepFake voice recognition
23. Puah YH (2021) Deep fake images
24. Davide-Coccomini (2022)Combining-EfficientNet-and-Vision-Transformers-for-Video-Deepfake-Detection
25. Elliott J (2020) DeepFake detection challenge
26. GREATGAMEDOTA (2020) FaceForensics
27. Khan ZY, Niu Z (2021) CNN with depthwise separable convolutions and combined kernels for rating prediction. Expert Syst Appl 170:114528
28. He K, Zhang X, Ren S, Sun J (2016) Identity mappings in deep residual networks. In: Computer Vision–ECCV: 14th European Conference, Amsterdam, The Netherlands, October 11–14, Proceedings, Part IV - 14, pp 630–645
29. Rafique R, Nawaz M, Kibriya H, Masood M (2021) Deepfake detection using error level analysis and deep learning. In: 4th International Conference on Computing & Information Sciences (ICCIS), pp 1–4
30. Al-Qizwini M, Barjasteh I, Al-Qassab H, Radha H (2017). Deep learning algorithm for autonomous driving using googlenet. In: IEEE Intelligent Vehicles Symposium (IV), pp 89–96
31. Ilhan HO, Sigirci IO, Serbes G, Aydin N (2020) A fully automated hybrid human sperm detection and classification system based on mobile-net and the performance comparison with conventional methods. Med Biol Eng Comput 58:1047–1068
32. Jabber B, Lingampalli J, Basha CZ, Krishna A (2020) Detection of covid-19 patients using chest x-ray images with convolution neural network and mobile net. In: 3rd International Conference on Intelligent Sustainable Systems (ICISS), pp 1032–1035
33. Kumar PR, Kiran R, Singh UP, Rathore Y, Janghel RR (2022) Rice leaf disease detection using mobile net and inception V.3. In: IEEE 11th International Conference on communication systems and network technologies (CSNT), pp 282–286
34. Basha C Z, Bhavana N, Bhavya P, Sowmya V (2020) Rainfall prediction using machine learning & deep learning techniques. In: International Conference on Electronics and Sustainable Communication Systems (ICESC), pp 92–97
35. Graves A, Graves A (2012) Long short-term memory (2012). In: Supervised sequence labelling with recurrent neural networks, pp 37–45
36. Patel Y, Tanwar S, Bhattacharya P, Gupta R, Alsuwian T, Davidson IE et al (2023) An improved dense CNN architecture for DeepFake image detection. IEEE Access 11:22081–22095
37. Kohli A, Gupta A (2021) Detecting DeepFake, faceswap and face2face facial forgeries using frequency CNN. Multimed Tools Appl 80:18461–18478
38. Heryadi D, Hampton S (2019) Characterizing performance improvement of gpus. In: Practice and Experience in Advanced Research Computing on Rise of the Machines (learning), edn, pp 1–5
39. Kramer O, Kramer O (2016) Scikit-learn. In: Machine learning for evolution strategies, pp 45–53
40. Zhong Y, Liu F, Huang G, Zhang J, Li C, Ding Y (2024) Thermogravimetric experiments-based prediction of biomass pyrolysis behavior: a comparison of typical machine learning regression models in Scikit-learn. Mar Pollut Bull 202:116361
41. Yang X, Zhou Y, Liao W, He T, Yan J (2024) Pharotate: A rotation detection benchmark using TensorFlow. In: ICASSP IEEE International Conference on Acoustics, Speech and Signal Processing (ICASSP), pp 9481–9485
42. Gupta P, Bagchi A (2024) Introduction to NumPy. In: Essentials of Python for Artificial Intelligence and Machine Learning, Springer, pp 127–159
43. Oliphant TE (2006) Guide to numpy. vol. 1, Trelgol Publishing USA

Deepfakes: Turning Reality into Chaos

Somsourjo Roy and Sourodip Roy

Abstract "AI is like fire. It can cook your food, or it can burn down your house". What if your acquaintances suddenly start believing you to have committed actions which you have not? Words which you have not spoken? What if there is a clear video of you available online to the World, where you have done things which you have not? Would your World not turn upside down? Over the past few decades, technology has reached unprecedented levels—giving birth to new challenges humanity has never faced before—one of which, is deepfakes. Deepfakes can be images, videos, or even recordings which have been digitally altered in such a way that people are convinced that someone has done or said something which they have not. It has enabled the creation of realistic fakes through the use of artificial neural networks, data interpretation, and advanced AI techniques. While it is argued that deepfake algorithms may be used for more productive activities like classroom teaching, this paper intends to alert the general audience to the dangers that the technology poses to mankind and the probable legal solutions that one might look up to. Moreover, it also strives to provide a glimpse into the potential methods of mitigating these deepfake fabrications by developing an additional tool that functions as a countermeasure against deepfakes.

Keywords AI · Deepfakes · Digital face manipulation · Digital forensics · Generative AI · Biometrics

---

S. Roy (✉) · S. Roy
Department of Law, University of Engineering and Management, Kolkata, India
e-mail: somsourjo.roy2021.1@uem.edu.in

S. Roy
e-mail: sourodiproy148@gmail.com

M. Chakraborty et al. (eds.), *Proceedings of 5th International Ethical Hacking Conference*, Lecture Notes in Networks and Systems 1148,
https://doi.org/10.1007/978-981-97-8457-8_12

1 Introduction

The incept of Artificial Intelligence has brought a vast change in the prospects and the ideologies of the humankind it helped in accelerating the achievement of the desired goal more quickly similar to what any catalyst does. In this process, the greatest mind-sets and the budding ones forecasted their interest in discovering this field but with a different lens. The study of Deepfakes through this chapter will help the reader to get an understanding of how digital things work. The very basics of every Artificial Intelligence is machine learning which symbolizes that the machine contains the capacity of the human mind to read and produce the result according to one's desire and the search history actually learns through the search and interest. There are three main types of machine learning: supervised, unsupervised learning, and reinforcement learning. Supervised learning or predictive analytics involves building algorithms on governed sets of data to make predictive calls ranging from the identification of images to academic outcomes. Clustering helps in service segmentation, and it is also applicable to grouping customers and detecting intrusion. On the other hand, reinforcement learning enables the machines to learn by taking trial and error and when successfully reaching the decisions can be rewarded as in the case of gaming or training of automobiles. Machine learning is implemented in some of the industries where each of them utilizes its force. It is effective in finance in evaluating risks and trading and also client support. In the pharmaceutical industry, it improves most of the drug discovery procedures, and clinical trials, thus the production process. Companies use it to control the operation of their machines, check the quality of their products according to computer vision, and improve their processes. Machine learning is being used by insurance companies in the following ways; Insurance companies use machine learning for recommending activities and streamlining claim processes and the activities involved. Customer approach has systems that suggest items to customers depending on their interests along with having adequate supply and forecasting of demand. On the one hand, machine learning has several benefits such as the establishment of trends, forecasting, planning, and data processing in a timely and efficient manner; on the other hand, many challenges come with machine learning. These include the existing time and financial constraints the frequency with which predictions are wrong and the operational time and costs required to construct and establish machine learning models. These challenges notwithstanding, continuous efforts and innovations for the advancement of machine learning technologies continue forth. The future of machine learning seems bright and very promising with applications expanding in the AI sector and other industries. Alongside progress has already embraced the idea of exploiting machine learning in order to implement positive changes in efficiency, productivity, and decision-making that affect various sectors. From the above implications, it can be concluded that there are positive as well as negative impacts of machine learning. We will further deduce that the machine Learning System used in AI can cause harm to an individual by violating the privacy of the person involved. But if one turns to the "inaccuracy" of the AI market as applied to the criminal rhetoric then nothing can be changed. This industry is huge and diversely spread over the countries but one thing that is universal constant: poverty, illiteracy, and ignorance about it. This paper will also discuss how the deepfakes are created using machine learning process; with the deep learning method they use in their algorithm to learn from historical data processes they used to generate the new data by using similar kinds of features and seeing the previous works and focus on the build of strong fake version of original images and video. The word deepfake was first used in 2017 on Reddit by a person who recreated porn movies by using face swapping technology. This was used in the past in the adult entertainment industry. But as computers started to advance in terms of efficiency and capabilities, creating a convincingly believable scene through the help of AI became a reality. Autoencoder is the collector of deepfake technology. This is a type of computer application that picks up a person's face and body gestures and comprises them into an application. Then, it can also encode this and put it on someone else. This is how they can appear to put one face on another body. One of its other attributes of deepfake tech is something called a Generative Adversarial Network. This kind of network is a conglomerate of computers that collaborate to create the illusion of absolute reality. Another computer, called the manipulator, attempts to determine whether the "thing" created by the generator is legitimate. An additional computer is called the discriminator and it attempts to call whether it is a real or fake one. They repetitively do this until the artificial product looks super real. This is how deepfakes get so real by Bruce Metzger. Some are quite interesting, and they have the capability of creating a number of creative things; however, deepfakes have raised concerns because of their ability to create fake content—and that is something that cannot lead to anything but fraud and other similar forms of exploitation. Therefore, people have been stressing the need for the industry and even the government to intervene in order to find a way to detect and regulate deepfake technology. One of the units under discussion is called VAEs or Variation Autoencoder. In general, it can be said that this is the kind of AI algorithm that uses deep learning techniques to generate new things, locate unusual things, or to solve issues. These parts of machine learning algorithms called VAEs are very suitable for synthetic time series data generation to train other algorithms especially where the sources are the Internet of Things, Electroencephalogram, and financial data. Now this is where VAEs are important because they can detect deepfakes and protect people from the harmful side effects of safety measures, but at the same time, it is also essential that the industry and the government collaborate if this technology is used responsibly and ethically. These algorithms are pretty adept at writing text, creating images, and generating videos. However, this has a detrimental effect on individuals and society as it has brought about a major breach of privacy and security. A lot of pornography especially girls has been posted online and this has caused a lot of tension in society. It has left many people who are keen on fashion confused and scared Deepfakes have also been responsible for an unprecedented boost in the production of porn. Things have become so messed up that it is difficult to determine what has actually happened and what has been told or fabricated. It can also switch the voice of someone, legal laws to do with the originality of a document, and even tap someone's phone without them knowing. It's turned out to be hectic. There are some laws and regulations that are being proposed in an attempt to deal with these issues but they are not all the same, they effectively apply. Law enforcement is as imperfect as one can pursue it: from the CDA to copyright and from state regulations to the general legislation. They should be cautious and be informed about what is happening in the world around them.

2 Research Methodology

This paper used qualitative research methodology, which focuses on theoretical materials as both primary and secondary sources and has been thoroughly analyzed and evaluated through numerous books, articles, reports, statistics data, graphs, experiments, surveys, case studies, interviews, and observation. The researchers also visited the T.S Dutta Library at New Town and the Ramkrishna Mission Institution Library, Golpark, Kolkata for additional citations and references.

3 Literature Review

Deepfake technology has significant legal implications. While it fosters the sharing of false content for entertainment purposes, its applications and implications are much broader. In particular, deepfakes violate or significantly challenge the existing laws in several domains, such as the spreading of disinformation, privacy invasion, fraud, trust disruption, and many others [1]. As a result, multiple legal responses and remedies are encouraged by the scholars and the community. The proposed models include IP laws, privacy regulation, criminal statutes, and technology-focused regulation. However, there are significant concerns such as jurisdictional difficulties, ongoing technological advancements, detection and attribution confusion, and the need to balance free speech. As a result, a comprehensive approach is required to mitigate the aforementioned challenges that are based on technology, reform, legal practice, international cooperation, and public awareness to safeguard individuals' rights and preserve societal trust in the digital era. In this field, significant research is yet to be performed and this chapter will try to focus on this gap including the unchartered areas.

4 Discussion

This paper aims to delve into the intricacies of deepfake technology, shedding light on its underlying concepts and the mechanisms of data manipulation involved. It will explore how this technology poses a threat to individual privacy and security, potentially infringing upon their rights and freedoms. Additionally, the paper will dissect the AI-driven processes behind the creation of deepfakes, providing insights into the mindset of those involved in this industry and their covert operations. Lastly, the discussion will extend to proposed measures for mitigating this issue, emphasizing the role of legal frameworks in addressing the challenges posed by deepfake technology once victimized by this menace.

5 Objective of Study

The foremost objective of this research paper is to generate awareness of the consequences of deepfake technology and how it can affect daily life. It also explores the legal remedies available to the aggrieved person.

6 What is Machine Learning?

Machine learning algorithms can be broadly categorized into three main types: supervised learning, non-supervised learning, and reinforcement learning. Every algorithm type exists for particular special purposes, and solving other types of problems or tasks means employing other algorithm types. Foremost, we will see how machine learning works and what it is, the system of machine learning is a process through which a machine learns from the dataset given by the user which helps the machine to understand what the user desires and provides efficient and desirable results. Further, it is the integrated part of AI, the constant trial and error is been done to refine their algorithm settings and performance by getting near to perfection while providing human-like solutions to proposed challenges. The ML is a loop-like system in which the continuous circulation of information entered by the user helps in the filtration of information according to the needs of the user. It forms a friendly user interference where the user doesn't have to repeat each time to the search engine about how they want their results. Now through this interference, the Machine creates its predictive algorithm which on the other hand acts as a supplement to the entire process of machine learning. There are various ways through which we feed machines with data such as supervised, unsupervised, and reinforcement [2].

Supervised—In this format of learning the machine is given the data but, in a label, and set structure such as information related to a certain object is provided with the machine and to settle the system with the knowledge of that specific object some label associated with that object is provided to the machine this helps the machine to acknowledge that object with the label associate with it.

Unsupervised—In this form of learning the machine is given a set of data which helps the machine to predict according to the information provided to it but in this case, no label related to the data is provided to the machine. In this case, the machine predicts upon the dataset given.

Reinforcement—This process includes trial and error so the inputs are given to the system simultaneously the system provides you with an output if the output doesn't match the desired result the user will provide negative feedback to the system and load the system with more specific result near to the output the user desires now the system creates another output which is parallel to the out demanded by the user now the user give positive feedback to the system. This is how the reinforcement process functions.

Further, the concepts concerning ML are the types of algorithms they used under the various forms of learning starting with:

6.1 Supervised Learning:

6.1.1 Supervised Learning Algorithms

Saturated learning narratives are used for cases, where the trained data has labels in it, meaning its final output is known. Thus, these algorithms function by way of training a mapping that can be obtained from the input data to the output labels. Some common supervised learning algorithms include: Some common supervised learning algorithms include:

- Decision Trees: A decision tree is a class of algorithms that is recursive could split the input space using feature values only. They are organized in such a way that they have branches following the rules from data features. Every single node of the structure is a decision-making point, and the branches (nerve trees) lead to further nodes (leaves) which indicate what the result or output is. Decision trees are designed for both classification and regression tasks, but those of the latter kind must be numeric. They possess the ability to be easily understood by intuition, though they may be difficult to interpret when they grow out in deep trees and are prone to overfitting.
- Support Vector Machines (SVMs): SVMs (Support Vector Machines) provide the solution to a complex problem in the form of an optimum hyperplane that runs through different classes of data points in the input space. The intention is to influence the gap between the classes and thereby find support vectors, which are the two classes that are closest to each other. SVMs are superior for high-dimensional information attributes where it is possible to solve both linear as well as non-linear issues through kernel effects. These models have a high reputation and are good for classifications, but may be computationally inefficient and can be sensitive to the parameter setup.
- Logistic Regression: Logistic regression is a binary classification technique that, similarly to linear regression, fits a function that maps a set of input features to a considered binary response. It does this by using the logistic function which is used to keep the output from becoming negative or greater than 1, where these values would not be probabilities. Logistic regression is simple and easy to understand, thus being the favorite in binary classifications; however, it is based on the premise that there is a straight line between the variable experiencing the situation and the logarithmic probability of the outcome.

6.2 *Unsupervised Learning*

- Linear Regression: Linear regression is the most common approach employed for the prediction of the outcome variable(s) by the linear combination of the input(s). It takes a linear shape and cuts off the error square between the forecast and the real observations to get the least sum of squares. Linear regression makes linear relationships simple to understand and it can be used for a large proportion of regression problems but the relationship between variables may be not linear causing the regressor to make errors.
- K-Means Clustering: K-Means is an algorithm that nowadays we can call a "clustering method" which divides the input data into k clusters based on the similarity of the data points. It accomplishes this task by attributing each datum to the closest cluster center, then regenerating the centers by the average of the allotted points, in the iteration-wise manner that it finetunes the clusters. The K-Means method is well designed to do clustering, numeric data, and categorical data being among the parameters that it can execute satisfactorily but K and initial cluster centers may affect the results generally.
- Hierarchical Clustering: In hierarchical clustering architecture the hierarchic structure of clusters is generated either by merging them (conglomeration style) or by splitting them (division style) in accordance with their similarities. The process that can result in a tree-shaped diagram (dendrogram) that demonstrates how the clusters are arranged is called agglomerative clustering. Hierarchical clustering has the capability to deal with both numerical and categorical data and proffers a visualization for the organization of data. It does not imply specification of the number of clusters in advance but rather might be computationally taxing if the data does not meet demand [3].

6.3 *Reinforcement*

- Q-Learning: Q-Learning is a model-free method by which Q-values for each state-action pair are calculated to the reward that was expected in the future. It chimes in the pairs of states and actions in every iteration and takes into account the actual reward and the estimated future rewards from the Bellman equation. Q-learning can be applicable for both discrete and continuous state and action spaces and because of its simplicity, the practical implementation is also not a problem. Q-learning is able to cope with higher-dimensional spaces without an in-dept system.
- Deep Q-Networks (DQN): DQN applies double learning quatree with deep neural networks for dimensional state space treatment. It is the basis of a neural network with network-enhanced capabilities of the Q value function and learns from raw sensory inputs like images. DQN has achieved remarkable results in other application areas, including the Atari Video games; yet, it is a difficult procedure to train and superior computer resources are required.

6.4 Neural Networking

Neural networks are a particular class of the machine learning algorithmic paradigm that imitates the brain's working manner primarily. Our brain is based on neurons that are networked together to enable us to think and learn, neural network is also made of such interconnected nodes (which you would think of as artificial neurons) that help computers to learn from data.

It has included neural networks as its contents and explained its internal nerve circuits, its learning process, its type, and also the challenges facing the implementation of neural networks.

A neural network is layered, which includes the input layer, hidden layers, and the final (or output) layer. The beginning layer is where the request is being made, while hidden layers take this input data and with some patterns on it, the output layer shows the result.

Neural Networking and its application in real world [4]:

- **Image Identification**: Identifying objects, faces, or animals in pictures.
- **Natural Language Processing (NLP)**: Understanding and generating human languages, like chatbots and translation services.
- **Stock Market Predictions**: Predicting financial trends and stock prices.
- **Drug Discovery**: Helping researchers find new medicines by predicting how different compounds will behave.
- **Personal Assistants**: Powering virtual assistants like Siri and Alexa, which understand and respond to voice commands.

So, further, in neural networking, the weightage of the data is a property calculated to minimize the errors in the output results through specific tasks: feed-forward prediction, recurrent memory task, convolution of image recognition, and radial basic function for task recognition. There are certain disadvantages suffered like time consumption, high computation requirements, and challenges suffered in the financial and healthcare care systems where update updates misprediction, and changes in the condition of the patients are faced by Machine Learning systems including the financial market where there are unpredictable uncertainties in the different sectors which are typically integrated to each other affecting all the sectors under single event occurrence this creates a problem for the system to understand on which data to use in order to set the algorithm and which data no to use cause each contradicts with each other with a small change in the market entire predicts false lie, difficult to interpret decision due to "black box" nature. The entire process or the small grid of the entire pattern of how machine learning works, is the place where AI marks its control and takeover by manipulating the data and using it for its desired goals so this machine learning is also scrummed to AI and how this might have its effects in future will be discussed further in this paper.

7 How Deepfakes are Made?

Deepfakes are produced through the joint functioning of the encoder and decoder network systems within the purview of the GAN or Generative Adversarial Network frame. While the encoder does the primary operation of evaluating the source material to retrieve the important characteristic features (for instance, the facial features of a human face), the decoder does the secondary operation of receiving these data from the encoder and then converting them into manipulated data. Thus, a cycle is established which gets repeated again and again until the desired result is achieved. When the manipulated data in hand belongs to a video, the cyclic process is performed for every frame of the video. However, it should also be noted that a majority of artificial intelligence tools do not use the Generative Adversarial Network or the GAN framework but rather only the combination of encoder-decoder systems [5, 6].

This entire cycle can be compared with that of the biological processing of the human brain. When we see a material object, for instance, a horse, we use the word "HORSE" to assign a name to the object. We know that the object before us is indeed, a horse because that information is stored in our brains. There is a direct connection between the image and the encoded word "HORSE" processed and stored in our brains. [7].

Now, here is a question.

Can we close our eyes and imagine the picture of a horse racing through the grasslands? Yes, we can, but how do we do that? It is because the neural fibers in our brains have stored the data of the relationship between the image of a horse with that of the five-letter word "HORSE." That means we can retrieve this information at our will. Similar is the case for artificial intelligence. Once it stores the given data, for instance in this case, the image of a horse, after analyzing several such images, it would be able to produce its image of a horse. Thus, the difference between a human and an artificially intelligent device is smaller than ever. Frightening, but true [7].

8 Future Aspects of Learning

Now dealing with the future aspect of Machine Learning we must first have the evidence of the past and the past gave us a quite evident answer that the future of machine learning under the control of AI will take the next leap. The outlook of machine learning may develop immensely and take to various fields near the horizon as we continue to research it and advance the technology, as well as have plenty of data we are lucky to get. Due to the improvement of algorithms, models such as neural networks (transformers and Generative Adversarial Networks) and advanced models are allowing ML to set the limits of what it can do in the processing of natural languages, images, and others. An intersection of quantum computing and ML called quantum machine learning is promised to rapidly grow the capabilities of solving problems that are climate-related. Also, it can provide unprecedented insight into datasets due to the power and complexity that it can process. ML, on the other hand, is succeeding by working alongside technologies such as the IoT and edge computing, by processing data in real-time and making decisions on the spot, leading to a wiser home, cities, and industries. This integration plays a critical role in the implementation on particularly applications like autonomous vehicles, healthcare, and industrial automation. It is where enhanced personalization and user experience through ML will uncover new horizons of personalized medicine by providing predictive analytics as a part of personalized medicine, early diagnosis, and treatment planning. Advancements in the technology related to similar smart assistants such as Siri, Alexa, and Google Assistant will ensure that they are not only capable but also intuitive to know more complex queries. These ML applications will consist of tools to fight fraud, machine trading, risk management finance, predictive analytics for retail, and maintenance and quality control for manufacturing. Therefore, ethical AI and fairness will be a key consideration in the ongoing research, where bias in ML models is investigated and mitigated in order to guarantee fair and cognizant decisions. It will be vital to establish the right rules and regulations including ethics for the use of AI so as to avert non-responsible use. Security and privacy are also indispensable. We are taking the lead role in adversarial machine learning as it helps in the development of ML models to have higher resistance against attacks and privacy-preserving techniques such as federated learning and differential privacy that enable protection of individual data from being exposed but it brings us the question that in order to receive such enhance protection what are the requirements to be followed? The very first criterion is to surrender most of your data to the machine in order to receive what you demand but this brings our right to privacy into problem cause it becomes open to the world because the moment you enter any of the information relating to you into the server the very next moment it gets spread into the entire digital servers, even though the presence of the end-to-end encryption which unenabled so the servers of the user to open your private file/information yet there is no certainty that it won't happen. Along with such disadvantages, ML can provide us with vast facilities ML facilitates the progress of autonomy and hence endows devices such as cars and drones with the expertise to navigate through detailed terrain while ensuring the safety of occupants and users. NLP will make conversational AI that is even more natural and human-like through further improvements, that will impact the dialogue between people and machines. Conditional Generation will also evolve and thereby get the writers to write high-quality texts, now very prominent ones are the GPTs this efficient intelligence is the product of ML and AI. Thus, various tasks performed by humans will be enhanced through augmented intelligence, which aims to strengthen fundamental human capabilities rather than replace them and provide higher-level intelligence and efficiency. Machines and software that show human-centric responses would understand and rationalize the crop of human emotions and rationality, which will make collaboration more efficient. In research and development, continuous learning techniques will facilitate the models instead of forgetting what they already know, and AutoML will automate the development, training, and deployment of ML models, which will operate in a way that the average person will be able to use the technology makes it become a concept of our daily lives as opposed to something difficult to visualize. This entire concept is where today's industry is lying from the drown to dusk every single project or even in every step of human life AI is getting involved [8].

8.1 The Dual-Edged Sword of AI and Its Future Aspects

Now acknowledging the facts and the intent that this paper tried to portray is about Machine learning and Deepfakes which have brought chaos to society but with shadows of positive remedy it is providing to society in exchange for destroying it from within. The AI technology sector is a fast-growing industry through ML, neural networks, and the use of big data. AI solutions are being widely implemented in areas such as medicine, banking, commerce, manufacturing, and entertainment at the moment due to the fact that they either improve productivity or quality. Artificial intelligence is used in healthcare for predictive analytics, personalized medicine, and healthcare diagnostic imaging. In the field of finance, it helps to identify fraud and promote algorithmic trading, and risk management. AI is retail's currency through its application in individualized marketing and supply chain management, and in manufacturing, as seen in the growth in predictive maintenance and automation. It is also very promising to be able to develop more advanced models such as transformers and GANs in the future of AI, in addition to integrating AI with emerging quantum computing technologies that can solve complex problems. The efforts to develop ethical AI solutions, which primarily target the questions of transparency, preventing bias, and fairness, continue to gain more and more weight. In addition, AI has driven the integration of IoT and edge computing which creates conditions for making cognitive decisions in real-time.

The increasing trend of deepfakes is among the areas that AI is currently known for which means manipulated synthetic media using AI algorithms to reproduce someone's image and likeness on the existing image or video. The entertainment, education, and marketing industries find not only benefits but also difficulties when deepfakes are utilized in their domains. They are misinformation, political manipulation, and damage to one's reputation. The future of deepfakes lies in innovating strong recognition devices, structuring legislation, and encouraging mindful use that will bake in the problem. One of the goals of the introduced novel means for detecting deepfakes, and also legal and policy mechanisms in this respect is to respond to the threat of their abuse. Finding a balance between beneficial innovation and the right moral responsibility will be most critical for AI industries and deepfakes' s vision for the future, leading to a just use of the powerful devices which can be also used against society people. The auto generative artificial intelligence(genAI) and large language models (LLMs) through which the deepfakes are mostly formed and with advancement of these factors lead to the uproar. Now the creation of deepfakes involves certain steps most of it is done through the saved data that the user has/had been saving on the online platform over the years the AI reads through all that data to know you(the user), further in creation of the exact replica of a person, their voice, actions or even video of doing certain things which they have never done this all are done by using variation auto-encoder(VAN), through learning pixel intensity, face swapping methods, trial and errors, fine-tuning of the fake to the original, model training of the AI to accurately train and trade their function of creation of the fakes in align to the realistic movements of individuals. One of the prominent and recognized figures of today's technological industry Elon Musk stated that AI will become "one of the biggest threats to humanity." Amidst all these threats, issues, and challenges that AI and Machine Learning have brought to society which indeed happened due to the greediness of mankind to find the easiest way to complete any hurdles and jobs of life, they created the most significant danger or rather a poison or entity that can take over the smartest person and even the entire humankind within a second. The essence of which is rather tasted or rather experienced by the people of the society through deepfakes there is no silver bullet to solve this problem but combined efforts of both the people and the government can help to mitigate it to a certain level because the advancement of the precaution and remedies this metaverse will also advance itself with the help of deep learning [9].

9 How Does Deepfake Affect You and Me in the Society?

While deepfake is already here among us, awareness about this technology is shockingly absent. We, as human beings, have the habit of not bothering with something unless it involves us. So, the authors would specifically emphasize the word "you and me" here.

9.1 Sociological Impact

Indeed, the repercussions of such manipulated data if left unchecked, can prove to be detrimental to the functioning of a healthy society. When artificial intelligence is hired to produce convincingly incorrect data of people committing actions that they, in reality, have never said or done, such offenses shatter the balance of social trust in society. Innocent victims of such advanced manipulation get ridiculed by society while the real person guilty of such offense hides behind the shadows. The use of such technology can seriously affect the political state of a place, influence public opinion, and also the social cohesion of the community. The paper specifically highlights a few of these impacts through the following points:

9.2 *Political Manipulation*

Not only can such manipulations be misleading but also vicious to the harmony of the community. Imagine a well-renowned political figure saying or doing things that have been recorded on camera. The actions are such, that can potentially damage the social reputation of the individual. These political videos of manipulated data are created with the sole intention of adversely affecting the image of the political representative in society. In 2022, a deepfake video of the President of Ukraine, Volodymyr Zelenskiy came up in which he was seen to declare the Ukrainian troops "to lay down" their weapons and stop fighting against Russia. In response to this, Mr. Zelenskiy expressed himself through a post on his official Instagram account and debunked the video. However, the fake video had already been watched by 5 million users all over the globe. If the President of a country is not safe from such offenses, then needless to say, none of us are safe either [10].

9.3 *Privacy Concerns*

While the problem of privacy infringement remains unsolved, the rising cases of deepfakes have only accelerated the issue. Using the physical representations of a person without his consent to produce morphed videos of the individual is performed with the malicious intent of the offender to exploit the person. More often than not, such videos contain pornographic content and the use of the person's physical characteristics to create these, without their consent is outright derogatory for the victim. A report published in 2019 stated that more than 96% of all deepfake videos that are available online constitute non-consensual pornographic content [11]. About this statistic, the authors would highlight a very recent case of 2024 in which the Prime Minister of Italy, Giorgia Meloni had filed a suit claiming $100,000 over pornographic videos of her circulated through the net. Needless to say, these were made through deepfake technology and it is unfortunate how international leaders and celebrities (especially women) are victims of such latest inventions.

9.4 *Rise in Advanced, Organized Cyber-Crime*

In 2023, the India Today News published an article that read that a man from Kerala had lost Rs 40,000 to an advanced online deepfake fraud [12]. The offender had impersonated the voice and the physical features of the victim's friend to ask him for money as an excuse for a medical emergency. The victim, fooled by the convincing details of his friend over the video call, immediately donated the sum. The only loophole left by the criminal was that the call came from an unknown number. By altering and imitating, deepfake can change the voice and physical characteristics of a person.

9.5 Psychological Impact

A recent study published in Scientific Reports in September 2023 has given insight into how our brains react differently to original and technologically produced content. In the words of the author, Anna Eiserbeck, doctoral researcher at the Abdel Rahman Lab at Humboldt University of Berlin, when we look into the image of a smiling face, that is perceived as a fake source, then the neurons in our brains do not react strongly. However, if we look into the image of a smiling face, which is thought to be a genuine source, then there are strong neural reactions in our brains. In contrast to this observation, the same cannot be said to be true for angry faces. For aggressive images, the neural reactions remain the same irrespective of whether the person perceives the image to be true or fake [13].

The research proves that artificial intelligence has the potential to influence our brains and stimulate different responses, all through artificial sources. According to Professor Joel Pearson, a cognitive neuroscientist at the University of New South Wales, there are already substantial concerns about the impact of deepfake technology and AI on the human brain and mental health.

"AI is already affecting us and changing our mental health in ways that are bad for us" John Tarnovschi and Joshua Kalla, two authors of the experimental thesis titled: "Negative Consequences of Informing Voters about Deepfakes: Evidence from Two Survey Experiments" have shed light on the psychological repercussions of deepfakes through two experiments. At the end of these two surveys, the experiment concluded that informing and intimating the general public about deepfake technology does not help them differentiate between the original and the artificial. Instead, people become increasingly anxious and skeptical as to the authenticity of every material that they watch, which was the case for the participants in this experiment. They failed to recognize even the genuine videos and had mistaken them for a deepfake video [14].

Thus, it proves that deepfake source materials will only lead to increased mistrust and chaos in society.

9.6 Positive Impact of Deepfakes

When knowing about deepfake technology, the reader needs to be aware of not just the negative side, but also the positive side. A few of the most crucial benefits of AI's deepfakes are highlighted in the following lines.

9.7 Education

Deepfake has opened the doors to advanced classroom learning using synthetic media in education. Here is a question for you. You are learning about the discoveries of Galileo Galilei. What if you had the opportunity of directly with the legend himself? Through the help of deepfake algorithms, it is now possible for realistically animated historical figures to demonstrate their inventions, deliver speeches, and give presentations in the classroom. The result is a far more engaging session rather than purely teaching from textbooks.

For instance, Claude, an education-oriented AI design can imitate popular figures from the past and answer students in their native languages while demonstrating the concepts thoroughly. In the words of Drew Rose, CSO and founder of the cybersecurity firm Living Security, "educators can leverage deepfake to create immersive learning experiences."

9.8 Business and Marketing

Deepfakes have the possibility of finding usefulness in the business world as virtual proxies for various functions such as presentations, video conferences, and metaverses. These digital avatars can hide identities and avoid unattractive looks. Besides, customization of news updates and sports information with deepfakes would also allow ordinary people to exhibit clothing designs. Instead of relying solely on text-based chatbots, companies may offer customer support utilizing personalized deepfake avatars crafted for individual customers. AI-driven avatars are capable of showcasing products like garments in digital formats thereby avoiding expensive photo shoots. What is more, it promotes dynamic product demonstrations and virtual shopping experiences. This approach presents new opportunities for innovative and cost-efficient marketing and advertising strategies.

It is evident that Deepfake technology has both a bright side and a dark side just like any other coin; however, through their article's focus on darker aspects of Deepfake algorithms, the authors' reasoning seems to be swayed toward the latter position rather than the former one. Any mishandling of this technology could lead to far-reaching implications for society [15].

10 Precautions that Can Be Taken into Account by the People

Proper knowledge is one of the mandated things for the people before opting to use any of the metaverse rather than any platform where significant disclosure of informant is present.

In the discourse of this research will also want the people to understand the need to read the T&C of every site, and app before releasing any important information and try to provide as much less information as possible. Try to not avoid such websites which require significant information about you.

Then comes the zero-trust mindset this is the foremost important aspect that today's generation has to keep in mind because, with the advent of various platforms, it has become difficult for people to keep anything personal. For example, on Facebook, Instagram, and WhatsApp platforms people are continuously sharing their every moment without being aware that these pictures, videos can be used and manipulated by AI creating something that will be disastrous for them as well as for society.

Then to cross-check every detail before releasing any data to any software, people must evidently access the privacy setting and undergo all the necessary counter measures which in turn will help the users to keep their profile safe and secure.

Individuals should always assess the regulations and direction provided by the authorities on the verge of creating a protective shield for the individuals. Precautions like using a watermark, not to give every clear picture or video not to release very sensitive content on media.

The combined efforts of the people and the government will help in realizing this global concern of deepfakes and discrepancies in the metaverse. Taking AI as the greatest weapon against AI will be an effective means to protect the people and society as a whole.

The existing laws that are present and which are trying their best to restore and hold the matrix by fulfilling their duty as the guardian of the society plus protecting its aggrieved children (individuals).

11 Legal Remedies

11.1 Defiance Act

- Proposed federal bills granting an option to victims who can file suits against the creators of the deepfakes who either knowingly or willfully ignored the victim's lack of consent, giving civil actions for those that they can be identified in "digital forgeries."

11.2 Copyright Laws

- Power civil actions against generating AI content that infringe on copyrighted materials without their consent, that helps defend original works from unapproved use.

11.3 Trademark Laws

- Give victims the chance to file claims in situations where their image or identity is either swindled for commercial purposes or used for false endorsements. This way, they will be protected against such scenarios.

11.4 State Laws

- Laws enforced in the states of California, Florida, Texas, and Virginia put aforementioned malicious actors who produce or disseminate deepfakes without consent into miles while imposing sanctions and disclosing requirements

11.5 IT Act, 2000 (India)

- Section 66E: Fines and prison sentences will be brought against people who publish, take, or transmit photos that violate privacy, thus preventing incidents like this from happening.
- Section 66D: Unmask impersonation using intercepting communication through computer devices or imprisonment and fines established.

11.6 Digital Personal Data Protection Act, 2023 (India)

- Protect citizens' personal data, including AI-generated data, by fighting against misuse, thus, civilians' privacy is granted and people are away from being subjected to unauthorized access to their information.

11.7 Defamation Laws

- Deepfake victims have the right to sue for defamation of the character and seek appropriate compensation as small compensation will not do justice to the reputational damage inflicted by deepfakes.

11.8 Privacy Laws

- Private invading might be addressed by AI-generated non-consensual advertising, and the injured parties have the right to sue for privacy infringement.

12 Loopholes Constituting These Remedies

1. Ambiguity in law there is no proper functioning and the laws are not strong enough to address the grievances of the people. Further, there is no prominence in the laws that are able to address the sexualized crimes that are now being created by deepfakes using AI as a tool.
2. Law is unaware of whom to hold as guilty the developer of the creator of deepfakes i.e., there the question arises is AI the guilty entity or the developer who created such sexually explicit videos, images, and contents.
3. There is a lack of specific laws addressing the deepfakes and AI. India suffers paucity of laws to grapple with this high-end metaverse issues, where the creator and developer are critical to identifying.
4. Then proper administration of the laws is not committed, even the legal cells are not fully knowledgeable about how to implement the laws in resolving the issues related to deepfakes and generative artificial intelligence(genAI).
5. Responsibility should be taken by the factories that design and develop technology, the platforms where these tools are mentioned, and the websites that create the tools to develop deepfake content. Implying safety by design and putting people instead of people in need according to the goal is necessary to block the misuse of social media technologies (deepfakes).
6. Ethical guidelines and responsible creation of the software, the developer should be very careful and responsible while developing any software or app that might pose potential threat to the user if negligence is done and security is lost. Further, the provider must notify all the dos or don'ts related to the software and apps. Most importantly strict regulations are not imposed on this industry which are always running after benefits, profits, and most common excuse used by this industry is growth to the economic structure of the society. Irrespective of everything the people of the society and their security must be kept as top priorities to this tech companies.
7. In light of the fact that the matter of deepfakes is transnational, global cooperation, proper plans, and responsible for ethics are needed for conflicts and to let the content develop properly. A global approach to achieving this is the main step that should be taken to deal with this abuse and its implications.

Legal remedies are available for deepfake and AI-driven punishments in India. Although these are not specifically targeted by the existing laws, they can be taken up under several contemporary legal procedures that provide criminal or civil lawsuits.

The IT Act, 2000 (IT Act), provides key provisions: A person is liable under section 66E if he is caught capturing, publishing, or transmitting private images. The punishment extends to three years in jail or a fine amounting to ₹2 lakh. Similarly, the perpetrator who has utilized the communication device in a malicious manner for impersonation or cheating can face the maximum penalty of three years in jail and/or be fined up to copyright law attainments give the unique opportunity, in the intellectual realm, for individuals to file civil suits against deepfakes exploiting protected materials—thereby, through the Digital Millennium Copyright Act (DMCA) provisions, compelling the internet sites to remove the infringing content. Consequently, celebrities may secure protection through both mark and trade laws under the Lanham Act by filing false claims for "false endorsement" in the event deepfakes deceive the public into believing celebrities have indeed received the portrayed product. In the United States, most states have laws that directly address deepfakes and regard this technology as an active measure to fight against the dissemination of pornographic deepfakes. For example, the states of Hawaii, Texas, Virginia, and Wyoming declare pornographic deepfakes illegal. Furthermore, both Texas and California are similar to each other in that they allow civil action lawsuits against deepfakes were influence political campaigns. The Federal government put forward the Deep Fakes Accountability Act which seeks to establish a new crime wing and civil penalties for the act of creating deepfakes. These penalties may allow victims to seek compensation from the creators to alleviate the psychological and long-term effects. Such legal structures tailored for the offenses with AI approach which shall be cybersecurity requirements met can be used to reinforce personal rights and privacy as well as bridge the bridges posed by the technology challenges [16].

What should be emphasized more than anything else in the context of mitigating the damage offered by deepfakes and AI offenses besides the application of current legal provisions, is the need for strong and explicit ones. It is, therefore, high time for individual efforts to be involved in this proactive solution. It is no longer a point of discussion that keep oneself updated on what exactly is passing through the legal system with respect to the digital age and protecting the rights and privacy of individuals.

13 Statistical Records of Deepfake Algorithms

(i). Despite the grim threat that the misuse of deepfake technology lurks over the heads of every member of society, the general public seems to be completely uninformed. According to a survey in 2022, made by iProov, in which more than 16,000 participants were taken into account, it was discovered that about 71% of people are not aware of what the term "deepfake" means. Only one-third of the participants had reported being informed of the existence of such technologies [17].

(ii). A recent report by Sumsub has discovered that the number of crimes associated with the use of deepfake has risen to 2X since 2022 and in the first quarter of 2023. The findings are pretty alarming because AI deepfake is still at its nascent stage and is yet to advance over the coming years.

(iii). Financial security in 2023 faced an unprecedented challenge as deepfake attempts rose by 31 times over the previous year, implying a gigantic surge of 3000 percent. This has been highlighted in Onfido Identity Fraud Report 2024—indicating that deepfakes have become a major concern for identity and biometric verification processes. It is important to note that this prevalence was driven by increasingly available generative AI and face-swap applications that have made deepfakes more sophisticated and easier to create. This alarming trend underlines the need for financial institutions and crypto exchanges (which traditionally relied on routine KYC checks) to upgrade their security measures in light of evolving digital deception [18].

(iv). The research carried out and published in PLOS One, covering 529 individuals, established that about a quarter of the subjects were unable to identify deepfake audio from authentic ones. This is troubling information as it shows the possibility of misuse of such technologies for fraud purposes and dissemination of false information. In other words, a significant percentage of societies are vulnerable to deceptive practices concerning deepfake audios; therefore, public awareness should be raised to create more advanced detection methods [19].

(v). According to Sumsub Identity Fraud Report, an authoritative digital identity verification firm, India, Bangladesh, and Pakistan are among the ten most affected countries in Asia Pacific by identity theft through the use of Deepfake technology. The U.K.-based verification platform has reportedly noticed a substantial increase in these crimes during 2023 with projections showing a further rise next year [20].

(vi). The report of Sumsub Identity Fraud Report further indicated that the cryptocurrency industry is especially vulnerable to deepfake scams, making up 88% of cases in 2023, while fintech accounted for 8%. In addition, prevalent fraudulent methods in 2023 encompassed money muling, counterfeit IDs, account hijacking, and coerced verification, the latter experiencing a 305% surge compared to 2022.

(vii). A recent Sapio Research study [21] found that up to 85% of security experts acknowledge that AI machine learning technology is directly responsible for this rise in a number of cyber frauds [22].

(viii). According to the deepfake phishing statistics in the Global Incident Response Threat Report [23], approximately 78% of deepfake phishing attacks are sent via email.

(ix). In a report published by Sumsub about deepfake crime stats [20], it says that the online media sector is projected to be the most targeted industry constituting 4.27% while coming next were Professional Services with 3.14%, Healthcare with 2.41%.

(x). Deeptrace AI (presently known as Sensity AI) carried out deepfake statistics which showed, that a staggering 96% of all online deepfake videos belonged to non-consensual pornography; and by December 2020, there had been at least identified over 85,000 deepfake videos.

Figure 1 is an international statistical report that sheds light on the awareness of the general public on the harmful consequences of deepfake technology. While it is clear from the graph that the public is not well versed in the potential dangers of possible AI manipulation, it is shocking to see that a good percentage of thousands

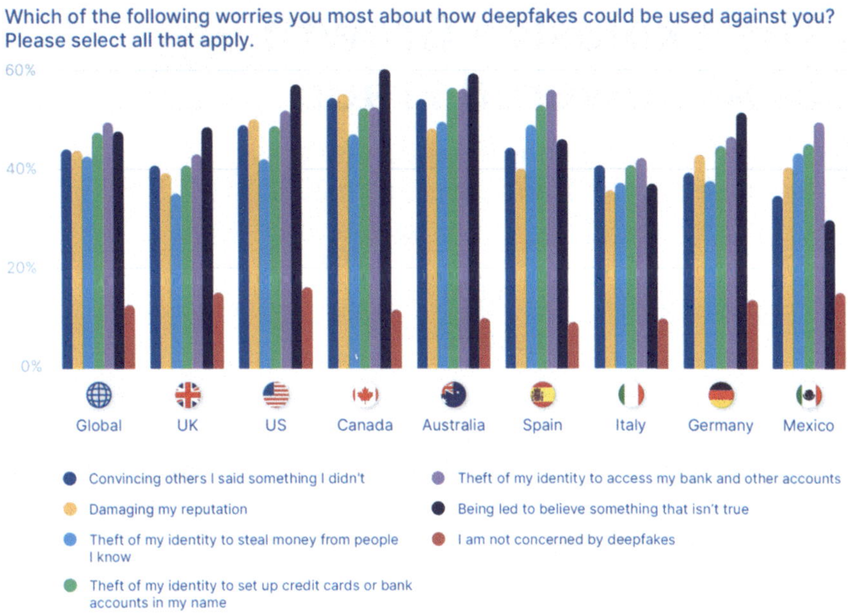

Fig. 1 Statistical data related to the growth of AI and Deepfakes

of people do not bother themselves with the impending risks of deepfakes. It is also worth noting that the given data has been collected from developed countries that have a high literacy rate. This implies that the statistics would be much more alarming for countries that are considered to be in a developing economic phase. This is the reason that most of the South East Asian countries are reportedly stated to be at a higher risk of AI manipulation [20].

Figure 2 depicts the statistical data of more than 2 million crime cases of deepfake by the official Reports of Sumsub Identity Fraud Report, 2023 [20].

Fig. 2 The explosive growth of AI-powered fraud

References

1. Amit KS (2024) Deepfake: phenomenon, impacts and challenges. Medium
2. S (2023) Understanding the machine learning process: key steps and more. Simplilearn.com. https://www.simplilearn.com/what-is-machine-learning-process-article
3. Machine Learning Algorithms - Javatpoint (n.d.) www.javatpoint.com. https://www.javatpoint.com/machine-learning-algorithms
4. What is a Neural Network? IBM (n.d.) https://www.ibm.com/topics/neural-networks
5. Catherine B, Dominic R (2022) How easy is it to make and detect a Deepfake? Carngie Mellon University. https://insights.sei.cmu.edu/blog/how-easy-is-it-to-make-and-detect-a-deepfake/
6. Independent (2023) What are deepfake videos, how they are made, and how to spot fake ones? Here is all you should know. Economic Times. https://m.economictimes.com/news/how-to/what-are-deepfake-videos-how-they-are-made-and-how-to-spot-fake-videos-here-is-all-you-should-know/articleshow/105031370.cms
7. Lutz F (2022) Overview of how to create deepfakes—it's scarily simple. Forbes. https://www.forbes.com/sites/lutzfinger/2022/09/08/overview-of-how-to-create-deepfakesits-scarily-simple/

8. Zysman J, Nitzberg M (2024) Generative AI and the future of work: augmentation or automation? SSRN Electron J. https://doi.org/10.2139/ssrn.4811728
9. Access NCBI through the World Wide Web (WWW) (1995) Molecular biotechnology 3(1):75–75. https://doi.org/10.1007/bf02821338
10. Hancock J (2021) The social impact of deepfakes. Natl Sci Found 24:150–151
11. Coleman K (2023) How deepfakes are impacting culture, privacy, and reputation. Status Labs. https://statuslabs.com/blog/what-is-a-deepfake
12. Neil J (2024) Deepfakes and their impact on society. CPI OpenFox. https://www.openfox.com/deepfakes-and-their-impact-on-society/
13. Eiserbeck A (2023) Deepfake smiles matter less—the psychological and neural impact of presumed AI-generated faces. Sci Rep 13:16111
14. Anna S (2024) Artificial intelligence has psychological impacts our brains might not be ready for, expert warns. Australian Broadcasting Corporation News. https://www.abc.net.au/news/health/2024-05-01/artificial-intelligence-ai-psychology-mental-health/103753940
15. David P (2024) Deepfakes for good? How synthetic media is transforming business. Tech Informed. https://techinformed.com/deepfakes-for-good-how-synthetic-media-is-transforming-business/
16. AI-Generated Deepfakes and the Emerging Legal Landscape (2024) JD Supra. https://www.jdsupra.com/legalnews/ai-generated-deepfakes-and-the-emerging-9345756/
17. iProov (2022) How to protect against deepfakes—statistics and solutions. https://www.iproov.com/blog/deepfakes-statistics-solutions-biometric-protection
18. Onfido (2024) Identity fraud report 2024. https://onfido.com/landing/identity-fraud-report/
19. Griffin L (2023) Warning: Humans cannot reliably detect speech deepfakes. PLoS ONE 18:8
20. Sumsub (2023) Sumsub research: Global deepfake incidents surge tenfold from 2022 to 2023. https://sumsub.com/newsroom/sumsub-research-global-deepfake-incidents-surge-tenfold-from-2022-to-2023/
21. Sensity AI (2020) Reports. https://sensity.ai/reports/
22. Sapio Research (2023) AI in market research: revolutionising insights. https://sapioresearch.com/opinion/ai-in-market-research/
23. VMWare (2022) 2022 Global Incident Response Threat Report. https://www.vmware.com/content/dam/learn/en/amer/fy23/pdf/1553238_Global_Incident_Response_Threat_Report_Weathering_The_Storm.pdf
24. G (2023) Introduction to support vector machines (SVM). GeeksforGeeks. https://www.geeksforgeeks.org/introduction-to-support-vector-machines-svm/

AI and Cybersecurity

TLMMChain: Blockchain-Based Hierarchical Security Management Scheme for Three-Level Mobility Model in Next-Generation Networks

Mohuya Chakraborty and Sudip Kumar Palit

Abstract This paper focuses on the security aspects of the Three-level Mobility Model (TLMM) in the next-generation networks. Prior research has indicated that hierarchical architectures outperform other network topologies in terms of handoff delay, location update, and signaling overhead in mobile networks. The optimality test of the time-bandwidth product proved the supremacy of the three-tiered architecture of mobile networks in literature. However, earlier works of TLMM have not addressed the security issues. Designing a security protocol for TLMM is therefore the goal of this research. Our solution to the current research gap is a lightweight blockchain-based security protocol that can be used for local, regional, and global mobility. The work's uniqueness resides in its proposal of a protocol to establish a peer-to-peer network where users can communicate with the data stored in a secure distributed ledger, reducing the amount of duplicate information that users must transfer. The implementation of blockchain technology is done using Hyperledger Composer and Hyperledger Fabric. Burrows-Abadi–Needham (BAN) logic has been used in the formal security verification of the suggested system to provide mutual authentication among the participants. Furthermore, the suggested model performs noticeably better than the other current systems by providing thirteen distinct security characteristics, requiring less processing time (about 8 ms) and communication overhead (around 2832 bits).

Keywords TLMM · Authentication · Security · Network attacks · Blockchain · Hyperledger · BAN logic

M. Chakraborty (✉)
Cybersecurity Centre of Excellence, Institute of Engineering and Management, University of Engineering and Management, Kolkata, India
e-mail: mohuyacb@iem.edu.in

S. K. Palit
Department of Computer Science and Engineering, Institute of Engineering and Management, University of Engineering and Management, Kolkata, India

M. Chakraborty et al. (eds.), *Proceedings of 5th International Ethical Hacking Conference*, Lecture Notes in Networks and Systems 1148,
https://doi.org/10.1007/978-981-97-8457-8_13

1 Introduction

Significant progress has been made in wireless communication technologies, which has fueled the development of Next-Generation Networks (NGNs) to meet the various mobility needs of contemporary applications. The Three-Level Mobility Model (TLMM) proved to be a crucial framework to handle the complexities of mobility in IP-based networks among the wide range of mobility management solutions available. TLMM's history, development, key components, and contributions to improving mobility management in wireless networks are needed to understand this work.

The evolution of IP-based mobility technologies and a discussion of the foundation for the development of TLMM is discussed in the study [1, 2]. A discussion on protocols like Mobile IP (MIP), TeleMIP, Hierarchical Mobile IP (HMIP), and others is covered in these studies. The mobility challenges of the 4G network along with its advantages and disadvantages are described thoroughly here. TLMM is modeled as a complete system managing the convergence and control of micro, macro, and global mobility requirements and successfully controlling the mobility in IP-based next-generation networks in an orderly way [3]. TLMM proposes a Three-Level Mobility Model that builds on well-known protocols like Mobile IPv4 and TeleMIP, that leverages dynamically auto-configured Care-of-Addresses for efficient packet routing. To enhance mobility management and reduce packet loss at mobile nodes, the study [4] added the bi-casting and Fast Handover for MIPv6 (FMIPv6) features to TLMM. Furthermore, in the research [5], Integrated Services (IntServ) and Differentiated Services (DiffServ) architectures are combined to conduct a systematic analysis of the Quality of Service (QoS) structure of TLMM. This study mainly focuses on the capacity to which TLMM's dynamic handoff policy guarantees suitable quality of service when interdomain roaming is present. Furthermore, to enhance the ability to manage incorrect handovers and to optimize the network utilization of TLMM, an integrated QoS framework is offered in the study [6].

In this paper, we introduce a unique strategy to cover the security aspects of TLMM architecture for NGNs. Blockchain technology is adopted to design the authentication protocol to enhance the security infrastructure for this purpose. Due to the absence of a decentralized network structure, conventional authentication techniques have weaknesses with respect to the security and integrity of mobile users' and mobile network operators' identities and data. In this paper, we aim to develop a robust authentication strategy for TLMM that enhances security and trust in the network by adopting the inherent consistency, decentralization, and transparency properties of blockchain. Our approach incorporates creating a distributed ledger where login credentials and user identities are securely stored and managed. Enforcing authentication standards and enabling secure communication among mobile units and network components are two uses for smart contracts. Through this blockchain-based authentication mechanism, we address the security concerns associated with identity management, ensuring secure and seamless communication in next-generation wireless networks.

The main contributions of the paper include.

1. Design of a novel hierarchical blockchain-based two-way authentication and session key consensus protocol (TLMMChain) to provide secure and measured admittance to the confidential information of mobile users in intrasubdomain (Local), intradomain (Regional), and interdomain (Global) mobility in next-generation networks.
2. Burrows–Abadi–Needham (BAN) logic is used in formal security assessments to confirm that the protocol is secure.
3. Informal security analysis to support the validity of thirteen security features of the proposed protocol.
4. A comparative performance analysis of computational time and communication overhead of TLMMChain with other current schemes.

This is how the remaining part of the paper is structured. A brief description of the preliminary work, including NGNs, their security, and authentication, followed by a brief overview of TLMM and blockchain, is provided in Sect. 2. This is followed by the literature review in Sect. 3. Section 4 proposes the novel security management protocol in TLMM followed by security analysis in Sect. 5. Section 6 is reserved for performance analysis. The paper completes with concluding remarks and future scope in Sect. 7.

2 Preliminaries

In this section, the various terminologies and concepts that are required in the current research are discussed as per literature review of the current topic.

2.1 NGN and TLMM

NGN is a dynamic, IP-based telecommunications infrastructure that combines voice, data, and multimedia services into a single platform [7]. It offers increased scalability, efficiency, and versatility, promoting quicker data transfer, supporting multiple devices, and providing strong security. This paradigm shift toward a more flexible and interconnected telecommunications ecosystem is reflected in NGN. NGN requires advanced security and authentication, particularly in 5G networks [8]. These networks use encryption techniques for data integrity and SIM-based devices as well as user authentication. Device security solutions address IoT device growth, while network slicing security meets service requirements. Strong monitoring, incident response systems, and secure APIs ensure a secure telecoms environment.

TLMM is a mobile communication architecture featuring key components such as Home Agent (HA) for maintaining location information of Mobile Node (MN), Foreign Agent (FA) offering a persistent regional CoA, and Gateway Mobility Agent

(GMA) providing global CoA of MN [3]. The MN seamlessly transitions between subnets, retaining three assigned CoA: local, regional, and global. This three-tiered architecture optimizes mobility management, reducing latency through the strategic use of GMA and enabling private addressing within the provider's domain.

2.2 Blockchain

Blockchain is a decentralized technology revolutionizing traditional transaction systems by securely, transparently, and impenetrably maintaining records across a network of computers [9]. Its foundation is made up of a number of blocks, each containing a list of transactions that are cryptographically linked, time-stamped, and transparent. Its uses in supply chain management, banking, and healthcare enhance security by cutting out intermediaries.

Distributed ledger features can be built in the modular architecture of Hyperledger Fabric, which provides a permissioned blockchain platform [10]. Hyperledger Fabric is appropriate for enterprise-grade blockchain systems and offers features like pluggable consensus methods, support for smart contracts, and privacy protection. On the other hand, Hyperledger Composer helps to build blockchain applications by providing a simplified framework [11]. To enable developers to quickly prototype and construct business networks, Hyperledger Composer provides powerful programming tools on top of Hyperledger Fabric. Hyperledger Composer and Hyperledger Fabric together perform to create scalable, secure blockchain applications for a variety of uses.

Hyperledger Composer consists of the following basic components to streamline the applications of blockchain-based development:

- Assets: Assets are either immaterial or tangible things that are managed and stocked in a blockchain. These may be any financial product or paper, digital items, etc. Assets are defined using Hyperledger Composer's modeling language, which includes attributes, an identity, and occasionally a connection with other assets.
- Participants: Any organization or network members are participants that communicate with blockchain to access, modify, or exchange assets. As part of the definition of a business network, participants are defined. These can be people, companies, as well as smart contracts. Every participant has a distinct identity with corresponding access control guidelines.
- Transactions: The operations that participants do when interacting with assets on the blockchain are called transactions. These tasks could involve the creation, modification, transfer, or querying of assets. Smart contracts, sometimes referred to as chaincode, create transaction functions that carry out transactions. These functions provide the business logic regulating how the blockchain network behaves.

These elements combine to support Hyperledger Composer-based applications, allowing for the blockchain's use in business network modeling, asset and participant requirements, and transaction execution to accomplish desired business goals.

3 Literature Survey

Researchers have discovered several weaknesses and made several breakthroughs in authentication techniques in recent years. An authentication technique based on one-way hash functions was proposed by Lee et al. [12], but shortcomings in several attacks were discovered by Madhusudhan and Suvidha after additional research [13]. Park et al. [14] further analyzed the Lee et al. [12] approach and highlighted its weaknesses in offline password guessing and user impersonation attacks. In 2018 and 2019, Wu et al. and Lu et al. presented schemes that used hash functions and elliptic curve multiplication operations [15, 16]. A novel lightweight authentication scheme was presented by Madhusudhan and Shashidhara in 2019 [17]. Palit and Chakraborty proposed an authentication scheme for global mobility networks in 2019 [18]. Global mobility-related drawbacks were discussed and carefully handled in this scheme. In 2020, Yu et al. presented a unique authentication scheme [19]. Moreover, in 2020, hash functions, symmetric encryption and decryption functions, and elliptic curve multiplication operations were used in the novel authentication schemes of GLOMONET proposed by Sohail et al. and Nikooghadam et al. [20, 21]. In 2021, Rahmani et al. and Kang et al. introduced lightweight authentication schemes utilizing only one-way hash functions [22, 23]. These works also highlight the different ongoing research, address weaknesses, and then offer new mechanisms to withstand the weaknesses. In the year 2022, Palit et al. proposed a mutual authentication and session key agreement scheme, which utilized lightweight hash functions as cryptographic primitives [24].

However, some research works used blockchain technology to design authentication schemes for mobile networks. Most of these researchers worked on 5G architecture. In 2020, Hojjati et al. presented an authentication scheme based on the blockchain framework to withstand security vulnerabilities of 5G networks including Denial of Service (DoS) attacks [25]. Again in 2020, Haddad et al. proposed a blockchain-based strategy specifically for the 5G network that is resistant to multiple known threats in addition to Distributed Denial of Service (DDOS) and DoS attacks [26]. Furthermore, Palit et al. [27] introduced a user authentication protocol based on blockchain in 2021. Unlike previous schemes, this paper is not based on 5G architecture. Moving forward to 2022, Palit et al. developed a lightweight authentication scheme specifically tailored for the 5G network. This lightweight scheme focused on efficiently addressing security concerns while minimizing computational overhead [28].

4 TLMMChain: The Proposed Solution

The principle of operation of the hierarchical authentication architecture of the proposed solution, TLMMChain is depicted in Fig. 1. It works in two different scenarios. The first scenario occurs when a User Equipment (UE) gets service from an FA as it moves out of the scope of its HA in the same domain (Regional). In the other scenario, when a UE moves out of the scope of its HA and obtains service from an FA, under a different domain (Global). In this case, a gateway agent (GA) comes in between the two domains. As shown in the figure, a network is segmented into various domains, each managed by a distinct organization (A, B, or C), which could be public or private entities. These organizations oversee their respective wireless networks, with UE accessing services within their designated domains. Within each domain, subdomains are overseen by FAs e.g., FA3B is responsible for providing wireless services to the UEs in subdomain 3 under organization B.

Before granting service, the FA (e.g., FA3B) authenticates the UE, and vice versa, within its HN (e.g., organization B). This authentication involves a two-way authentication plus session key consensus between UE and FA phase (Regional).

As a UE moves beyond its HN to another network (e.g., domain A of organization A), it can access services from a FA (e.g., FA1A) within the subdomain 1

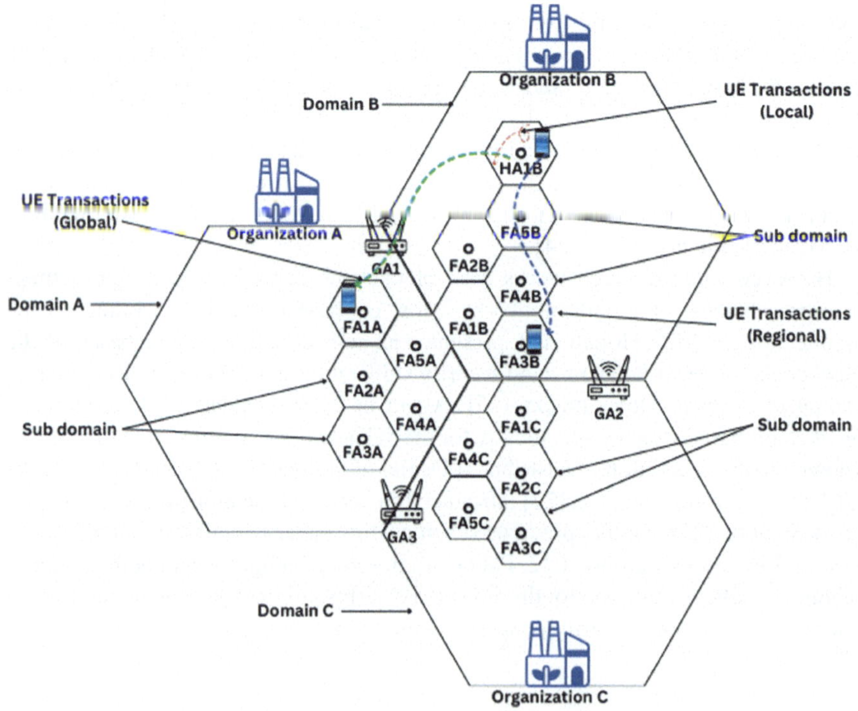

Fig. 1 Hierarchical authentication architecture

under domain A. When crossing network boundaries, a GA (e.g., GA1) facilitates communication between the FA (FA1A) of the visiting network and the HA (HA1B) of the home organization (organization B). Consequently, the FA (FA1A) can provide services to the UE with support from the GA (GA1) and the HA (HA1B) of organization A. In this context, the authentication process involves a two-way authentication plus session key consensus between UE and FA phase (global), reflecting interaction across different network domains.

In the proposed model, there are a total of three phases needed to execute authentication plus a session key agreement scheme within a domain. These are the UE registration phase (Regional), FA registration phase (Regional), Two-way authentication plus session key consensus between UE and FA phase (Regional). The participants in this case are UE, FA, and HA. On the other hand, there are a total of three phases required to execute authentication plus a session key agreement scheme between different domains. These are the UE registration phase (Regional), FA registration phase (Global), Two-way authentication, and session key consensus between UE and FA phase (Global). The participants in this case are UE, GA, FA, and HA.

The descriptions of the phases for both the scenarios are described below. The notations used in the phases are shown in Table 1.

Table 1 Notations

Symbol	Description
idX	The uniqueness of participant X
pwUE	Password of UE
skHA	A secret key of HA
skFA	A secret key shared between HA and FA
skGA	A secret key shared between HA and GA
skFAg	A secret key shared between FA (global) and GA
skfuu	Session key between FA and UE, computed by UE
skfuf	Session key between FA and UE, computed by FA
skhgh	Session key between HA and GA, computed by HA
skhgg	Session key between HA and GA, computed by GA
skgfg	Session key between GA and FA, computed by GA
skgff	Session key between GA and FA, computed by FA
M‖N	Chain operation between M and N
$M \oplus N$	XOR operation between M and N
H()	Hash function
SHR_Asset_X	Stake the asset Asset with Participant X
$ST_{(p1,p2,..)}$Asset	In the asset Asset, create and save (p1, p2,..)
$UPD_{(p1,p2,..)}$Asset	Update (p1,p2,..) in the asset Asset
$AD_{(p1,p2,..)}$Asset	Incorporate (p1,p2,..) into the current asset Asset
ACS_Asset	Admittance of the asset Asset

Algorithm 1	

Transaction1 ueSecretGen ():
UE inputs idUE, password pwUE of UE and identity idHA of HA
Creates a random number uRand;
Set aue = H (idUE$\oplus$pwUE) $\oplus$ uRand;
pidUE = H (idUE $\oplus$ uRand);
Performs ST$_{(idUE,\ pwUE,\ aue,\ pidUE,\ uRand)}$UserEquipmentSecret,
ST$_{(idUE,\ pidUE,\ idHA)}$UserEquipmentSent,
SHR_UserEquipmentSent_HA;
Transaction2 ueRegistrationRequest ():
HA ACS_UserEquipmentSent and inputs identity idHA of HA;
A random number (bRand) is generated;
Fix bue = H (pidUE $\oplus$ bRand);
cue = H (pidUE||bRand);
	due = H (bRand||pidUE);
	Performs ST$_{(pidUE,\ bRand,\ cue,\ idHA)}$UserEquipmentRegistrationData,
	ST$_{(bue,\ cue,\ due,\ bRand,\ idHA)}$HomeAgentSent,
	SHR_HomeAgentSent_UE;
	Transaction3 haRegistrationReply ():
	UE ACS_HomeAgentSent,
	ACS_UserEquipmentSecret;
	Set eue = bue $\oplus$ H(pwUE);
	fue = cue $\oplus$ H(pwUE); gue = due $\oplus$ H (pwUE); Performs AD$_{(eue,\ fue,\ gue,\ bRand)}$UserEquipmentSecret;
	Therefore, the asset UserEquipmentSecret stores
	(aue, idUE, pwUE, pidUE, uRand, eue, fue, gue, bRand);

Fig. 2 UE registration phase (regional)

4.1 *Authentication Protocol for TLMM*

4.1.1 UE Registration Phase (Regional)

This phase is the subject of three deals (transactions). These are the functions ueSecretGen(), ueRegistrationRequest(), and haRegistrationReply(). UserEquipmentSecret, UserEquipmentRegistrationData, UserEquipmentSent, and HomeAgentSent are the four assets used. Figure 2 depicts the operation of this phase in Algorithm 1.

4.1.2 FA Registration Phase (Regional)

This phase is the subject of two deals (transactions). These are the functions sdrSecretGen() and sdrRegistration(). ForeignAgentSecret is the sole asset used here. Figure 3 depicts the operation of this phase in Algorithm 2.

Algorithm 2	
Transaction1 sdrSecretGen (): FA inputs identity idFA Generates a random number fRand Set afa = H (idFA||fRand); ST$_{(idFA,afa)}$ForeignAgentSecret SHR_ForeignAgentSecret_HA	**Transaction2 sdrRegistration ():** HA ACS_ForeignAgentSecret; Set skFA = H (afa$\oplus$skHA); Performs AD$_{(skFA)}$ForeignAgentSecret SHR_ForeignAgentSecret_FA

Fig. 3 FA registration phase (regional)

Algorithm 3	
Transaction1 faSecretGen ():	**Transaction3 gaSecretGen ():**
FA inputs identity idFA	GA inputs identity idGA
Creates a random number fgRand	Generates a random number gRand
Set afag = H (idFA‖fgRand);	Set aga = H (idGA‖gRand);
$ST_{(idFA,afag,idHA)}$ ForeignAgentgSecret	$ST_{(aga,\ idGA)}$GatewayAgentSecret
SHR_ForeignAgentgSecret_GA	SHR_GatewayAgentSecret_HA
Transaction2 faRegistration ():	**Transaction4 gaRegistration ():**
GA ACS_ForeignAgentgSecret;	HA ACS_GatewayAgentSecret;
Set skFAg = H (afag⊕skGA);	Set skGA = H (aga⊕skHA);
Performs $AD_{(skFAg)}$ForeignAgentgSecret	Performs $AD_{(skGA)}$GatewayAgentSecret
SHR_ForeignAgentgSecret_FA	SHR_GatewayAgentSecret_GA

Fig. 4 FA registration phase (global)

4.1.3 FA Registration Phase (Global)

This phase is the subject of four deals (transactions). These are the functions faSecretGen(), faRegistration(), gaSecretGen (), gaRegistration (). The three assets used are ForeignAgentgSecret, GatewayAgentSecret, and ForeignAgentGatewaySecret. Figure 4 depicts the operation of this phase in Algorithm 3.

4.1.4 Two-Way Authentication Plus Session Key Consensus Between UE and FA Phase (Regional)

In this phase, there are five deals (transactions). These are the functions ueAuthenticationSent(), faAuthenticationSent(), haAuthenticationReceived(), faCommunicateUe() and ueConfirmFa(). The nine assets used are UserEquipmentSecret, UeSentFa, ForeignAgentSecret, FaSentHa, HaPermissionFa, UserEquipmentRegistrationData, FaSentUe, SkfuFa, SkfuUe. Figure 5 shows the operation of this phase in Algorithm 4.

4.1.5 Two-Way Authentication and Session Key Consensus Between UE and FA Phase (Global)

There are seven transactions used to implement this phase. These are ueAuthenticationSent(), faAuthenticationSent(), gaAuthenticationSent(), haAuthenticationReceived(), gaCommunicateFa(), faConfirmUe() and ueConfirmFa(). There are seventeen assets used. These are UserEquipmentSecret, UeSentFa, FaSentGa, GaSentHa, HaToGa, GatewayAgentSecret, ForeignAgentgSecret, HaPermissionFa, GaToFa, UserEquipmentRegistrationData, FaSentUe, SkhgHa, SkhgGa, SkgfGa, SkgfFa, SkfuFa, SkfuUe. Figure 6 shows the algorithm for this phase in Algorithm 5.

Algorithm 4

Transaction1 ueAuthenticationSent ():
UE ACS_UserEquipmentSecret; Inputs idUE, pwUE of UE and idHA of HA;
Set aue1 = H (idUE ⊕ pwUE);
if aue = aue1 then
Set pidUE = H (idUE ⊕ uRand); bue = eue ⊕ H (pwUE); cue = fue ⊕ H (pwUE); due = gue ⊕ H (pwUE); Selects a random number nmRand; Set cue1 = cue ⊕ bRand; pidUE1 = pidUE ⊕ nmRand; Performs ST$_{(cue1,\ nmRand,\ pidUE1,\ idHA)}$UeSentFa, SHR_UeSentFa_FA;
end

Transaction2 faAuthenticationSent ():
An arbitrary (random) number nfRand is picked by FA; ACS_ForeignAgentSecret
Set bfa = H (nfRand||skFA); afa1 = afa ⊕ nfRand; Performs ST$_{(bfa,\ afa1,\ nfRand)}$FaSentHa;
SHR_FaSentHa_HA;

Transaction3 haAuthenticationReceived ():
HA ACS_ForeignAgentSecret, ACS_FaSentHa; Set afa = afa1 ⊕ nfRand;
Discovers skFA corresponding to afa from ForeignAgentSecret;
if bfa = H (nfRand||skFA) then
HA authenticates FA; HA sets HaPermissionFa.status = "TRUE" of asset HaPermissionFa; Performs SHR_HaPermissionFa_FA
end

Transaction4 faCommunicateUe ():
FA ACS_HaPermissionFa;
if HaPermissionFa.status = "TRUE" then
FA ACS_UeSentFa; Set pidUE = pidUE1 ⊕ nmRand;
ACS_UserEquipmentRegistrationData with respect to pidUE;
if cue = cue1 ⊕ bRand then
FA authenticates UE; Set bue = pidUE ⊕ bRand; due = H (bRand||pidUE);
Chooses random numbers b1Rand, nf1Rand; bue1=pidUE ⊕ b1Rand, cue1= H(pidUE||b1Rand), due1=H(b1Rand || pidUE)
performs UPD$_{(b1Rand,cue1)}$UserEquipmentRegistrationData
Set due11 = due ⊕ bue; bueNew = bue1 ⊕ bue; cueNew = cue1 ⊕ bue; dueNew = due1 ⊕ bue; bRandNew = b1Rand ⊕ bue; Set session key skfuf = H (due||nf1Rand||nmRand); Performs ST$_{(skfuf)}$SkfuFa, [Here, Asset SkfuFa created to store the Session key between FA and UE, computed by FA]
ST$_{(due11,\ bueNew,\ cueNew,\ dueNew,\ bRandNew,\ nf1Rand)}$FaSentUe;
SHR_UserEquipmentRegistrationData_UE, SHR_FaSentUe_UE;
end
end

Transaction5 ueConfirmFa ():
UE ACS_UserEquipmentSecret, ACS_FaSentUe, Computes due = gue ⊕ H (pwUE);
if due = due11 ⊕ bue then
UE authenticates FA; Set bue1 = bueNew ⊕ bue; cue1= cueNew ⊕ bue; due1 = dueNew ⊕ bue; b1Rand = bRandNew ⊕ bue; eue1 = bue1 ⊕ H (pwUE); fue1 = cue1 ⊕ H (pwUE); gue1 = due1 ⊕ H (pwUE); Sets session key skfuu = H (due||nf1Rans||nmRand); Performs UPD$_{(eue1,\ fue1,\ gue1,\ b1Rand)}$ UserEquipment, ST$_{(skfuu)}$SkfuUe;
end

Fig. 5 Two-way authentication and session key consensus between UE and FA phase (regional)

Algorithm 5

Transaction1 ueAuthenticationSent ():
UE ACS_UserEquipmentSecret; Inputs idUE, pwUE of UE and idHA of HA;
Set aue1 = H (idUE ⊕ pwUE);
if aue = aue1 then
Set pidUE = H (idUE ⊕ uRand); bue = eue ⊕ H (pwUE); cue = fue ⊕ H (pwUE);
due = gue ⊕ H (pwUE); Selects a random number nmRand; Set cue1 = cue ⊕
bRand; pidUE1 = pidUE ⊕ nmRand; Performs $ST_{(cue1,\ nmRand,\ pidUE1,\ idHA)}$UeSentFa,
SHR_UeSentFa_FA;
end

Transaction2 faAuthenticationSent ():
FA picks a random number nfRand; ACS_ForeignAgentgSecret.
Set bfa = H (nfRand || skFAg); afa1 = afag ⊕ nfRand;
Performs $ST_{(bfa,\ afa1,\ nfRand)}$FaSentGa; SHR_FaSentGa_GA;

Transaction3 gaAuthenticationSent ():
GA chooses a random number ngRand;
ACS_ForeignAgentgSecret, ACS_FaSentGa, ACS_GatewayAgentSecret
if bfa==H(nfRand || skFAg), then
Choose random ngRand, Set bga = H (ngRand || skGA); aga1 = aga ⊕ ngRand;
Performs $ST_{(bga,\ aga1,\ ngRand)}$GaSentHa; SHR_GaSentHa_HA;
end

Transaction4 haAuthenticationReceived ():
HA ACS_GatewayAgentSecret, ACS_GaSentHa; Set aga = aga1 ⊕ ngRand;
Finds skGA corresponding to aga from GatewayAgentSecret;
if bga = H (ngRand||skGA) then
HA authenticates GA; HA sets HaPermissionFa.status = "TRUE" of
asset HaPermissionFa; Performs SHR_HaPermissionFa_GA, Generates random
number nhRand, Computes nhg=H(nhRand||skGA), ah=H(skGA||nhRand)
Create a session key skhgh=H(nhg||nhRand||ngRand), Perform $ST_{(nhRand,\ ah)}$HaToGa,
$ST_{(skhgh)}$SkhgHa, SHR_HaToGa_GA
end

Transaction5 gaCommunicateFa():
GA ACS_HaPermissionFa_GA, CS_HaToGa, ACS_GaSentHa, ACS_ForeignAgentgSecret,
Computes nhg=H(nhRand||skGA)
if ah=H(skGA||nhRand), then
Computes session key skhgg=H(nhg||nhRand||ngRand), Creates random number
ng1Rand, Computes ag=H(skFAg||ng1Rand), ngf=H(ng1Rand||skFAg)
Computes session key skgfg=H(ngf||ng1Rand||nfRand), $ST_{(ng1Rand,\ ag)}$GaToFa,
$ST_{(skgfg)}$SkgfGa, $ST_{(skhgg)}$SkhgGa, SHR__HaPermissionFa_FA,
SHR_GaToFa_FA
end

Transaction6 FaConfirmUe ():
FA ACS_HaPermissionFa
if HaPermissionFa.status= "True", then
FA ACS_UeSentFa; ACS_ForeignAgentgSecret, ACS_GaToFa, ACS_FaSentGa
if ag=H(skFAg||ng1Rand), then
Computes ngf=H(ng1Rand||skFAg), Computes session key skgff= H(ngf ||
ng1Rand||nfRand), Performs $ST_{(skgff)}$SkgfFa, Set pidUE = pidUE1 ⊕ nmRand;
ACS_UserEquipmentRegistrationData with respect to pidUE;
if cue = cue1 ⊕ bRand then
FA authenticates UE; Set bue = pidUE ⊕ bRand; due = H (bRand||pidUE);
Chooses random numbers b1Rand, nf1Rand; Set due11 = due ⊕ bue;
Compute bue1 = pidUE ⊕ b1Rand, cue1=H(pidUE || b1Rand),
due1= H(b1Rand||pidUE), Delete (pidUE, bRand, cue)
$UPD_{(pidUE,\ b1Rand, cue1)}$ UserEquipmentRegistrationData,
bueNew = bue1 ⊕ bue; cueNew = cue1 ⊕ bue; dueNew = due1 ⊕ bue;
bRandNew = b1Rand ⊕ bue; Set session key skfuf = H (due||nf1Rand||nmRand);
Performs $ST_{(skfuf)}$SkfuFa, $ST_{(due11,\ bueNew,\ cueNew,\ dueNew,\ bRandNew,\ nf1Rand)}$FaSentUe;
SHR_UserEquipmentRegistrationData_UE, SHR_FaSentUe_UE;
end
end
end

Transaction7 UeConfirmFa():
UE ACS_UserEquipmentSecret, ACS_FaSentUe,
Computes due = gue ⊕ H (pwUE);
if due = due11 ⊕ bue then
UE authenticates FA; Set bue1 = bueNew ⊕ bue; cue1= cueNew ⊕ bue; due1 =
dueNew ⊕ bue; b1Rand = bRandNew ⊕ bue; eue1 = bue1 ⊕ H (pwUE);
fue1 = cue1 ⊕ H (pwUE); gue1 = due1 ⊕ H (pwUE); Sets session key skfuu =
H (due||nf1Rans||nmRand); Performs $UPD_{(eue1,\ fue1,\ gue1,\ b1Rand)}$
UserEquipmentSecret, $ST_{(skfuu)}$SkfuUe;
end

Fig. 6 Two-way authentication and session key consensus between UE and FA phase (global)

5 Security Analysis

We examine TLMMChain's security in this section of the article. We employ formal methods, such as BAN logic for rigorous security verification. Additionally, we provide an informal security analysis to identify potential vulnerabilities and threats. This combined approach offers a comprehensive assessment of TLMMChain's security, ensuring its resilience against various attacks and risks.

5.1 Formal Security Analysis Using BAN Logic

We have conducted a formal security analysis using BAN logic [29], a method commonly employed for evaluating authentication and key distribution protocols. Initially, we outlined the various symbols and principles of BAN logic. Then, we utilized this framework to demonstrate that our protocol ensures mutual authentication between the FA and the UE.

BAN logic utilizes the following symbols:

- P $|\equiv$ M: Principal P believes statement M, or has the authority to believe M.
- P $\triangleleft$ M: Principal P sees message M, meaning M has received, read, and can repeat M.
- P $|\sim$ M: Principal P has previously stated or sent a message to M, indicating trust in M at that time.
- P $\rightarrow$ M: Principal P has jurisdiction over statement M, implying authority or control.
- #M: Formula A is fresh, meaning it has not been sent in any previous message before the current protocol execution.
- $< M1 >_{M2}$: Combination of M1 and M2; M1 is kept secret, and the presence of $< M1 >_{M2}$ establishes the identity of the sender.
- $\{M1\}_{K1}$: Encryption of formula M1 using secret key K1.
- P1 $\overset{K1}{\leftrightarrow}$ P2: Principals P1 and P2 can communicate using the same key K1, ensuring secure communication accessible only to them or to principals trusted by both.

Here are the following principles used in BAN logic:

1. Message Meaning Rule: $\frac{P1|\equiv P1 \overset{K1}{\leftrightarrow} R1, P1 \triangleleft \langle A \rangle_{K1}}{P1|\equiv R1|\sim A}$ If principal P1 believes that principal R1 possesses a secret key, K1, and P1 sees that message A has been encrypted with K1, then P1 will infer that someone, say Q1, once asserted A.
2. Session Key Rule: $\frac{P1|\equiv \#(A), P1|\equiv R1|\equiv A}{P1|\equiv P1 \overset{K1}{\leftrightarrow} R1}$ If principal P1 believes that session key A is fresh and also believes that principal R1 believes in A, then P1 concludes that they are providing R1 with access to session key A.
3. Freshness Concatenation Rule: $\frac{P1|\equiv \#(A)}{P1|\equiv \#(A,B)}$, If principal P1 believes that A is fresh, then P1 will also believe that the combination of A and B is fresh.

4. Belief Rule: $\frac{P1|\equiv(A)}{P1|\equiv(A,B)}$ If principal P1 believes in statement A, then P1 will also believe in the conjunction of A and B.
5. Nonce Verification Rule: $\frac{P1|\equiv\#(A),P1|\equiv R1|\sim A}{P1|\equiv R1|\equiv A}$ If principal P1 believes that A is fresh and also believes that principal R1 once communicated with A, then P1 will believe that R1 believes in A.
6. Jurisdiction Rule: $\frac{P1|\equiv R1\rightarrow A,P1|\equiv R1|\equiv A}{P1|\equiv A}$ Principal P1 believes in statement A if P1 believes that principal R1 has jurisdiction over A and P1 believes that R1 believes in A.

5.1.1 BAN Logic Implementation for the Two-Way Authentication and Session Key Consensus Between UE and FA Phase (Regional)

To demonstrate the security of the proposed Two-way authentication and session key consensus between UE and FA phase (Regional), we adhere to the following steps:

1. Specify the goals of our scheme.
2. Describe the assumptions we have about our scheme's initial state.
3. Convert the suggested scheme into an idealized representation of BAN logic.
4. To achieve the goal stated in Step 1, apply the BAN logic principles, the idealized form from Step 3, and the beginning assumptions from Step 2.

1. Goals of the scheme:
   - G1 : $\mathrm{FA}|\equiv\left(FA\overset{skfu}{\leftrightarrow}\mathrm{UE}\right)$ [Since skfu = skfuf = H (due||nf1Rand||nmRand), we use a common term skfu in place of skfuf and skfuu throughout this section.]
   - G2: $\mathrm{UE}|\equiv\left(\mathrm{FA}\overset{skfu}{\leftrightarrow}\mathrm{UE}\right)$
   - G3: $\mathrm{FA}|\equiv\mathrm{UE}|\equiv\left(\mathrm{FA}\overset{skfu}{\leftrightarrow}\mathrm{UE}\right)$
   - G4: $\mathrm{UE}|\equiv\mathrm{FA}|\equiv\left(\mathrm{FA}\overset{skfu}{\leftrightarrow}\mathrm{UE}\right)$
2. Here are the underlying assumptions of our suggested plan:
   - AS1 :$\mathrm{FA}|\equiv\left(\mathrm{FA}\overset{due}{\leftrightarrow}\mathrm{UE}\right)$
   - AS2 :$\mathrm{UE}|\equiv\left(\mathrm{FA}\overset{due}{\leftrightarrow}\mathrm{UE}\right)$
   - AS3 : $\mathrm{FA}\ |\equiv \#\mathrm{nmRand}$
   - AS4 : $\mathrm{UE}\ |\equiv \#\mathrm{nf1Rand}$
   - AS5 : $\mathrm{UE}\ |\equiv \#\mathrm{nmRand}$
   - AS6 : $\mathrm{FA}\ |\equiv \#\mathrm{nf1Rand}$
   - AS7 : $\mathrm{UE} \lhd \langle \mathrm{skfu}\rangle_{\mathrm{due}}$
   - AS8: $\mathrm{FA}\lhd \langle \mathrm{skfu}\rangle_{\mathrm{due}}$
3. Here is our suggested protocol in its idealized version:

- M1: UE → FA: (skfu: $\langle \text{nmRand} \rangle_{\text{due}}$)
- M2: FA → UE: (skfu: $\langle \text{nf1Rand} \rangle_{\text{due}}$)

4. Proof through use of BAN logic:

After receiving M1 we could get

- S1: FA $\triangleleft$ (skfu: $\langle \text{nmRand} \rangle_{\text{due}}$)

From *message meaning rule,* S1, and AS1 we are able to write

- S2: HA $|\equiv$ GA $|\sim$ nmRand

From *nonce verification rule,* S2, and AS3 we are able to write

- S3: FA $| \equiv$ UE $| \equiv$ nmRand, where nmRand is part of the session key skfu.

From freshness concatenation rule and AS6 we are able to write

- S4: FA $|\equiv$ #skfu [as nmRand is a part of skfu]

From S3, S4, and *session key rule* we can say that

- S5: $\text{FA}|\equiv \left(\text{FA} \overset{skfu}{\leftrightarrow} \text{UE}\right)$ **[G1]**

From *message meaning rule,* AS1, and AS8 we are able to write.

- S6: FA $|\equiv$ UE $|\equiv\sim$ skfu

From *nonce verification rule,* S4, and S6 we are able to write

- S7: $\text{FA}|\equiv\text{UE}|\equiv \left(\text{FA} \overset{skfu}{\leftrightarrow} \text{UE}\right)$ **[G3]**

After receiving M2 we can write

- S8: UE $\triangleleft$ (skfu: $\langle \text{nf1Rand} \rangle_{\text{due}}$)

From *message meaning rule,* S8, and AS2 we can get

- S9: UE $|\equiv$ FA $|\sim$ nf1Rand

From *nonce verification rule,* S9, and AS4 we can write

- S10: UE $|\equiv$ FA $|\equiv$ nf1Rand

From AS5 and *freshness concatenation rule* we are able to write

- S11: UE $|\equiv$ #skfu [As nf1Rand is a part of skfu]

From S10, S11 and *session key rule* we can write

- S12: $\text{UE}|\equiv \left(\text{FA} \overset{skfu}{\leftrightarrow} \text{UE}\right)$ **[G2]**

From AS2, AS7 and *message meaning rule* we are able to write

- S13: UE $|\equiv$ FA $|\equiv\sim$ skfu

  From S11, S13 and *nonce verification rule* we can say that

- S14: UE$|\equiv$FA$|\equiv\left(\text{FA} \overset{skfu}{\leftrightarrow} \text{UE}\right)$ **[G4]**

5.1.2 BAN Logic Implementation for the Two-Way Authentication and Session Key Consensus Between UE and FA Phase (Global)

In the suggested approach, for the two-way authentication and session key consensus between UE and FA phase (Global) phase, BAN logic can be applied to prove the mutual authentication between the participants: UE and FA, FA and GA, and GA and HA for the session keys established by them. Similar to earlier sections, mutual authentication between these pairs is confirmed using BAN logic. The process mirrors the one discussed previously for regional phase authentication between UE and FA. Specific session keys, such as skgfg and skgff for GA and FA, and skhgg and skhgh for GA and HA, are utilized to confirm mutual authentication between respective entities.

5.2 Informal Security Analysis

In this segment, we show how the suggested TLMMChain system may effectively handle a number of security issues, including replay attacks, insider assaults, DOS, DDOS, and mutual authentication.

5.2.1 Perfect Mutual Authentication

In the global TLMM phase, FA and GA engage in mutual authentication using the private key skFAg, while GA and HA authenticate each other on skGA. Authentication of UE by FA, GA, and HA is performed by verifying the equation cue = cue1 $\oplus$ bRand, where bRand and cue are stored in the blockchain alongside UE's pidUE. Initially accessible to HA and UE, these data become available to FA upon mutual authentication between (UE, FA), (FA, GA), and (GA, HA) pairs. Hence, the relation (cue = cue1 $\oplus$ bRand) ensures a genuine UE is authenticated by the rest of the participants. Similarly, UE authenticates FA, GA, and HA by verifying the relation due = due11 $\oplus$ bue, computed using private data accessible only to UE. FA derives due11 = due $\oplus$ bue using pidUE and bRand from the blockchain, allowing only valid UE to decrypt FA's encrypted relation.

In the regional TLMM phase, FA and HA mutually authenticate using the secret key skFA, while FA and HA authenticate UE reciprocally, following the same relation and logic as in the global TLMM, with the only distinction being the absence of GA in this phase.

5.2.2 Compromised Server Attack

Hackers find it more difficult to breach TLMMChain, a blockchain-based authentication system that stores transaction data on distributed ledger nodes. By preventing attempts to remove or alter data and resisting attacks from compromised systems, the blockchain network guarantees the accuracy of recorded data.

5.2.3 Replay Attack

In TLMMChain, messages exchanged within each session are encrypted using unique secret values cue and bue, which change with every session and are known only to valid participants. Additionally, new random values such as nfRand, nmRand, ngRand, nhRand, ng1Rand, b1Rand, and nf1Rand are utilized for computing all messages in each session. These measures effectively prevent replay attacks within our scheme.

5.2.4 DOS and DDOS Attack

DOS and DDOS attacks cannot be carried out due to the distributed and decentralized nature of TLMMChain's blockchain-based structure. Because there isn't a single point of control and each node maintains a copy of the ledger, attackers find it extremely difficult to disrupt the network. The cryptographic protections built into blockchain technology further make TLMMChain unassailable from tampering and attempts at forgery. Blockchain technology offers continuous and secure data storage.

5.2.5 Man-In-The-Middle Attack

TLMMChain prevents man-in-the-middle attacks by safeguarding secret keys and parameters, stored either in the blockchain or accessible only to authorized participants. For instance, parameters like (pidUE, bRand, cue) for UE are stored in the blockchain by HA but accessible only to genuine UE, HA, and FA. Similarly, secret values of UE like (idUE, pwUE, aue, pidUE, eue, fue, gue, uRand, bRand) are accessible only to UE. The communication between FA and HA is encrypted and decrypted using the secret key skFA, similarly, the secret key skFAg is used between FA and GA, and skGA is used between GA and HA. As a result, TLMMChain effectively prevents man-in-the-middle attacks.

5.2.6 Single Point of Failure

A single point of failure is not possible in the blockchain-based TLMMChain approach since records are stored on a distributed ledger.

5.2.7 Insider Attack

As all parameters are accessible via application calls made during message transfers between participants, insider attacks are successfully protected with TLMMChain. This ensures that privileged access to the database is only attainable through authorized means. Parameters exchanged between participants are computed using secret keys and passwords such as pwUE, skFA, skFAg, and skGA, which are either inputted by the participants themselves (e.g., pwUE for UE) or stored securely in the blockchain (e.g., skFAg, skGA, skFA). Consequently, the nature of TLMMChain makes insider attacks impractical.

5.2.8 Anonymity and Untraceability

Encryption is a technique used by TLMMChain to guarantee the privacy of the user equipment's identity during communication. The encrypted identity (pidUE), which is produced during UE registration under HA as pidUE $=$ H(idUE $\oplus$ uRand), where uRand is a randomly selected number by the UE, is created to guarantee that idUE is kept secret from HA. pidUE is also encrypted with a session-specific random number (nmRand) during the two-way authentication plus session key consensus between UE and FA phases in order to preserve UE anonymity between sessions. Furthermore, each session's parameters (e.g., nfRand, nmRand, ngRand, nhRand, ng1Rand, b1Rand, and nf1Rand) are generated as random numbers in order to ensure untraceability and prevent third parties from tracing message senders.

5.2.9 Impersonation Attack

User impersonation within TLMMChain is rendered infeasible due to the requirement of knowing the parameter cue1, where cue1 $=$ cue $\oplus$ bRand. Only legitimate participants can access cue and bRand since they are safely kept in the blockchain and vary with every session. Moreover, impersonating FA, GA, or HA is impractical as their communication relies on secret keys such as skFA, skFAg, and skGA, ensuring the integrity of their interactions.

5.2.10 Forward Secrecy

Despite the possibility that Æ may obtain UE's long-term secret key, pwUE, as well as the shared secret keys, skFA, skFAg, and skGA, intercepting all communication between participants is unattainable since messages are stored in the blockchain. Consequently, Æ cannot compute any previous session keys such as skfuf, skfuu, skhgh, skhgg, skgfg, skgff, skfuf, and skfuu. Thus, the proposed scheme effectively implements perfect forward secrecy.

5.2.11 Local Password Verification

Before sending the authentication request through FA in TLMMChain, UE inputs their password (pwUE) and identification (idUE) to prove their legitimacy. Without knowledge of these secrets, an unauthorized user Æ cannot compute aue1 = H(idUE ⊕ pwUE). Consequently, subsequent verification checks, such as (aue1 = aue), would fail. TLMMChain effectively prevents unauthorized access by implementing local password verification.

5.2.12 Offline Password Guessing Attack

Æ may attempt to guess the identity as idUEA and the password as pwUEA of a UE, enabling them to compute aueA = H(idUEA ⊕ pwUEA). However, to satisfy the equation (aueA = aue), Æ must correctly guess both the identity and password of the UE. This renders offline password guessing attacks impractical since correctly guessing both identity and password simultaneously is highly unlikely. TLMMChain effectively mitigates offline password guessing attacks through this mechanism.

5.2.13 Stolen Smartcard Attack

If Æ manages to steal or obtain UE's smartcard, they could potentially access all confidential parameters loaded within it using a power analysis attack, as described by Kocher et al. [30] and Messerges et al. [31], and subsequently utilize them for unauthorized access. However, since all of UE's sensitive data is kept on the blockchain, our solution does not require a smartcard. Since only legitimate UE can access this kind of data, there is no chance that a stolen smartcard attack could be successful.

6 Performance Analysis

This section delves into the efficacy of TLMMChain with respect to performance across three crucial parameters: (1) Security features, (2) Computation time, (3) Communication overhead. In this section, a performance comparison of TLMMChain with four related schemes [12, 19, 22, 23] is shown.

6.1 Security Features Comparison

In this section, we have examined the performance of TLMMChain regarding several well-known security features. Then a comparison of the proposed scheme and other related schemes [12, 19, 22, 23] concerning these security features has been

Table 2 Security features comparison

Schemes	S1	S2	S3	S4	S5	S6	S7	S8	S9	S10	S11	S12	S13
[12]	×	×	×	×	✓	✓	×	✓	×	✓	×	×	✓
[19]	✓	✓	✓	×	–	✓	×	✓	✓	–	×	–	✓
[22]	✓	–	✓	×	✓	✓	×	✓	✓	✓	×	✓	✓
[23]	✓	–	–	×	–	✓	×	×	✓	✓	×	✓	✓
Proposed	✓	✓	✓	✓	✓	✓	✓	✓	✓	✓	✓	✓	✓

S1: Perfect mutual authentication, S2: Offline password guessing attack, S3: Local password Verification, S4: DOS and DDOS attack, S5: Man-in-the-middle attack, S6: User anonymity and untraceability, S7: Not a single point of failure, S8: Stolen smartcard attack, S9: Perfect forward secrecy, S10: Insider attack, S11: compromised server attack, S12: Impersonation attack, S13: Replay attack
✓: secure, ×: insecure, –: does not concern

conducted. TLMMChain is successful in mitigating security attacks like replay, offline password guessing, stolen smartcards, and impersonation attacks. Furthermore, our proposed scheme provides perfect forward secrecy, guarantees anonymity, and is untraceable. A remarkable achievement of TLMMChain is that it can withstand DDOS and DOS attacks. As the blockchain framework used in the TLMMChain, it is resilient to the compromised server attack and there is no single point of failure. However, from Table 2, we can see that most of the other schemes are susceptible to most of these attacks.

6.2 Computation Time Comparison

Table 3 illustrates the computation overhead of the protocols, focusing solely on the two-way authentication plus session key consensus between the UE and FA phases at the regional level. We exclude the global phase as no other scheme has addressed this scenario in their research, making it impossible to include this for comparison. Additionally, registration phases are omitted from the comparison since they execute only once, unlike the repeated execution of the two-way authentication plus session key consensus between the UE and FA phases. Using the notation T_H to denote the time required for executing a one-way hash operation, we assume a one-way hash function execution time of 0.5 ms based on experimental results from [32]. The execution times in milliseconds for the schemes [12, 19, 22, 23] are 14, 11, 09, and 15 ms, respectively. Our proposed model, TLMMChain, demonstrates the minimum execution time at 8 ms compared to other schemes [12, 19, 22, 23]. However, it is worth noting that the computation time for the two-way authentication plus session key consensus between the UE and FA phases (Global) in TLMMChain is 15 ms, which is higher than the regional phase due to the involvement of an additional participant, the Gateway Agent (GA).

Table 3 Computation time comparison

Scheme	[12]	[19]	[22]	[23]	TLMMChain (Regional)
UE	$10T_H$	$8T_H$	$6T_H$	$13T_H$	$10T_H$
FA	$8T_H$	$7T_H$	$4T_H$	$5T_H$	$5T_H$
HA	$10T_H$	$7T_H$	$8T_H$	$12T_H$	$1T_H$
Total	$28T_H$	$22T_H$	$18T_H$	$30T_H$	$16T_H$
Execution time (ms)	14	11	9	15	8

The notation T_H signifies the execution time for a one-way hash function

6.3 Communication Overhead Comparison

The size of various parameters used during interactions among members in the two-way authentication plus session key consensus between the User Equipment (UE) and Foreign Agent (FA) phase are calculated as part of the TLMMChain communicational cost evaluation. The parameters considered include random numbers, entity identities, entity passwords, Boolean flags, outputs of SHA-256, outputs of MAC functions using HMAC-SHA-256, and timestamps. The lengths of these parameters in our proposed protocol are as follows: Password of an entity: 256 bits, Identity of an entity: 160 bits, Random number or nonce: 128 bits, Output of a one-way hash function using SHA-256: 256 bits, Timestamp: 32 bits, Boolean flag: 16 bits, Output of MAC function using HMAC-SHA-256: 256 bits.

Table 4 presents a comparative study of communication overhead in terms of the number of bits during the two-way authentication plus session key consensus between the User Equipment (UE) and Foreign Agent (FA) phase (Regional). It compares the communicational overhead between our scheme and schemes used in previous literature [12, 19, 22, 23]. In scheme [12], the number of communicated messages is 5, while for the remaining schemes, including TLMMChain, the number of communicated messages is 4. Furthermore, the total size of messages in bits for schemes [12, 19, 22, 23] is 3872, 4000, 3232, and 5856, respectively. In contrast, TLMMChain incurs a total size of 2832 bits during the communication being considered.

The reason the suggested approach has a lower communication cost than other schemes is that TLMMChain maintains assets in the distributed ledger, where users may access them immediately. As a result, the protocol is made to enable authorized individuals to obtain the information needed for authentication from the blockchain. Furthermore, our technique is very effective for quick communication because of

Table 4 Comparison on communication overhead

Scheme	[12]	[19]	[22]	[23]	TLMMChain (Regional)
No of messages	5	4	4	4	4
Total size in bits	3872	4000	3232	5856	2832

the decreased number of messages transferred (four), which helps reduce communication overhead. During the two-way authentication plus session key consensus between the User Equipment (UE) and Foreign Agent (FA) phase (global), the communication overhead is 4256 bits. Hence the proposed scheme is suitable for global mobility management in next-generation wireless networks using three-level architecture [33, 34].

7 Conclusion

This paper addresses a gap in research on TLMM in NGN. It introduces a lightweight blockchain-based security protocol specifically designed for TLMM, enabling direct user interaction with secure distributed ledger data. The protocol uses Hyperledger Composer and Hyperledger Fabric for blockchain implementation, employing BAN logic for robust mutual authentication. The results show superior performance with thirteen security characteristics, minimal processing time (approximately 8 ms for regional, 15 ms for global), and reduced communication overhead (around 2832 bits for regional and 4256 bits for global). Future work should enhance the proposed blockchain-based security protocol for TLMM, focusing on scalability, real-world implementation, user experience, adaptation to emerging technologies, and industry standards collaboration.

References

1. Chakraborty M, Misra IS, Saha D, Mukherjee A (2003) A comparative study of existing protocols supporting IP mobility. In: 2003 World Wireless Congress 2003 May 27, vol. 75
2. Saha D, Mukherjee A, Misra IS, Chakraborty M (2004) Mobility support in IP: a survey of related protocols. IEEE Netw 18(6):34–40
3. Chakraborty M, Misra IS, Saha D, Mukherjee A (2007) TLMM: MIP-based three level mobility model. Comput Commun 30(6):1244–1257
4. Chakraborty M, Misra IS, Saha D, Mukherjee A (2004) A novel fast handover technique for TLMM: three level mobility model for next generation wireless IP-based networks. In: Proceedings of the IEEE INDICON 2004. First India Annual Conference. IEEE, pp 327–332
5. Chakraborty M, Misra IS, Saha D, Mukherjee A. (2004) QoS support in TLMM: Three level mobility model for IP-based networks. In: Distributed Computing-IWDC 2004: 6th International Workshop, Kolkata, India, December 27–30, 2005, Proceedings. Springer Berlin Heidelberg, pp 192–197
6. Chakraborty M, Misra IS, Mukherjee A (2008) Integrating QoS management in TLMM: three level mobility model for next generation wireless IP-based networks. 中国通信: 英文版. 2008(3):125–39
7. Gupta U, Pantola D, Bhardwaj A, Singh SP (2022) Next-generation networks enabled technologies: challenges and applications. Next Gener Commun Netw Ind Internet Things Syst 16:191–216
8. Nguyen DD, Sood K, Xiang Y, Gao L, Chi L, Yu S (2023) Towards IoT node authentication mechanism in next generation networks. IEEE Internet Things J 10:13333

9. Zheng Z, Xie S, Dai HN, Chen X, Wang H (2018) Blockchain challenges and opportunities: a survey. Int J Web Grid Serv 14(4):352–375
10. Hyperledger fabric retrieved from https://www.hyperledger.org/projects/fabric. Accessed 15 June 2023
11. Hyperledger Composer retrieved from https://hyperledger.github.io/composer/latest/. Accessed 15 July 2023
12. Lee C, Lai Y, Chen C, Chen S (2017) Advanced secure anonymous authentication scheme for roaming service in global mobility networks. Wirel Pers Commun 94:1281–1296
13. Madhusudhan R, Suvidha KS (2017) An efficient and secure user authentication scheme with anonymity in global mobility networks. In: 2017 31st International Conference on Advanced Information Networking and Applications Workshops (WAINA), pp 19–24
14. Park K, Park Y, Park Y, Reddy AG, Das AK (2017) Provably secure and efficient authentication protocol for roaming service in global mobility networks. IEEE Access 5:25110–25125
15. Wu F, Li X, Xu L, Kumari S, Sangaiah A (2018) A novel mutual authentication scheme with formal proof for smart healthcare systems under global mobility networks notion. Comput Electr Eng 68:107–118
16. Lu Y, Xu G, Li L, Yang Y (2019) Robust privacy-preserving mutual authenticated key agreement scheme in roaming service for global mobility networks. IEEE Syst J 13(2):1454–1465
17. Madhusudhan R (2018) A secure and lightweight authentication scheme for roaming service in global mobile networks. J Inf Secur Appl 38:96–110
18. Palit SK, Chakraborty M (2018) A mobile user authentication technique in global mobility network. In: Proceedings of International Ethical Hacking Conference 2019: eHaCON 2019, Kolkata, India, Springer Singapore, pp 171–190
19. Yu S, Lee J, Park Y, Park Y, Lee S, Chung B (2020) A secure and efficient three-factor authentication protocol in global mobility networks. Appl Sci 10(10):3565
20. Sohail MM, Hassan M, Mansoor K, Ghani A, Jawad K (2020) An improved authentication protocol for global mobility network (GLOMONET). In: 17th International Bhurban Conference on Applied Sciences and Technology (IBCAST), pp 401–406
21. Nikooghadam M, Amintoosi H, Kumari S (2020) A provably secure ECC-based roaming authentication scheme for global mobility networks. J Inf Secur Appl 54:102588
22. Rahmani AM, Mohammadi M, Lansky J, Mildeova S, Safkhani M, Kumari S, Karim SHT, Hosseinzadeh M (2021) AMAPG: advanced mobile authentication protocol for GLOMONET. IEEE Access 9:88256–88271
23. Kang D, Lee H, Lee Y, Won D (2021) Lightweight user authentication scheme for roaming service in GLOMONET with privacy preserving. PLoS ONE 16(2):e0247441
24. Palit SK, Chakraborty M, Chakraborty S (2022) MASKA: mutual authentication and session key agreement protocol in global mobility networks. In: Applications of machine intelligence in engineering. CRC Press, pp 309–321
25. Hojjati M, Shafieinejad A, Yanikomeroglu H (2020) A Blockchain-based authentication and key agreement (AKA) protocol for 5G networks. IEEE Access 8:216461–216476
26. Haddad Z, Fouda MM, Mahmoud M, Abdallah M (2020) Blockchain-based authentication for 5G networks. In: IEEE International Conference on Informatics, IoT, and Enabling Technologies (ICIoT), pp 189–194
27. Palit SK, Chakraborty M, Chakraborty S (2022) AUGChain: blockchain-based mobile user authentication scheme in global mobility network. J Supercomput 78(5):6788–6816
28. Palit SK, Chakraborty M, Chakraborty S (2023) Performance analysis of 5GMAKA: lightweight mutual authentication and key agreement scheme for 5G network. J Supercomput 79(4):3902–3935
29. Burrows M, Abadi M, Needham RM (1989) A logic of authentication. Proc R Soc Lond A Math Phys Sci 426:233–271
30. Kocher P, Jaffe J, Jun B (1999) Differential power analysis. In: Advances in Cryptology—CRYPTO'99: 19th Annual International Cryptology Conference Santa Barbara, California, USA, August 15–19, Proceedings. Springer Berlin Heidelberg, pp 388–397

31. Messerges TS, Dabbish EA, Sloan RH (2002) Examining smart-card security under the threat of power analysis attacks. IEEE Trans Comput 51(5):541–552
32. Banerjee S, Odelu V, Das AK, Chattopadhyay S, Kumar N, Park Y, Tanwar S (2018) Design of an anonymity-preserving group formation-based authentication protocol in global mobility networks. IEEE Access 6:20673–20693
33. Saha Misra I, Chakraborty M, Saha D, Mukherjee A, Deb B, Chatterjee B (2003) Global Mobility management: A three Level Architecture for next generation wireless networks. In: Proceedings Distributed Computing—IWDC 2003: 5th International Workshop, Kolkata, India, Dec 27–30, pp 184–193
34. Saha Misra I, Chakraborty M, Saha D, Mukherjee A, Deb B, Chatterjee B (2001) A three level hierarchical mobility management architecture for IPv4 networks. In: Proceedings of International Conference IST Mobile and Wireless Summit, pp 358–362

Permissioned Blockchain-Based Multifactor Authentication Scheme for Internet of Things Devices Using Physical Unclonable Functions

Abhilash Kayyidavazhiyil and Mohuya Chakraborty

Abstract This paper presents a novel multifactor mutual authentication scheme for authenticating Internet of Things (IoT) devices using the power of both the permissioned blockchain smart contract and Physical Unclonable Function (PUF). The authentication capability of an IoT device can be enhanced by combining a permissioned blockchain-based smart contract, a computer program that runs automatically as and when the terms of the contract or agreement are met, and PUF, which gives a response or a unique identifier for a given input and challenge. The work's novelty is in the method it recommends for building a peer-to-peer network that enables every user to unswervingly communicate with the data stored in the safe distributed ledger. As a result, a participant's need to submit relevant data is much diminished under this proposed approach. Security analysis of the proposed framework using Automated Validation of Internet Security Protocols and Applications (AVISPA) shows its potential to combat numerous attack scenarios like man-in-the-middle, spoofing, and replay attacks along with better authentication due to the uniqueness of the physical microstructure of PUF. Comparative analysis of the schemes using Elliptic Curve Cryptography (ECC) and without using ECC have been made, which shows the efficiency of the former regarding security.

Keywords IoT · Blockchain · Smart contract · AVISPA · PUF · Multifactor authentication · ECC

A. Kayyidavazhiyil (✉)
BITS Pilani, Dubai Campus, Dubai, UAE
e-mail: p20230909@dubai.bits-pilani.ac.in

M. Chakraborty
Cybersecurity Centre of Excellence, Institute of Engineering and Management, University of Engineering and Management, Kolkata, India

M. Chakraborty et al. (eds.), *Proceedings of 5th International Ethical Hacking Conference*, Lecture Notes in Networks and Systems 1148,
https://doi.org/10.1007/978-981-97-8457-8_14

1 Introduction

The Internet of Things (IoT) provides a way to link commonplace items that are utilized daily such as sports gadgets, coffee machines, mobile phones, watches, spectacles, toasters, coke vending machines, shirts, and other related devices to other objects and applications of the cloud, in such a way to make them intelligent (smart) and interactive. These "smart" devices are used to gather physical information from the environment to make our lives easier and healthier and help to optimize the use of threatened possessions. Looking at the evolution of IoT, we find that prior to the advent of IoT, we were exposed to human-to-human communication with the fixed telephony and SMS services, Internet of Content such as www, which is a huge repository of linked information across the globe, Internet of Services like e-commerce and Internet of People such as social media (Skype, Facebook, YouTube, etc.). We are now in the era of IoT also called Machine-to-Machine communication, where objects talk to each other without human intervention resulting in huge gathered data in the cloud. IoT devices consist of tiny sensors embedded in ordinary devices to gather information.

There are several security challenges and issues of IoT devices such as lack of authentication and encryption, insufficient testing and updating, brute forcing, and risk of default passwords, malware, ransomware, and IoT botnet attacks on cryptocurrency [1, 2]. Authenticating IoT devices is of prime importance as any weakness in this process may lead to data theft, financial loss, and exposure to sensitive personal information such as medical data. Researchers across the globe have worked in this area using various technological innovations, especially for multi-server environments using remote user authentication [3–13]. The paper proposes a revolutionary multifactor authentication method that combines two cutting-edge technologies—Blockchain and Physical Unclonable Functions (PUF). Blockchain smart contracts give us an edge over traditional contracts as these are computer programs that are run automatically on the blockchain when certain conditions are met. They are also immutable, transparent, and decentralized. Conversely, PUFs are a type of hardware security solution that leverages the inherent differences in a device to produce a unique, unclonable response to an input. A PUF can be compared to biometrics for people on a higher level because they are inborn, distinctive identifiers for every silicon component. Hence this two-step verification process also called multifactor authentication would enable us to extract the exact identity of the IoT device more safely rather than relying on usernames and passwords alone.

The key contributions of the paper are as follows.

1. A multifactor authentication mechanism that combines blockchain and PUF.
2. Informal security analysis presenting resistance to ~ ten attacks.
3. Formal security analysis using AVISPA.
4. Performance analysis of the proposed scheme with and without ECC.

The paper is structured as follows. Section 2 presents a summary of relevant studies after the introduction in Sect. 1. The foundation and first information about

PUF and blockchain smart contracts are provided in Sect. 3. Performance analysis and the suggested solution are presented in Sects. 4 and 5, respectively. Section 6 provides some highlights on future scopes as the paper comes to a close.

2 Related Work

The Internet of Things (IoT) has proved to be one of the most significant technologies in recent years due to its wide range of application areas. The variety of applications causes a significant number of users' private information to diffuse, which is extremely dangerous for their security. In the IoT environment, user authentication is important because it enables secure communication between the user and the device. An IoT device must both confirm its identification to other networked devices and the identities of all other networked devices. Secure data retrieval and effective access control should be ensured by the integration of authentication systems with IoT [14]. Literature survey shows that researchers across the globe have worked on various authentication schemes for global mobility networks [15] and IoT devices [16]. Some of these schemes are Cloud-Based IoT Authentication [17], Lightweight Authentication [8, 14, 16, 17], Decentralized Blockchain-Based Authentication [18, 19], Biometrics-Based Remote User Authentication Schemes [20–22], PUF-Based Authentication [23–25], etc. However, there are several security issues of IoT authentication schemes such as limited computation power and memory storage, energy requirement causing some IoT devices to switch to battery saver mode, expandability in the number of devices, mobility, heterogeneous device support, physical theft, security and privacy of stored IoT sensor data [26].

Considering Cloud-Based IoT Authentication, identity and access management (IAM) enables authentication methods for cloud infrastructure and makes them available. It is also possible to employ a shared access signature (SAS) key or single sign-on (SSO) certification function, which greatly decreases attack surfaces by not being exposed to the web and not relying on open ports. Even in cases when a device is compromised or a piece of connected hardware is tampered with, infrastructure can usually protect the stored data. The rules and procedures for customizing and setting up IAM/active dictionary (AD) vary across different cloud platforms, though. Developers responsible for creating and maintaining the platforms for their various organizations will need to thoroughly research and test these new developments. If the error is set on IAM, a role, or a policy, it may result in additional errors in other places [26]. The need for individual customization of the products makes shipment more difficult. A lightweight authentication scheme is therefore needed to combat the limitations of Cloud-based IoT authentication schemes. The rise of linked devices increases the need for more adaptable user authentication methods. Lightweight ciphers' performance benefits offer reduced block and key sizes as well as easier key scheduling. Because of the reduction in computational complexity relative to the more well-known heavy cryptographic techniques, hardware implementation of lightweight algorithms has a disadvantage in terms of security.

We then look toward decentralized blockchain-based IoT authentication methods as a result of this. IoT size and other features make it impossible to create a reliable centralized authentication solution. Integration of blockchains smart contracts simplifies the sharing of IoT services and resources as well as authorizes the automation cryptographically [27]. Biometric-based remote user authentication comprising User Registration Phase, Login Phase, Authentication Phase, and Password change phase, poses less computational cost but fails to provide mutual authentication, and it is vulnerable to server spoofing attacks as well [28]. Using PUF-based IoT authentication, an attacker cannot assume the uniqueness of a altered device whose key has been recovered by taking advantage of the device's exclusive arbitrary patterns for safe key creation and key agreement. The PUF key can only be generated when it is needed for a cryptographic operation and can be instantly erased later [29]. In this work, we have suggested a PUF-based multifactor authentication strategy that leverages the capabilities of blockchain smart contracts, taking into account the benefits and drawbacks of current solutions.

3 Groundworks

In this section we present the preliminaries and groundwork related to the present research work. Overview of blockchain and smart contract is presented in section A followed by that of PUF in section B.

3.1 Overview of Blockchain Smart Contract

Asset tracking and transaction recording inside a corporate network are made easier by blockchain, a decentralized, immutable database. An asset can be anything from a house, automobile, money, or technology to intellectual property like patents, copyrights, or brand names. Assets can even be virtual. The mysterious minds of bitcoin, known as Nakamoto et al., published a highly-publicized document in 2008 that contains the first known mention of blockchain technology (2008). On a blockchain network, almost anything valuable can be registered and exchanged, lowering risk and increasing efficiency and security for everybody involved. A smart contract is simply a digital contract that has blockchain security coding. It functions similar to a conventional contract while also being automatically enforced by computer programs that run exactly as they are configured (coded, programmed), and is enforced by code. It is basically a business logic-encoding program that runs on a dedicated virtual machine embedded in a blockchain or distributed ledger. The principle of operation of smart contact is as follows: (i) initially two users create the smart contract; (ii) all contract terms are written as a code or computer program' (iii) the smart contract is then stored in the Blockchain' (iv) smart contracts execute automatically when the event is triggered as per the written code. Smart contracts have several advantages over conventional contracts in terms of efficiency, speed, trust, security, transparency, and accuracy. Applications of smart contract include supply chain management, insurance claim, safeguard of medication efficacy, improvement of retailer-supplier relationship, record storing, enhancement of speed and efficiency of international trade and streamlining the way in which data is collected and processed in IoT devices etc.

3.2 Overview of Physical Unclonable Functions

A unique "fingerprint" or trust anchor is assigned to a physical entity via a physical unclonable function (PUF), a tool that leverages the inherent randomness established during creation. These devices can be used for several purposes, such as identity, authentication, anti-counterfeiting, and key generation. It is also feasible to implement sophisticated protocols like key renovation, key exchange, oblivious transfer, and virtual proof of existence.

PUFs rely on their physical microstructure's distinctiveness. Instead of encoding a single cryptographic key, PUFs use challenge-response authentication to evaluate this microstructure. The complicated interplay of the physical stimulus with the device's physical microstructure causes the structure to react to a physical stimulus in an unpredictable (but repeated) manner. This specific microstructure is dependent on uncontrollable physical elements that are introduced during manufacturing (like a fair coin). The administered stimulus serves as the challenge, and the PUF's response is referred to as the response. A challenge-response pair, or CRP, is made up of a particular challenge and its associated answer. The characteristics of the microstructure itself serve to identify the device. Such a gadget is immune to spoofing attack because the challenge-response process does not immediately reveal this structure. PUFs can be implemented with a lot less hardware than other cryptographic primitives, like pseudo-random functions, that give unpredictable input/output behavior. PUFs can occasionally even be created using pre-existing hardware that satisfies the necessary requirements. PUFs come in two varieties: strong and weak. The difference has to do with how many possible responses there are. A powerful PUF can produce many challenge-response pairings, but a weak PUF can only produce a few, frequently just one. It is difficult to put a strong PUF into practice. PUFs also struggle to control the noise when producing the answers. In the event that the underlying PUF—an Arbiter PUF—with a sizable challenge-response pair space is not present, the PUF-FSM design exclusively employs error-free responses that are meticulously determined on demand, eliminating the need for error correction logic and associated computation, storing the helper data, and loading it on-chip. It validates enhanced security, especially to reliability-based attacks and supports a wide range of applications from authentication to more advanced cryptographic applications built upon shared keys [30].

So, with this overview of blockchain smart contract and PUF, the proposed solution toward IoT device authentication integrating both the technologies have been discussed in Sect. 4.

4 Proposed Solution

In this section, we provide a multifactor authentication method that safeguards knowledge flow and integrity on Internet of Things networks. It is based on the combination of PUF and permissioned blockchain smart contracts. The integral part of the scheme is storing of k challenge-response pairs for each IoT device as discussed in sub-sect. 4.4. The traditional way of storing them is subject to theft thereby leading to authentication failure. In this paper we have proposed storing them in permissioned blockchain, which produces almost zero authentication failure.

4.1 Architecture

The network's participants fall into one of four categories. These are serving network (SN), blockchain network, home network (HN) and user equipment (UE). The UE is initially registered with the HN through Home Agent (HA). If it moves out of the scope of HN and enters an SN, it has to re-register itself with the SN with the permission of the HN through the Server Agent (SA). The SN may also be regarded as the foreign network and accordingly the SA may be treated as a foreign agent. There are four phases in this scheme. These are:

1. UE registration with HN
2. SN registration with HN
3. SN registration with the permissioned blockchain network to store challenge-response pairs generated by the PUF mechanism.
4. Multifactor authentication and session key agreement between UE and SN

4.2 UE Registration with HN

In the registration phase UE registers with HN. The transactions of this phase are UESecretGen(), UERegistrationRequest(), and HARegistrationReply(). The assets used are UESecret, UERegistrationData, UESent and HomeAgentSent. Two types of parameters are used here. They are assets and transactions. Assets are the records that are stored in the blockchain network whereas transactions are the processes that are called by the participants like HA, UE, and SA by inputting the essential data in the Hyperledger Fabric. The UE registration Phase comprises three transactions as stated in Algorithm 1 and pictorially depicted in Fig. 1.

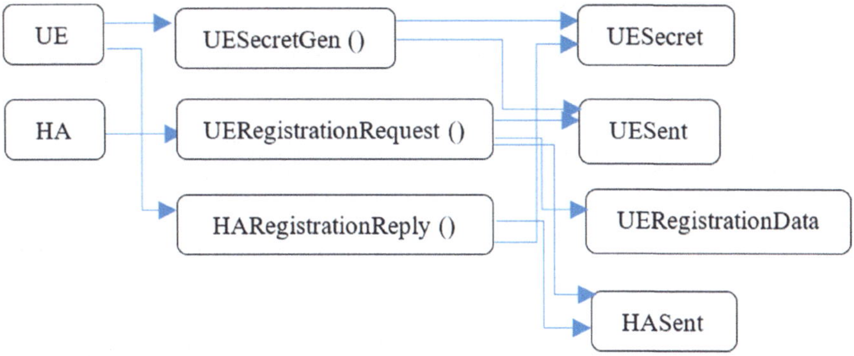

Fig. 1 UE registration phase

Transaction 1 UESecretGen():
UE inputs idUE, password pwUE, and identity idHA of the HA.
Creates a random number mRand.
aUE = H(idUE⊕pwUE) ⊕mRand and pidUE = H(idUE⊕mRand).
Generates a UESecret asset with the identity idUE.
Adds (idUE, pwUE, aUE, pidUE, mRand) to UESecret .
Adds (idUE, pidUE, idHA) to the newly created asset UESent with identity idUE.
Shares UESent with HA.

Transaction 2 UERegistrationRequest():
HA uses UESent to get (idUE, pidUE, idHA) and compares own identification with idHA.
If a match is found, inputs idUE and idHA.
Creates a random number bRand.
bUE = H(pidUE⊕bRand), cUE = H(pidUE||bRand), and dUE = H(bRand||pidUE).
Adds (pidUE,bRand,cUE, idHA) to UERegistrationData with identity pidUE.
Creates HomeAgentSent asset with the identity idUE.
idHA, bUE, cUE, dUE, bRand, and idUE are added to HomeAgentSent.
Shares the HomeAgentSent asset with UE.

Transaction 3 HARegistrationReply():
HA calculates
eUE = bUE H(pwUE), fUE = cUE H(pwUE), and gUE = dUE H (pwUE).
Adds (eUE,fUE,gUE,bRand) to the asset UESecret.
UESecret contains (idUE, pwUE, aUE, pidUE, mRand, eUE, fUE, gUE, bRand).

Algorithm 1 UE Registration with HN

4.3 SN Registration with HN

In this phase, SN registration with the HN of the UE takes place in order to avail roaming services when the UE goes out of the scope of HN. The transactions of this

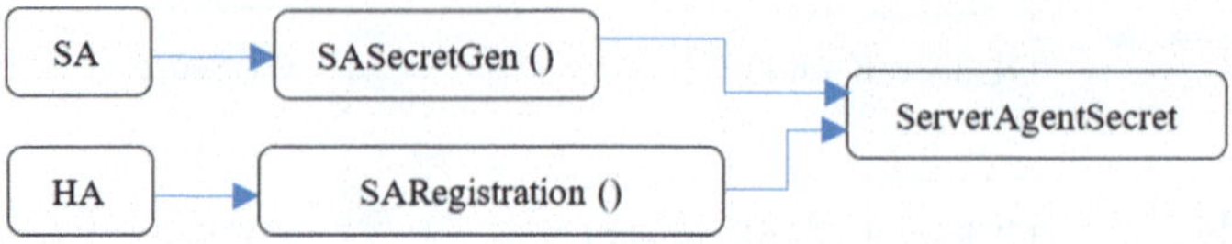

Fig. 2 SN registration phase

phase are SASecretGen() and SARegistration() and the asset is ServerAgentSecret. There are two transactions in this phase as given in Algorithm 2 and depicted in Fig. 2.

Transaction 1 SASecretGen():
SN Inputs idSA and idHA.
Creates a random number fRand.
Creates aSA as a ServerAgentSecret asset with the identity H(idSA||fRand) computed earlier.
The asset ServerAgentSecret is stored with (aSA, idSA, and idHA).
Shares the ServerAgentSecret asset with HA.

Transaction 2 SARegistration():
HA accesses the ServerAgentSecret asset (aSA, idSA, idHA), matches the returned idHA to its identity.
If they match,
Determines SA's secret key, skSA = H(aSA skHA).
Adds SA to the ServerAgentSecret asset and share it with SA.
The dataset is contained in the asset ServerAgentSecret (aSA, idSA, skSA, idHA).

Algorithm 2 SN Registration with HN

4.4 *SN Registration with Blockchain Network*

The SA issues the UE a series of k challenges. Each challenge is sent through the UE's PUF mechanism to receive the appropriate response from an IoT device. A secure channel is used to send SA the k returned responses. The entire process takes place by multifactor authentication and key agreement protocol described in sub-sect. 4.5. The secure channel in this scheme is the permissioned blockchain network using Hyperledger Fabric. The server is responsible for the secure storage of these k challenge–response pairs for each IoT device. The SN has to register itself with the secure Hyperledger Fabric blockchain peer-to-peer network with the message SNreg to securely store the k challenge-response pairs for each IoT device created above. The Hyperledger Fabric blockchain network acknowledges the registration with the message AckSN. It is fed to the smart contract of the permissioned blockchain network for validation. This is depicted in Fig. 3.

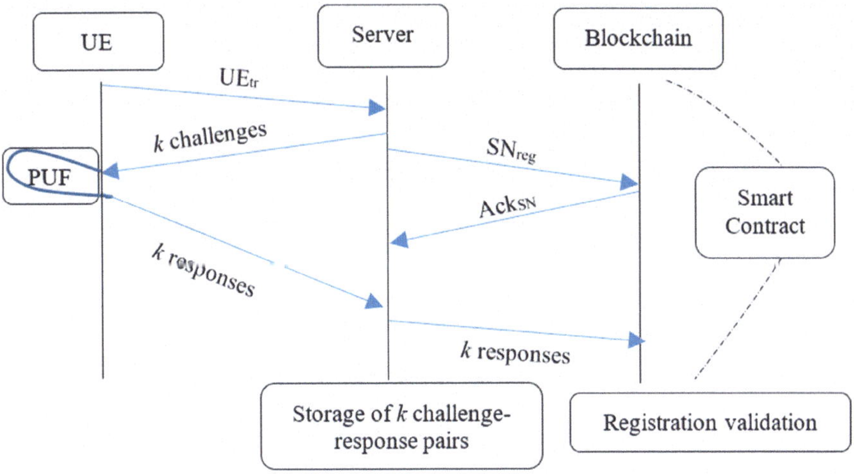

Fig. 3 Multifactor authentication and session key agreement phase

4.5 Multifactor Authentication and Session Key Agreement Between the UE and the SN Phase

In this phase UE requests for roaming service to SA. The transactions of this phase are UEAuthenticationSent(), SAAuthenticationSent(), HAAuthenticationReceived(), and SAAutheticationSentUE(). The assets used are UESentSA, ServerAgentSecret, SASentHA, and HAPermissionSA, Algorithm 3 explains the four transactions of this phase.

Transaction 1 UEAuthenticationSent():
UE sends a request to SN for transaction with the message UEtr.
From its own database of saved challenges, the server randomly chooses two C1, C2 challenges that are connected to UE. The value is then determined by hashing the corresponding responses R1=PUF(C1), R2=PUF(C2) with the timestamp *t*.

$$\delta = H(R1||R2||t)$$

Once more, *t* is made secret from prying eyes by being XORed with the sum of C1 and C2's answers to produce *t'*. After assessing the PUF of C1 and C2 upon getting this communication from the server, the UE derives *t* from *t'*.

$$t' = t \oplus (PUF(C1)||PUF(C2))$$

Computes idUE by hashing the responses of the challenges with the obtained *t*.

$$idUE = H(PUF(C1)||PUF(C2)||t)$$

The hash of the XOR of C1 and C2 is used to determine P, which is the next step.

$$P = H(C1 \oplus C2)$$

A randomly chosen value x multiplied with P results in PPUB, the public key of UE and the corresponding private key PPRV is given as the product of x and idUE as in Elliptic Curve Cryptography (ECC). ECC's core operation is based on scalar multiplication. Here x is an integer and P an PPRV are points on an elliptic curve [25, 31] as given below.

$$PPUB = x*P$$
$$PPRV = x*idUE$$

The UE should check whether δ is similar to idUE. Then, in order to ensure the integrity, an XOR hash function of the various acquired values is computed, yielding the value *d*.

$$d = H(PUF(C1) \oplus PUF(C2) \oplus idUE \oplus P \oplus PPUB \oplus t)$$

UE then creates an asset UESentSA with the message containing idUE, P, PPUB, *d* and adds it in the asset UESentSA. Shares the asset UESentSA with SA after that.

Transaction 2 SAAuthenticationSent():
Server verifies UE's identity by comparing idUE to the previously computed value and *d*'s accuracy.
SA uses this asset to retrieve (aSA, idFA, skFA, and idHA) from ServerAgentSecret.
SA inputs aSA. Creates random number nfRand.
bSA = H(nfRand||skSA), aSA1 = aSA ⊕ nfRand.
Produces SASentHA asset with identification bSA.
Adds datasest (bSA, aSA1, and nfRand) to the SASentHA asset.
Shares SASentHA with HA.

Transaction 3 HAAuthenticationReceived():
HA accesses the asset SASentHA and retrieves (bSA, aSA1, and nfRand)

Algorithm 3 Multifactor Authentication between UE and SN

Fig. 4 Blockchain smart contract logic/pseudo code

```
Smart Contract PUFBCAuth()
if (SA and UE authentication succeeds)
  create SkmfSA and SkmfUE
  initiate roaming
else
  terminate
```

Inputs bSA.
Access the ServerAgentSecret asset and retrieves (aSA, idSA, skSA, idHA).
Finds skSA from the asset ServerAgentSecret that corresponds to aSA by computing
aSA = aSA1 ⊕ nfRand.
Checks whether received bSA matches with H(nfRand||skSA).
If not, terminates the session.
else
HA confirms SA's identity.
A flag status of HAPermissionSA.status =}} TRUE" is put on an asset called HAPermissionSA
Gives SA access to the asset HAPermissionSA.

Transaction 4 SAAutheticationSentUE():
SA accesses the assets HAPermissionSA, UESentSA, SASentHA and
Retrieves all the datasets from these assets.
Checks for validity of HAPermissionSA.status =}} TRUE''
If not, terminates the session.
else
Sends this authentication data and public session key to UE.

Algorithm 3 Multifactor Authentication between UE and SN

Assuming SA and UE are sincere participants, they use Diffie-Hellman Key exchange algorithm to authenticate one another and generate a special session key. The Diffie-Hellman protocol's goal is to make it safe for two users to exchange a secret key that will later be used to encrypt messages and is restricted to the key exchange [31]. For future use, this session key is kept in the permissioned blockchain as two distinct assets with the names SkmfSA and SkmfUE for SA and UE, respectively. The result is further sent to the blockchain smart contact for validation and initiation of the roaming service state of the UE in the SN. Figure 4 gives the logic and pseudo code of the blockchain smart contract used for this purpose.

5 Performance Analysis

We go over the suggested protocol's informal security analysis in this section. The Automated Validation of Internet Security Protocols and Applications (AVISPA) formal security analysis comes next. Additionally, we conduct a comparative analysis of the suggested approach with and without ECC.

5.1 Informal Security Analysis

In this section, we discuss the informal security analysis of the proposed protocol.

1. Blockchain application Hyperledger Fabric ensures the immutable storage of all transactions and records in the distributed ledger, as well as the integrity of the messages transacted by all participants. As a result, it guards against compromised server attacks.
2. The protocol's integrity is maintained at all stages since the messages contain a hash function that includes the message or its by-products.
3. Since blockchain is distributed, Denial-of-Service (DoS) and Distributed DoS (DDoS) threats are also removed.
4. The use of a blockchain network and smart contracts assures that there is no single point of failure.
5. UE authentication contains hash values comprised of responses to challenges that can be created by UE having the PUF. Together with the PUF idea, the use of smart contracts of the blockchain ensures multifactor authentication.
6. Mutual authentication of UE with HA, UE with SA, and SA with HA occurs, resulting in flawless mutual authentication of all participants and zero compromise.
7. Resistance to spoofing, man-in-the-middle, and replay attacks is ensured since the sender's authenticity is confirmed by the receiver at each step of the protocol.
8. Spoofing attack: There are two sorts of spoofing attacks- IP spoofing and website spoofing. In IP spoofing, an attacker sends a communication from an incorrect "from" address to a target party. Website spoofing is accomplished by creating a fake website. IP spoofing or website spoofing are not allowed in the proposed system since members connect with the network through the Hyperledger blockchain portal. Before taking part, participants must register. Spoofing attacks do not affect the portal because only registered valid participants use it.
9. Man-in-the-middle-attack: All participants in the proposed scheme keep an individual secret key or values. Participants use these keys and values to validate the authenticity of the participants from whom they receive messages. As a result, because each participant can detect changes in the messages, man-in-the-middle attack is remediated in the proposed approach.
10. Replay attack: A replay attack is possible if the attacker replays a previous message to any participant and the victim is unaware that an earlier message has been repeated. In the proposed system, UE and SN keep a secret parameter h, which varies with each session. This parameter h is a secret value that UE keeps in the blockchain. If SN is real, then access to h is granted for that session. As a result, if the attacker repeats an old message, the value of h will be an old value of h that differs from the current value of h, and the condition CUE = CUE' h, at the SN site will fail. As a result, the suggested approach easily detects replay attacks.

5.2 Formal Security Analysis

The proposed protocol's formal security is validated using the widely used automated validation of Internet security protocols and applications (AVISPA) simulation tool [32]. This simulation tool is widely used in formal security validation to explain if authentication strategies are stable in the face of a man-in-the-middle attack and a replay attack. AVISPA is implemented using a role-based high-level protocol simulation language (HLPSL) [33]. The HLPSL2IF translator (IF) converts the HLPSL source code into an intermediate format. Lastly, the intermediate format (IF) is translated to the output format (of) by one of four backend mechanisms: the tree automata-based protocol analyzer, SAT-based model-checker (SATMC), CL-based attack search engine (CL-AtSe), or on-the-fly model-checker (OFMC) (TA4SP).

The results of the simulation of the proposed protocol using AVISPA are described below. CL-AtSe backends were employed to ensure the security of the proposed method. The security protocol animator SPAN [34] describes the scheme's outcome. Replay attacks are discovered using the CL-AtSe backend. The passive attacker is permitted to take part in select sessions with other authorized participants to accomplish this. The CL-AtSe backends also determine a scheme's vulnerability to man-in-the-middle attacks. To do this, we use the DolevYao (DY) intruder model, in which the attacker is given access to all crucial shared parameters of the suggested protocol before being given authorization to hold several sessions with other authorized participants. As a result, an attacker is aware of everyone's identities as well as the PUF, generator for elliptic curve point multiplication, and hash functions used in the simulation. The experimental finding shows that this simulation can track an attack if an attacker knows the secret key shared by any two participants. Fortunately, no assaults were discovered when the suggested method was simulated. The executive summary of simulations using SPAN 1.6 shows that the suggested system is safe for a limited number of sessions and meets the objectives set forth by both the OFMC and CL-AtSe backends, and with and without ECC, hence thwarting man-in-the-middle attacks and replay assaults as depicted in Fig. 5. Table 1 shows the simulation result parameters obtained.

Table 2 shows the performance analysis vis-à-vis design parameters such as number of rounds, total size in bits, overall feature vectors, total number of operations, and total execution time [35, 36]. Simulation results show that the number of rounds, size in bits, execution time have increased in the proposed model with ECC in comparison to that without. The particulars of overall feature vectors such as size in bits and abbreviation are enumerated in Table 3. Table 4 depicts computation time during execution.

Here, TH, TPUF, and TECC represent the execution time of the one-way hash function, PUF function, and elliptic curve multiplication respectively [37]. The use of ECC will enable us to adopt asymmetric key cryptography giving us two keys: private (PPRV) and public (PPUB), whereas the model without using ECC will use symmetric cryptography giving us a single key for both encryption and decryption. Symmetric key cryptography is easier to use and implement but the level of security

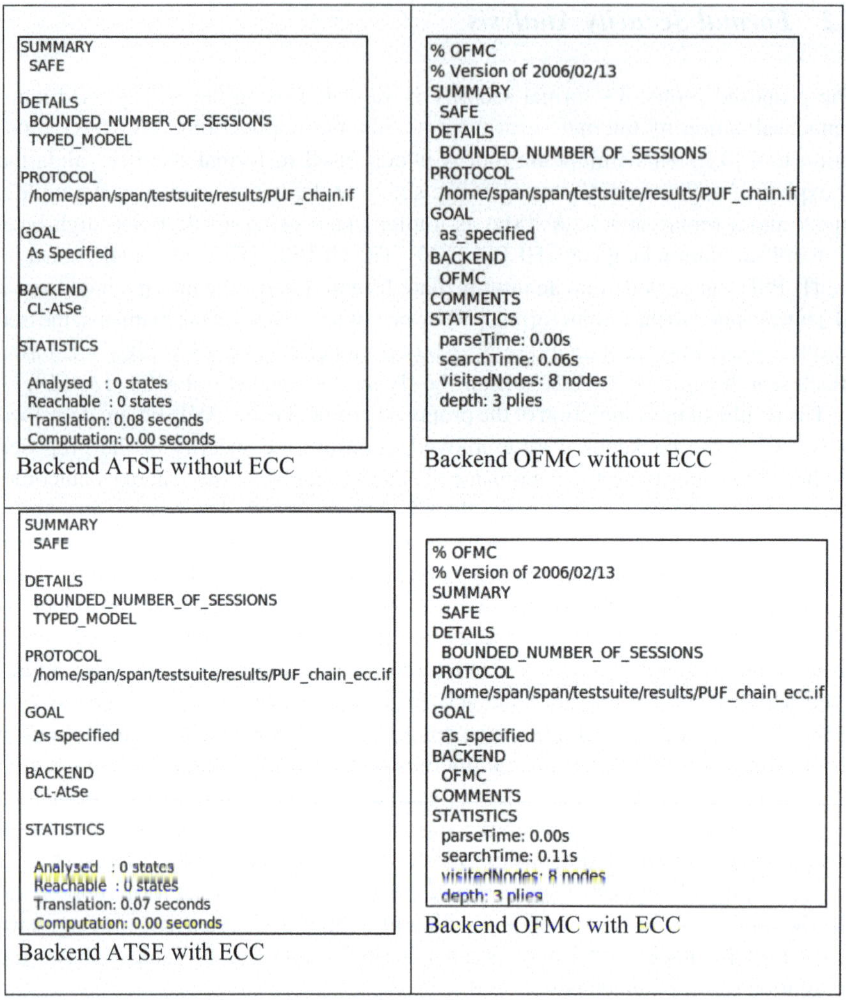

Fig. 5 Simulation result summary using SPAN 1.6

Table 1 SPAN 1.6 Simulation result parameters

Proposed protocol with backend OFMC			Proposed protocol with backend CL-AtSe		
	With ECC	Without ECC		With ECC	Without ECC
Status	Safe	Safe	Status	Safe	Safe
Parse time	0.00 s	0.00 s	States analyzed	0	0
Visited nodes	8	8	Translation time	0.07 s	0.08 s
Search time	0.06 s	0.06 s	Computation time	0.43 s	0.00 s
Depth	3 piles	3 piles	States reachable	0	0

Table 2 Performance Analysis vis-à-vis Design Parameters

Design parameters	Protocols	
	Proposed solution (with ECC)	Proposed solution (without ECC)
No. of rounds	6	6
Overall feature vectors	12H + 3Xrand + 1idX + 2XECC + 2Rx + 2Cx	12H + 3Xrand + 1idX + 2Cx + 2Rx
Total size (in bits)	4480	3840
Total No. of operations	20TH + 4TPUF + 4TECC	20TH + 4TPUF
Total execution time (microseconds)	13,694	504

Table 3 Overall feature vectors

Feature vectors	Abbreviation	Size (in bits)
An arbitrary number or a nonce	Xrand	128
Identity of an entity	idX	160
Password of an entity	pwX	256
The output of a one-way hash function using SHA-256	H	256
Timestamp	t	32
Challenge	Cx	64
Response	Rx	48
Output of an ECC point multiplication	XECC	320

Table 4 Computational execution time

Function	Execution time (in microseconds)
TH	25.1
TPUF	0.5
TECC	3297.5

is less as compared to that of asymmetric key cryptography. However, the implementation cost, execution time, and space complexity are more for asymmetric key cryptography. So, the protocol designer may make a trade-off based on his application and needs. The proposed model with ECC is expected to outclass other related protocols that have not used a combination of blockchain and PUF. This is because blockchain provides security and PUF provides uniqueness of the physical microstructure of PUF providing better authentication.

The proposed scheme has been implemented in the Hyperledger composer tool using the Hyperledger fabric framework used to design permissioned blockchain. The performance analysis of the Hyperledger composer in general shows that it is very much scalable under asset (records) read, and write operations. If the number of UE increases from 10 to 50, and then to 100, the asset creates time change and asset read timer change that is expected to take place would be nominal [18]. Hence, Hyperledger composer if used would show scalability under transaction execution time, throughput, and an average latency of the said model. So, it may be concluded that if the number of IoT devices increases, the system should give a stable performance.

5.3 Discussion

The proposed system faces a few challenges, including small-sized IoT devices to incorporate PUF causing area allocation problems, vulnerability to diverse dependability problems with occasional alteration of the PUF answers with time and under varied operating circumstances, and security issues in storing challenge-response pairs in a public permissionless blockchain. However, keeping in mind the strengths and opportunities of IoT including efficiency in connectivity, data collection, and automation, and reduction in cost with improved services, the current body of research shows widespread benefit to mankind.

6 Conclusion

This paper has depicted a novel multifactor authentication of IoT devices by the integration of two innovative technologies—PUF and Permissioned Blockchain smart contracts. The proposed model is more efficient than Public Key Infrastructure (PKI) based Authentication Mechanism in IoT. This is mainly because a Certificate Authority (CA) software runs on one computer, which requires backups, maintenance, etc. in PKI. If we integrate PKI with blockchain so that the CA runs on a group of connected computers and when the code is accessible to anyone on a blockchain instead, PKI will be even more robust and trustworthy as blockchain eliminates the need for a trusted third party. But even then, it will not outclass our proposed protocol with PUF. Since each node individually processes each transaction in blockchain, the number of nodes has a negligible impact on the supported rate of transactions per second. Performance and scalability issues of blockchain may be resolved by using side chains to offload the transaction processing from the main chain. Blockchain is slower than centralized system. However, they are resistant to malicious activities. Lastly, with the incorporation of PUF, the model becomes much more error free and secured. Performance analysis of the proposed solution shows that it has the capability to withstand the security issues arising in IoT authentication and outperforms the related solutions with respect to the uniqueness of the physical microstructure of PUF as well as immutable, decentralized, distributed, secure nature of permissioned blockchain used here by the integration of the two technologies. The issues discussed above with regard to small-sized IoT devices to accommodate PUF and security of storing challenge response pairs in public permissionless blockchain are open research areas that may be addressed in future.

References

1. Kavianpour S, Shanmugam B, Azam S, Zamani M, Narayana Samy G, De Boer F (2019) A systematic literature review of authentication in internet of things for heterogeneous devices. J Comput Netw Commun 2019:14. https://doi.org/10.1155/2019/5747136
2. Kayyidavazhiyil A (2023) Combined tri-classifiers for IoT botnet detection with tuned training weights. Int J Image Graph. https://doi.org/10.1142/S021946782550007X
3. Kayyidavazhiyil A, Kaipacheri S (2022) Trusted execution environments for internet of things devices: a recent study. Tech Innov Eng Res 6:79–95
4. Liao Y-P, Wang S-S (2009) A secure dynamic id based remote user authentication scheme for multi-server environment. Comput Stand Interfaces 31(1):24–29
5. Kayyidavazhiyil A, Jain S (2017) Authentication using a client device and a mobile device. US Patent 9:805
6. Sood SK, Sarje AK, Singh K (2011) A secure dynamic identity-based authentication protocol for multi-server architecture. J Netw Comput Appl 34(2):609–618
7. Li X, Xiong Ma YJ, Wang W (2012) An efficient and security dynamic identity-based authentication protocol for multi-server architecture using smart cards". J Netw Comput Appl 35(2):763–769
8. Leu J-S, Hsieh W-B (2014) Efficient and secure dynamic id-based remote user authentication scheme for distributed systems using smart cards. IET Inf Secur 8(2):104–113
9. Shunmuganathan S, Saravanan RD, Palanichamy Y (2015) Secure and efficient smart-card-based remote user authentication scheme for multiserver environment. Can J Electr Comput Eng 38(1):20–30
10. Zhu H (2015) Flexible and password-authenticated key agreement scheme based on chaotic maps for multiple servers to server architecture. Wireless Pers Commun 82(3):1697–1718
11. Li X, Niu J, Kumari S, Liao J, Liang W (2015) An enhancement of a smart card authentication scheme for multi-server architecture. Wireless Pers Commun 80(1):175–192
12. Irshad A, Ahmad HF, Alzahrani BA, Sher M, Chaudhry SA (2016) An efficient and anonymous chaotic map based authenticated key agreement for multi-server architecture. KSII Trans Internet Inf Syst 10(12):5572–5595
13. Maitra T, Islam SKH, Amin R, Giri D, Khan MK, Kumar N (2016) An enhanced multi-server authentication protocol using password and smart-card: cryptanalysis and design. Secur Commun Netw 9(17):4615–4638
14. Amin R, Kumar N, Biswas GP, Iqbal R, Chang V (2018) A light weight authentication protocol for IoT-enabled devices in distributed cloud computing environment. Futur Gener Comput Syst 78:1005–1019
15. Palit SK, Chakraborty M, Chakraborty S (2022) MASKA: Mutual authentication and session key agreement protocol in global mobility networks. In: Applications of Machine Intelligence in Engineering. https://doi.org/10.1201/9781003269793-35
16. Feng W, Qin Y, Zhao S, Feng D (2018) AAoT: lightweight attestation and authentication of low-resource things in IoT and CPS. Comput Netw 134:167–182
17. Zhou L, Li X, Yeh K-H, Su C, Chiu W (2019) Lightweight IoT-based authentication scheme in cloud computing circumstance. Futur Gener Comput Syst 91:244–251
18. Palit SK, Chakraborty M, Chakraborty S (2022) AUGChain: blockchain-based mobile user authentication scheme in global mobility network. J Supercomput 78(3):1–29. https://doi.org/10.1007/s11227-021-04139-y
19. Hammi MT, Hammi B, Bellot P, Serhrouchni A (2018) Bubbles of trust: a decentralized blockchain-based authentication system for IoT. Comput Secur 78:126–142
20. Li C-T, Hwang M-S (2010) An efficient biometrics-based remote user authentication scheme using smart cards. J Netw Comput Appl 33(1):1–5. https://doi.org/10.1016/j.jnca.2009.08.001
21. Kaur Dhillon P, Kalra S (2017) A lightweight biometrics-based remote user authentication scheme for IoT services. J Inf Secur Appl 34(2):255–270. https://doi.org/10.1016/j.jisa.2017.01.003

22. Khan MK, Zhang J (2007) Improving the security of 'a flexible biometrics remote user authentication scheme. Comput Stand Interfaces 29(1):82–85
23. Aman MN, Chua KC, Sikdar B (2017) Mutual authentication in IoT systems using physical unclonable functions. IEEE Internet Things J 4(5):1327–1340. https://doi.org/10.1109/JIOT.2017.2703088
24. Kong J, Koushanfar F, Pendyala PK, Sadeghi A-R, Wachsmann C (2014) PUFatt: embedded platform attestation based on novel processor-based PUFs. In: Proceedings of the 51st ACM/EDAC/IEEE Design Automation Conference (DAC), San Francisco
25. Braeken A (2018) PUF based authentication protocol for IoT. Symmetry 10:352. https://doi.org/10.3390/sym10080352
26. https://www.trendmicro.com/vinfo/hk-en/security/news/internet-of-things/cloud-based-iot-solutions-responding-to-traditional-limits-and-security-concerns
27. Christidis K, Devetsikiotis M (2016) Blockchains and smart contracts for the internet of things. IEEE Access 4:2292–2303
28. The pros and cons of biometric authentication | TechTarget
29. Ruhrmair U, Holcomb DE (2014) Pufs at a glance. In: Proceedings of the Conference on Design, Automation & Test in Europe (DATE '14), Dresden, Germany
30. Gao Y, Ma H, Al-Sarawi SF, Abbott D, Ranasinghe DC (2018) PUF-FSM: a controlled strong PUF. IEEE Trans Comput Aided Des Integr Circuits Syst 37(5):1104–1108. https://doi.org/10.1109/TCAD.2017.2740297
31. Li N (2010) Research on Diffie–Hellman key exchange protocol. In: 2nd International Conference on Computer Engineering and Technology, pp V4-634–V4-637. https://doi.org/10.1109/ICCET.2010.5485276
32. AVISPA: automated validation of internet security protocols and applications (ercim.eu)
33. HLPSL-Tutorial.pdf (unimi.it)
34. SPAN - Security protocol animator for AVISPA (irisa.fr)
35. Sadhukhan D, Ray S, Biswas GP, Khan MK, Dasgupta M (2020) A lightweight remote user authentication scheme for IoT communication using elliptic curve cryptography. J Supercomput. https://doi.org/10.1007/s11227-020-03318-7
36. Majumder S, Ray S, Sadhukhan D et al (2021) ECC-CoAP: elliptic curve cryptography based constraint application protocol for internet of things. Wireless Pers Commun 116:1867–1896. https://doi.org/10.1007/s11277-020-07769-2
37. Liu F, Yan Y, Sun Y, Liu J, Li D, Guan Z (2022) Extremely lightweight PUF based batch authentication protocol for end-edge-cloud hierarchical smart grid. Secur Commun Netw 2022:1–14

AI-Driven Security in Banking: Boon or Bane

Manoggnya Reddy Kanjula and Jammula Sravya

Abstract "The science and engineering of making intelligent machines, especially intelligent computer programs, related to the similar task of using computers to understand human intelligence, but AI does not have to confine itself to methods that are biologically observable" (McCarthy in John McCarthy's Home Page. Stanford University, 2024)—definition of Artificial Intelligence as provided John McCarthy who is an important proponent to the field of artificial intelligence. AI technology was always meant to aid humans in achieving their goals more efficiently; has evolved and advanced in a heartbeat and now is facing a threat to job security in the modern world. Currently, AI holds an influence in not one field of an economy like Transportation, Communications, Service industry, Healthcare, Banking, and immeasurably more. This paper aims to cover the effect of AI in banking, the Ethical implications, and its role in preventing fraudulent acts in the financial field of banking. The impact of use of AI in banking industry. Including the challenges that AI has caused and the threats it poses to the financial sector. In this article, we explore the question around the impact of AI on banking operations focusing on core stars of credit risk analysis and assessment. This question also explores the possibility of AI in manipulating the data collected that can eat better credit risk predictions. Fraud prevention and risk management are considered paramount among the various important pillars of the modern financial panorama. AI is highly adaptable, it proves to be a formidable strength in detecting anomalies, and potential fraud transpiring before it takes place, ultimately fulfilling the gaping hole of inability left open by the traditional fraud detecting systems. DigitalOcean (Understanding AI fraud detection and prevention strategies, 2023) Fraud tactics evolve with time and technology, which can be predicted and ensured by the Self-learning and evolving system of AI. But, the organizations that are securing the services of AI should ensure the potential harm that such indulgence of AI may lead to. Hence, AI possesses the ability to act as a show of dominance and also as a potential hazard, the only way to employ it as a strength is by making a proper application of the system and the purpose behind the same.

M. R. Kanjula (✉) · J. Sravya
ICFAI Law School Hyderabad, Hyderabad, India
e-mail: kmanoggnyareddy@gmail.com

M. Chakraborty et al. (eds.), *Proceedings of 5th International Ethical Hacking Conference*, Lecture Notes in Networks and Systems 1148,
https://doi.org/10.1007/978-981-97-8457-8_15

Keywords Artificial intelligence · Banking · Fraud prevention · Risks

1 Ethical concerns

2 Introduction

Artificial Intelligence (AI) has revolutionized the field of data science, leaving its mark on numerous industries such as medical diagnostics, fraud detection, education access, and of course, and banking. In the banking sector specifically, AI has contributed to risk management and even stimulated consumption to some extent, similar to its impact on productivity. However, the implementation of AI in core banking functions in India has been limited due to regulatory concerns. AI has been commonly used in Indian Banks, mostly for customer service, as Chatbots. (Emerj) The most recent advancement in such use includes, AI technology for facial recognition in real time for the purpose of capturing customer feedback. (Emerj).

Other minimal uses include:

- The usual Fraud detection along with Credit risk management. (Emerj)
- Automated back-office workflows and minimized turnaround time. (Emerj)
- Personalized wealth management and investment advisory services. (upGrad)

The RBI, otherwise known as the Reserve Bank of India, has not officially considered the inclusion of AI in the banking sector. Despite this AI is widely used in customer service and the management of the same within the banking industry. The primary rationale for this hesitation in India is the perceived threats posed by AI, which seem to outweigh the said benefits. To fully exploit the benefits of AI, it is impertinent for policymakers and regulators to collaborate with banking institutions to develop a proper and robust framework for regulating the concerns while developing and fostering digital innovation.

Striking the balance between innovation and risk mitigation itself poses challenges as sometimes prioritizing one over the other seems necessary. The introduction of AI or technology in the Banking Sector has already taken place through the rise of automatic teller machines and online banking. The core functions and the Decision-making process within financial institutions have not been affected or underground changes due to AI given the emergence of this particular information technology the question of impact on Bank operations remains of particular importance due to the increasing complaints potential of AI and optimizing the operations and enhancing the efficiency [1]. However, the growing influence of AI is very much undeniable, warranting careful consideration by policymakers and regulators. The development of Cyber-attacks into more sophisticated and ubiquitous is notable. The adaptation of AI by cybercriminals for the purpose of evasion has noticed considerably great consequences. Though as of 2021, Research in the domain of cyber security has not reached the level of sophistication of these types of AI-powered cyber-attacks [2].

These kinds of Cyber-attack techniques pose a severe danger to Enterprises and Organization Infrastructures with the power to wipe away data and create significant damage. An Avant-Garde generation of Cybercriminals one that is settled and secretive has emerged and this new generation of threats is said to be smarter and independent with the inclusion of the assistance of AI.

One of the recommendations that can be applicable to such threats and potential dangers include, The Indian government will need to adopt policies and regulations accordingly for the purpose of facilitating the significant digital transition. And this development of regulatory infrastructure mandates the inclusion of Technology experts along with the policy maker for better understanding, managing, and mitigating the risks of implementing AI into the financial sector.

As a public interest statement, AI is essential for the development of future prospects for banks. It is proven as a powerful tool for analyzing the creditworthiness of credit applicants in many other countries that have implemented the use of AI and the financial industry [3]. It has also proven to contribute to the establishment of economic and social inclusion in the banking Industry.

This research paper uses a qualitative methodology, preliminary through reviews of existing academic research, legal case laws, and documents. The scope of the paper is to focus on the impact of the use of AI in banking industry. The paper analyzes the challenges that AI has caused and the threats it poses to the financial sector. Finally, in conclusion, the paper provides for the approach to be taken by professionals, regulators, and policymakers about the way to tackle the situation and adapt to the present increasing technology innovations.

3 Use of AI in the Financial Sector

AI is an encompassing term that covers various approaches and strategies for replicating refined functions like language and decision-making [4]. Within the field of AI, machine learning and deep learning are subfields that focus on identifying patterns and neural networks within data in order to facilitate complex information processing and sound decision-making. It has been observed that banking organizations can create customized services for their clients by using AI-powered systems to learn about their preferences [5]. The financial sector is becoming increasingly competitive, and banks are expected to adapt by implementing innovative security measures suitable for the emerging dangers.

For instance, Biometric authentication—like fingerprints, facial recognition, and voice recognition for the purpose of verification of identification of customers instead of sole dependence on passwords (Farhad Ahamed).

Block chain technology for the purpose of securing financial transactions and records, as its nature can enhance data integrity and security (Mhlanga).

But when it comes to implementing these new technologies incumbent banks find it inherently disadvantaged as compared to the more established financial institutions, which may create additional security risks due to the incompatible security software systems with innovative technology solutions [6]. These disadvantages include a lack of interoperability and integration of multiple systems and their data's (Bhargav). Facing challenges in relation with failed connectivity, lack of consistent and accurate data is also a part of the problem (Bank of Baroda).

Though AI provides many benefits like improved risk management and profit margins constant regulatory changes and perspectives of critiques have been playing their role as hurdles.

This cautiousness in regards to the implementation of AI beyond automated minimal tasks like generating predictions is justified.

Comparing developed countries' AI technology implementations in the USA and Europe offers insights into benefits and threats, enabling better understanding and risk mitigation.

Hence AI has been used in other countries, in the manner stated below:

- One of the most prominent applications of AI in US banking is undeniably that of customer service. Both the Indian and US banking companies use AI power chart boats and virtual assistants to provide 24/7 customer support and interaction.
- They use natural language processing and automated machine learning to understand the general questions that are enquired by customers by accessing relevant data provided priory to provide accurate and kind responses.
- Moreover, AI is used in the detection of fraud and risk management. By analyzing the huge amount of available transaction data mated algorithms tend to study and identify patterns that are suspicious to detect fraudulent activity in real time. This approach proactive in nature helps in mitigating financial losses and protects the assets of the customer.
- Additionally, driven credit scores are becoming more prevalent using sources of data and machine learning for accurate creditworthiness.

The scenario is almost similar in the European banking market where AI is widely adopted to enhance various aspects of their financial operation such as fraud detection and anti-money laundering, personalized customer service, credit risk assessment, process automation, and Personalized financial advice [7]. Lewis [8] they have successfully managed to use generative AI to make traditional technology more accessible and understandable for easier use and supervisory Technology [9].

European banks are seen to be collaborating on fraud-fighting tools such as fraud patterns and anomaly detection [10].

Despite the numerous benefits, the adoption of AI in the financial sector of the USA presents grave concerns about data privacy, algorithmic bias, and job displacement. There is a need for robust governance along with ethical guidelines.

Hence from the information mentioned above answers our Research question: Is the banking sector making profit from the use of AI? We may without a doubt draw the conclusion that, like other businesses, the financial services have benefited somewhat from the deployment of AI technology. Similar to many other technologies AI also provides us with both potential benefits and consequences. But as we stated at the beginning of the article the main issue here is that the consequences outweigh over the benefits of the use of AI in the banking industry as the consequences are not limited to micro level but on macro level.

4 AI and Threat of Cybercrimes

AI may offer powerful technology for preventing cybercrime threats, such as artificial neural networks which can be used to detect any kind of irregularities and also suggest methods of counting a measure. Thus, AI has been widely employed by businesses to improve traditional technique analysis and create countermeasures against cyber-attacks. We may divide the application of AI in the banking industry into three primary categories: detection, response, and protection.

Losses suffered due to these cybercrimes are classified as two separate. Namely, direct and indirect losses. Direct refers to the actual theft of money and data, whereas indirect refers to customer relations and increased customer frustration. Considering the economic losses due to this committing of cybercrimes are estimated at a few billion, it becomes important and favorable to investigate the disruptions caused by AI in the banking industry.

AI-based fraud is a growing concern in the financial sector and more particularly in the online banking sector. Cyber fraudsters have adopted advanced technologies and sophisticated tools for scams and committing fraud. Some of the AI-based fraud committed are seen in the following forms.

(i). Synthetic Identity Fraud

A combination of real and fabricated information generated through the use of AI is known as synthetic identity. These identities are to open fraudulent customer accounts and transact with those particular accounts and can generate realistic personal details which make it difficult to detect through traditional fraud detection.

It becomes an increasingly difficult proposition to recognize these synthetic identities as they are awesomely created using Data that is stolen and legitimate while the false information usually includes names, addresses, or other personal information. Using AI algorithms fraudsters have become capable of creating highly believable synthetic identities which pose difficult to recognize as forgery through traditional verification checks. The main purposes these identities are used for include opening bank accounts, applying for credit cards, and making fraudulent transactions or purchases posing to the prevention of money laundering and causing significant financial losses to the original owner of the legitimate information [11].

In the case of Suffolk County District Attorney's Office in New York, 13 individuals were held guilty for being a part of Synthetic identity scheme (Gates). Fake synthetic identities were created by stealing Social Security numbers of children, migrants, elderly, and including deceased people who were not having any credit history (Gates). By using these, many shell companies were created to report these fake identities as good customers to increase their credit history boosting their credit scores. They then used these to obtain loans and credit cards from several financial institutions which were never repaid causing huge monetary loss (Gates).

(ii). Deep Fakes

AI can create deep fakes, such as fake videos or audio recordings. In the financial sector, cybercriminals can use these fake videos or audio to Imitate executives or high-level managerial personnel to authorize any kind of unauthorized or illegal transactions, especially in online banking. This also includes manipulating stock prices and the financial market. These deep fakes use both machine learning techniques and have a high potential to defraud. This mechanism includes identity theft and manipulation.

A finance worker, in February 2024, at an MNC in Hong Kong was deceived into paying $25 million to scammers who used deep fake to pose as a company's CFO in a video conference call (Heather Chen).

(iii). Automated Hacking

Automated hacking leverages AI and machine learning to carry out brute force attacks through which the ultimate purpose is to try every possible combination to outrun and crack the password of the security system. Not only it is used to crack but also to automate the process of Malware these AI algorithms are also used to carry out sensitive data breaches or to discover loopholes in the banks' security systems.

(iv). Social Engineering

Artificial Intelligence can be used to analyze a person's social media footprints and other internet activities which are then used for the purpose of creating personalized phishing emails. This manipulation, influence, or deceive users often results in identity theft. For example: $100 million spear phishing scam that targeted Facebook and Google by impersonating a vendor (Check Point).

These are only a few of the illustrations of the potential threats posed to institutions in the financial sector through the misuse of artificial intelligence that needs serious retrospection to counter AI-based Banking frauds in India. Though the digital revolution is prompting banks and Financial Institutions to stay updated on these technologies it is important to invest in advanced security measures to aid in countering threats. The main AI threats that the banking industry may face in the near future are: Data security and privacy risks, Algorithmic bias and discrimination, Job displacement, and lack of transparency and explainability. By proactively addressing these threats banks can find ways to mitigate and protect themselves while protecting their customers too.

5 RBI and Regulatory Compliance

The Central Bank of India has played a significantly crucial role in the purpose of making and implementing regulations and compliance with the financial sector. In recent years, RBI has introduced several measures to minimize regulatory oversight and promotion of compliance with the guidelines. RBI has notably focused on the improvement of regulations in regard to cyber security for banks and other financial entities. It has issued guidelines to mitigate and robust data security standards across the sector.

RBI's mandates include implementing prevention of money laundering regulations and terrorist financing to prevent the leaking of funds from the financial system. These include monitoring transactions, increased transparency of the transaction, conducting due diligence on customers by way of KYC norms, reporting suspicious activities, and collaborating with Global agencies.

In short, RBI's implementation strategy of AI in the banking sector is being dealt with careful consideration of both benefits and consequences. This is being done by engaging with various stakeholders to formulate appropriate regulations that aim to leverage the benefits of AI while trying to deal with its associated risks for a trustworthy banking ecosystem.

6 Conclusion

There cannot be any preference given to one stakeholder at the expense of others in the regulation of AI. AI regulation has to be applied consistently and equally to all Financial Institutions that want to implement the usage of AI for the providing of financial Services or regulatory complaints or for monitoring. Legal compliance loses its meaning when the compliance transparency and reduces accountability along with endangering complacency.

References

1. Arrieta AB et al (2019) Explainable artificial intelligence (XAI): concepts, taxonomies, opportunities and challenges toward responsible AI. Inf Fusion 2019:82–115
2. Goncalves AJ (2023) Artificial intelligence—based cyber security in the context of industry 4.0—a survey. Electronics
3. Genovesi S et al (2023) Standardizing fairness-evaluation procedures: interdisciplinary insights on machine learning algorithms in creditworthiness assessments for small personal loans. AI Ethics 4:537–553
4. Lundberg SM, Erion G, Chen H, DeGrave A, Prutkin J, Nair B, Katz R, Himmelfarb J, Bansal N, Lee S-I (2020) From local explanations to global understanding with explainable AI for trees. Nat Mach Intell 2:56–67
5. Muizu WOZ, Diana S (2019) Improving employee performance through organizational culture, leadership, and work motivation: survey on banking organizations in Southeast Sulawesi. Jurnal Bisnis dan Manajemen
6. Erik D, Kristina B (2023) Beyond the safeguards: exploring the security risks of ChatGPT. ArXiv: arXiv:abs/2305.08005
7. Yu X, Guanjun L, Chungang Y, Changjun J, Mengchu Z (2023) Time-aware attention-based gated network for credit card fraud detection by extracting transactional behaviors. IEEE Trans Comput Social Syst 10:1004–1016
8. Lewis M, Yinhan L, Naman G, Marjan G, Abdel-Rahman M, Omer L, Veselin S, Luke Z (2019) BART: denoising sequence-to-sequence pre-training for natural language generation, translation, and comprehension. In: Annual Meeting of the Association for Computational Linguistics

9. Hacker P, Engel A, Mauer M (2023) Regulating ChatGPT and other large generative AI models. In: Proceedings of 2023 ACM Conference on Fairness, Accountability, and Transparency
10. Venkata RS, Swetha B, Bhagawan G, Logeshwaran J (2023) Fighting insurance fraud with hybrid AI/ML models: discuss the potential for combining approaches for improved insurance fraud detection. In: 2024 4th International Conference on Communication, Computing and Industry 6.0 (C216)
11. Atif U, Nasir N, Saima M (2023) Intelligent anti-money laundering fraud control using graph-based machine learning model for the financial domain. J Cases Inf Technol 25:1–20
12. DigitalOcean (2023) Understanding AI fraud detection and prevention strategies. Web Article. June 2024
13. McCarthy J (2024) John McCarthy's Home Page. In: 12 November 2007. Stanford University. June 2024

Criminal Liability of Artificial Intelligence-Generated Deepfakes in India

Shambhu Prasad Chakrabarty and Shrabana Chattopadhyay

Abstract The ongoing advancement of technology has initiated a race to use artificial intelligence in every domain of knowledge, including our personal spaces. Artificial Intelligence (AI) is a generic term used for many applications in different levels of intervention on humanity mainly referred to by its acronym "AI". One of the applications of AI has mastered to personalise us in the digital world. Deepfakes refer to one of these manifestations to deceive humans with the aid of AI, wherein machine learning algorithms, specifically generative adversarial networks (GANs), are employed to produce synthetic media that bears resemblance to human beings, albeit with some elements of the content being altered. In contemporary times, the task of differentiating authentic information generated by human agents from that produced by Deepfakes algorithms has become increasingly arduous. Deepfakes are socially disturbing and so manipulative to the human brain because of their ability to create a public deception with regrettable results. Hence, both society and nations must possess systems capable of autonomously detecting and assessing material without human intervention. Regrettably, AI has yet to be acknowledged as an independent legal discipline. In this scenario, can per se AI be held responsible for its act especially when AI is not recognised as a legal entity? The objective of this study is to ascertain the regulatory framework governing AI as perpetrators of Deepfakes crimes and to establish the criminal liability of AI entities involved in criminal activities within India. Furthermore, it elucidates the potential significance of Deepfakes in the realm of criminal justice administration. The paper will additionally examine whether the gaps in the current legislation to address these concerns. The paper will culminate in providing recommendations for criminal justice administration of AI in India.

Keywords Deep learning · Machine learning · Artificial intelligence · Deepfakes · Crimes · Liabilities and Responsibilities

---

S. P. Chakrabarty · S. Chattopadhyay (✉)
IEM's International Institute of Juridical Sciences (IIIJS) (Department of Law), University of Engineering and Management, Kolkata, India
e-mail: shrabana.uem@gmail.com

M. Chakraborty et al. (eds.), *Proceedings of 5th International Ethical Hacking Conference*, Lecture Notes in Networks and Systems 1148,
https://doi.org/10.1007/978-981-97-8457-8_16

1 Introduction

The emergence of AI has given rise to a sophisticated algorithm known as deepfake technology. Deepfake refers to an algorithm that enables users to replace any other features in photos or videos with those of others. Deepfake refers to an artificial intelligence-driven technique that uses machine learning algorithms, specifically Generative Adversarial Networks (GANs), to produce mock media including photos, videos, and sounds. The objective of deepfake technology is to produce exceptionally authentic mocked media that closely resemble actual individuals, although with certain elements of the content altered. The term "deepfake" was originally introduced in late 2017 when an unidentified user on the social media platform Reddit shared movies in which the faces of pornographic actresses were digitally replaced with the faces of celebrities [1]. The Internet rapidly propagated the promise of this technology, making it widely available to everybody [2].

Deepfake technology is founded upon two methodologies, specifically, deep learning and GANs [3]. Deep learning is a kind of machine learning that uses algorithms based on the anatomy and functioning of the brain, referred to as artificially generated neural networks, to handle and examine vast quantities of data. Deep learning has been utilised in several domains like computer vision, natural language processing, speech recognition, and robotics. GANs are a specific class of deep-learning frameworks that employ two neural networks, namely a generator and a discriminator, to undergo training on a given dataset. The ultimate objective is to produce novel mocked data that closely reflects the characteristics of the original dataset. The generator produces counterfeit samples, while the discriminator evaluates the genuineness of both the created sample and the real sample from the trained dataset. The two networks have been taught in an adversarial fashion, with the generator attempting to produce samples that deceive the system of discrimination, while the discriminator aims to accurately differentiate between the created samples and the authentic ones. This procedure persists till the generator is capable of generating exceedingly authentic mocked data [4]. The whole process is summarised hereunder:

The Fig. 1 shows how deepfake works in creating a fake image of the concerned person and the Fig. 2 presents the outcome of such deepfake process [4].

2 Research Methodology

This research study analyses the available materials on the criminal consequences of deepfakes generated by AI in India. The objective is to detect developing patterns and assess research methodologies, to offer valuable insights for the development of future studies. The text engages in a thorough examination of many secondary sources. The necessary information for this research was obtained primarily from online publications and articles. This study encompassed various online English

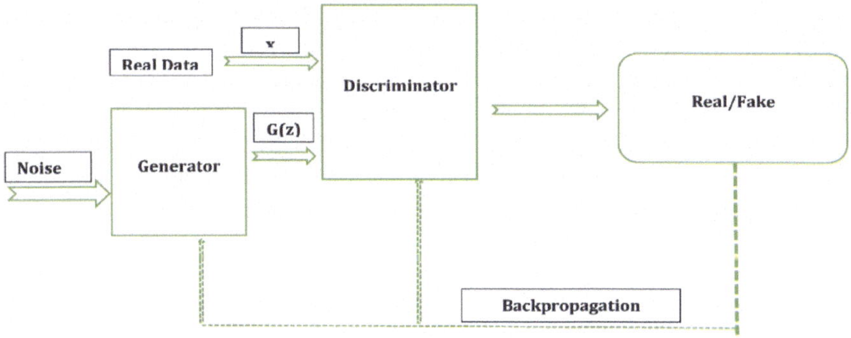

Fig. 1 Process of Deepfake technology

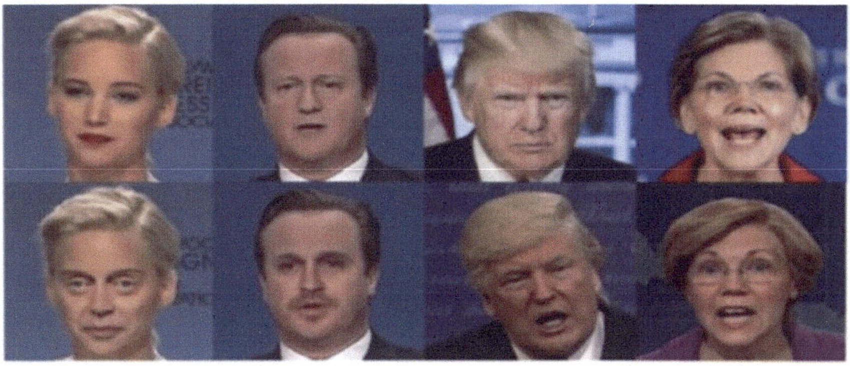

Fig. 2 Example of Deepfake image [5]

newspapers such as The Economic Times, The Wire, The Times of India, and Times Now, have been referred to assess the subject areas.

3 Subject Matter Under the Criminal Justice System of India

In India, the legal subjects who can perform legal acts are limited to individuals and legal entities. Every individual is subject to the law and can engage in legal actions or establish legal relationships, which must be adhered to within the bounds of legal capability and legal authority. The current application of Criminal law maintains the belief that humans are primarily responsible for doing criminal acts [6]. The Criminal law also governs criminal activities performed by individuals who share duties with others [6].

The rationale behind these restrictions exclusively targeting human beings is rooted in the notion that only humans possess a cognitive state, possess the capacity to be accountable for their conduct, and are susceptible to the impact of criminal regulations. Primarily, awareness of situations and the ability to possess intentions are exclusive to humans. Consequently, only humans have the capability to act deliberately, carelessly, or negligently, thereby committing crimes. Furthermore, the retribution for the perpetrator of the offence necessitates that the individual be held legally accountable. Only those who possess the cognitive capacity required for rational decision-making are capable of being held legally accountable for their actions. Furthermore, the application of criminal law assumes that the intending party is capable of being influenced by legal directives and the accompanying penalties, as criminal law aims to deter undesired behaviour. Given that non-human creatures cannot understand the relevance of norms, sanctions, and the societal implications of undesired actions, their creative behaviour and conflicts cannot be regulated by criminal rules.

An individual is considered subject to the law only when they may be held accountable for their illegal actions under the existing legal framework. Based on the evidence presented during the trial, the defendant, who has been accused by the public prosecutor of committing a criminal offence, must be proven to be physically and mentally sound, capable of comprehending and responding to all questions posed to them, to be held responsible for their actions. Corporate entities can also play a role in this scenario. For example, according to section 34 of the Companies Act, 1956, a "company" refers to a body corporate, which encompasses both legal entities such as firms and other associations of individuals, as well as non-legal entities. In other words, the firm has acquired legal status and can consequently be held liable for criminal offences [7].

The identification of the corporation as an entity and its accountability for its drive must prioritise the advancement of the corporation itself. Corporations can only carry out actions with the involvement of administrators. Therefore, the external forces (actus reus) in corporations are contingent upon the interaction between the organisation and its stakeholders. Corporations that engage in illegal activities are always subject to penalties. In this scenario, the corporation becomes implicated in the commission of a crime. Corporations can act, but they do not possess the capacity to employ and frequently violate the law. This responsibility falls on individuals such as directors and administrators [6].

Criminal law represents the most potent form of socio-legal control in contemporary society. AI entities are concerned that several cases arise from the perception that they are exempt from legal obligations, particularly in the realm of criminal law. If a crime is committed by someone who lacks mental capacity, such as children or animals, the specific offender can be assumed innocent due to their inability to consciously conduct the offence. However, in the case where an innocent individual receives a directive from another person, if the owner of the animal commands an attack on someone, then the owner may face legal consequences. This can also be likened to artificial intelligence (AI) lacking mens rea, or the intention to commit a crime, and hence not being legally accountable. However, programmers or users of AI can be held criminally liable if the AI does a criminal act. If an individual takes the help of AI to commit any offence, for example where by using AI, a robot is converted into an offender, such an action satisfies the requirement of actus reus. This is because the robot carries out the crime of murder, and the person who created the AI must be held accountable for their actions. According to the maxim "actus non facit reum nisi mens sit re", an action is considered offensive only when there is a guilty intention behind it.

If the legal system opts to particularly govern AI crimes, it might be asserted that AI crimes hold special legal significance within the legal system. Both humans and AI systems can be held responsible for regulating crime by AI. Two methodologies can be employed to examine offences carried out by AI.

(i). The significance of criminal law regulating AI does not mandatorily bind AI systems. Only humans, in their roles as users, developers, and disseminators, will be subject to criminal standards and consequences. If individuals contribute to the criminal behaviour of AI systems, they may face criminal punishment. They might have to provide compensation and pay fines, as well as restrict future usage of AI systems, in response to prohibitions on their deployment, requirements to deactivate them, or the need to reprogram their components.
(ii). AI systems will be held accountable under criminal law, meaning they will be directly impacted by legal consequences for any crimes they commit. The legal response may take the shape of punitive measures, such as imposing penalties on individuals or implementing punishments, such as reprogramming AI systems that pose a threat. This alternative method positions AI as a legal entity [8]. Therefore, while AI is now used as a tool for legal regulation, according to modern law, AI does not own its own legal identity and is instead regarded as an entity that is subject to legal regulation, meaning it is owned and managed by a person or legal entity.

Under the existing legal system in India, AI cannot be subjected to execution under criminal law. An individual or a legal organisation must assume responsibility in the case of criminal conduct that ensues as a consequence. The user, along with the legal entity represented by the director of the company or foundation authority, must bear this duty. Humans can control AI to carry out their desired actions, often to engage in specific criminal activities. In such instances, the most apparent course of action is to ensure that the individual responsible for influencing the AI is held accountable. However, the approach to taking responsibility may differ based on the amount of accuracy of the tools employed by the AI. Nevertheless, the individual responsible for AI is not solely confined to its utilisation, but it is crucial not to overlook the significant role played by the AI inventor. This AI developer develops artificial intelligence systems that are utilised by AI users. The way the inventor of the AI must bear the accountability of any legal actions performed by the AI that they have developed, they are equally liable for any wrongful actions too. AI users lacking familiarity with AI science will face a disadvantage. The government can establish and enforce specific regulations about AI for its legal accountability. These regulations would govern the rights and responsibilities of both AI users and AI creators, and establish boundaries for liability in AI-related matters [9].

To prevent AI from being manipulated by hidden actors to achieve their desired outcomes, AI systems need to possess a higher level of sophistication that enables them to comprehend the situation and make appropriate decisions. At this juncture, there are also potential instances where the AI executes activities that surpass its initial objective. For instance, autonomous vehicles are intended not to harm humans, but instead, a situation may come that an automated vehicle, operated by AI cause the death of individuals due to the wrong operation of AI. In such a situation, the outcome deviates from the expected outcome for humans, and it could be advantageous to apply the idea of assigning responsibility depending on forecasts and the possibility that the intended illegal conduct will lead to the actual violation. Accomplices or instigators can be held criminally liable in these circumstances if they could have reasonably foreseen the many consequences that may arise as a potential outcome of the intentional act [10]. Sections 34 and 149 of the Indian Penal Code, 1860, govern the involvement of multiple individuals in the commission of a crime. To ensure accountability, it is necessary to hold not only the person who directly conducts a criminal act but also anybody who participates in it, liable and subject to conviction.

4 Legal Liability of AI for Commission of Deepfake Crimes

The presence of AI has had a profound impact on society, causing significant disruption. Artificial intelligence systems possess distinct characteristics that set them apart from typical computer algorithms. AI possesses the capability to autonomously acquire knowledge, gather expertise, and generate diverse solutions by analysing numerous scenarios without relying on the programmer's intentions. It is capable of functioning independently [11]. The intelligent course of action is to engage in decision-making and exercise choice. If AI possesses human-like intelligence and cognitive abilities, it raises problems regarding its legal accountability for its acts.

AI has greatly benefited human existence in various aspects. The current status of AI has significant legal implications for society, particularly with regard to legal accountability. It is worth noting that there is currently a lack of strong regulations on the responsibility for AI's actions in the event of a crime. To address this issue, it is important to take into account many factors, particularly when discussing legal responsibility. The topic of legal culpability becomes complex to debate due to the consideration of legal subjects' capacity for responsibility. To determine if AI can be subjected to legal proceedings, several conceptual factors need to be reconsidered. These criteria include whether AI possesses the legal capacity to perform legal acts, whether it has the authority or ability to act by the law, whether it can act as a rights advocate authorised by law, and whether it has legal rights and obligations [43].

Regarding the criminal culpability of AI in the commission of criminal acts, it is imperative to initially address the issue of criminal liability. Criminal culpability is the process of establishing the responsibility of a suspect or defendant for criminal conduct that took place. Put simply, criminal culpability is a legal concept that determines whether an individual can be found guilty or acquitted. As already mentioned the principle of culpability in criminal law considers that a criminal offence cannot be established if there is no fault, barring a few statutory exceptions [43].

Humans, as legal entities, possess complete awareness when performing legal action, whereas AI is a human-created tool that lacks consciousness. Consequently, AI cannot be considered a legal entity capable of being held accountable under criminal law. Negligence is the suitable framework for attributing criminal responsibility to accidental activities that arise in the realm of programming or routine tasks carried out by AI. This occurs when the AI agent fulfils its obligations without any malfunction. In this context, the responsibility lies with the designer or operator to proactively address and prevent unforeseen outcomes that may arise during the normal functioning of the AI system, outcomes that the programmer or user should have anticipated. Autonomous vehicles are a reality today. Regulations need to be formulated earlier to avoid any future complications.

5 Legal Framework for Deepfakes in India

India lacks any explicit legislation or regulations that expressly address the issue of deepfakes. Nevertheless, to tackle the issue of misusing or abusing deepfake technology, there are various legal grounds within our existing laws that can potentially be used or expanded to be applicable. The following is a concise overview of the provisions outlined in the existing legislation in India, relevant to understand the techno-legal position:

5.1 Right to Privacy

Privacy right is considered to be the utmost entitlement of every individual that originates from Article 21 in Part III of the Constitution of India, 1950. In the case of Justice K. S. Putuswamy v. UOI [12], the nine-judge bench of the Apex Court acknowledged the fundamental importance of the right to privacy including the concept of 'informational privacy'. This means that individuals can regulate the distribution of private data, including digital privacy, which is a subset of informational privacy [13]. Consequently, employing personal data like images or audio-visual recordings in deepfake films without the explicit permission of the person amounts to a violation of their basic privacy right. In addition, section 66E of the Information Technology Act, 2000 (IT Act) specifies the penalties for invading someone's privacy of any person by intentionally or knowingly taking, broadcasting, or spreading images of their private body parts without their consent. The punishment for this offence can include imprisonment for up to three years, a fine of up to two lakh rupees, or both.

5.2 *Offences Concerning Computers*

Engaging in the improper or malicious use of deepfakes can result in computer-related offences under the IT Act of 2000. Section 67 of the Act stipulates that the dissemination or communication of obscene content in electronic format is subject to punishment [13]. Additionally, section 67A specifies penalties for distributing or transmitting material that contains sexually explicit acts or actions in electronic form [14]. Additionally, the act of publishing or transmitting electronically any information that portrays children engaging in sexually explicit acts or conduct is subject to punishment under section 67B [15].

If someone's digital autograph, login credentials, or any additional unique identifier is used to make a deepfake movie to perform fraud or dishonesty, the individual responsible will be subject to punishment for the crime of identity theft as outlined in section 66C of the IT Act [16]. Moreover, the act of cheating through impersonation by utilising computer resources or other communication device is subject to punishment as per section 66D of the IT Act [17].

In addition, the Central Government has the authority to instruct the intermediary to restrict public access to any deepfake content that is created, transmitted, received, stored, or hosted on any computer resource. This action can be taken if the government deems it necessary or expedient to protect the country from any other consequences concerning it [18].

5.3 *Defamation in India*

Defamation can have both criminal and civil penalties [19]. In civil defamation, damages will be due to the person who was defamed if an action for defamation is filed and defamation is proven to have occurred [20]. Criminal defamation includes published visible representations that either damage or will damage the person's reputation [21]. Section 500 of the Indian Penal Code, 1860 (IPC) lays down the penalty for such offence as simple imprisonment for up to two years or fine or both, but these are insufficient to address the many ways deepfakes occur [22].

Cyber defamation was before specifically covered by IT Act section 66A [23]. The clause forbade sending any material via computer resource that is unpleasant or menacing, or that is intended to irritate, inconvenience, impede, insult, injure, criminally intimidate, incite animosity, hatred, or bad will. The Supreme Court overturned this clause, though, in Shreya Singhal v. Union of India [24]

5.4 Other Criminal Offences

Deepfake videos are primarily manipulated reproductions of the original content and might potentially be seen as acts of forgery. Forgery, as defined in section 468 of the Indian Penal Code (IPC), is the intentional act of creating a false electronic record to damage the image of a party or with the knowledge that it has the potential to be utilised for that purpose. Those found guilty of this offence can be sentenced to imprisonment for a maximum of three years and may also be required to pay a fine [25]. If a deepfake video incites or attempts to incite hatred, contempt, or disaffection towards the legally established Government in India, it will be subject to punishment for the crime of sedition under section 124 of the Indian Penal Code (IPC) [26].

Additionally, the act of threatening a person included in a deepfake video with harm to their reputation or property, to cause alarm or coerce them into committing an illegal conduct, establishes the crime of criminal intimidation [27]. Section 506 of the Indian Penal Code (IPC) stipulates penalties for the aforementioned offence, which encompasses the act of utilising photographs or recordings to instil fear or intimidate an individual, as well as their possessions or reputation. In addition, depending on the potential impact of deepfake content, the accused may face charges of intentional insult to incite a breach of peace [28], encouraging animosity between communities based on race, ethnicity, birth location, place of residence, language, and other factors; and doing things that hurt the peace [29], or for deliberate and malicious acts aimed at outraging the religious sentiments of any particular group by insulting their religion or religious beliefs [30].

5.5 Violation of Copyrights

Deepfake videos are typically altered or counterfeit renditions of the audio or visual recording extracted from a copyrighted movie or film. According to section 14 of the Copyright Act, 1957, the owner of the copyright for a cinematographic movie and sound recording has the sole right to create or authorise the creation of copies of the film. This includes taking photographs of any images within the film or creating other sound recordings that include it [31].

The Delhi High Court in Amarnath Sehegal v. UOI, acknowledged the author's moral right in his work [32]. The author has the right to seek compensation for any violation of their moral right, specifically in cases where their work has been altered in a way that could harm their reputation or honour [33]. The copyright holder can be held legally responsible for civil penalties, such as injunctions, financial compensation, and other remedies, as provided by law, for the violation of their rights [34].

Furthermore, individuals who are aware of and intentionally violate or assist in the violation of copyrights in a work of art, or any other rights granted by the Act, may be subject to imprisonment for a maximum of three years and a fine of up to two lakh rupees [35]. However, copyright remedies may not be the optimal approach in situations when the individual targeted by the deepfake may not have the copyright. In cinematographic films, the right to reproduce is held by the film producers rather than the actors [36]. Similarly, in photographs, the copyright may be held by the photographer rather than the individual depicted in the photo [37]. Therefore, the recourse for infringement of copyright may not be accessible to the intended recipient [38].

5.6 Theft of Identities and Deepfakes

Identity theft is the act of dishonestly acquiring someone else's private identifying information and exploiting it for personal gain. This might include activities like creating bank accounts or obtaining government benefits using the victim's name. As per a recent news report in the United States, the CEO of an undisclosed energy company situated in the United Kingdom was tricked into believing that he was conversing with the CEO of his German parent company. The imposter CEO told him to transfer €220,000 ($243,000) to a Hungarian supplier. The convincing nature of the Deepfaked voice led to the successful execution of the fraudulent transaction [39].

Moreover, the case of reporter Rana Ayub serves as evidence that our legal system fails to sufficiently safeguard individuals who fall victim to revenge porn. Rana Ayyub was the victim of a manipulated pornographic video where her image was modified to falsely portray her as the main participant. This brutal attack took place shortly after she started lobbying for the victim of sexual assault in Kathua. An eight-year-old girl was subjected to repeated sexual assault and then murdered. BJP officials in the Jammu area have spearheaded a campaign advocating for the accused, contending that they were unjustly targeted due to their religious connection. Rana Ayyub faced online abuse and hate speech when she expressed her views, but she fought back by utilising Twitter. Several digitally manipulated the tweets depicting Rana were being shared on the microblogging platform, and she noticed that she was included in some of them [39].

5.7 Content Revisions and Intermediate Responsibilities

The internet poses a significant threat to individuals' right to privacy. Online service providers bear a significant responsibility to safeguard the privacy of their clients' sensitive data. The 2016 US Elections put significant pressure on digital intermediaries because of the possible catastrophic consequences of deepfakes in disseminating disinformation.

In 2019, Facebook, Amazon, and Microsoft established the 'Deepfake Detection Challenge' to promote the development of dependable technology for detecting

Deepfakes [40]. The victorious team in the competition achieved a system with a detection accuracy of only 65.18 per cent for identifying deepfakes. Hence, social media companies such as Facebook, Reddit, and Twitter have enforced laws to combat the issue of deepfakes.

Facebook's policy mandates that fact checks be conducted by independent third-party fact-checkers, who collaborate with over 50 partners and work in more than 40 languages. Given the inefficacy of fact-checking in countering deepfakes, these approaches are excessively limiting. Although fact-checking can be effective in reducing the dissemination of false information, it is not sufficient to address the complex issues arising from deepfakes, such as privacy breaches and intellectual property infringements.

According to section 79 of the IT Act, Indian intermediaries are required to delete illegal information once they become aware of it or get a court order to do so as discussed in Myspace Inc v. Super Cassettes Industries Ltd. [41]. Currently, most content moderators utilise a hybrid approach that integrates artificial intelligence (AI) with human reviewers. Complying with the new Intermediary Rules may be unattainable in the case of deepfakes, since both human assessors and technology cannot identify them with absolute precision.

6 Are There Any Loopholes in Existing Laws?

"The current laws are insufficient as they were not specifically formulated with these emerging technologies in mind", states Shehnaz Ahmed, the fintech leader at the Vidhi Centre for Legal Policy in Delhi. However, she warns that implementing incremental changes to legislation is not the correct approach. "There is currently a moral panic stemming from these recent well-known cases, but we appear to be diverting our attention from the question that is more significant—which should be India's regulatory strategy regarding emerging technologies such as AI?" she asserts. She emphasises the necessity of establishing a regulatory framework for AI technology that is grounded in comprehensive market research evaluating the various types of harm caused by it. "In addition, it is crucial to establish a strong enforcement mechanism, as the effectiveness of laws depends not only on their design but also on the ability of institutions to implement them", she emphasises [42].

Highlighting a deficiency in the current IT Rules, she argues that it solely deals with cases where the unlawful material has already been posted and the resulting damage has been experienced. Instead, there should be greater emphasis on preventive actions, such as informing users when they are viewing a manipulated image.

Mr. Gupta acknowledges the necessity of overhauling the current laws. He highlights that the existing regulations solely concentrate on either censoring or criminally prosecuting online takedowns. Nevertheless, they lack a thorough comprehension of the operation of GNAs and the substantial harm that it has the potential to cause. The legal provisions impose full responsibility on the person who is victimised to initiate a complaint. According to him, many people have an unsatisfactory experience with local police stations when it comes to investigating crimes or ensuring that the culprits face appropriate consequences.

7 Efforts at the International Level to Combat Deepfakes

All the major countries are developing laws to combat the adverse effect of the deepfake technology. However, there is a coalition of twenty-nine nations, who have united to mitigate the potential "catastrophic harm, whether intentional or unintentional", resulting from the escalating adoption of artificial intelligence. Among all those countries, US, UK, India, the European Union, China, and South Korea have taken the most visible steps and come out with specific laws in this regard. The Declaration establishes a framework for nations and their governments to work together and engage in joint efforts to address the current and future threats associated with artificial intelligence (AI). It outlines a set of objectives and priorities to be pursued.

1. Identifying dangers in the field of artificial intelligence;
2. Developing risk-based policies in different nations to enhance transparency among private entities involved in the advancement of cutting-edge AI technologies. Countries that have implemented proactive measures to combat the threat of Deepfakes [43]

7.1 United States of America

In October 2023, President Joe Biden of the United States enacted a comprehensive executive order concerning artificial intelligence (AI) to effectively address its various hazards, including those related to national security and privacy. The Department of Commerce has been assigned the responsibility of formulating criteria for labelling AI-generated content, which will facilitate the identification process, commonly referred to as watermarking. California and Texas have enacted legislation that makes it illegal to broadcast and distribute deepfake films to manipulate the intention of manipulating election results. The dissemination of non-consensual deepfake pornographic material in Virginia is subject to criminal penalties as mandated by the law.

The Deep Fakes Accountability Bill, 2023, which was just presented in Congress, mandates that creators must identify deepfakes on internet platforms and notify users of any modifications made to a film or other content. Not labelling these 'malicious deepfakes' would result in criminal punishment. The United States has urged the Department of Homeland Security (DHS) to create a specialised task force dedicated to tackling the issue of digital content forgeries, commonly referred to as "deepfakes". Several states have implemented their laws to address the issue of Deepfakes [44].

7.2 United Kingdom

The UK government intends to establish nationwide regulations for artificial intelligence companies to assess the enforcement of legislation mandating transparent labelling for AI-generated photographs and films.

7.3 European Union

The European Union (EU) has enhanced its Code of Practice on Disinformation to mandate prominent digital media platforms like Google, Meta, and Twitter to detect and categorise deepfake content, with the potential consequence of facing significant monetary sanctions. The Code was originally established as a voluntary self-regulatory system in 2018. However, it now receives backing from the Digital Services Act which intends to toughen the oversight of online platforms to combat different types of abuse. Furthermore, as per the EU AI Act proposal, suppliers of deepfake technology would be required to adhere to transparency and disclosure standards. The European Union has implemented the Digital Services Act, which mandates social media companies to comply with labelling requirements, promoting transparency and assisting users in verifying the legitimacy of media [45]. On March 13, 2024, the EU has come up with a new law called EU Artificial Intelligence Act, 2024 to promote innovation and assure safety and adherence to fundamental rights.

7.4 China

In January, 2024, the Cyberspace Administration of China enforced updated regulations to control the usage of deep synthesis technology and reduce the dissemination of incorrect information. The rule guarantees that if any information gets altered by the use of technology, it must be identified and its source should be able to be recognised. Providers offering deep synthesis services are required to adhere to local rules, maintain ethical standards, and follow the principles of "proper orientation in governance as well as appropriate public opinion orientation" [44].

7.5 South Korea

South Korea on January 30, 2024, has amended its existing law, namely Public Official Election Act which provided for criminalising the dissemination of deepfakes that have the potential to harm public interest. Violators may be subject to a maximum prison sentence of five years or fines reaching 50 million won (about 43,000 USD) [44].

8 Conclusion & Suggestion

The utilisation of deepfake technology is expected to gain significant popularity in the forthcoming years. Despite the recent introduction of special legislation in many nations to address the issue of deepfake, none have been successful in effectively implementing these laws and holding perpetrators accountable under criminal statutes. Therefore, in connection with deepfakes, it is worth considering implementing ex-ante regulation. This could involve either banning the development, providing, consumption, or possession of deepfake technology in the consumer market, or implementing an obligatory ex-ante legitimacy test. This test would need to be conducted before publication of any information and/or dispersed by citizens. Before introducing material in court, it may be necessary to conduct an ex-ante test. Enacting specific legislation to hold individuals criminally accountable is of similar significance.

In response to the public outcry surrounding the deepfake video of Mandana, Ashwini Vaishnaw, the Union Minister of Electronics and Information Technology, conducted a meeting on November 23 with social network platforms, companies operating with AI, and industrialists. During the meeting, he recognised the emergence of a new crisis caused by deepfakes and acknowledged the absence of a parallel verification system to address this issue for a significant portion of society. In addition, he declared that government officials will present preliminary regulations, which will be available for public feedback, in the next 10 days to tackle the problem. The regulations would establish responsibility for both content providers and intermediaries of social media. He also stated that all social networking businesses have reached a consensus on the imperative need to affix labels and watermarks to deepfakes.

Nevertheless, Rajeev Chandrasekhar, the Minister of State for Electronics and Information Technology (MeitY), asserts that the current regulations are sufficient to address deepfakes, provided they are rigorously enforced. According to him, a designated person known as a Rule 7 officer, would be assigned to closely supervise any breaches, and an internet-based platform will be established to facilitate affected users and people in submitting First Information Reports (FIRs) for deepfake offences. Social media corporations were reminded of their need to promptly remove content that violates the regulations, as stated in section 66D of the IT Act and Rule 3(1)(b) of the IT Rules. A piece of advice was issued to them to ensure compliance. Mr. Gupta highlights that the advice made by the MeitY lacks legal authority and significance. It is crucial to demonstrate a certain level of response, considering the ethical anxiety regarding generative AI that was triggered when the clip featuring Rashmika Mandanna went viral. It fails to consider the possibility that deepfakes may be transmitted through channels other than social media platforms [42].

In addition, certain specialists have formulated specific scenarios about the liability of artificial intelligence under criminal law. The most discussed is Gabriel Hallevy's concept. Hallevy [45] developed an AI criminal responsibility scenario by including the concepts of actus reus (activity) and mens rea (condition) from Israeli criminal law. Hallevy subsequently categorised it into the following groups:

1. Actus reus refers to the situation where the individual in question either does an action that is prohibited by the law or fails to carry out a conduct that is mandated by the law.
2. Mens rea refers to the mental state of a person in a legal context. It encompasses those who possess knowledge or awareness, as well as those who lack knowledge, are unaware, or act mistakenly.

According to the aforementioned classification, criminal acts perpetrated by artificial intelligence can manifest in three potential scenarios. Firstly, illegal acts occur when legal entities intentionally employ AI to subvert the law. Secondly, illegal acts occur when legal entities intentionally employ AI for purposes that are not in compliance with the law. Furthermore, illegal activities may occur as a result of the legal subjects' negligence in utilising AI. Hallevy presented three accountability models for illegal activities done by AI: the perpetration-via another model (PVM), the natural-probable consequence liability model (NPCLM), and the direct liability model (DLM) [45].

8.1 Pvm

According to this approach, AI is regarded as an agent that cannot engage in criminal behaviour. In this scenario, artificial intelligence (AI) serves just as a tool, whereas the individuals responsible for creating and employing AI to commit illegal activities are the true culprits. This concept establishes accountability through severe liability. In other words, criminal culpability is attributed without the need to evaluate the presence of intent or negligence. The deliberate actions of the culprit in building and using AI for illegal objectives can be deemed perilous to society.

8.2 Npclm

The NPCLM model specifically targets illegal behaviours resulting from malfunctions or failures of AI systems. Regarding such instances, all people involved (including the proprietors, designers, or supervisors) might be held directly criminally liable if they knew the risks. They may also be held responsible if they are unaware of the vulnerabilities of the artificial intelligence they employ. Proportional culpability can be applied when there is a varying cause-and-effect relationship among different individuals involved in a criminal act.

8.3 Dlm

This concept applies the same standards of criminal responsibility to AI as it does to humans. AI is not regarded as a passive entity but rather possesses a level of consciousness or cognitive state when engaging in illegal behaviour. AI possesses the capacity to exhibit responsibility. There is significant controversy over the validity of this strategy. Instances such as software defects or mistakes committed by AI in self-defence that lead to violations of accountability will also be considered equivalent to the concept of excusable causes under criminal law. Granting AI the status of a legal entity entails the inclusion of all parties involved in the planning, production, implementation, or ownership of the technology. These parties would be held legally responsible for any unlawful actions, regardless of the extent of their ownership stake.

The existing laws in India about cyber offences perpetrated by deepfakes are insufficient to comprehensively tackle the problem. The absence of explicit rules in the IT Act, 2000 about machine learning, artificial intelligence, and deepfakes poses challenges in properly governing the implementation of these technologies. To enhance the regulation of offences by employing deepfakes, it might be imperative to revise the IT Act, 2000 by incorporating measures that explicitly deal with the utilisation of deepfakes and the corresponding sanctions for their improper use. Possible measures might involve imposing harsher sanctions on persons who produce or disseminate deepfakes with harmful intent, along with implementing more robust legal safeguards for any person whose photos or a similar are exploited without obtaining explicit permission [46].

It is crucial to acknowledge that the creation and utilisation of deepfakes is a worldwide concern, and it will probably necessitate global collaboration and cooperation to adequately control their usage and prevent infringements on privacy. Meanwhile, it is crucial for both people and organisations to acknowledge the possible hazards linked to deepfakes and remain cautious in confirming the genuineness of material encountered on the internet.

Legal progress lags behind technological advancements. The inclusion of rules about data about an individual, the right to be forgotten, and extraterritorial applicability is vital in the Digital Personal Data Protection Act, 2023. Concurrently, it is necessary to create an artificial intelligence system that can detect and prevent the usage of deepfakes on many platforms, such as government authorities, social media, and large corporations. It is crucial to consider that the creation and use of deepfakes present a worldwide problem that will probably necessitate cooperative endeavours to control their usage and safeguard privacy rights. An excellent strategy for protecting oneself and one's business from the danger of deepfakes is to practice prudence and thoroughly authenticate online content.

Meanwhile, the governments can undertake the below-mentioned actions:

(a). Firstly, the censorship technique involves restricting general access to misinformation through the issuance of directives to intermediates and publishers.

It aims to restrict middlemen and publishers from disseminating misleading information to the public

(b). The second option is the punitive approach, which holds individuals or groups responsible for creating or spreading false information.

(c). The third one is known as the intermediary regulation approach by which the online intermediaries will be asked to promptly eliminate wrong information from online portals. In case of any failure to do so, they may be held liable under Sections 69A(33) and 79(34) of the IT Act, 2000 [47].

References

1. Europol (2020) In: Malicious uses and abuses of artificial intelligence. Available via Europol. https://www.europol.europa.eu/cms/sites/default/files/documents/malicious_uses_and_abuses_of_artificial_intelligence_europol.pdf. Last accessed 15 Feb 2024
2. Anderson KE (2018) Getting acquainted with social networks and apps: combating fake news on social media. Libr Hi Tech News 35:1–6
3. Jia WS et al (2022) A comprehensive overview of deepfake: generation, detection, datasets and opportunities. 513 Science Direct. https://doi.org/10.1016/j.neucom.2022.09.135
4. J. Thies et al (2016) Face2Face: real-time face capture and reenactment of RGB videos (CVPR 2016, Las Vegas, June 2016). <https://www.cvfoundation.org/openaccess/content_cvpr_2016/papers/Thies_Face2Face_RealTime_Face_CVPR_2016_paper.pdf> Last accessed 6th Feb 2024
5. A Review of Image Processing Techniques for Deepfakes—Scientific Figure on ResearchGate (2024) Available from: https://www.researchgate.net/figure/Examples-of-original-and-deepfake-videos_fig3_361422044 [accessed 19 Jun 2024]
6. Rahman RA, Habibulah R (2019) The criminal liability of artificial intelligence: is it plausible to Hitherto Indonesian criminal system? Legality Jurnal Ilmiah Hukum 27(2):147–160
7. Sari AP, Harwika DM (2022) Legal liability of artificial intelligence in perspective of civil law in Indonesia. Int J Social Sci Res Rev 5(2):57–60
8. Lagioia F, Sartor G (2020) Ai systems under criminal law: a legal analysis and a regulatory perspective. Philos Technol 33(3):433–465
9. Lodder AR, Oskamp A, Duker MJA (2018) AI & criminal law: past, present & future. JURIX
10. Lima D (2017) Could AI agents be held criminally liable: artificial intelligence and the challenges for criminal law. SCL Rev 69:677
11. Ravizki EN, Yudhantaka L (2022) Artificial Intelligence Sebagai Subjek Hukum: Tinjauan Konseptual dan Tantangan Pengaturan di Indonesia. Notaire 5(3):351
12. Justice KS (2017) Putuswamy v. Union of India. 10 S.C.C. 1
13. Section 67, Information and Technology Act (2000)
14. Section 67A, Information and Technology Act (2000)
15. Section 67B, Information and Technology Act (2000)
16. Section 66C, Information and Technology Act (2000)
17. Section 66D, Information and Technology Act (2000)
18. Section 69A, Information and Technology Act (2000)
19. Pradeep T, Rajan A (2018) A critical study on cyber defamation and liability of ISPS. Int J Pure Appl Math 119(17):1717–1719
20. Shivi (2016) Defamation laws and judicial intervention: a critical study. ILI Law Rev pp 170–175
21. Section 499, Indian Penal Code (1860)
22. Section 500, Indian Penal Code (1860)

23. Section 66A, Information and Technology Act (2000)
24. Shreya Singhal v. Union of India A.I.R. (2015) S.C. 1523
25. Section 469, Indian Penal Code (1860)
26. Section 124, Indian Penal Code (1860)
27. Section 503, Indian Penal Code (1860)
28. Section 504, Indian Penal Code (1860)
29. Section 153A, Indian Penal Code (1860)
30. Section 295A, Indian Penal Code (1860)
31. Section 14(d) and Section 14(e), The Copyright Act (1957)
32. Amarnath Sehegal v. Union of India 117 (2005) DLT 717
33. Section 57, Copyright Act (1957)
34. Section 55, Copyright Act (1957)
35. Section 63, Copyright Act (1957)
36. Section 2(d)(v), The Copyright Act (1957)
37. Section 2(d)(iv), The Copyright Act (1957)
38. Ciccatelli AG (2019) (https://www.ipwatchdog.com/author/amanda-ciccatelli/), In: Honor of April Fools' Day: Diving Into Deepfakes, IP Watchdog, 01/04/2019, available at: https://www.ipwatchdog.com/2019/04/01/honor-april-fools-day-diving-deepfakes/id=107878/, Last accessed 10 Feb 2024
39. Aditya M (2024) Dissecting the framework of deep fakes in India—a Glaring Lacuna, cell for law and technology, National Law Institute University, Bhopal, available at https://clt.nliu.ac.in/?p=887, Last accessed 6 Feb 2024
40. Gong LY, Li XJ (2024) A contemporary survey on deepfake detection: datasets, algorithms, and challenges. Electronics 13:585. https://doi.org/10.3390/electronics13030585
41. My Space Inc. vs Super Cassettes Industries Ltd., 23 Dec 2016
42. Aaratrika Bhaumik, Regulating deepfakes and generative AI in India | Explained—The Hindu, December 04, 2023 06:57 pm | Updated 08:14 pm IST, available at https://www.thehindu.com/news/national/regulating-deepfakes-generative-ai-in-india-explained/article67591640.ece, Last accessed 6 Feb 2024
43. Vikrant R, Anuradha G, Rachita T (2023) Deepfakes and breach of personal data—a bigger picture. LiveLaw, 24 Nov 2023 12:16 PM available at https://www.livelaw.in/law-firms/law-firm-articles-/deepfakes-personal-data-artificial-intelligence-machine-learning-ministry-of-electronics and info, Last accessed 6 Feb 2024
44. Hailtik AGE, Afifah W (2023) Criminal responsibility of artificial intelligence committing deepfake crimes in Indonesia. Asian J Social Hum 2:776–795
45. Hallevy P (2018) Dangerous robots–Artificial intelligence vs. human intelligence. Dangerous Robots–Artificial Intelligence vs. Human Intelligence (11 Feb 2018)
46. Sousa Antunes H et al (eds) Multidisciplinary perspectives on artificial intelligence and the law, law, governance and technology series, vol 58. https://doi.org/10.1007/978-3-031-41264-6_19
47. Sections 69A(33) and 79(34) of the Information Technology Act (2000)

Adaptive Machine Learning-Based Intrusion Detection Systems for IoT Era

Rahul Vadisetty

Abstract The spread of Internet of Things (IoT) devices encompasses enormous security flaws in their network. The integration of Intrusion Detection Systems (IDS) that rely on adaptive machine learning (AML) into IoT networks serves to address the distinct security issues that arise from the variety and widespread use of IoT devices. In this paper, a hybrid AML-based IDS using deep learning models, such as LSTM and CNN + GRU to detect anomalies and threats such as DDoS, DoS, and U2R has been adopted to evaluate its anomaly detection capability in a variety of datasets, such as UNSW_NB15, CICIDS2017, NID, and TON_IoT. When tested on these datasets, the accuracy of 98.9%, 99.5%, 98.7%, and 99.9% are obtained on the UNSW_NB15, CICIDS2017, NID, and TON_IoT datasets respectively by the Max Voting mechanism. Furthermore, the MSE and R2_score for the hybrid model for all the datasets were found to be ~ 4×10^{-5} and ~ 0.99 respectively, with ROC-AUC plots of ~ 0.9, which further proves its supremacy. The study in addition highlights the supremacy of AML-based IDS in protecting IoT networks and the potential of ML methods to improve cybersecurity through a comparative performance analysis with related works. The results point out the need for IDS to strike a balance between reducing false alarms and having a high detection rate with the goal of protecting IoT devices.

Keywords IoT networks · Adaptive machine learning · Intrusion detection systems · Cybersecurity · Anomaly detection · Network security

1 Introduction

IoT (Internet of Things) consisting of various devices such as simple sensors to general-purpose computing devices, covers different domains such as health, agriculture, and home automation [1]. The heterogeneity, as well as ubiquity, of such IoT

R. Vadisetty (✉)
Wayne State University, Detroit, Michigan, USA
e-mail: rahulvy91@gmail.com

M. Chakraborty et al. (eds.), *Proceedings of 5th International Ethical Hacking Conference*, Lecture Notes in Networks and Systems 1148,
https://doi.org/10.1007/978-981-97-8457-8_17

devices, limited but non-trivial computing power, and their diversity in applications, pose unique security challenges not well-handled by traditional security mechanisms [2]. This adaptive system uses machine learning (ML) algorithms to detect malicious traffic flowing in the computer networks, like malware and unauthorized login attempts, by analyzing the pattern and behavior of network traffic [3]. This addictiveness comes from their capability of automatically updating their learning to new network behaviors to detect new threats without having to write extra rule enforcement or reach human beings [4].

Features of this cyber system include anomaly detection, horizontal scalability, real-time processing, resource-efficient processing, and self-adaptive operations [5]. Anomaly detection, representing the capability of the cyber system to identify deviations from the normal network behavior, is important to mitigate security breaches, especially from intrusion [6]. In turn, scaling can ensure that the system can keep pace with the rising number of such devices, real-time processing can help mitigate or shut down threats before they spread widely, adaptability can allow a system to react to a changing environment and a new crop of threats, and efficiency can minimize computational overheads on IoT devices themselves [7].

An intrusive sensor can waste network resources, compromise the network, spy on communications, transmit fake messages, and change the authenticity of data [8]. One of the most frequent and serious risks to Wireless Sensor Network (WSN) security is the denial-of-service (DoS) attack [9]. DoS attacks continue to be a serious problem in modern times. This attack comes in different forms, and its principal objective is to stop operations that WSNs deliver [10]. DoS attacks are a serious and highly risky attack type since current technology is dependent on data. When processed, data also referred to as raw information becomes meaningful [11]. The amount of data consumed has expanded dramatically since the dawn of the computer age [12]. Since then data consumption has skyrocketed. Big data can contain system events, network traffic logs, and logs from individual system components [13]. Together, artificial intelligence and big data algorithms can examine network data from the past and present to assess whether the current traffic is malicious or normal [14].

Certain systems can identify actual time incidents and notify the appropriate sensors in the nodes of the networks that are susceptible to attacks. Intrusion Detection System (IDS) is the name given to these systems [15]. An IDS is an assertive tool for detecting attacks that quickly finds and categorizes unauthorized access, assaults, and safety policy violations [13]. The IDS monitors network traffic or the activities of particular hosts continuously, analyzes trends, and detects anomalous or abnormal activity [16]. The most popular IDS is set up over an internet connection and operates in a networked manner inside a network system to monitor traffic movement for attacks [17]. The implementation of a Network IDS for WI-FI-based WSNs has generated hype for several machine learning and deep learning techniques, including SVM, Perceptron, Convolutional Neural Networks (CNN)[5] K-Nearest Neighbors (KNN), Feedforward Neural Networks (FFNN), Deep Neural Networks (DNN) [14] IoT security can be improved through adaptive machine learning-based IDS, however, challenges remain regarding data privacy, understanding of models, and the need for examples that function well in a range of scenarios involving IoT

[18]. These systems play an indisputable role in protecting the globalized world against emerging cyber threats, which makes them indispensable for guaranteeing the safety of the Internet of Things in the future [19].

Here are the key contributions of this study include.

- Use of LSTM and CNN + GRU models to effectively identify a variety of cyber threats in IoT networks, including DDoS, DoS, and U2R.
- Incorporating continuous adaptive detection capabilities to new and evolving attack patterns.
- Increasing the effectiveness of network security by achieving better detection rates and fewer false positives.
- Demonstrating efficacy by outperforming cutting-edge models in cyber threat detection and mitigation.

The remaining portion of this work is organized as follows. Section 2 describes the related work, The proposed methodology is described in Sect. 3 followed by a discussion of the results in Sect. 4. The paper concludes in Sect. 5 with closing remarks and future scope.

2 Literature Review

The swift expansion of IoT devices across multiple industries highlights the urgent requirement for strong cybersecurity protocols. To defend these networks against sophisticated cyber threats, recent literature has focused on further development and improvement of ML-based IDS (ML-IDS). This section provides an in-depth literature review of related works followed by research gaps.

2.1 Related Works

In [20], a deep learning algorithm has been developed to identify network intrusions by analyzing large collections of network activity data. In the Google Colaboratory (Colab) setting, PySpark with Apache Spark is being used for this investigation. In the study, Scikit-Learn and Keras libraries are used, and "TON_IoT" and "CICIoT2023" datasets are utilized to assess and instruct the prototype. The technique of correlation ensures that significant features are included in the tests by reducing the number of characteristics in the datasets. A hybrid algorithm for deep learning is made up of CNN in one size and LSTM in another. Ten methods for deep learning and machine learning were contrasted with the developed approach. Evaluation of the effectiveness of the simulation was performed by measuring accuracy, precision, recall, and F1 score parameters.

In [21], a novel intrusion detection system leveraging ML and deep learning-based architecture was developed using the Bot-IoT dataset, which was released in 2019 and addresses the issue of unbalanced classes. Three distinct subsets were created with this data set by choosing various sets of features. The Argus sequence number (seq) and dates of the first subset are different from those of the other two. Timestamps were eliminated from the second subset with the expectation that the model would remember these details.

Aravamudhan [20], presents a novel approach to combating cyberattacks within the IoT ecosystem by developing an advanced IDS using deep learning techniques. Through the use of Gradient Implementing Boosting and the Fast Region-Based Convolutional Neural Network (Fast R-CNN) for optimization, the study presents a hybrid model that greatly improves detection efficiency and accuracy. By utilizing dimensionality-reducing methods like PCA and SVD for dataset structuring, the suggested model can identify cyber threats with an astounding 99.5% accuracy rate.

The two parts of the FlowGuard algorithm—the flow handler and the flow filter were created in [21]. The flow filter stream is responsible for detecting fraud that would not be detected by circular fluctuations. Its goal is to remove these liquids under the flow handler's created sorting rules. The flow handler uses CNN and LSTM, two ML models that have been established, to recognize and separate fraudulent flows. The efficacy of the FlowGuard technique was assessed using the CICDDoS2019 dataset, yielding accuracy rates of 99.9% for attack detection and 98.9% for target type recognition.

A hybrid deep learning algorithm was evolved using feature selection in [22]. Ten alternative variable arrangements, including grouping size, the number of filter units, and the number of BiLSTM units, were used to assess CNN and BiLSTM-based deep learning models. Using the CICDDoS2019 dataset for training, assessment, and verification allowed the system to reach a 94.52% peak accuracy. Three different models of the deep learning architecture used in [23] to detect DDoS attacks consisted of CNN, RNN, and DNN. CICDDoS2019 dataset was used to train each model, and TON-IoT dataset was used to evaluate each model. The main task of the dataset used here is that it consists of multiclass and binary problems. To compare binary versus multiclass classification, three different datasets were preprocessed. The datasets used were: Dataset_2_class, Dataset_7_class, and Dataset_13_class. The accuracies obtained by the CNN algorithm of binary classification, 7-class classification, and 13-class classification were 99.95%, 95.90%, and 95.12%, respectively.

Zakariah et al. [18] solves the issue of encrypted data, conventional network firewalls, and identification tools based on ML. The article introduces a deep neural network architecture that is more complex than used before for IoT network invasion identification. The new approach accurately models the data flow's ordered properties using the attention mechanism and LSTM network, enhancing anomaly recognition. Furthermore, ADASYN, which extends the volumes of minority-class samples to enhance the model's binary-threat classification accuracy, is applied. The experimental outcomes on the NSL–KDD dataset illustrated the technique's strong outcome, where the MLP classifier exhibited high accuracy of up to 87%, precision of 89%, recall of 87%, F1 score of 89%, and 88% of AUC score. The model even performs better in the all-class classification task as noted by the reported AUC of 0.94, accuracy and precision of 87% each, recall of 84%, and F1 score of 83%. The results indicated the model's good performance and comprehensive detection ability, as well as its significant improvement concerning the current methods, making it a feasible method for addressing IoT network security problems.

Saba et al. [24] sheds light on the fact that the Internet of Medical Things (IoMT) is subject to cyberattacks and puts forward a novel method of intrusion detection, fuzzy-based self-tuning LSTM, to provide greater security. This complex system has perfected the craft of avoiding underfitting and overfitting via early stopping and dynamic epoch adjustment. A huge amount of testing has shown that the suggested model is superior to existing IDS solutions for the IoMT, retaining greater accuracy, reducing false positive rates, and enhancing the detection rate. Moreover, research shows the significant disadvantage of static settings of deep learning models and, thereby, the benefits of flexibility, making new useful contributions to forming IDS models that are much more efficient in IoMT conditions.

Saba et al. [24], aims to tackle a hard problem, that of weaknesses of IoT devices. Security remains a demonstrated rid, with the expansion of IoT to multiple domains. Traditional security is not sufficient here and newer approaches must be considered. ML is particularly advantageous when it is required to find patterns where the other approaches do not work well As a result, the presented approach can analyze all possible dependencies between the network flows. The perfect performance of this model when used in the NID and BoT-IoT datasets, which gave the accuracy of 99.51% and 92.85%, respectively, proves how effective it can be and how one can spot attacks or unusual behaviors for the betterment of IoT security.

The summarized work [19] offers an overview of IDS suitable for IoT networks, focusing on the special constraints posed by the reduced processing energy and energy accessibility of IoT devices. In general, traditional safety solutions are often infeasible due to the constraints, resulting in frequent assaults such as DoS and routing disruptions. Even though IDSs have previously discovered limited achievement utilizing similar surroundings such as WSN, Mobile Ad hoc Networks, and Cyber-Physical Systems (CPS), this study addressed the potential problems and performance of applications in IoT settings. To evaluate their applicability for IoT systems, the discussion covers existing IDS approaches created especially for IoT in addition to methodologies from MANET, WSN, and CPS. The study suggests that current research may soon produce IDS solutions that are better suited to the drawbacks of IoT devices and outlines prospective directions for future studies aimed at improving the security of IoT networks.

New research shows creative ways to use cutting-edge deep learning techniques to improve IDS in the IoT ecosystem [6]. One method uses a mixed design that blends Gradient Boosting and Fast R-CNN to achieve 99.5% accuracy in identifying threats. PCA and SVD are used for the data structure. Using adaptive manufactured sampling and an attention-enhanced LSTM network, another study demonstrates a significant improvement in anomaly detection with high-performance metrics across multiple evaluations. Furthermore, a novel fuzzy-based self-tuning LSTM system is proposed for the IoMT, exhibiting improved accuracy and flexibility through swiftly stopping and changing training epochs [25]. Finally, a CNN-based model for IoT security greatly outperforms conventional security systems by utilizing the interconnectedness of devices to analyze network traffic effectively. When taken as a whole, these studies point to a move toward more advanced, machine learning-based security solutions that can handle the particular difficulties presented by IoT environments.

2.2 Research Gaps

There are several research gaps in the development of IDS for IoT networks based on Adaptive ML (AML), despite the literature survey depicting a variety of deep learning techniques for these systems. Firstly, the adaptability of these systems against complex attack vectors like adversarial attacks where hackers manipulate input data to avoid detection, advanced persistent threats (APTs) where the attackers establish an undetected presence in the network to steal information, and zero-day exploits, where the attackers take advantage of errors in codes or software that the target has yet to discover to steal sensitive data are becoming more and more common, is where a critical gap exists. Secondly, IoT devices being small have limited computational capability, memory, and battery life. Incorporating lightweight and energy-efficient AML mechanisms to mitigate IoT threats has received little attention, despite the exploration of architectures such as CNNs and LSTM networks. Thirdly, enhancing real-time threat detection and mitigation using AML techniques is the need of the hour. This paper addresses the above-mentioned problems by designing a hybrid AML-based IDS using LSTM, CNN, and GRU algorithms. The model has been tested on four publicly available datasets and compared with the related works.

3 The Proposed Methodology

In this section, we propose an AML-based IDS by using a hybrid model of the combination of LSTM and CNN + GRU algorithms to enhance IoT network security. Performance analysis of the system proves its efficiency in identifying, detecting, and mitigating complex threats like DDoS, DoS, and U2R attacks. Furthermore, comparative analysis with related works proves its supremacy.

3.1 Dataset

Four publicly available datasets such as UNSW_NB15, CICIDS2017, NID, and TON-IoT have been used in this study. These datasets represent IoT network traffic that includes normal traffic and attack traffic. The attack traffics include ransomware

Table 1 Dataset details

References	Dataset	Anomaly
[22]	UNSW_NB15	DoS
[23]	CICIDS2017	DDoS, PortScan, Botnet, Infiltration, Web Attack including SQL Injection, and XSS
[26]	NID	DoS, U2R, R2L, Probing
[27]	TON_IoT	DoS, DDoS, and ransomware against web applications, IoT gateways, and computer systems on IoT/IIoT networks

attacks, SQL injection, cross-site scripting (XSS), botnet attacks, distributed denial-of-service (DDoS), denial-of-service (DoS) attacks, and absorption attacks. It also covers other types of infiltration, like user-to-root (U2R), root-to-local (R2L), and probing. This large dataset is a critical tool for the creation and analysis of adaptively implemented ML-supported IDS specially designed for the dynamic and evolving threat environment of IoT-based networks. The details of the datasets used are shown in Table 1.

3.2 Preprocessing

We specifically extracted DDoS, DoS, and U2R attack vectors from the above-mentioned resources and made our training dataset. In the preprocessing step, all missing values and duplications were removed.

In practice, preprocessing methods are required to convert unprocessed sensor data and network traffic into a form suitable for analysis. They also encompass solving such issues as class imbalance by using over-sampling or under-sampling techniques, managing missing values, and general data normalization to make feature scales the same. Moreover, one more important area of preprocessing known as feature engineering is a method of forming or compiling accurate characteristics of data to distinguish fundamental patterns between network actions as shown in Fig. 1. As depicted in the figure, the proposed methodology starts with data analysis or preprocessing followed by handling missing values, feature selection and subsequently segregating the features into categorical and numerical parts. The former requires label encoding before being normalized whereas the latter is directly fed to the normalization block. Thereafter the normalized data is split into the training dataset (80%) and the testing dataset (20%) before finally arriving at the output.

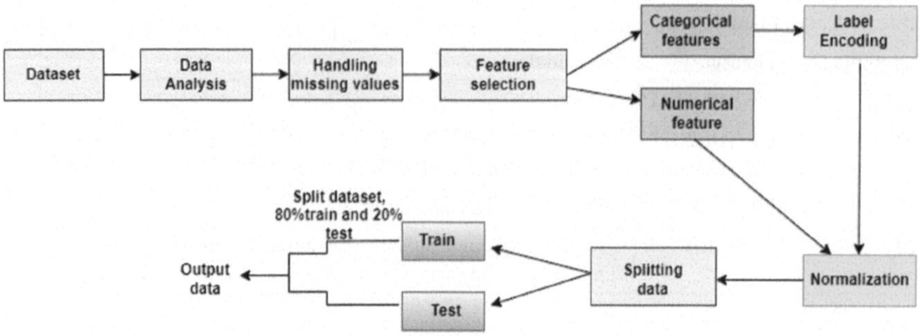

Fig. 1 Preprocessing of data

3.3 Handling Missing Values

To make sure the detection models will be reliable and precise, one should treat missing values delicately during preprocessing. Strong strategies are much needed to manage the effect of missing values in intermittent IoT data streams, such as the issues of connectivity or sensors not working.

$$x_i = \frac{\sum_{j=1}^{n} x_j}{n} \tag{1}$$

Observed values are presented as x_j, non-missing values as n, while the missing value in each entry is an imputed value presented as x_i as shown in Eq. (1). On the other hand, median imputation provides flexibility against outliers by replacing all missing observations of the characteristic with the mean, including the non-missing ones.

3.4 Handling Duplication

While dealing with duplicate data, the detection models are required to always work accurately and stably. Duplicate data instances can result in the model's performance being biased and swaying the learning process. It is considered a common practice to go over every instance in the dataset and pinpoint the dataset's duplicates. This is done by simultaneously comparing each instance to every other instance. If both instances have the same value for all features, then one of them is deleted from the dataset conclusion.

3.5 Convolutional Neural Network

Due to the complexity of the flow of data packets originating from IoT devices, Convolutional Neural Networks (CNNs) can easily recognize both trends and abnormalities. They apply their power of deep learning to identify the difference between possible attacks and standard network behavior on their own. This differentiates CNN-based IDS for the guarantee of more effective security performance. The element of adaptability safeguards and eliminates the possibilities of new threats by adjusting new strategies to minimize risks in the dynamic IoT network environment. Such systems occupy a proactive stance in the cybersecurity environment and thus guarantee the security and integrity of the system's devices and data.

3.5.1 Input Layer

CNN uses its input layer as a gateway to handle network traffic data. IoT devices send raw packet data including payload data, protocol types, source and destination IP addresses, etc. at this layer after preprocessing the data. The preprocessing steps comprise converting the data to numeric followed by normalization and feature extraction. This layer helps CNN identify understated patterns by encoding complex information about network activity, which may indicate anomalies or intrusions and thereby precisely distinguish potential attacks from normal traffic.

3.5.2 Convolutional Layer

Spatial patterns are used by the convolutional layer to extract features from network traffic data that indicate unusual packet patterns or traffic spikes by employing convolution operations. CNN becomes effective by using a series of convolutional layers that gradually extract more complex features to distinguish between malicious and legitimate network traffic.

3.5.3 Relu

Rectified Linear Unit (ReLU) serves as an activation function incorporating nonlinearity into a model by converting the output of a convolutional layer into a more usable format to identify complex patterns. It also aids in detecting irregularities in network traffic data. Several advantages of RELU include.

- **Uncomplicatedness**: ReLU has an unpretentious mathematical representation.
- **Training Speed**: The design of ReLU allows for faster convergence during model training compared to other activation functions.
- **Gradient Preservation:** ReLU helps mitigate the vanishing gradient problem that can occur when training deep neural networks. The problem arises when

gradients become extremely small. It makes it difficult for the model to learn. Updating its parameters effectively becomes a challenge.

By incorporating ReLU as the activation function in an IDS based on CNNs we can anticipate enhanced performance in terms of anomaly detection. The system's classification accuracy will see improvement. This approach promises significant benefits. Relu is represented by Eq. (2).

$$\text{Relu}(x) = \max(0, x) = \{x \text{ for } x > 0;\ 0 \text{ for } x \leq 0\} \tag{2}$$

3.5.4 Pooling Layer

CNN reduces the number of calculations needed while still keeping important information. This is achieved by downsampling feature maps. It uses a pooling layer. Merging nearby features such as selecting the highest or middle values pooling enhances CNN's capability to concentrate on crucial patterns. It can also disregard minor changes, which is evident in network traffic data.

3.5.5 Flattened Layer

The flattened layer is super important in a CNN because it helps transition from the convolutional to fully connected layers. It converts complex feature maps from convolutional layers into one-dimensional vectors. This simplifies the task for the CNN. It does this by transforming the extracted features into a single continuous input. And is particularly beneficial for classifying network traffic data.

3.5.6 Fully Connected Layer

The fully connected layer is a vital part of CNN in handling high-level feature representation and classification. It functions by connecting all neurons in the layer above to those in the layer below. This connection facilitates complex feature combinations. It also enhances pattern recognition. This layer plays a key role in accurate intrusion detection by identifying complex relationships in network traffic data. It uses the features learned from earlier layers.

3.5.7 Output Layer

The flattened layer is extremely important in CNN because it helps to transform the convolutional layer into fully connected layers by taking the complex feature maps from convolutional layers and transforming them into a one-dimensional vector making it easier for the CNN to process the extracted features as a single continuous input. This is handy and thus helps in classifying network traffic data. Equation (3) gives the mathematical expression of the output size.

$$\text{Output size} = \frac{N - F + 2P}{S} + 1 \quad (3)$$

3.6 *The Proposed Model*

3.6.1 Long Short-Term Memory (LSTM)

Long Short-Term Memory (LSTM) type recurrent neural networks (RNNs) excel in analyzing data that occur in sequence over time (i.e., time-series data) by remembering long-term patterns in data sequences. This utility renders them particularly valuable for identifying anomalies in various fields. These include detecting financial fraud and enhancing network security. They are also useful for predicting when equipment might need maintenance LSTMs excel at detecting long-term patterns within data and hence can identify behaviors that deviate from the normal ones indicating malicious activities. This approach particularly excels in scenarios where past trends precast unusual events, making it possible to identify potential issues early on before they escalate.

LSTM can remove gradient issue that results from the effect of activities inverting gradients gradually degrading during computing. The basic building blocks of LSTM are the memory units known as cells. The LSTM algorithm is shown in Fig. 2. It comprises three different components: a forget gate f(t), an input gate i(t), and an output gate o(t). Input and output gates represent the data inputs and outputs at time t. To decide whether or not data will be overlooked, the forget gate compares real-time data inputs with the historical data state.

The relationship between gates in an LSTM cell is expressed by the following calculation:

$$i_t = Sigm(W_{xi}x_t + U_{hi}h_{t-1} + b_i) \quad (4)$$

$$o_t = Sigm(W_{xo}x_t + U_{ho}h_{t-1} + b_O) \quad (5)$$

$$f_t = Sigm\big(W_{xf}x_t + U_{hf}h_{t-1} + b_f\big) \quad (6)$$

$$\hat{a}_t = \tanh(W_{x\hat{a}}x_t + U_{h\hat{a}}h_{t-1} + b_{\hat{a}}) \quad (7)$$

$$c_t = f_t \times x_{t-1} + i_t \times \hat{a}_t \quad (8)$$

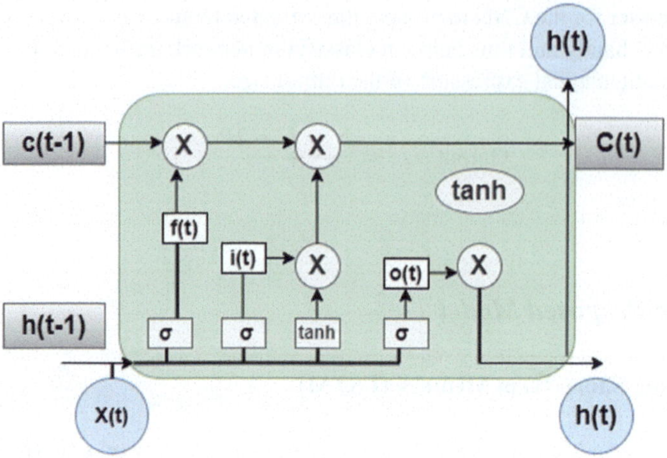

Fig. 2 LSTM basic model

$$h_t = o_t \times \tanh(c_t) \quad (9)$$

Equation (4) calculates the input gate activation i_t, which determines the relevance of the new input x_t to be stored in the cell state. Equation (5) gives the output gate activation o_t and is computed to control the flow of information from the cell state to the hidden state. Equation (6) computes the forget gate activation f_t whereas Eq. (7) computes the candidate cell state $\hat{a}_t$, representing the new information to be added to the cell state. The cell state c_t is updated, combining the relevant information from previous cell states and new information from input and candidate cell states in Eq. (8). Finally, the hidden state h_t in Eq. (9) represents the output of the LSTM cell and carries information to subsequent layers or the final classification layer. Table 2 represents the details of the parameters used in the proposed LSTM model.

Table 2 Proposed model parameters

Parameter	Binary	Multiclass
Activation function	Sigmoid	Softmax
Loss function	Binary cross entropy	Sparse Categorical Cross entropy
Optimizer	Adam	Adam
Learning rate	0.002	0.002
Epsilon	1e-08	1e-08
Schedule decay	0.004	0.004
Epochs	50	50
Dropout	0.1	0.1

3.6.2 Proposed CNN + GRU

In this section, we present the hybrid model by combining CNN and Gated Recurrent Unit (GRU) to effectively detect deviations and attacks in IoT network traffic. Figure 3 illustrates how the CNN portion of the model uses convolutional layers to extract features and spatial patterns from raw network traffic data.

The model can discriminate between possibly harmful and benign activity. It does this by setting specific parameters such as activation functions, strides, and filter sizes for these convolutional layers. Pooling layers are also included to increase computational efficiency and downsample feature maps. Four different datasets are used to train and assess the architecture of the model: UNSW_NB15, CICIDS2017, NID, and TON_IoT datasets. These datasets cover a broad spectrum of network activity situations. This enables the CNN + GRU model to pick up on and adjust to different patterns and irregularities found in IoT networks. As an adjunct to CNN, the GRU algorithm continuously updates the representations of network traffic data based on its inherent motion and framework by using gating mechanisms to selectively update their hidden state at each time step. The algorithm can learn automatically

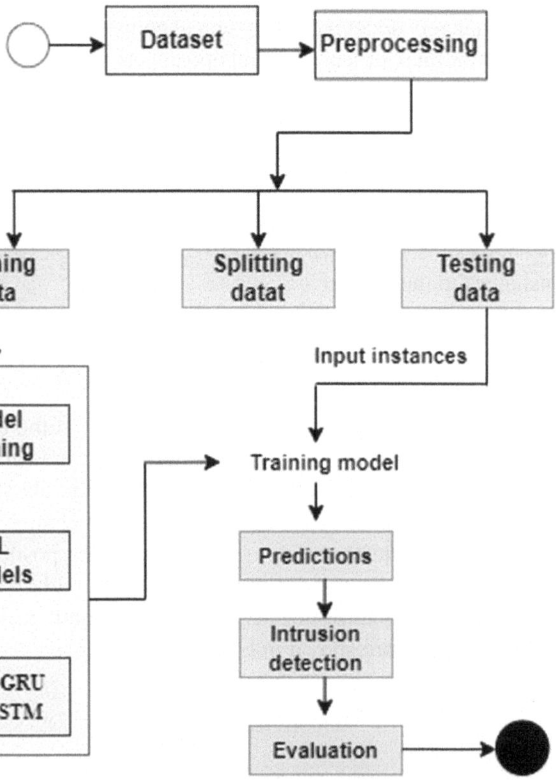

Fig. 3 Proposed CNN + GRU model

Table 3 Parameter details of the proposed CNN + GRU model

Model	Parameters	Values
	Convolutional layers	3
CNN	Filter sizes	3
	Activation function	Relu
	Pooling Operation	Max Pooling
	Pool size	2*2
	Stride	2*2
	GRU layers	2
GRU	Learning rate	0.001
	Epochs	50
	Activation function	Relu, Softmax
	Learning rate	0.001
	Batch size	128
	Loss function	Cross-entropy
	Optimization method	Adam

and identify new dangers or deviations in typical network behavior by parameterizing the GRU layers with rates of learning and epochs, as shown in Table 3.

3.6.3 Evaluation Matrices

The accuracy, precision, recall, specificity, sensitivity, and F1 score are used to validate the expected results. As shown in Eq. (10), accuracy is the percentage of correctly classified instances in the test set.

$$\text{Accuracy} = \frac{\text{TP} + \text{TN}}{\text{TP} + \text{TN} + \text{FP} + \text{FN}} \tag{10}$$

Examples in the test set that are truly positive and that the classifier accurately labels as positive are known as true positives (TP). Instances in the test set that are truly negative but are accurately classified as such by the classifier are called true negatives (TN). Instances in the test set that are truly negative but are mistakenly classified as positive by the classifier are known as false positives (FP). Instances that are truly positive in the test set but are mistakenly classified as negative by the classifier are known as false negatives (FN). The mathematical expressions of sensitivity and specificity are shown in Eqs. (11) and (12) respectively.

$$\text{Sensitivity} = \frac{\text{TP}}{\text{TP} + \text{FN}} \tag{11}$$

$$\text{Specificity} = \frac{\text{TN}}{\text{TN} + \text{FP}} \tag{12}$$

As shown in Eq. (13), precision is the proportion of instances that are truly positive out of all instances predicted to be positive by the classifier. The estimation of precision involves the division of the true number of triumphs by the number of successes predicted by the classifier.

$$\text{Precision} = \frac{\text{TP}}{\text{TP} + \text{FP}} \tag{13}$$

As shown in Eq. (14), Recall is the proportion of true positive instances out of the total number of positive instances in the test set.

$$\text{Recall} = \frac{\text{TP}}{\text{TP} + \text{FN}} \tag{14}$$

As shown in Eq. (15), the F1 score is the harmonic mean of precision and recall, providing a combined measure of precision and recall.

$$\text{F1 score} = 2 * \frac{\text{Precision} * \text{Recall}}{\text{Precision} + \text{Recall}} \tag{15}$$

3.6.4 Experimental Setup

TensorFlow and Python programming are used for the implementation of the model described in the present research. The complete experimental setup was done in Python using Anaconda. Keras libraries are used to build, compile, and test the model. TensorFlow was used to develop the models as a backend, as was Python 3.12 on a computing environment used in these experiments with a 2.7–2.90 GHz Intel(R) Core (TM) i7-7500U CPU and 8GB RAM.

4 Results and Discussion

Experimental evaluation results of the proposed model for intrusion detection and discussions are presented in this section.

4.1 Results

The research showcases the efficacy of IDS based on AML in bolstering the security of IoT networks. Using four datasets the models employ advanced techniques like CNNs and GRU algorithms. The model demonstrates low false positive rates. High precision of detection and efficient resource utilization are also observed. The study emphasizes the importance of hybrid models integrating supervised and unsupervised learning techniques to combat risks posed by cyberspace in IoT settings.

4.1.1 Lstm

The evaluation of various datasets for the LSTM model for IDS in IoT Networks shows maximum values of the following: accuracy of 99.20%, precision of 99.32%, recall of 98.7%, and F1 score of 98.99% proving the reliability in intrusion detection tasks as shown in Fig. 4. These datasets provide a solid foundation for developing and evaluating these systems. They ensure effective threat detection in dynamic environments.

Implementing an LSTM model to detect anomalies in network traffic, particularly for IoT environments has yielded the following performance metrics spanning several datasets as shown in Table 4.

All datasets showed that the LSTM model was highly accurate. This indicates that it is capable of capturing the temporal dependencies and patterns common to data

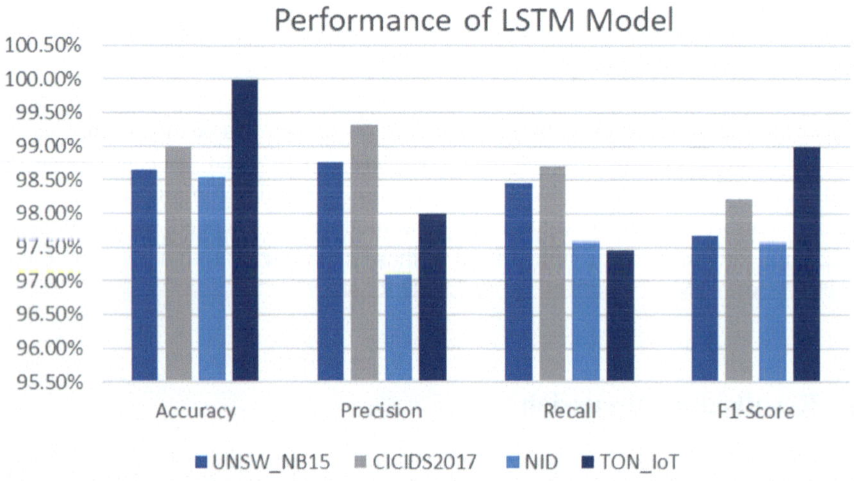

Fig. 4 Performance metrics of the proposed LSTM model

Table 4 Results of LSTM across all datasets

Dataset	Accuracy (%)	Precision (%)	Recall (%)	F1-Score (%)
UNSW_NB15	98.65	98.75	98.45	97.67
CICIDS2017	98.99	99.32	98.70	98.22
NID	98.54	97.10	97.56	97.56
TON_IoT	99.99	97.99	97.45	98.99

on network connections. This is important for anomaly detection. Event sequences frequently indicate malicious activity. The LSTM model successfully integrates the recognition of true anomalies with minimizing false alarms. This is evidenced by the high precision and recall across datasets. The intrusion detection system's dependability and credibility depend heavily on this balance. Since LSTM can learn long-term dependencies it is especially well-suited to time-series data frequently encountered in network traffic. This feature enables it to identify complex target sequences that emerge over time. More straightforward, non-sequential models might fail to detect this.

Unlike conventional algorithms for ML that require fixed-length inputs, LSTM is capable of handling variable-length patterns of network data because of the diversity of network traffic. LSTM installation in resource-constrained environments is a common feature of many IoT devices. It may present challenges due to their high computational demands. LSTM sensitivity is affected by parameter settings. These include the learning rate number of units per layer and number of layers. Determining the optimum setup requires extensive testing. Fine-tuning can be a real resource constraint sometimes because LSTMs are known to be complex and prone to overfitting. It means they are good at modeling a specific set of data, but might not generalize well to unknown data. This is true, especially when the model has been trained on smaller or less diverse datasets. It is important to use regularization techniques like dropout or early stopping to help reduce that risk. The LSTM model is also capable of spotting network outliers across a variety of different datasets making it quite reliable. It could even help improve the security of the IoT networks. However, the research gap remains for achieving perfect results even in situations where there is limited computing power including real-time IoT applications.

4.1.2 Cnn + gru

The evaluation of various datasets for the CNN + GRU model for IDS in IoT Networks shows maximum values of the following: accuracy of 99.55%, precision of 99.5%, recall of 99.6%, and F1 score of 99.55% proving the reliability in intrusion detection tasks as shown in Fig. 5. This hybrid model is premeditated to detect several attacks such as DoS (Denial of Service), DDoS (Distributed Denial of Service), and U2R (User-to-Root). Table 5 shows the performance metrics spanning across several datasets under consideration for this hybrid approach.

The high accuracy score suggests that the combined strengths of CNN for spatial feature extraction and GRU for temporal sequence learning are highly effective in detecting intrusions in IoT networks. These strengths allow us to identify complex patterns that might signal malevolent actions in the network. All the datasets we have seen so far have high precision and recall scores. This shows that our model is good at catching illicit events without accidentally missing anything important. It also shows its capacity for correctly recognizing positive samples. In the field of cybersecurity, where it can be extremely costly to fail to detect an attack, this is crucial. The CNN layers do a good job of extracting key features from the input data,

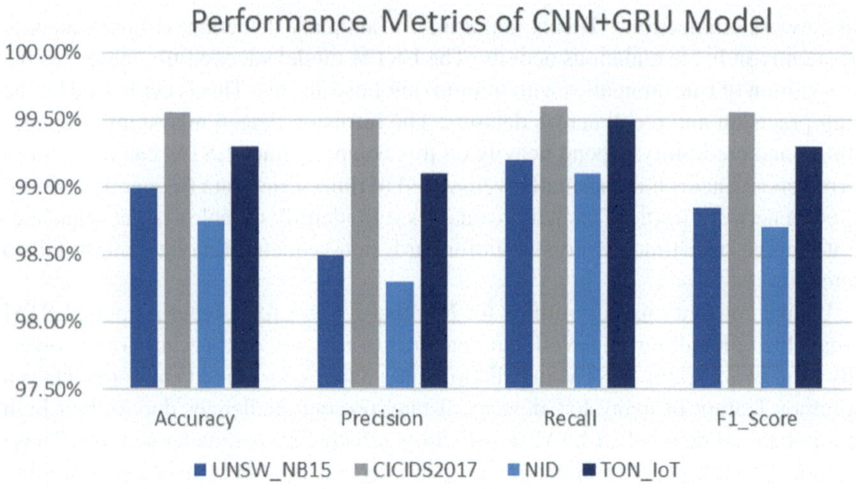

Fig. 5 Performance metrics of the proposed CNN + GRU model

Table 5 Results of CNN + GRU across all datasets

Dataset	Accuracy (%)	Precision (%)	Recall (%)	F1-Score (%)
UNSW_NB15	98.99	98.5	99.2	98.85
CICIDS2017	99.55	99.5	99.6	99.55
NID	98.75	98.3	99.1	98.70
TON_IoT	99.30	99.1	99.5	99.30

which is essential for identifying minute irregularities in big, complicated datasets. In contrast, dependencies in time-series practical data within system traffic are handled by the GRU layers. Because anomalies, or attacks, are much less frequent than normal data in intrusion detection, the architecture is strong enough to handle imbalanced data. Methods such as dropout and data preprocessing (e.g., scaling and balancing) have likely contributed to this robustness. Figure 5 shows the performance metrics of the proposed CNN + GRU model for IDS.

This discussion points out how much deep learning models can help improve cybersecurity in the IoT. But we still need to do more testing and tweaking to ensure these models fit each network's specific needs and stay up-to-date with the latest threats.

4.2 *Discussion*

On a variety of different datasets, both the CNN + GRU and LSTM models perform well in identifying anomalies. When deciding between them, one should keep in mind the specific execution settings and requirements these specifications can vary in what kind of information they are dealing with (spatial–temporal or just plain temporal), how much computing power is available, and if one needs to crunch numbers right away or if one can take their time with it. Depending on the situation, one might have to use simpler models or modify these architectures a bit to make them work better with limited resources. For applications where one needs to dig deep into complicated data and pull out lots of specific details, the CNN + GRU combo is the way to go. But if one just needs to make predictions or classifications based on simple time-series data, then the LSTM might be the better choice. The data shows how intrusion detection methods have improved over time. SVM, LSTM, and CNN have all done well, with really high accuracy rates. In 2024, SVM was the champion, hitting an accuracy of 93%. LSTM was not too far behind at 89%, and CNN did even better with 94.5%. A proposed hybrid model that combines LSTM, CNN, and GRU nailed it with the highest accuracy of 99.9% as shown in Table 6 which makes a performance comparative analysis with related works. To build the proposed hybrid model we have defined each of the LSTM model and the combined CNN + GRU model 10 times resulting in a combination of a total of 20 learners. Then finally, the Max Voting Classifier method was used where the class which had been predicted mostly by the learners had been the final class prediction of the ensemble model. We have used cross-validation, a technique used in machine learning to evaluate the performance of a model on unseen data. This process is repeated multiple times, each time using a different fold as the validation set. Finally, the results from each validation step are averaged to produce a more robust estimate of the model's performance. Here a tenfold cross-validation has been used at the final stage. The Receiver Operating Characteristic—Area Under Curve (ROC-AUC) plots (~ 0.9) for the hybrid model are shown in Fig. 6a. ROC is a probability curve and AUC represents the degree or measure of separability. An excellent model has AUC near to the 1 which means it has a good measure of separability. A poor model has an AUC near 0 which means it has the worst measure of separability. The Mean Squared Error (MSE) and R2_score of the model for all the datasets were found to be ~ 4×10^{-5} and ~ 0.99 respectively, all of which prove its supremacy as shown in Table 7. MSE is used to evaluate the average of the squares of errors between estimated values and actual values and R2-score is used to evaluate the performance of the model in predicting the variation in the output (dependent) variable from the input (independent) variables. It ranges between 0 and 1, with 1 indicating a perfect fit and 0 indicating no predictive capability. The table further shows accuracy, AUC, and confusion matrix. The concept of the confusion matrix is depicted in Fig. 6b. The confusion matrix shows the number of true values and predicted values. The number depicting true normal predicted as normal are called true negatives (TN); that depicting true attack predicted as attack are true positives (TP); that depicting true attack predicted as normal are called false negatives (FN); and the number depicting true normal predicted as attack are called false positives (FP). The proportion of accurately predicted instances to all instances is known as accuracy.

Table 6 Comparative analysis of the proposed model with related works

Year	Technique	Dataset (%)	Accuracy (%)
2024 [28]	SVM	KDD99 dataset,	93
		UNSW-NB15	93.3
2023 [18]	LSTM	NSL–KDD	89
2021 [29]	CNN	CICDDoS2019	94.5
Proposed model	LSTM + CNN + GRU	UNSW_NB15	98.9
		CICIDS2017	99.5
		NID	98.7
		TON_IoT	99.9

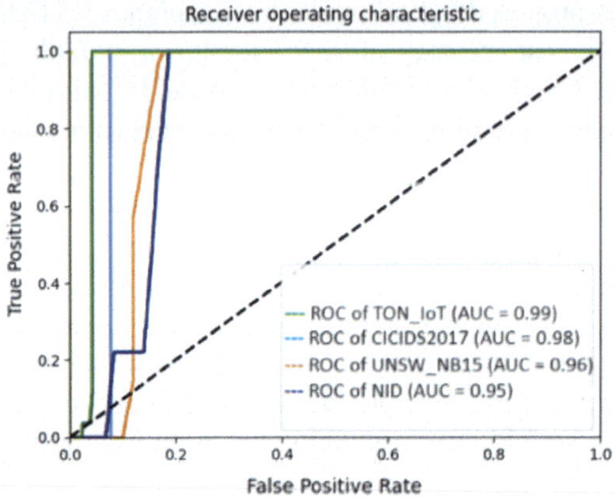

True Normal	TN	FN
True Attack	FP	TP
	Predicted Normal	Predicted Attack

Fig. 6 **a** ROC-AUC plots of the proposed hybrid model; **b** Concept of confusion matrix

Table 7 Performance metrics of the proposed hybrid model

Dataset	AUC	R2_Score	MSE	Accuracy (%)	Confusion matrix
TON_IoT	0.99	0.9999	4.2×10^{-5}	99.9	$\begin{bmatrix} 74737 & 0 \\ 0 & 40524 \end{bmatrix}$
CICIDS2017	0.98	0.9972	4.17×10^{-5}	99.5	$\begin{bmatrix} 99987 & 4 \\ 11 & 73043 \end{bmatrix}$
UNSW_NB15	0.96	0.9955	3.99×10^{-5}	98.9	$\begin{bmatrix} 33835 & 16 \\ 33 & 10524 \end{bmatrix}$
NID	0.95	0.9945	3.95×10^{-5}	98.7	$\begin{bmatrix} 99955 & 91 \\ 56 & 73176 \end{bmatrix}$

4.3 Limitations

The resource constraints inherent variability in data streams and the diverse and dynamic nature of IoT environments present challenges for AML-based IDS in IoT networks. Continued research is required to create scalable and portable AML methods. Although the model exhibits strong performance on various datasets, the distinct attributes of each dataset (such as the frequency distribution of features and the nature of deviations) may necessitate customized adjustments to the model's architecture and hyperparameters. To optimize performance for a particular dataset, one could investigate various dropout rates or add extra layers. The model's capacity to process data in real time has not been specifically tested in this study. Given the heavy computing needs of those CNN and GRU layers, running things in real-time could be a bit tricky, especially on the often not-so-powerful hardware you find in lots of IoT situations. To find out how good the model does against new or unknown kinds of attacks, we need more research. Looking into ways to keep learning could help the model stay strong against new dangers. We need to make sure it keeps getting smarter to handle whatever comes its way.

5 Conclusion

The present body of research deals with the design and performance evaluation of an AML-based IDS for safeguarding IoT networks from intruders. The proposed system shows prodigious performance at detecting and preventing all kinds of advanced cyberattacks, like DDoS attacks, DoS attacks, and U2R attacks in IoT networks by using deep learning models such as LSTM and CNN + GRU. By using a hybrid approach where LSTM, CNN, and GRU are combined we get a more accurate and more precise performance in detecting intrusions in IoT networks based on training four datasets as discussed in the manuscript. The hybrid approach does so by adapting to scenarios where new and unknown attacks come into existence thereby increasing the resiliency of the IoT network defense system. The experiential learning discloses that the proposed AML-based IDS performs well at keeping false positives low compared to traditional models, while also scoring higher in detection rates. Furthermore, our system beats out some of the latest models to show its supremacy when LSTM is combined with CNN + GRU through performance comparative analysis of accuracy metric. This proves that it is effective in preventing cyberattacks by keeping IoT devices safe and secure in an increasingly digital and connected society. In the future, edge computing and AML-based IDS are anticipated to converge, allowing for quicker threat detection and more efficient data processing. Furthermore, by facilitating the growth of federated learning, AML-based IDS will gain from collaborative learning across several IoT nodes without compromising data privacy. This just goes to show that more and more people are working on creating all-in-one solutions for detecting intrusions in IoT networks of all shapes and sizes.

References

1. Neto ECP, Dadkhah S, Ferreira R, Zohourian A, Lu R, Ghorbani AA (2023) CICIoT2023: a real-time dataset and benchmark for large-scale attacks in IoT environment. Sensors 23:5941
2. Raptis G E, Katsini C, Alexakos C (2021) Towards automated matching of cyber threat intelligence reports based on cluster analysis in an internet-of-vehicles environment. In: IEEE International Conference on Cyber Security and Resilience (CSR), pp 366–371
3. Chartuni A, Márquez J (2021) Multi-classifier of DDoS attacks in computer networks built on neural networks. Appl Sci 11:10609
4. Amouri A, Alaparthy VT, Morgera SD (2020) A machine learning based intrusion detection system for mobile internet of things. Sensors 20:461
5. Ibrahim Hairab B, Aslan HK, Elsayed MS, Jurcut AD, Azer MA (2023) Anomaly detection of zero-day attacks based on CNN and regularization techniques. Electronics 12:573
6. Kumar SS, Selvi M, Kannan A (2023) A comprehensive survey on machine learning-based intrusion detection systems for secure communication in internet of things. Comput Intell Neurosci CIN 2023:8981988
7. Albulayhi K, Sheldon FT (2021) An adaptive deep-ensemble anomaly-based intrusion detection system for the internet of things. In: IEEE World AI IoT Congress (AIIoT), pp 0187–0196
8. Boahen EK, Frimpong SA, Ujakpa MM, Sosu RNA, Larbi-Siaw O, Owusu E et al (2022) A deep multi-architectural approach for online social network intrusion detection system. In: IEEE World Conference on Applied Intelligence and Computing (AIC), pp 919–924
9. Satam P, Hariri S (2020) WIDS: an anomaly-based intrusion detection system for Wi-Fi (IEEE 802.11) protocol. IEEE Trans Netw Serv Manag 18:1077–1091
10. Batchu RK, Seetha H (2021) A generalized machine learning model for DDOS attacks detection using hybrid feature selection and hyperparameter tuning. Comput Netw 200:108498
11. Wajahat A, He J, Zhu N, Mahmood T, Nazir A, Ullah F et al (2024) Securing android IoT devices with GuardDroid transparent and lightweight malware detection. Ain Shams Eng J 15:102642
12. Al S, Dener M (2021) STL-HDL: a new hybrid network intrusion detection system for imbalanced dataset on big data environment. Comput Secur 110:102435
13. Cil AE, Yildiz K, Buldu A (2021) Detection of DDoS attacks with feed forward based deep neural network model. Expert Syst Appl 169:114520
14. Mahmood T, Li J, Saba T, Rehman A, Ali S (2024) Energy optimized data fusion approach for scalable wireless sensor network using deep learning-based scheme. J Netw Comput Appl 224:103841
15. Kasongo SM, Sun Y (2021) A deep gated recurrent unit based model for wireless intrusion detection system. ICT Express 7:81–87
16. Pande P, Mathur H, Gupta LK (2024) Machine learning-based intrusion detection system using wireless sensor networks. In: Fourth International Conference on Advances in Electrical, Computing, Communication and Sustainable Technologies (ICAECT), pp 1–10
17. Asharf J, Moustafa N, Khurshid H, Debie E, Haider W, Wahab A (2020) A review of intrusion detection systems using machine and deep learning in internet of things: challenges, solutions and future directions. Electronics 9:1177
18. Zakariah M, AlQahtani SA, Al-Rakhami MS (2023) Machine learning-based adaptive synthetic sampling technique for intrusion detection. Appl Sci 13:6504
19. Khan ZA, Herrmann P (2019) Recent advancements in intrusion detection systems for the internet of things. Secur Commun Netw 2019:1–19
20. Aravamudhan P (2023) A novel adaptive network intrusion detection system for internet of things. PLoS ONE 18:e0283725
21. Jia Y, Zhong F, Alrawais A, Gong B, Cheng X (2020) Flowguard: an intelligent edge defense mechanism against IoT DDoS attacks. IEEE Internet Things J 7:9552–9562
22. Moustafa N, Slay J (2025) UNSW-NB15: a comprehensive dataset for network intrusion detection systems (UNSW-NB15 Network Dataset). In: Military Communications and Information Systems Conference (MILCIS), pp 1–6

23. Sharafaldin I, Habibi Lashkari A, Ghorbani AA (2019) A detailed analysis of the CICIDS2017 data set. In: Information Systems Security and Privacy: 4th International Conference, ICISSP 2018, Funchal-Madeira, Portugal, pp 172–188
24. Saba T, Rehman A, Sadad T, Kolivand H, Bahaj SA (2022) Anomaly-based intrusion detection system for Iot networks through deep learning model. Comput Electr Eng 99:107810
25. Hidayat I, Ali MZ, Arshad A (2023) Machine learning-based intrusion detection system: an experimental comparison. J Comput Cogn Eng 2:88–97
26. Hamid Y, Balasaraswathi VR, Journaux L, Sugumaran M (2018) Benchmark datasets for network intrusion detection: a review. Int J Netw Secur 20:645–654
27. Tareq I, Elbagoury BM, El-Regaily S, El-Horbaty E-SM (2022) Analysis of ToN-IoT, UNW NB15, and edge-IIOT datasets using Dl in cybersecurity for IoT. Appl Sci 12:9572
28. Geetha C, Johnson SD, Oliver AS, Lekha D (2024) Adaptive weighted kernel support vector machine-based circle search approach for intrusion detection in IoT environments. Signal Image Video Process 18:1–12
29. Alghazzawi D, Bamasag O, Ullah H, Asghar MZ (2021) Efficient detection of Ddos attacks using a hybrid deep learning model with improved feature selection. Appl Sci 11:11634

AI and Law

Is India Refraining from technology?—An Abridgment on Technical Aspects Vis-à-Vis Ethics in Legal Meadow

Priyadarshini Bandyopadhyay

Abstract "Cases like Carlill… peculiar facts become the law or any judgments where the interest is retained. It's not the subject, it's the point of interest that is required to retain on it by students", as stated in one of his speeches by Hon'ble former judge of the apex court Rohinton Fali Nariman. We are now in a veritable golden knowledge democracy! The world today is massively impacted by technological advancement in any of the genres and this calls for upgradation of them and judicial implications are not left behind. Even though the presence of scope, technological advancement is relatively slow to the judiciary as there is an urge to "adhere to old patterns". According to the Chief Justice of India, Hon'ble Dr. D.Y. Chandrachud, the Indian judiciary is resistant to including technology, and the very possible reasons prevailing could be that, if the world witnesses the court proceedings, it can automatically call for unanimous opinions from many or there is a very strict tendency to adhere old school methods, despite the changes around us. From the inclusiveness of National Law schools and conventional universities conducting legal studies to the digitization of court proceedings including digital platforms that support the legal meadow and have already segmented towards a lot of positive impacts and a "clear table" that includes preservation as well as challenges. We are in a digital platform-enabled world, where ordinary people are able to challenge knowledge, breaking down in-group and out-group barriers. If the law and legal system are controlled by computers, how much would it matter?—the ethical issues like accessibility, productivity, diversity, and autonomy. Technology should increase the specialization of one's role and improve productivity as well as mobility that increases mobility and autonomy and a rise in personal time vis-à-vis, reduction in individual efficacy and a lack of human involvement in work as we are alienated and focused on wider knowledge and being put into "silos" by technology.

Keywords Technology · Ethics · Silos

P. Bandyopadhyay (✉)
The University of Burdwan, Bardhaman, India
e-mail: barshabanerjee2@gmail.com

M. Chakraborty et al. (eds.), *Proceedings of 5th International Ethical Hacking Conference*, Lecture Notes in Networks and Systems 1148,
https://doi.org/10.1007/978-981-97-8457-8_18

1 Introduction

Alfred Schutz describes a world, in which digital platforms have enabled ordinary people to challenge a knowledge elite, breaking down the in-group and out-group barriers.

Covid-19 proved to be a game changer and a historic event that not only qualifies to be called as said but also is an unanticipated and impactful era that changed the vision of the entire world in multiple aspects, that was hampered due to the lacuna created which automatically turned out to be restrictions which got imposed due to occurrence of rare scenarios, to be more precise, there have been questions on India' testing strategy for the transmission rate of the disease among individuals, as we were greatly focused on the passengers arriving in India from other countries only and lost on actual nature of outbreak. Such machine-based tests are the base for gearing up our country India, as well as several other countries in the world, against the pandemic clubbed with taking decisions on other necessary steps to curb out the health emergency. This created a lesser human interaction and a rise in technological touch and thus, we are slowly being put into silos, by technology.

There is a very spectacular and serious rise and development in the implementation of Artificial Intelligence in various fields and even various strategies are in the process for handling and accelerating the performance of the same at a higher pace.

Software development, use of artificial intelligence (AI) mainly now for transcription and translation for the courts for removing the language barrier while discoursing, Blockchain and Digital accessibility specifically are in emphasis. "*..we are conscious of the fact that artificial intelligence has the flip side as well, for instance we would be very difficult and defeated in allowing artificial intelligence to tell us what sentence to hand down following a conviction in a criminal case, many courts across the world have experimented with it; I don't think that we as judges would ever want to see our discretions which we exercise on sound judicial lines in terms of sentencing policy and at the same time artificial intelligence is replete with possibilities*".—Hon'ble CJI.

The world is massively impacted upon technological advancement, and upgradation of judicial implications is not to be left behind, even though the presence of the scope of such advancement is slow yet happening at its own pace.

As per reports, 4.32 cr. total pending cases, 69 thousand in the Supreme Court, and a backlog of about 59 lakhs across twenty-five High courts. Our judge-to-population ratio is not in correct proportionate in a country like ours and added, according to the United Nations, the ratio of police to population is less than 150 per 1 lakh.

Our judicial structure is a colonial one, borrowed from the British era, which reflects the prejudice of faith and belief in the system vide its doctrinal and empirical way. The highest collegium for judges is constantly deciding the flag bearers of Indian judiciary, based on numerous parameters that includes the no discriminatory rule when it comes in terms of gender perspective or minority. A constant strict evaluation towards the service and seniority is a point of concern too. The independence of the judiciary is considered to be of cardinal value and any kind of outside forces are not entertained as it shall take away the meaning of being truly independent, even though it is opened for public criticism purposively for transparency.

Thus, a logical way of looking into the current era goes straight to digitalization of judiciary which is constantly being questioned if it could not only help with backlog of cases but also how much more the involvement of AI in judicial system can bear fruit.

2 Legal Reasoning—Digitization of Judiciary is a Persuasion or Anarchy?

Aristotle stated that "dialectics" mean the collation of two partially correct striving ideas to come up with a third idea which is considerably more correct as it consumes the accurate portions of the two-half completed, collided views. Furthermore, he affirmed that the expert opinion in a certain area, working on a regular basis, is of high value. The third idea is the outcome of the two competing elements which are the view of the experts in the respective, to come up with more accurate and this is the exact pictorial representation of Human Behaviour that states mostly of "Opinion" and less of "Knowledge" of what should be and not be something else.

To understand the Science behind Law, the methodologies and observations of natural sciences are to be taken into consideration. Nature is governed by law and may likewise work as an essence with respect to humanity and law. Such vast ideas and layers of this involve formulation of a Hypothesis (which are further examined between hypothesis and the actual material reality), examination of it, and thus, asserts or denies it via experimentation.

As stated by Karl Popper—*Law is a science in dialectical sense, as any observation of the dialectic of plaintiff and defendant (thesis and antithesis) before the judge (synthesis), should show.*

That, the life of law is more of experience than logic, Jensen is absolutely right in saying that formal logic is not the base for the kind of decisions made by judges, the real query lies in whether there is availability of some kind of logical expositions that "compel" or "force" judges to make decisions in a certain way. If logic is considered as a set of principles in the way where we evaluate the authenticity of arguments, the existing question that still lies is whether the law really provides a bunch of principles on the basis of which we can evaluate the validity of the legal supposition and arguments.

The new Globalization challenges that intertwine today's world with respect to the development of Law and justice. The rise in knowledge and circulation of information cross-borders, demands a legally unified approach in along with equality and access to law, added, fairness and transparency, along with accountability and lined-up legal system with inclusivity of legal proforma and a higher outreach of stabilized international law. This has a vital role with respect to the inclusion of artificially injected tools in the legal system that is now slowly adhering in the entire process of justice delivery.

In our current technological era, which is going ahead by providing us with lots of information as well as acts as an aid in our almost every genre of work, along with even working as our replacement, which is automatically throwing us on a verge of being alone or in *Silos*, thus, have created an alarming scenario where we have no idea of, how much a human brain is capable of processing infinite amount of information via technology, at one go? Or how much information is too much information?

During the evolving era of technologies, i.e., almost 1960, the primary objective for machines was to perform a certain task faster and now we are more concerned about whether they are becoming our threat instead of a help!

When we talk about empirical work in legal research, we are always focused on the beneficiary partnership bestowed by technology, which includes conceptual, planning, collection or empirical, interpretation, and the diffusion phases.

If we take the real example of DABUS (**Device for Autonomous Bootstrapping of Unified Sentience**), a technology that is in talks for its patentability claims, where the inventor Dr. Thaller claims it to be human-like as well as sentient and develops ideas, as it is neither created for problem-solving nor having been trained with exclusive data', thus, having a human-based approach in understanding situations and act as required along with the capability to foresee, qualifying it to claim invention in its (DABUS) own name (Fig. 1).

As per reports, over 4.7 crore cases are pending across Indian courts. Almost 87.4% are at a pending juncture with district court and about 12.4% are pendency with High courts and about 70,000 cases are pending with the Supreme Court. Based on the same works as various aspects of technology and ethics in the legal meadow, the Generative AI that does the "lawyerly works" and blows winds towards the discussion on the pros and cons of over-inclusiveness or under-inclusiveness of the

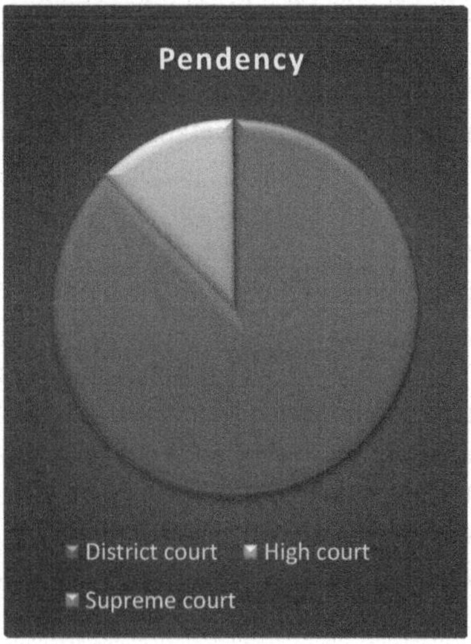

Fig. 1 Pendency of cases

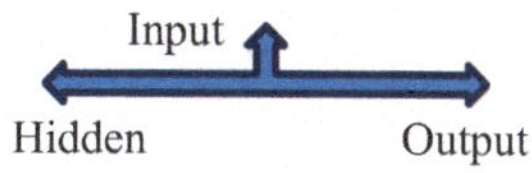

Fig. 2 Node layers (neural network)

machine learning and digital tools that are now being used in the judicial system and everyday playing pivotal role.

To talk about—adding values to digital transformation, these are hard and digital transformations are harder. The facets of organization that add value to the work are in association with external partners to come up with a new operating model. Firms have engaged in digital conversion by coordination, automation and outsourcing. Organizations have changed for good and basic, that targets to the direction towards the creation of values and that too from external partners.

2.1 *Man vs. Machine*

Let us understand a bit about the mechanism of computers! No, not like software engineers but like a lawman with a finite knowledge (Fig. 2).

For example, if we look into Neural network, it reflects human brain behavior, allows computer programs to recognize a particular pattern, and solves common problems in the field of AI/machine learning. Data sets are passed in these Node layers and they rely on training data.

We have learned in school the mathematical model of linear regression, which is used to predict future scenarios; now, if we think of these nodes or this artificial neural network (ANN), as its self linear regression model, then the weight of the connection that is between the nodes, determines the amount of influence each input has on the very output. Now, the data is transferred from one layer in the neural network to the other and is fed forward network. One premediated single node in the neural network decides, whether we should go surfing, and the decision to move on or not with it is exactly our predicted outcome.

Let us look into the real implementation of them moving on.

2.2 *Adaption of Artificial Intelligence by the Apex Court*

Since 1950, more than thirty thousand judgments are given by the Supreme Court of India. Our Chief Justice is on a mission to make these judgments easily accessible and free of cost. The motto of digitization of Indian courts is to be executed in such a manner that these judgments are available to the public at large not only Indian citizens but also anybody across the world, via official Supreme Court reports that are accessible as well as having its free judgment text portal.

AI and Machine learning are used even to reach out by making available the judgment in the language that the citizens of the country understand. India is a diversified and multi-lingual country, where the medium of communication varies.

A team of experienced software developers worked towards the process of developing this particular software that helps in the digitization or translation of judgments in a constitutionally valid manner which is now known as Vidhik Anuvaad software (SUVAS).

In furtherance to the motto, the apex court's intelligence committee also launched its portal named Supreme Court Portal for Assistance in Courts Efficiency (SUPACE), which particularly being the idea of a tech expert team led by Manthan Trivedi, who is of the opinion that,

> "*Keeping that in mind—a very farfetched idea to be very honest—we designed a system that was to assist the human*".

Live transcripts of proceedings are available and being adapted by the apex court, when the lawyer is arguing his case, a complete text is made available on screen in the courtroom, automatically making it a court of record and accountable to its citizens. According to lawyer and tech expert Vikas Mahendra, who assisted in the transcription through AI, it is of the opinion that this tool could be beneficiary to the district judiciary that records the greatest number of evidences and such a tool will improve the efficiency of the court. Added to accurate and speedy disposal of cases, issue of pendency of cases will be tackled with ease.

The digitization of the court proceedings with the help of Supreme Court Interact (SCI) software, will help in the reduction of paperwork by making it paperless and "clean table", that includes preservation. From e-filling of cases, Live-hearing of the trial and arguments, Tele-law which is an e-interface and pre-litigation mechanism, Nyay Bandhu which is a *pro-bono* legal service portal and a few others play a crucial role in the mechanism for techno-legal advancement.

- *Impending Areas of Closer Assistance by AI*

If we look into the various genres of a repetitive arena in cases of law, we may find certain areas that are exactly the same in each and every scenario when the cases are being filed before the court. Such data and patterns when analyzed, could be seen to have a repetitive nature. Automation could be introduced to help in disposing off the cases sooner.

Added, from the perspective of para-legal jobs, legal analytics and robotics could be an inclusive help in doing the same or similar kind of jobs on a daily basis which will ultimately lead to speedy disposal of cases, as it also helps in fetching of dates of hearing and avoids pendency and back logging of any particular case.

Fair justice is not rabbit run but tortoise moves—to which nowadays everyone won't agree but are more interested in moving towards drawing fast conclusions where we are more likely to make mistakes. It is open to readers to draw a perception of two separate contentions, one being positive and the other negative.

In a scenario of the same genre and lesser merit-based cases, if we think of coming up with software that is based on algorithms and machine learning, for a repetitive scenario based-judgment that saves time and effort, as well as deciding the questions in law, may prove to be a speedy remedy and could be viewed as the conception of online dispute resolution without human intervention.

Taking positive inference into account, we can take note of the Automated Number Plate Recognition for the real time identification of vehicles considered as a main tool for security and transportation. Such a system is designed to be used for the extraction of information and providing aid to the Road authorities, Parking authorities, and the legal system as well. This particular automated tool is said to be adopted in India as well. This may be a positive mechanized way of introducing new areas of alternative dispute resolution for saving the court's time which may be accompanied with settlement.

Inference can be drawn in a negative approach as well, in the example of implementation of AI tools in direct legal machinery for imparting justice, which is of huge concern if we look into the inefficacy side of it; AI tools used in the U.S.A. COMPAS (Correctional Offender Management Profiling for Alternative Solutions) helps judges in terms of risk assessment by determining factors such as criminal history, social and economic background, and mental health to anticipate the presence of the habit of relapsing into crime, have proved to make mistakes and creating a serious ambiguous discrimination in black and white offenders as well as most likely picking up wrong offenders, neglecting potential risk factors, unimpressive explanations and so on. Thus, putting their liberty at risk, whose mispredictions may even lead to injustice and threaten lives. This tool has raised eyebrows of many and is a massive flaw.

Hence, any technological inclusion is warmly welcomed until it affects our basic principles of delivering justice and keeping our faith intact upon judiciary.

2.3 *AI Vis-à-Vis Ethics*

We are in a veritable golden knowledge democracy! Not only the demographic boundaries but also the man-made differences of any sort that are yet prevailing are being curbed out by the technology. The accessibility of AI through various devices that are available in any form to almost every one of us is creating a huge difference in the way of life that we thought of not less than a decade ago. Yet, on the other hand, if we look into, the marginalized community lack access to technology and internet that is automatically widening the disparity between access of those with and people with no access.

In a parallel domain, we are coming up even with a concern about whether an AI may replace lawyers? It has boomed almost every stakeholder's mind concerned, but such prediction in favor would be to a great extent incorrect as we have to understand that a Generative AI tool is there to help with the quality of the work by even saving time, in association with its human user. As per the survey report of LexisNexis, almost eighty-four percent of legal professionals believe that Generative AI will increase their efficiency as well as revolutionize the practice of law.

And to understand the impact further, we are supposed to categorize AI into primarily Weak and Strong AI.

For robots, they need a sharp foresight and an existing enough detail about the environment as well as an updated complete model of a given scenario. The results of a planned action can be given but are largely dependable upon changing scenarios as well as the robot's condition in a given situation. Such could be specific and applicable for a given scenario while another conception could be derived from a traditional view.

On a view towards the counter side, if we are concerned about certain scenarios like, if giving up our privacy to connect with others, how much will it matter? Or, if law and the legal system are controlled by computers, how much would it matter? Whether a software can be allowed to make life-changing decisions? How better can they do it?

This exact scene calls for understanding Ethical Issues which include in itself Accessibility, Productivity, Diversity, and Autonomy.

In looking into the usage of AI in the criminal justice system, reference could be derived from the case of Justice K.S. Puttaswamy (Retd.) and Another v. Union of India and Others [1], the Supreme Court of India held that right to privacy is a fundamental right under the Indian Constitution. This decision could potentially be used to challenge the use of AI in the criminal justice system if it is found to violate the right to privacy.

In the case of Pradyuman Bisht v. Union of India, which focuses on the gun firing incidents of courts, that led to a question on security and thus installation of CCTV cameras was demanded. Digitization of Indian courts for future should include tools which can be used in all procedures of trials and determination of offenders and offenses. One such segment used in U.S.A. is COMPAS (Correctional Offender Management Profiling for Alternative Solutions) which helps judges in terms of risk assessment by determining factors such as criminal history, social and economic background, and mental health to anticipate the presence of habit of relapsing into crime.

In the case of United States v. Loomis [2], the Ninth Circuit Court of Appeals held that the use of a risk assessment algorithm which is generally used to determine the risk of escape for offenders or violence and such information is used for security and programming, where in a pretrial detention decision violated the defendant's right to Due process. The court found that the algorithm was not sufficiently transparent and that it was not clear how the algorithm was making its decisions.

Along with the various upsides of using AI and certain black boxes AI tools, whose usages are restricted up to the system, yet certain evidentiary benefits that incorporate in law enforcements, viz: the predictive nature of AI for improved public safety, Automatic license plate scanner that detect the vehicles associated with criminal activities and thus raising the efficacy of Motor Vehicles rule implementation along with reduction in crimes due to early detection, Recognition of faces or finger prints can eventually save the amount of time procured to detect an offender, if he is a new one or a recidivist. The discussions on benefits in courts are already made above.

Recently, there has been a positive impact on Bengaluru's traffic conditions. From the time, Bengaluru traffic police adapted the AI-based tool known as Actionable Intelligence for Sustainable Traffic Management (ASTraM), from the month of January 2024, there has been an effective monitoring of traffic flow patterns. This tool not only provides traffic jam alerts but also reports situations that delay traffic via mobile applications, along with Traffic Management Center (TMC). Additionally, they have also identified via this technology that vehicle breakdowns, as per data considered to be 71% liable for a total of over 2462 road incidents, being one of the major reasons for road congestion.

- *Ethical Concerns*

The challenges include vulnerable data availability and ultimately leading to compromise in data security, lack of understanding and various Privacy concerns which are further pathways to Ethical concerns.

In an overview, Generative AI can lead to concerns that include the creation of fake content which can be gradually used for misinformation or fraud, a serious attack on Transparency and Data privacy implications in case of handling personal data, and added potential biases in any primary or training data which can be further enhanced or amplified in the final or generated content. It even includes any unauthorized use of copyrighted elements as well.

(a). Predictive Algorithm can lead to serious discrimination when it comes to rendering justice. Algorithms can falsely reflect certain data that create concerns among defendants. In many other countries like the U.S.A., such also leads to accidental grievous racial discrimination, and thus, be difficult to hold AI accountable for its deeds.
(b). All cases are unique when it comes to factual aspects and for rendering justice, personal involvement with respect to the application of mind by the judges is highly expected. With the AI, the greater chances lie with interference regarding the judge's discretion as when it comes to any ambiguous laws, such are removed as and when required to the extent of ambiguity, such work is impossible by AI as the algorithms have already freezed the data.
(c). Influence of third party is highly expected as well because such models can be biased, lack transparency, and if being faulty in nature by itself from inception, it might come up with faulty decisions that would automatically disrupt fairness, as they are risky.

These negative ethical concerns automatically interfere with the Due process and the procedure established by law, leading to question our fundamental right to a fair trial. We are using the tools and heading towards putting ourselves in *Silos* by these technologies. But how much is judiciary getting affected by it is the question of the hour.

3 The Flip Side of the Coin—An Anticipatory View on Inclusivity of AI

Indian Judiciary as an institution along with its stakeholders is resistant to include technology and the probable reasons are certain to prompt and may cause long-term hurdles that might require an urge for rethinking, for the transformation to a digital era.

The concern may rise with regards to the world witnessing the court proceedings, it can call for unanimous opinions from many and there is a high wish for adherence to old-school methods despite the changes around. Such practice could be seen in the lower courts or among the district bar and bench, as the tendency to follow a convenient way is always acceptable than a wish for upgradation via tech methods.

We are extremely dependable upon our judicial institutions as they call for the most realistic and ideal savior when it comes to any of our crisis that are substantial in nature.

The torch holders of this phenomenon are chosen very carefully and only after a rigorous training, they are qualified to be the upholders of our constitution.

In such a scenario, what concerns the most is the "over inclusion" or "under inclusion" of AI. This is to be looked minutely and in a detailed manner because such will lead to a concerning factor that questions on "application of mind by the judges", or cuts down the autonomous behavior of the judiciary, which may eventually create a doubt in the mind of the citizens that might led to a lower faith in the judicial system.

Various strategic studies have been published that discuss about how and where AI is going and leading us along. Tech giant Infosys' owner Mr. Narayan Murthy, once said, on being asked if software engineers will lose their jobs for Generative AI (ChatGPT), and he replied.

> *"... no no! it will never happen... because of a very important reason and that is the human mind is the most flexible instrument. It can adapt very well and all that happened was people started solving bigger and bigger problems...we use ChatGPT as the base and then show our creativity, show our smartness, show our innovation".*

The focus on AI usage includes administrative efficacy as well as expanding decision-making process. Recently, a well-researched paper has come up with a cumulation of ideas that mainly deal with the digitization pattern of AI in the judiciary. It has documented three important questions viz:

(a). Whether machine learning tools can assist in the intelligent scheduling of cases;
(b). Whether an AI-enabled program can extract the exact position of law from a pile of precedents referred;
(c). Whether the robots decide the question of law;

We are at the stepping stone with regards to the usage of AI. The question regarding the inclusivity of AI may pertain to be looked into the global arena that has already laid down certain tools to improve procedural efficacy and aid in decision-making phenomena. With a goal to eradicate the arbitrariness in human decision-making, predictive justice tools are already in work with the larger justice system. The versatility of the impending technologies is already in discussion as per global surveys and even the marginal challenges that procure on a short-term or a long-term basis, inside the judicial institution.

To answer the first scenario above, the administrative work done by the judges could be curtailed a lot that is done along with the judicial work and will reduce in time consumption. The basic technical hurdles while accessing the database are to be looked into and solved. The accessible data is required to be seen and made available to be used by the public at large and is expected to be glitch-free.

Remedies like legal bots can come as a savior to prospective litigants with upgraded and informed decision-making with respect to their legal rights and with readily and cheap access to basic legal services. The workloads could be categorically looked into and further be divided accordingly.

Any aid to decision-making process in the judiciary primarily saves time and cuts off pendency and so, AI-driven integration is welcomed. The data' are accessible by not only Bars and Benches but also the public at large. The computational tools for justice delivery systems made with AI have been already experienced in the process automation with regards to traffic challan cases through online payment mechanisms. Generally, motor vehicle compensation claims are largely calculated and are based on the established methodology that hardly involves interpretation, such matters if looked at by AI and machine learning tools, can prove as an aid to judge. Added, the predictive justice tools are also surrounded with concerns which are logical in nature; as these do not require human interference and should aid in the judiciary instead of itself being the judiciary, such may lead to thoughts with regards to societal biasness, cloudy decision-making, curtail down the judicial autonomy, etc.

The huge and already established segments for basic and frequent that are even similar questions of law could be answered using AI tools. The legal services that emerge out of either pro-bono or paid could be made readily available via the tools. We still have many such operating websites that are already established and known to the general public who deal with similar scenarios.

3.1 AI and Process in Law

The "Process" is about the discussion on both the doctrines of Due process in Law and Procedure established by law, in a nutshell.

Legal procedure established by law means that it is going to govern all the facets and no arbitrariness is allowed. This means that when a law is duly enacted, it is required to be followed in the exact manner as it is provided it is not taking away any rights bestowed by the constitution. The focus is mainly on following a strict process of law and enacting, "Law as it is" or "Law as it ought to be".

Due process in law primarily refers to the constitutional guarantee to protection of individual rights of citizen and even to see that the law is fair, just, and as well as non-arbitrary. The government should mandatorily follow the fair and just procedures before depriving an individual and as well as required to be heard "*audi alterum partem*" along with evidences to be seen for ensuring reasonability and natural justice. This includes both procedural and substantive as to whether constitutional rights are not overlooked in the conviction and proper procedure is followed and also questions the government process or enactment if found ambiguous.

The doctrines are to be ensured in application in both pre-trail and trail stages and no such deprivations should occur. It puts a check on coercive or arbitrary powers of State as well and calls for the right to fair trail along with access to justice with properly obtained evidence as well as review when it comes to miscarriages of justice.

Judicial decisions are required to be transparent, consistent, and scrutinized. The involvement of AI in an adverse way can really disrupt the entire procedure and might call for mayhem.

3.2 Moral Compass

From the readings of an international bestseller, I appreciated the work of Hayao Miyazaki, a Japanese director, who in almost all his films showed humans, fantasy, technology, and nature being initially in a state of conflict and in the conclusion coming together. Additionally the authors came up with the view that the Japanese are skillfully efficient when it comes to bringing the nature and technology together, significantly not man versus nature but man and nature together.

Despite the other questions of law (set of rules), we are addedly focused on the implementation of AI in law, which raised the question of the basic jurisprudential idea of law and morals, where even though there are no clear differences between them yet, the general values that be buoyant among the people, automatically creating a correspondence between the two ideas, whereas, many eminent jurists were mostly of the view of keeping law and morality in consonance, added to consider the ultimate intention is societal maintenance with fewer conflicts, morality too considered as ends of law as well.

The fast-rising technology is slowly too becoming an integral part of the world's largest parliamentary democracy, and so, the concern lies in its positive sustainability, in a justified manner, even though legislation like Information technology has shaped India.

Every law is made for the public good and achieving justice, equality, reasonableness, and order, and according to natural law theory any such law in contravention held ultra vires. In an equivalent manner, we are concerned with the inclusivity of AI. Though law and morality are deeply connected, we may have the same morality but we differ in mentality!

If we are concerned about the machines themselves and expect them to act morally then we should also not forget the man behind the machine who actually fed the mechanism in it for-making it work in a particular way. We somehow may be bothered about the thesis of morality that the man behind is bearing.

3.3 *Conclusion*

This is of no debate that digitization is a boon and has eased litigation and also made it both time and cost-effective on the part of every stakeholder in it. Humans as well as technologies are frail; benefits and challenges goes hand-in-hand, and we have to be careful about the operational preparedness on each one's part. Upgradation of system and constant review of them along with raising the security values of them are to be viewed as a need to carry on the positive effect of the machines in our lives. It is to be looked into, the non-failure of technology and a standardized system to be implemented that is non-fallible rather than talking about the failure of the courts, as the huge shifting process requires and involves strong and poignant measures to be taken by the concerned parties. There is no such information about AI replacing judges, and so it is strongly advised to not draw a strict negative apprehension against the faith of judiciary in its motto of digitization but to be conscious about the process.

References

1. Puttaswamy JKS (Retd.) & Anr. vs. Union of India & Ors (2017) 10 SCC 1, AIR 2017 SC 4161
2. State v. Loomis, 881 N.W.2d 749 (Wis. 2016)

Analyzing Copyright Infringement Liability in Web Scraping Cases—A Critical Examination of Recent Global Trends

Arunabha Banerjee

Abstract Using copyright information from protected literary or artistic copyrighted works is on the rise as suggested by an influx of global cases on this subject matter. For instance, in Anderson v. Stability AI, et al. the plaintiffs alleged before the United States District Court of Northern California, San Francisco Division, that JPEG files generated by the defendants' Stable Diffusion and Mid Journey Product AI have been derived from the protected artwork of the plaintiffs. But the claims were set aside by the Court on the ground that the plaintiffs failed to allege adequate facts or substantiate their claims with sufficient evidence. In cases where wholesale copying could not be established through adequate evidence, the courts in the US have refused to find infringement. The issue merits an investigation into how web scraping can be tackled more effectively under the existing international copyright regime. Accordingly, the paper seeks to address the following questions: (a) Whether AI-generated content based on copyrighted works qualifies as derivative works? (b) Can web scraping be protected under the fair use exception? (c) In the event of infringement, can the AI owner be held liable for the activities of AI in using copyrighted information?

Keywords Artificial intelligence · Web scraping · Copyright infringement · Fair dealing · Authorship

1 Introduction

An artificial intelligence ('AI') programme may have looked at 10,000 copyrighted paintings online to generate its own version of painting. This unauthorized act of 'web scraping' [1] has, in recent times, attracted infringement suits from the copyright holders in the accessed works, prompting a series of intriguing questions under copyright law. Firstly, do these generated works constitute derivatives of the original protected works? Secondly, if the question of infringement is answered in affirmative,

A. Banerjee (✉)
Assistant Professor, Faculty of Law, National Law University, Jodhpur, India
e-mail: arunabha.nlu@gmail.com

M. Chakraborty et al. (eds.), *Proceedings of 5th International Ethical Hacking Conference*, Lecture Notes in Networks and Systems 1148,
https://doi.org/10.1007/978-981-97-8457-8_19

can the owner of AI be entitled to the defence of fair dealing? Thirdly, can the AI be considered as the author of a generated work, and thus, by rule, the first owner of the work or should such an ownership vest with the creator of the AI? The legally recognized owner must not only command rights but also bear liabilities in relation to the generated work. For AI to claim ownership and become liable for infringement, it must possess the requisite legal personality. If the ownership goes to the creator of the AI, the law must determine whether such owner had 'caused' the infringement because the infringing act can only be directly traced to the AI and not its owner who had simply programmed that AI. The present paper analytically explores these three questions and seeks to find a solution within the existing framework of copyright laws in the United States of America ('US') and the United Kingdom ('UK'). The choice of the jurisdictions has been influenced by the sudden increase in web scraping copyright disputes in both these countries in recent times. The paper has been divided into four sections. Section 1 introduces the research questions and outlines the approach to be taken by the author in finding the appropriate solution. Section 2 provides a gist of the ongoing relevant copyright disputes in the US and UK. Section 3 identifies and explains the core copyright principles of infringement, fair dealing, and authorship in the US and UK. Using the understanding developed under Sect. 3, the paper, in Sect. 4, will include an in-depth critical analysis of the issues identified in Sect. 2 and offer solutions for each issue through the corresponding sub section on conclusion. The flow of ideas running through this paper may be depicted as under:

Section 1: Research questions and Roadmap

⇩

Section 2: Cases and relevant arguments in the US and UK

⇩

Section 3: Theoretical Understanding of Copyright Infringement, Fair Dealing and Authorship

⇩

Section 4: Critical Analysis and Corresponding Points of Conclusion

2 Recent Cases on Web Scraping: A Global Review

Recently, multiple copyright infringement suits alleging web scraping have been filed in the United States of America by persons holding copyright in protected works available online. For example, in Anderson v. Stability AI, et al. [2] the plaintiffs including the likes of Jeff Koons, Mc.Kernan, Andy Warhol, and Ortiz filed a suit in the United States District Court of Northern California, San Francisco Division, and contended that the JPEG files generated by the defendants' AI programmes Stable Diffusion have been derived from the protected artwork of the plaintiffs. The plaintiffs contended that the other defendant DeviantArt.com has been persistently violating the copyright of the various authors who have uploaded their paintings on the website. Deviant Art's actions represented a violation of their own terms of service. The works generated by the users of Deviant Art constituted unauthorized derivations of the plaintiff's protected works. The infringing works were reflections of the plaintiffs' creations and the resulting infringement allegedly happened at three levels:

(a). Copying of protected images into Stable Diffusion to create the training data set by employing Large Artificial Intelligence Open Network ('LAION') through the method of 'diffusion'.
(b). Distribution of Stable Diffusion to users of defendant websites.

Thus, it was argued by the plaintiffs that the output images generated by Stable Diffusion ought to be treated as derivative works of the copyrighted images belonging to the plaintiffs. But the claims were set aside by the Court on the ground that the plaintiffs failed to allege adequate facts or substantiate their claims with sufficient evidence. While granting leave to amend, the court noted that it was unclear how the five billion protected images could be compressed into Stable Diffusion. Stability, the AI developer, successfully established that Stable Diffusion used mathematical and statistical models to compress conceptual elements from the images. Thus, the plaintiffs failed to prove how the incorporation of the concepts from the protected images and allowing access of the AI to the users amounted to unauthorized reproduction. The court also refuted the contention of the plaintiff that substantial similarity need not be proved in case of direct copying. As opposed to the ruling in Range Rd. Music, Inc. v. E. Coast Foods, Inc. [3] wholesale copying could not be established. The court drew support from Litchfield v. Spielberg [4] and Authors Guild v. Google, Inc. [5] where it was observed that a derivative infringing work must be substantially copied from the copyrighted work. The protection would extend only to the 'expressive content' of the work and not its general and unprotectable aspects.

A similar suit was also been filed against the defendant in Getty Images (US), Inc. v. Stability AI, Inc. [6] before the Delaware District Court on May 12, 2023, where Getty Images ('Getty') alleged that the Stability had indulged in unauthorized use of its copyrighted images to train 'stable diffusion', the AI programme owned by Stability. According to Getty, this act of training not only violated Getty's copyright on the images used for training but also defeated several licence agreements executed by Getty with numerous technology innovators. An unfair advantage, as a result, was conferred on Stability.

In Getty Images (US) INC. and others v. Stability AI Ltd. [7] contested before the English High Court, Getty also claimed secondary infringement under sections 22, 23, and 27 of Copyright, Designs and Patents Act, 1988, ('CDPA'). Pursuant to amendment applications filed by Getty in October and November, the final judgement was pronounced by Mrs. Justice Smith on December 1, 2024. The court refused to find merit in Stability's arguments that no infringement had taken place in the UK and that Getty had no real prospect of success. Pointing to the remarks made by Mr Emad Mostaque, founder and chief executive officer of Stability AI, in several media interviews, the court observed that both these arguments were contradicted by Stability's conduct in electronic media. A summary judgement finding no infringement could not be issued at this juncture without having a full-fledged trial. The evidence placed on record by Stability did not justify a dismissal of the infringement suit at such a preliminary stage. Likewise, the court preferred to have a more extensive trial to determine whether infringing 'articles' under sections 22, 23, and 27 of the CDPA can have an intangible nature. In March, 2024, Stability submitted its first statement of defence outlining its process of generation of images by Stable Diffusion. The arguments were twofold. Firstly, the alleged infringing acts took place in the US, where the servers of Stability were located, hence preventing courts in the UK from having jurisdiction in the current suit. Secondly, assuming that the UK courts did have jurisdiction, the generated images did not result in infringement as their nature was dictated by the user's prompt rather than the content copied by Stability. This 'eccentric and contrived use' lent itself to the application of the fair dealing exception provided under section 30A of the CDPA. Section 30A protects infringing use which constitutes a mere 'imitation of style' from a 'multitude of elements' in protected works. Stability depicted its content generation process through the following diagram:

Based on the process outlined in Fig. 1 [8], Stability argued that the content generation process by Stable Diffusion did not involve a verbatim reproduction or use of substantial elements from the protected works of Getty. The original works were encoded with 'noise' to aid creation and storage of additional copies, and then decoded to make sure that Stable Diffusion could match the user's prompts with the subject matter of the protected work and produce the desired result for that user. The programming of the software in Stable Diffusion allowed each user to generate synthetic images based on Stable Diffusion's comparison of the protected works and decoded copies. The encoded model weights in the software also differed markedly from the copyrighted images in the training data because these model weights only contained data structures and not the data itself [9].

In J. Doe 1 v. GitHub, Inc. [10], the California District Court refused to find infringement in a case where the plaintiffs alleged that the defendant used the copyright protected code to train its AI programmes Copilot and Codex. The plaintiffs

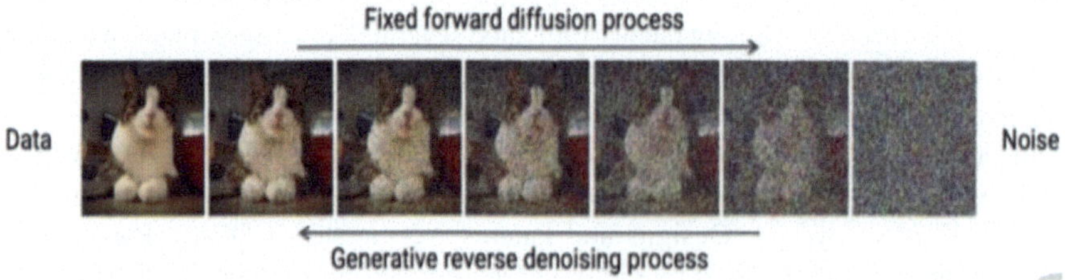

Fig. 1 Content generation process in stability AI

had shared the code on GitHub under an open-source licence where it could be used subject to the conditions that the user must attribute the use of the code in his or her work along with a copyright notice and the relevant licence agreement. The Court could not find sufficient evidence to establish that the defendants had actually used the plaintiff's code, however, it granted the plaintiff's motion seeking an injunction on the use of the code by the defendant based on probable harm in the future. The plea for monetary compensation was declined as the Court did not consider the harm to be sufficiently 'concrete, imminent or substantial'.

The factual complexities of web scraping cases are perfectly illustrated by Thomson Reuters Enterprise Centre GMBH v. Ross Intelligence Inc. [11] where the Delaware District Court declined summary judgement on the purpose and character of use, copyrightable character of the work, transformative nature of the allegedly infringing work, and effect of relevant market. The court opined that except for access of the defendant to the plaintiff's database which could be determined as a matter of law, the above four elements of the fair use defence raised triable issues as to fact and would require a jury trial. At this stage, it remains unclear as to how much copyright protection the headnotes to the judgements uploaded on Westlaw would enjoy, how much of that work has been copied by the defendant, whether the defendant's use of the headnotes differs materially from the protected work and what could be the potential commercial consequences.

In spite of the evidentiary issues involved in proving similarity between the protected and infringing work, extent of copying, transformative nature of defendant's works, the flurry of lawsuits continues unabated. Notably, New York Times sued Microsoft and Open AI in the United States District Court, SDNY, on December 27, 2023. [12]

Both Microsoft and Open AI were also on the receiving end of plaints filed in the Manhattan Federal Court by the Intercept, Raw Story, and AlterNet on February 28, 2024. The plaintiffs object to the rampant use of copyrighted material used by Microsoft and Open AI in training ChatGPT by stripping the content of the digital copyright information under Digital Millenium Copyright Act. Microsoft and Open AI continue to defend their position stating that their works are not 'derivative' and constitute 'new, different and innovative products'.

3 Core Copyright Principles on Infringement, Fair Dealing, and Authorship

3.1 Determination of Copyright Infringement

The following factors must be looked at while determining copyright infringement:

(a). Similarity—Did the copying alter the nature of the plaintiff's work?
(b). Extent of Copying—Did the defendant reproduce a substantial part of the plaintiff's work?

(c). Misappropriation—Did the defendant take unfair advantage of plaintiff's work? [13]

The test of similarity in item (a) requires the court to answer the question whether the infringing copy is that which comes so near to the original as to suggest to the mind of every person seeing it. [14] For a successful infringement suit, the plaintiff must prove access to the work by the defendant and establish similarity. If the resemblance between the protected and the infringing work is sufficiently great, the court will draw an inference of copying. The onus will be on the defendant, in such a case, to disprove copying by adducing necessary evidence [15].

What constitutes 'substantial' copying under item (b) cannot be determined mathematically. It is a question of face and degree whether the copied amount is significant enough to be considered substantial [16]. To determine the substantiality of copying, the court must look at the quality of the copied portion rather than the quantity. For instance, if the defendant copies four pages out of forty pages from the plaintiff's literary work, it would still county as substantial copying if those four pages possess independent merit and contribute fundamentally to the overall value of the protected work [17].

A finding of misappropriation under item (c) would depend on the plaintiff's ability to show that the defendant has appropriated the skill and labour of the plaintiff and the similarity is not simply due to the origin of both the works from a common source [18].

A defendant incurs no liability if he copies only the 'ideas' or 'concepts' used in the protected work. To infringe, the defendant must also reproduce sufficient amount of the plaintiff's expression to render the two works 'substantially similar' [19]. When the plaintiff does not have direct evidence of copying, may prove it circumstantially by showing that the defendant had access to the original work and that the two works share similarities suggestive of copying [20]. The inverse ratio rule provides that the stronger the proof of access, lesser is the requirement of similarities between the two works to prove infringement. On the other hand, more compelling similarities supporting lessen the need for evidence of access. If the similarities are 'striking' enough, the product would have, in all probability, not been independently created and the defendant must have had access to the plaintiff's work [21].

3.2 *Fair Dealing—Exception to Copyright Infringement*

The importance of balance in intellectual property rights was summed up eloquently by Judge Kozinski in White v. Samsung Electronics America, Inc. [22] in the following words:

> *"Intellectual property law is full of careful balances between what's set aside for the owner and what's left in the public domain for the rest of us:...life of copyrights; copyright's idea expression dichotomy; the fair use doctrine; the prohibition on copyrighting facts...All of these diminish an intellectual property owner's rights. All let the public use something*

> *created by someone else. But all are necessary to maintain a free environment in which creative genius can flourish.*"

Historically, copyright began taking roots in our society by creating a monopoly in favour of the author of a work similar to that of a property. In the sixteenth and seventeenthth centuries, philosophers like John Locke and Raymond Astbury campaigned against this monopoly and urged for its curtailment. Unlimited grant for unlimited duration was seen as counterproductive and would halt the learning processes. The right should therefore be treated as a form of private property as well as an instrument of public policy to encourage creation and dissemination of knowledge. However, doctrines like fair use, which were earlier intended to facilitating the sharing of knowledge, ran into trouble as replication became much easier with newer forms of technology coming into the market [23].

The Supreme Court of Canada in Society of Composers, Authors and Music Publishers of Canada v. Bell Canada [24] observed that fair dealing permits users to indulge in some activities that might otherwise amount to copyright violation. The test for fair dealing was formulated in two steps by the Canada apex court in CCH Canadian Ltd. v. Law Society of Upper Canada:

(a). Whether the dealing is for the allowable purpose of research or private study, criticism or review, or news reporting.
(b). Whether the dealing is 'fair' in light of the following factors: purpose, character, and amount of the dealing, existence of alternatives, the nature of the work, and the effect of the dealing on the work [25].
(c). The Copyright Act of 1976 in the US requires the following four factors to be satisfied for invoking the 'fair use':

    a. purpose and character of the use, whether such use is of a commercial nature or is nonprofit;
    b. nature of the copyrighted work;
    c. amount and substantiality of the copying; and
    d. the effect of the use upon the potential market for the copyrighted work [26].

The section owes its foundation to the seminal case of Folsom v. Marsh, where Judge Story observed that a question of copyright infringement requires the court to ascertain:

> "*…the nature and objects of the selections made, the quantity and value of the materials used, and the degree in which the use may prejudice the sale, or diminish the profits, or supersede the objects, of the original work*" [27].

Judge Pierre Leval famously reasons in his seminal piece published in the Harvard Law Review that fair use of a copyrighted work transforms the original, and adds value by 'creating new information, new aesthetics, new insights and understanding' [28]. The United States Supreme Court recognized the transformative use doctrine in Campbell v. Acuff Rose and observed that 'transformative uses' augment the number of original works of authorship, whereas reproduction simply increases the number of copies to the detriment of the author, whose profits decrease and whose incentive also diminishes [29].

Transformative use synthesis of fair use has had widespread appeal, because it supports the underlying goal of copyright law by increasing our fund of knowledge and information. A transformative work is less likely to impose economic harm on the copyright owner. If an infringing work has the same purpose, meaning, or effect, as does the original work, the infringing work is more likely to displace the need for the original work. Alternatively, if the infringing work is ostensibly different in purpose, meaning, and effect, the copyright holder's conventional business might remain fully unharmed despite the new, unlicensed use [30].

In Sony Corp. of America v. Universal City Studios [31]. Rather than limiting fair use to transformative use, the Supreme Court viewed fair use in much broader terms, as an equitable rule of reason determined case-by-case. In upholding the fair use defence, the Court focused on the economic impact of the use on the incentives to produce copyrighted works: whether the use was commercial or nonprofit.

A couple of English decisions on the point of fair dealing are equally informative. In Hubbard v. Vosper [32] the question was whether Mr. Vosper's treatment of Mr. Hubbard's books amounted to fair dealing for criticism or review. At the outset, the court cautioned that it is not fair dealing for a person to take copyrighted material and use it for his own benefit. For example, 'The Times' publishing a letter on USA by Rudyard Kipling.

The following important points/questions emerge for consideration:

(a) Though difficult to define, fair dealing must be a question of degree.
(b) For criticism or review, one must look at the number and extent of the quotations and extracts.
(c) Are the extracts too many and too long?
(d) What is the end use of these extracts?

If they are used as a foundation for the entire critical commentary, it may qualify as fair dealing. However, it must not convey the same information as the author for a rival purpose. Similarly, taking long extracts and attaching smaller comments may be unfair. Here what counts ultimately is the impression conveyed by the allegedly infringing work. This must be decided by a tribunal of fact [33].

The Chancery Division observed in Ashdown v. Telegraph Group [34] that it is difficult to coin a precise definition of what fair dealing is. It is mostly a matter of 'fact, degree and impression'. The key question to be answered is whether the infringing work is commercially competing with the owner's exploitation of the copyrighted work? If yes, the fair dealing defence will fail. Secondly, whether the work has already been published or otherwise exposed to the public. If not, it will be treated as a breach of confidence by the courts rather than fair use. Though, the court conceded that sometimes it could be necessary for legitimate public controversy to utilize leaked information. Thirdly, the most important factor is the amount and importance of the protected work that has been copied. Taking of an excessive amount, or a small amount on a regular basis, would nullify fair dealing.

3.3 Authorship—the Key Point of Liability in Copyright Infringement Cases

A detailed discussion on the concept of 'author' can be found in the decision of the England and Wales Court of Appeal (Civil Division) in Kogan v. Martin and Ors. [35] The dispute arose regarding the authorship of a screenplay written about the last few years in the life of Florence Foster Jenkins, the celebrated American figure who was known for her love of music. The court looked into the definition of 'author' under section 9(1) of the CDPA, which explained the concept as a person who creates the copyrighted work. Section 10(3) of CDPA creates the fiction that all references to 'author' in case of such collaborative works would be considered as references to each 'joint author'. The observations of Laddie J. in Cala Homes (South) Limited v. Alfred McAlpine Homes East Limited [36] were relied upon by the court to identify the essential ingredients of authorship. Laddie J. concluded that an author should:

(a). Significant amount of the skill and labour, and
(b). Take responsibility for the contents.

Laddie J. however remarked that the second limb of the above test overstepped the mark and it was sufficient if the concerned person satisfied the first limb. This approach was also approved by Floyd LJ in Kogan. Floyd LJ observed that an author should expend skill and effort in creating, selecting or gathering together the detailed concepts or emotions affixed by the words in writing. Ability to accept responsibility for the work was not an essential requirement under the statute. The focus should be on who contributed to the creation of the work.

The other aspect of authorship is the nature of contribution. The contribution must be an authorial one. Simply doing some work or expending labour would not turn a non-author into an author. In the words of Prof. Ginsburg,

> "*Author is a human being who exercises subjective judgment in composing the work and who controls its execution.*"[37]

In Painer v. Standard Verlags GmbH [38], the Court of Justice for the European Union ('CJEU') further explained that according to Recital 17 in the Preamble to Directive 93/98, the intellectual creation should be traceable to the author's personality. The work must reflect this personality. During the process of making the protectable work, the author must have freely exercised a wide range of creative choices. It is this element of choice which adds the author's 'personal touch' to his or her creation. For instance, in Funke Medien NRW GmbH v. Bundesrepublik Deutschlandan [39], the CJEU held that authorship had to be determined by answering:

(a). Whether the authors exercised free and expressive choices?
(b). Whether these choices resulted in conveying 'original' content to the consumers of the work?
(c). Whether the choices resulted in an 'intellectual creation' of the author?

While the US Copyright Act of 1976 codified under 17 U.S.C. § 102 et seq. does not spell out a specific definition for 'author', 17 U.S.C. § 102(a) makes it clear that copyright protection could only subsist on an original work of authorship where the expression has been affixed on a tangible medium [40].

An accurate explanation of the concept of 'author' under the US copyright law can be found in Community for Creative Non-Violence v. Reid, where Justice Thurgood Marshall observed that:

> "*...the author is the party who actually creates the work, that is, the person who translates an idea into a fixed, tangible expression entitled to copyright protection.*"[41].

In other words, an author is a person who should contribute something which is capable of being protected under the law of copyright. It refers to the creator of the copyrighted work. It may also be a person who records the ideas of mind or supplies those ideas. The author is solely responsible and therefore deserves exclusive credit for the creation of a unique work. The Court of Appeals for the Ninth Circuit observed in Aalmuhammed v. Lee [42] that the word 'author' is generally used to signify the 'originator or the person who causes something to come into being'. The author therefore is the person with 'creative control' and supervises the creation of the work by exercising this control. The Ninth Circuit affirmed the definition of 'author' coined in Burrow-Giles Lithographic Co. v. Sarony [43] where the United States Federal Supreme Court defined 'author' as 'he to whom anything owes its origin'. The author was termed as the 'master mind' who executes a basic idea. The Ninth Circuit discussed all these authorities in Cindy Lee Garcia v. Google, Inc. and Ors. [44] and concluded that at no point did Garcia exercise any creative control over the script or her performance. The control rested with the director Mr. Youssef who had supplied the script, equipment, and given the necessary directions to the actress. Garcia could not be considered the originator of the idea or concept shown in the film. Hence, she could not claim authorship over her performance. The term 'author' thus refers to the creator of the copyrighted work. It may also be a person who records the ideas of mind or supplies those ideas. The author is solely responsible and therefore deserves exclusive credit for the creation of a unique work.

4 Critical Analysis and Concluding Remarks

4.1 Test of Copyright Infringement

4.1.1 Whether Web Scraping of Protected Works by AI Satisfies the Test of Copyright Infringement?

As on date, there is a lack of clarity on whether the use of copyrighted works to train AI and licensing that AI to website users amount to copyright infringement. Most of the disputes on this issue are either ongoing or have at best yielded interim orders pending final adjudication. The major problem in these kinds of cases is that the unauthorized use of the copyrighted works does not directly convert into infringing products. The original works undergo several stages of coding, decoding, and scaling within the AI software before they become capable of producing the alleged derivative output. Thus, any adjudicating authority will have to break through layers of algorithm before concluding on a causal link between the training data and the output being produced by the AI. Any defendant in such proceedings would point to these intervening stages between the input and out as proof of the altered nature and therefore the transformative character of the defendant's work making it dissimilar to that of the plaintiff's work. Additionally, a defendant could also argue that the repeated processing of the input reduced the proportion of the copyrighted works used in the overall scheme of the output generated by the AI.

4.1.2 Conclusion

The author therefore believes that an adjudicating authority must be careful in examining the evidence to ensure that any act of deceptive copying does not go unnoticed and unpunished. More emphasis must be laid on the qualitative aspect of the plaintiff's works copied and how they lend value to the works of the defendant. As pointed out by the Court of Appeals in Franz Hansfstaengal v. Baines [45] the court, in addition to focusing on the leading features and details of the infringing work, must also determine whether the general effect of the infringing work is such that the degree of identity or similarity would lead an observer to say that the infringing work is a copy of the protected work. Even if minimum content is copied and subject to extensive data processing within the AI, the adjudicating authority must discern whether the copied content is fundamental to the quality and financial value of the defendant's works.

The recent decision by the Autorité de la concurrence, the competition authority in France, to slap a fine of €250 million ($271 million) on Google for breaching the settlement with news agencies and publishers represents a step in the right direction [46]. However, the unauthorized use of news items to train Google's service does not address the issues surrounding copyright infringement. The legal as well as the technological community will wait with baited breath to receive the much-required clarity from the ongoing cases before the various courts in the US and UK.

4.2 *Feasibility of Application of Fair Dealing Doctrine*

4.2.1 Assuming There is Infringement, Can the Owner or Creator of the AI Invoke the Defence of Fair Dealing to Escape Liability of Copyright Infringement?

The use of fair dealing as a defence against copyright infringement could either stem from a list of permitted uses prescribed in the copyright statutes or whether it satisfies the general rule identified in a provision and/or judicial pronouncement. As on date, copyright statutes around the world do not recognize web scraping as a form of permitted use of copyrighted works. Thus, to check the legality of web scraping under copyright law, the adjudicating authority must determine whether the activity complies with the general 'rule' of fair dealing under the relevant copyright statute or a doctrine identified in a case precedent. Given the complex nature of the evolving AI technology and the regulatory challenges identified in Sect. 4.1.1, it would make more sense to go with a rule-based approach in tune with the unique facts and circumstances of each case.

4.2.2 Conclusion

The author believes that if substantial copying can be proved by the plaintiff, the infringing work would automatically cease to be transformative. The difficulties of proving substantial copying have been illustrated under Sect. 4.1.1. Thus, for a proper detection of infringement, the adjudicating authority will need to closely examine whether the owner or creator of AI is attempting to commercially benefit from the protected works. The nature and character of use and effect on the commercial market ought to be scrutinized more rigorously to identify instances of copyright infringement.

4.3 *Point of Liability for Infringement Resulting from AI Generated Content*

4.3.1 If Infringement is Proved, Can the Creator or Owner of AI Escape Liability Based on Lack of Authorship of the Infringing Works?

A potential argument from the creator or owner of the AI could be that the infringing copies were generated by the AI in an autonomous way, thus exempting the owner of creator from any liability. However, copyright law, in its present state only recognizes human authorship.

For instance, on February 14, 2022, the Review Board of the United States Copyright Office ('Review Board') refused to provide copyright for the art work produced by AI due to the traditional view that only a human being can be considered as an author of a copyrightable work. Dr. Stephen Thaler had filed a suit for an AI generated work under 'work for hire for the owner of the creative machine'. The art work was labelled as 'A Recent Entrance to Paradise'. Dr. Thaler claimed that the work has been autonomously created by the algorithms of the computer. The Review Board however refused to grant copyright protection due to the lack of human authorship. The Board further explained that longstanding principles of Copyright could not be ignored simply based on new innovations. Copyright could only be given for the intellectual labour of a man. An author is a person who gives effect to an idea. The Review Board accordingly concluded that human authorship was a fundamental component of copyright protection. AI has not yet reached a stage where it is self-reliant enough to be called an 'author' [47]. In its decision titled 'Cancellation Decision re: Zarya of the Dawn' by the U.S. Copyright Office on Feb. 21, 2023, the Office determined the test for authorship of copyrightable works created through AI generated content and human inputs. It was observed that where the input of AI is more than de minimis, the applicant must prove that the human creator of the AI had 'creative control' over the manner in which AI was used to generate content. If the content generation is solely mechanical, the application for copyright would be rejected as copyright could only subsist in a work of human authorship [48]. The decision has been referred in the latest 'Copyright Registration Guidance: Works Containing Material Generated by Artificial Intelligence' in March, 2023. While the guidance released by the Office represents an encouraging futuristic outlook in scrutinizing AI generated works more consciously and cull out human inputs, the Office continues to stick to its traditional stance on human authorship. The decision of granting copyright to an AI generated work, according to a policy statement released by the Office will be circumstantial depending on the manner of use of AI tool by the human creator [49]. The position has been confirmed yet again in a notice for comments issued by the Office on August 30, 2023 [50]. The notice came close on the heels of Judge Howell's decision in Thaler v. Perlmutter [51] on August 18, 2023, which reviewed the copyrightability of 'A Recent Entrance to Paradise' as determined earlier by the Copyright Office. Speaking for the United States District Court, District of Columbia, Judge Howell emphatically declared that human authorship is the foundation on which copyright rests and thus the Creativity Machine of Dr. Thaler could not qualify as an author in the copyright application. Neither does such an AI have the capability of enjoying rights in the protected work nor can it confer ownership on its developer simply by being possessed by the developer.

4.3.2 Conclusion

Thus, it is concluded that AI can neither qualify an author, not can it legally transfer ownership to its creator or owner. These cases would be fit for 'deemed authorship' as seen in computer generated works where authorship is vested on the person who 'caused' those works to be created by making the 'necessary arrangements' (e.g. Section 9(3) of CDPA) [52]. Accordingly, all liabilities attached to the ownership of the AI generated content must be answered by the creator or owner of AI only.

References

1. Gallese C (2023) Web scraping and generative models training in the directive 790/19. i-lex – Rivista di Scienze Giuridiche, Scienze Cognitive ed Intelligenza Artificiale. https://doi.org/10.6092/issn.1825-1927/18871
2. Case No. 3:23-cv-00201-WHO
3. 668 F.3d 1148 (9th Cir. 2012)
4. 736 F.2d 1352, 1357 (9th Cir. 1984)
5. 804 F.3d 202, 225 (2d Cir. 2015)
6. Case No. 1:99-mc-09999 (D. Del. Jan 01, 2023)
7. Getty Images (US) INC. and others v. Stability AI Ltd. [2023] EWHC 3090 (Ch)
8. Waters P, Getty images vs stability AI, https://www.lexology.com/library/detail.aspx?g=051387ba-b805-4342-b7c1-2b98cb4f9b1c
9. Davies CW, Getty images v stability AI: the implications for UK copyright law and licensing, https://www.pinsentmasons.com/out-law/analysis/getty-images-v-stability-ai-implications-copyright-law-licensing
10. 4:22-cv-06823, (N.D. Cal.) (2023)
11. Thomson Reuters Enterprise Centre GMBH v. Ross Intelligence Inc., No. 1:20-cv-613-SB (25 Sept 2023)
12. Case 1:23-cv-11195, Filed 27 Dec 2023
13. Narayanan P (2002) Law of copyright and industrial designs. P. 160, Eastern Law House, Kolkata (2002); See also Glyn v. Weston [1916] 1 Ch. 261, Williamson v. Pearson, [1987] FSR 97
14. Ladbroke (Football) Ltd. v. William Hill (Football) Ltd. [1964] 1 All ER 465 (HL)
15. LB Plastics v. Swish [1979] RPC 551 (per Lord Wilberforce)
16. Wham-O-Mfg Co. v. Lincoln Industries [1982] RPC 281
17. Leslie v. J. Young & Sons [1894] AC 335
18. Halsbury's Laws of England, 4th edn, vol 9, Para. 912
19. Mattel, Inc. v. MGA Entertainment, Inc., 616 F.3d 904, 913–14 (9th Cir. 2010)
20. Baxter v. MCA, Inc., 812 F.2d 421, 423 (9th Cir. 1987)
21. Ty, Inc. v. GMA Accessories, Inc., 132 F.3d 1167, 1170 (7th Cir. 1997); See also Jacobus Rentmeester v. Nike, Inc. D.C. No. 3:15-cv-00113-MO (Filed 27 Feb 2018)
22. 989 F. 2d 1512 (Ninth Circuit, 1993)
23. The Chancellor, Masters and Scholars of University of Oxford and Ors. v. Rameshwari Photocopy Services and Ors. (2016) 16 DRJ (SN) 678
24. [2012] 2 S.C.R. 326
25. [2004] 1 S.C.R. 339
26. US Copyright Office, https://www.copyright.gov/fair-use/
27. 9 F. Cas. 342 (C.C.D. Mass. 1841) (No. 4901); Sag, M.: The Pre-History of Fair Use. 76 BROOKLYN L. REV. 1371 (2011)
28. Leval, P.N.: Toward a Fair Use Standard. 103 HARV. L. REV. 1105, 1111 (1990)
29. 510 U.S. 569 (1994)
30. Leaffer MA (2019) Understanding copyright Law. P. 877, Carolina Academic Press, Durham, North Carolina
31. 464 U.S. 417 (1984)
32. 1972 2 Q.B. 84

33. British Broadcasting Corporation v. British Satellite Broadcasting Ltd. [1999] Ch 141, 148 (per Scott J.)
34. 2002 1 Ch. 149
35. [2019] EWCA Civ 1645
36. FSR 818.
37. Ginsburg JC (2003) The concept of authorship in comparative copyright law. 52 DePaul L. Rev. 1064, 1067
38. ECDR 6.
39. Case C-479-17
40. 17 U.S.C. § 102(a), https://www.govinfo.gov/content/pkg/USCODE-2022-title17/pdf/USCODE-2022-title17-chap1.pdf
41. 490 U.S. 730, 737 (1989)
42. 202 F.3d 1227 (9th Cir. 2000)
43. 111 U.S. 53, 56 (1884).
44. MANU/FENT/2325/2014
45. AC 20.
46. Gold J, French regulator fines Google $271M over generative AI copyright issue. https://www.cio.com/article/2069449/french-regulator-fines-google-271m-over-generative-ai-copyright-issue.html
47. US Copyright Review Board, Second Request for Reconsideration for Refusal to Register A Recent Entrance to Paradise, Correspondence ID 1-3ZPC6C3; SR # 1-7100387071, 14 Feb 2022, https://www.copyright.gov/rulings-filings/review-board/docs/a-recent-entrance-to-paradise.pdf; See also US Copyright Office, Compendium of US Copyright Office Practices, § 306, The Human Authorship Requirement
48. U.S. Copyright Office, Cancellation Decision re: Zarya of the Dawn (VAu001480196), https://www.copyright.gov/docs/zarya-of-the-dawn.pdf (letter from the Office to applicant cancelling the original certificate and issuing a new one covering only the expressive material created by the applicant)
49. Kushman B, U.S. copyright office opens the door for AI-assisted work to gain copyright protection, https://www.lexology.com/library/detail.aspx?g=04d34f34-b97e-4e5d-837a-1782e498eb30
50. United States Copyright Office, Artificial Intelligence and Copyright, 59942 Federal Register, vol 88, issue 167, https://www.copyright.gov/ai/docs/Federal-Register-Document-Artificial-Intelligence-and-Copyright-NOI.pdf
51. Case 1:22-cv-01564-BAH (18 Aug 2023)
52. Bainbridge DI (2010) Intellectual Property. P. 87, Pearson Education Limited, Harlow, Essex, England

Leveraging AI in Prevention and Protection of Women Against Cybercrime in India: A Paradigm Shift of Criminal Law in the Making

Alice Dey and Sadhan Kumar Dey

Abstract Innovative strategies are required for the efficient prevention and defense of legal rights considering the serious problems that the growth of cybercrime has presented to established legal systems in recent years. The paper additionally explores how India's criminal law is evolving due to technological developments in artificial intelligence (AI), with an emphasis on preventing cybercrime and strengthening the laws that protect women's rights. India is among the digital economies that growing at the fastest pace, that necessitates to fortify its legal defenses against cyberattacks promptly. The expansive presence of technology inclusive of its uses and misuses poses a set of different challenges both for International and National legal frameworks to combat them. In this respect, technologies powered by AI provide promising substitutes by giving legal domain the state-of-the-art tools for threat detection, evidence processing, and pro-active crime prevention. With the use of AI-enabled technologies, enormous volumes of data from various sources—such as social media, online forums, and digital communication channels—would be analyzed to look for trends that can point to cybercrimes directed at women according to NCRB Report. Thus, to tackle cyber dangers and defend the rights of women in India, this abstract idea calls for the creation of progressive legislation that integrates AI technologies, data privacy measures, and gender-sensitive policies. In conclusion, a paradigm shifts in the battle against cybercrime and the defense of women's legal rights in India is represented by the incorporation of AI into the legal sphere.

Keywords AI · Threat detection · Cybercrime · Women's rights · Data privacy measures

---

A. Dey (✉)
IEM's International Institute of Juridical Sciences, University of Engineering and Management, Kolkata, Kolkata 700160, India
e-mail: alicedey.uem.edu@gmail.com

S. K. Dey
Department of ESM, RCC Institute of Information Technology, Kolkata 700015, India
e-mail: deysadhan.rcciit@gmail.com

M. Chakraborty et al. (eds.), *Proceedings of 5th International Ethical Hacking Conference*, Lecture Notes in Networks and Systems 1148,
https://doi.org/10.1007/978-981-97-8457-8_20

1 Introduction

The database management system used in Indian e-governance relies on recently developed digital technologies and web tools that are accessible over the Internet, which have drastically altered how the government conducts business [1]. Electronic media and computer-generated e-records are the main sources of information used by government agencies and professionals to continue the nation's daily e-governance. At the Federal and Diocesan levels of government in India, information technology presents novel and sophisticated ethical, legal, and extra-legal challenges to E-governance. Daily e-governance operations include, among other things, the concern of cybersecurity against information technology abuses and misuses directed toward the most vulnerable segments of the population including children and women [2].

A fresh analysis of the cybersecurity problem in respect to the privacy of Indian women is important given the current circumstances. The results of information technology have opened the door to a new world of online shopping, business networking, and electronic banking. This has made it possible to cut expenses and replace complex, lucrative transactions with a faster, more efficient, and time-saving system of commercial dealings. Prolific offenders such as hackers and attackers have also devised strategies and tactics to breach internet accounts. They have been successful in breaking into any computer system without authorization and pillaging valuable information.

Cyberspace is the term for the computer-generated environment that is the internet, and cyber laws are the regulations that control it. It entails tracking a person's online activities by sending out notifications on boards that they visit, breaking into their converse or block-apartments, sending them a barrage of emails, and so on. Generally, the hacker has no reason for communicating and only wants to cause emotional and mental distress to others. The digital age has opened previously unheard and unexplored avenues for creativity and connectedness, but it has also created a pathway for new kinds of criminal activities, especially in cyberspace.

Cybercrime is a serious threat to people and society everywhere, including India [3]. Cyber-bullying, cyberstalking, online harassment, and other types of digital abuse disproportionately affect women. Given the importance of addressing these concerns, there is rising consensus regarding the need to incorporate AI into cybercrime prevention measures. However, the connection between AI and women's legal protection in India is still mostly unknown and requires additional research [4].

2 Overview of the Issue

India is witnessing a surge in cybercrimes targeting women, which adversely impacts these persons' financial stability, mental and physical well-being, and both. Financial fraud, vengeance porn, internet stalking, online bullying, and digital harassment are just a few examples of the various forms that computer-based crimes against women may exhibit. One of the most prevalent types of computer-based crimes against women in India is online harassment. It entails sending offensive, threatening, or abusive messages or remarks to women via online channels. According to a National Commission for Women study, 54.8% of women said they have experienced cyber-bullying to begin with [5]. Feeling vulnerable and uneasy can cause them extreme psychological distress, anxiety, and fear in women who are the targets of cyber-bullying.

Today's flawed internet reflects the disorganized society in which women are routinely mistreated, harassed, compromised, tracked, and ignored. It investigates gendered metrics for digital offenses such as individual harassment, cyber-stalking, hacking, defamation, manipulated pornographic images, and electronic extortion [6].

Another form of cybercrime that affects women in India is cyber-stalking. It describes a recurring behavior of tracking, monitoring, or following someone online in order to harass them. Cyber-stalkers follow and harass their targets via a variety of digital channels, including social media, emails, and messaging applications. A report released by the Mumbai Police's Cybercrime Cell claims that the number of cyberstalking cases in India within the last year has increased by 91% [7].

2.1 Cybercrime Landscape in India

The landscape of computer-based crime in India has evolved significantly over time, bringing with it a complex web of challenges and opportunities in the digital age. India's increasing digital infrastructure and quick adoption of technology have led to an increase in cybercrime incidents [8]. Cyberthreats come in many different forms, from financial fraud to identity theft, and they represent a major risk to individuals, businesses, and governmental institutions. The intensification of social platforms and e-commerce platforms has pointed vulnerabilities which are intensely becoming higher these days, allowing malicious intended people with numerous avenues for exploitation. The Executives and the Governing bodies at times face challenges in their efforts to pinpoint and seize cyber offenders. This kind of situation is attributed as the fallacies in constraints of resource, strict Jurisdictional barrier, and clandestineness of the internet [9].

The prevalence of AI-driven computer-based crime in India is concerning, particularly when it comes to targeting women and children. These vulnerable groups face serious threats from the exploitation of AI technology, such as deepfake and AI-generated pornographic content. In deeply patriarchal nations like India, the proliferation of AI-generated sexual imagery is a cause or great concern because it can result in serious mental health problems, social stigma, and even physical assault [10].

2.2 Importance of Leveraging AI in Cybercrime Prevention

AI has the potential to substantially improve efforts aimed at preventing computer-based crime by promoting quicker threat identification, proactive risk reduction, and more efficient use of available resources. Machine learning algorithms can examine large data sets to identify patterns that suggest harmful behavior, enabling law enforcement and cybersecurity professionals to respond swiftly and efficiently to emerging threats. Furthermore, AI-powered solutions can increase the overall resilience of the judicial system by improving the capacity of legal institutions to investigate and prosecute cybercrimes [11].

The potential of Artificial Intelligence to produce realistic-looking but fraudulent photos and movies exacerbates already-existing social problems. For example, it is possible to use AI-generated sexual content as a weapon to damage people's reputations, which could result in social exclusion and even violent outcomes like honor killings. The lack of social and legal support available in India for victims of internet abuse exacerbates this issue [9].

2.3 Significance of Legal Protection for Women in India

Computer-based crimes are increasing in India for several reasons. Cybercriminals now have access to a wider attack surface and pool of potential victims due to the rapid widespread use of smartphones and internet connectivity. In addition, the growth of e-commerce, online banking, and digital payment systems has increased the potential for financial fraud and data theft. The problem is exacerbated by the general public's inexperience of cybersecurity, insufficient legal protections, and the ever-evolving tactics employed by cybercriminals.

The integration of Artificial Intelligence provides a colossal room to maneuver for identifying those particular hindrances faced by Indian women in reference to e-crime prevention theories. With the advent and utilization of AI-driven technology, governing bodies along with working executive officials may propose and constitute beleaguered temporizations to combat cyber-bullying, cyber-stalking, and other all other forms of cybercrimes against women. It is rightly advised and suggested at every workshop, training facilities, etc., to enhance digital literacy among women so that they can be safe in the internet community, but along with it would also be applaudable if the process of taking down of the reporting of the problem and providing resolution at a superspeed pace to them against gender-based computer-based crimes would be benignant [12]. However, the Governing Authorities are foresighting the systemization of the utilization of Artificial Intelligence in Judicial Proceedings which can shape up women's access to fair justice and equitable treatment.

3 Structure of the Paper

This paper envisages to answer three research questions that form the blueprint of this whole article.

The questions are:

(a). Is AI being actually used in today's time by the existing Legal framework to combat Cybercrime physically and superficially?
(b). Are the new penal provisions capable enough to contain them?
(c). And the last question rises to a point where misuse of AI is controlled by using AI as a tool to prevent one, but will there be any other Technology curated or technology-abled tool to rise as a phoenix to combat cybercrime?

The paper starts with an introduction briefing about the menace the digital platform has become. With the advent of technology in its highest form has brought people to a situation where in order to apply it as boon they are falling prey to its fallacies. The paper then studies the historical background of Cybercrime which takes us to address the vulnerabilities faced by women in India and Abroad.

A secluded study of the NCRB Report with data and chart has been curated in order to understand the rate of surge of Cybercrimes against women all over India, Cybercrimes in different forms and also Cybercrimes in metropolitan cities as well. In order to answer the second question of research we need to go through the International Laws and the National laws available to us to comprehend the necessity for furtherance of laws in the making. While in order to find the third answer for the time being we have to anticipate the Technological Advancement secured by our Indian Government which reflects on quick resolvent strategies of the cybercrime cases filed by women.

4 Vulnerabilities Faced by Women

In this technological advanced era, the internet has become an unfathomable altar for communication, trade & commerce, and most importantly a social interaction. While the online world parades immense opportunities for idiosyncratic and adept growth, but along with it also proffers significant risks, particularly for softer gender i.e. women. Women often encounter unique and exacerbated vulnerabilities in the ambit of cybercrime, ranging from online harassment and cyber-stalking to more heinous forms of exploitation such as cyber-pornography and cyber grooming to cyber-frauds. These threats not only pulverize their privacy and security which also have vehement impacts on the mental health, professional lives, and even personal relationships of the women. The inconspicuousness and the comprehensive spectrum of the internet empower offenders to target women with relative aegis. Cybercrimes specifically targeting women often exploit societal power imbalances, reinforcing existing offline discrimination and violence. Furthermore, the intersection of gender with other factors such as race, sexual orientation, and socio-economic status can exacerbate these vulnerabilities, creating compounded risks for marginalized groups.

In order to effectively address cybercrime against women, a culture of equality and respect must be promoted online in addition to strengthening legal frameworks and advancing AI-based technologies. Governments, businesses, and civil society organizations must work together to build a more secure and welcoming digital environment where women may prosper without worrying about risks from the internet. A thorough analysis of a National Crime Records Bureau report will take place when we go to the following set of subtopics.

Though all the above preluded understanding comes to a stagnant circumstance when the atrocities do not cease by such offenders. With the technological upliftment in nooks and corners of society and availability and affordability of cheaper wireless connection foster the misuse of these in furtherance of cybercrimes against women. Cybercrime against women in India can take many other various forms of vulnerabilities, including as below but these are not restricted to the following:

- **Deepfakes:** The term "deepfake" refers to the technique of creating artificial intelligence photographs of fictitious events using deep learning. Deepfakes has no bounds because it isn't confined to only visuals or sounds. According to Danielle Citron a law Professor of Boston University states, "Deepfake Technology is being weaponized against women. Beyond the porn there's plenty of spoof, satire and mischief" [13].
- **Cyber-Stalking:** In order to inflict mental suffering on the victims, it involves sending them unpleasant emails and texts on a regular basis, posting abusive posts on message boards, and making fruitless attempts to contact the ladies and teenagers via social networking sites.
- **Online Harassment:** Women and teenagers often face relentless online harassment, including offensive messages, derogatory comments, and threats, leading to psychological distress and trauma.
- **Cyber Defamation:** Through the use of manipulated photos, blackmail, and the revelation of personal information, the victim is defamed. It usually involves asking the target for sexual favors and extortion.
- **Cyber Hacking:** False news and information are more ubiquitous now than ever before. Once the ladies and youngsters clicked on fraudulent URLs, they ended up being the targets of cyberattacks. Malware on their phones generated through AI now-a-days triggered the microphone and camera, downloaded all of their personal data, and took private photos and videos. Then, thieves use these pictures and information to carry out other crimes, like extortion.
- **Pornography:** This illegal practice includes posting modified photographs of victims and utilizing them for pornographic content, occasionally demanding money to have the images taken down from social media.
- **Cyber-bullying:** This entails posting abusive, defamatory, and fraudulent remarks concerning the victims on social networking sites and asking money to have them taken down. Rape and death threats may also be sent to the victim. Additionally, it involves making offensive comments on the victim's posts. Examples of digital

or communication technology that is used for harassment and bullying include a computer, cell phone, or laptop.

- **Sextortion**: Cybercrime against women in the current period most commonly takes the form of sextortion. The perpetrators began using their victims altered or private photos as a kind of blackmail to demand money or sexual favors. The perpetrators used intimidation tactics to express their dissatisfaction toward women, demanding sexual advances or communications in return. Considering they were poor, they also felt emboldened to scare people with their Photoshopped photos in an attempt to get money out of them.

4.1 Emerging Trends and Challenges in Cybercrime

As technology develops, a multitude of new challenges and trends are emerging in the field of cybercrime. Cybercriminals are constantly modifying their techniques and creating new technologies in order to inflict harm and exploit women. They will not allow advancement in technological field for better future of the nation. Understanding these patterns and emerging concerns is essential for staying one step ahead of all kinds of cyber hazards. They are committing the crimes using Cryptocurrencies amalgaming AI, inventing advanced cybernetics methodologies which include techniques like spear phishing, bitexting, and sensitive messaging (like honey trap) catapulted at users which forces them into giving over their personal details or carrying out actions that endanger their security. India's computer-based crimes against women could be significantly reduced with the use of technology. To avoid and lessen computer-based crimes, technologies such as blockchain, artificial intelligence, and machine learning can be employed. It's crucial to remember, though, that technology cannot end computer-based crimes against women on its own.

4.2 Analysis of NCRB Data

The **National Crime Records Bureau (NCRB)** is an Indian government organization tasked with gathering and examining crime data as specified by the **Special and Local Laws (SLL)** and the **Indian Penal Code (IPC)**. The NCRB aims to establish and preserve safe, shareable databases about crimes and offenders for law enforcement and to encourage the use of these databases in the provision of public services. Additionally, it gathers and analyzes national crime statistics and serves as a clearinghouse for data on crime and offenders on a national and worldwide scale. A certain percentage can be calculated to assess the rate of cybercrimes against women based on the data sets provided by NCRB.

The reason a data set has been tapped up so that the paper makes it clear that what kind of Vulnerabilities women go through on a day-to-day basis in the digital world.

Let us analyze the graphs provided below which depict crimes against Women overall throughout the country. And then Cybercrimes against Women throughout the country and limiting it to Cybercrimes in Metropolitan areas of India to narrowing it down to cybercrimes against women in Metropolitan areas of India. This data has been collected from the National Crime Record Bureau (NCRB) site where these reports have been published widely. The timeline frame adheres from 2020 to 2022.

In Table 1 and Fig. 1; the NCRB Annual Report provides us with a picture where the sum total of 65,894 cases which were registered in 2022 under the categorization of Cybercrimes in India including the States & Union territories depicting a rise of 24% in registration over as compared to 2021's 52,973 cases. Rate of Crime in this category spawned from 3.9 in 2021 to 4.8 in 2022. Throughout 2022, 64.8% of cybercrime cases were registered with reference to, mainly E-Fraud calculating till 42,710 cases followed by Extortion with 5.5% i.e. 3648 cases and Sexual Exploitation with 5.2% i.e. 3434 cases [14].

Table 1 Cybercrimes against women throughout india (State/UT-wise)—2020–2022

State/UT	2020	2021	2022
Andhra Pradesh	1899	1875	2341
Assam	3530	4846	1733
Bihar	1512	1413	1621
Chhattisgarh	297	352	439
Gujarat	1283	1536	1417
Haryana	656	622	681
Jharkhand	1204	953	967
Karnataka	10,741	8136	12,556
Kerala	426	626	773
Madhya Pradesh	699	589	826
Maharashtra	5496	5562	8249
Meghalaya	142	107	75
Odisha	1931	2037	1983
Punjab	378	551	697
Rajasthan	1354	1504	1833
Sikkim	0	0	26
Tamil Nadu	782	1076	2082
Telangana	5024	10,303	15,297
Uttar Pradesh	11,097	8829	10,117
Uttarakhand	243	718	559
West Bengal	712	513	401
Delhi	168	356	685
Jammu & Kashmir	120	154	173
Total all India	50,035	52,974	65,893

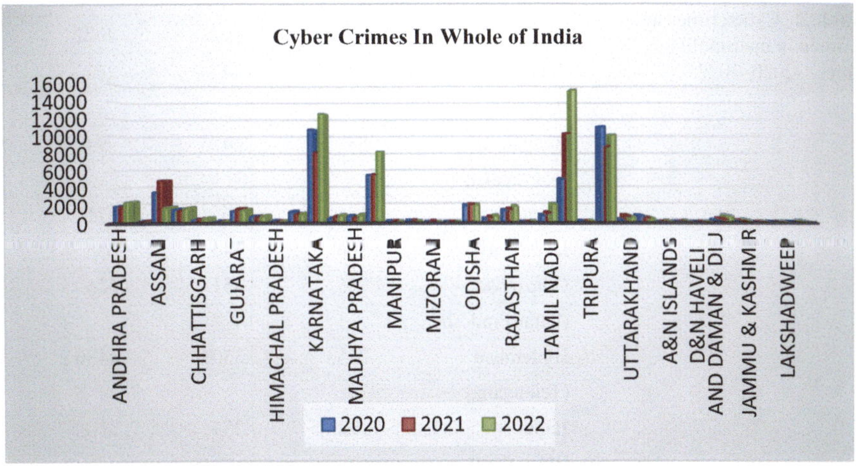

Fig. 1 A Graphical representation of the Above table cybercrimes against women in India (*Source*: NCRB Report)

In Table 2 and Fig. 2; the NCRB Annual Report provides us with a picture where the sum total of 24,420 cases which were registered in 2022 under the classification of Cybercrimes specifically committed in Metropolitan Areas of India portraying an increase of 42.7% i.e. 17,115 cases in 2021. The rate of crime surged from 15.0 in 2021 to 21.4 in 2022. It has been observed that the majority number of cybercrime cases is Computer-related offenses attracting Sec 66 of the Information Technology Act of 2000. As per the report a total of 12,213 cases formed the highest number of Cybercrimes accounting for 50.0% in the year 2022 [15].

Apart from studying the rate of Cybercrimes in India both at the National and the Metropolitan level we need to focus on the different types or modes of Cybercrime committed against women and how many are the victims.

In Table 3 and Fig. 3; depict the graphical representation of Different types of crime in different states. Maharashtra is the highest number in Sexual Exploitation of Women that is 787 cases followed by Uttar Pradesh with 542 cases and Karnataka with 338.

The next traumatizing and heinous crime lined up is that of Fraud in which Karnataka is ranking high with 11,205 cases followed by Maharashtra with 5515 cases and Uttar Pradesh with 4506 cases [15].

The next and the last set of tables and its graphical representation talk about some specific Cybercrimes against women in India which had made the lives of women a living hell. The problem is that the Indian Society presumes Women to be the guilty one in any offense against them and tags them directly which is known as Victim Shaming or Victim blaming; in which the women were never a part of.

In Table 4 and Fig. 4; the chart depicts Specific Cybercrimes against Women reported in the year 2022. There was a reporting of the highest number of Cyber Pornography cases in Uttar Pradesh is around 450, followed by Odisha with 269 and

Table 2 Cybercrimes against women in metropolitan cities—2020–2022

	2020	2021	2022
Ahmedabad	421	447	261
(Gujarat)			
Bengaluru	8892	6423	9940
(Karnataka)			
Chennai	186	76	271
Delhi City	166	345	685
Ghaziabad	756	451	626
(Uttar Pradesh)			
Hyderabad	2553	3303	4436
(Telangana)			
Jaipur	192	254	310
(Rajasthan)			
Kanpur	300	449	541
(Uttar Pradesh)			
Kolkata	172	220	75
(West Bengal)			
Lucknow	1465	1067	1134
(Uttar Pradesh)			
Mumbai	2433	2883	4724
(Maharashtra)			
Nagpur	243	192	211
(Maharashtra)			
Patna	312	304	222
(Bihar)			
Pune	238	225	357
(Maharashtra)			
Surat	204	296	371
(Gujarat)			
Total cities	18,657	17,115	24,420

then Karnataka with 234. Similarly, it also depicts a highest recording of other kinds of heinous cybercrimes cases amounting to 3669 cases in Karnataka, followed by 1836 in Maharashtra and 943 in Telangana [14].

All the above tables and figure chart are just a mere reflection of how cybercrimes are upsurging day by day and specifically with the AI-enabled applications. The women have fallen more easier target. The presence of mere laws would not suffice but acting and bringing evolving laws is the key to such detrimental changes for a safer future for women.

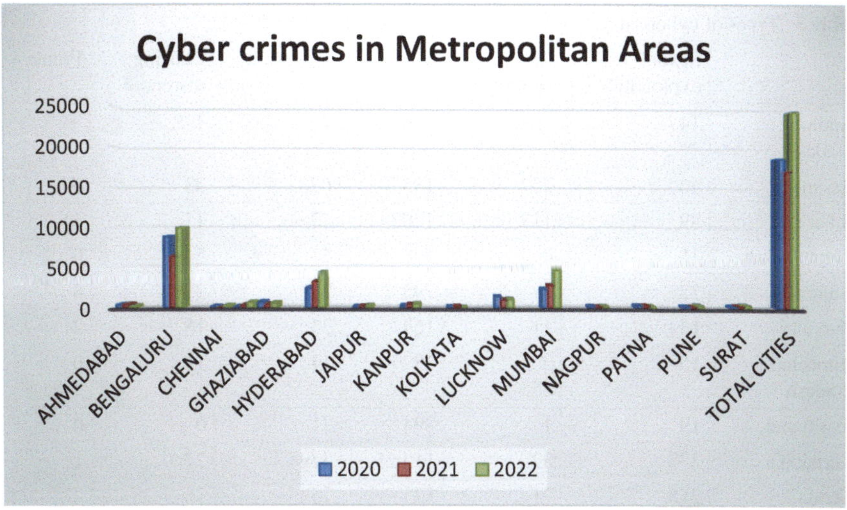

Fig. 2 Cybercrimes against women in metropolitan areas of India (*Source*: NCRB)

5 Laws Concerning Protection of Women Globally

The laws and the bills that have been passed and is of higher importance in order to render justice to women paving the way both for the world and India to make certain amendments in the changing structure of its prolonged existence.

William Perdue, Julia Spiegel along with others stated in California Law Review "Cyber-attacks have become increasingly common in recent years. Capable of shutting down nuclear centrifuges, air defense systems, and electrical grids, cyber-attacks pose a serious threat to national security. As a result, some have suggested that cyber-attacks should be treated as acts of war" [16].

Let's look onto the international arena of Laws for some better understanding of the situation our existent law is.

5.1 International Covenants on the Rights of Women in the Context of Cybercrime:

(a). Convention on the Elimination of All Forms of Discrimination against Women (CEDAW):

The Convention on the Elimination of All Forms of Discrimination Against Women (CEDAW) is an international convention which was adopted by the UN General Assembly in the year of 1979. It is always referred to as the universal women's bill of rights which is recognized worldwide. CEDAW pursues to elevate

Table 3 Types of cybercrime against women throughout India 2020–2022

	Sexual exploitation	Personal revenge	Fraud	Extortion	Causing disrepute	Prank
Andhra Pradesh	147	107	1463	47	1	3
Assam	46	257	185	628	33	53
Bihar	89	112	1207	77	15	11
Chhattisgarh	35	0	59	5	54	0
Gujarat	72	8	795	24	371	8
Haryana	134	11	164	35	15	2
Himachal Pradesh	15	3	28	14	10	0
Jharkhand	19	1	894	21	0	0
Karnataka	338	32	11,025	379	27	1
Kerala	218	71	323	25	36	2
Madhya Pradesh	102	4	343	11	251	13
Maharashtra	787	66	5515	129	31	0
Odisha	14	0	1832	0	0	0
Punjab	98	19	326	29	21	5
Rajasthan	271	26	621	98	101	2
Tamil Nadu	189	49	1650	31	17	1
Telangana	152	26	10,991	447	0	0
Uttar Pradesh	542	37	4506	1250	883	58
Uttarakhand	36	0	155	354	2	0
West Bengal	9	16	95	11	3	4
Delhi	54	2	278	8	0	8
Jammu & Kashmir	15	0	106	10	2	2

women's equal rights and possibilities transoceanically and cease all kinds of bigotry and intolerance against them in all realms of life. In addition, to interpreting discrimination against women, the covenant lays down a plan of execution for every Nation States to combat gender-based violence in the digital arena too. It focuses on a range of relatable subjects related to women, including their legal rights, work, political involvement, education, and healthcare. Member States that have adopted CEDAW are required to work on promoting and guaranteeing gender equality around the globe. CEDAW has been beneficial in the global advancement of women's rights and in the constitution of laws and policies that support gender equality and protection.

(b). Sustainable Development Goals (SDGs) for Women:

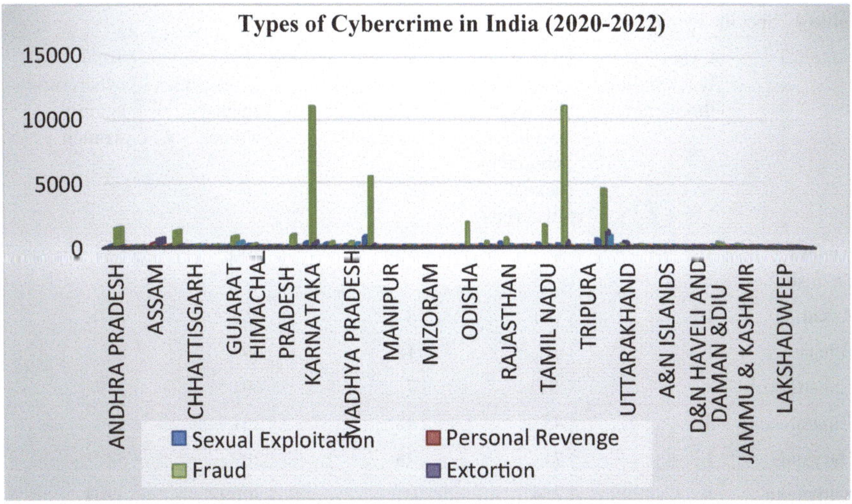

Fig. 3 Different types of cybercrime in India 2020–2022 (*Source*: NCRB)

The United Nations adopted a set of 17 global goals known as the Sustainable Development Goals (SDGs) in the year 2015 to combat the disparate social, economic, and environmental matters that the world is conflicting on contemporarily. Goal 5 articulates gender equality and aims to emancipate and liberate all women by acknowledging their rights toward sustainable development. Target 5.2 of Goal 5 directly undertakes the hindrances of each and every woman's public safety. This goal draws attention to the imperious urgency for managing the widespread concern of roughness and harassment against women even in cyberspace including public safety at streets, public transportation, and workplaces. In order to achieve Target 5.2, it is very important to strengthen the regulatory structures that prohibit and penalize violence against women, altering social norms that condone violence, increasing public awareness of the problem alongside offering support to victims, and ensuring that women have an important role in decisions affecting public safety and security are all essential.

(c). European Parliament Bill, 2023:

This bill of 2023 proposed for a regulation, i.e. Amendment 3–11, Recital 1–3 talks about presiding constraints on the development, marketing, and use of AI systems by Member States unless explicitly authorized by this regulation. And it also proposes to protect all natural persons with regard to the processing of personal data and provides for the adoption of rules on the protection of individuals with regard to the processing of personal data [17, 18].

Table 4 Specific cybercrimes against women in 2022

State/UT	Cyber blackmailing/ threatening	Cyber pornography/ hosting/ publishing obscene sexual materials	Cyber stalking/ cyber bullying of women (Sec.354D	Other cybercrimes against women	Total cybercrimes against women
Andhra Pradesh	16	89	129	400	637
Assam	10	101	3	455	610
Bihar	1	4	18	49	84
Chhattisgarh	3	200	7	69	279
Gujarat	3	43	48	251	352
Haryana	12	71	28	209	327
Karnataka	1	234	0	3669	3904
Kerala	0	122	45	203	381
Madhya Pradesh	8	96	82	201	390
Maharashtra	11	75	578	1836	2530
Odisha	0	269	0	32	574
Punjab	3	33	26	126	194
Rajasthan	16	136	79	253	498
Tamil Nadu	3	82	31	239	401
Telangana	23	14	279	943	1262
Uttar Pradesh	2	450	27	617	1101
Uttarakhand	10	21	17	30	161
West Bengal	0	9	11	119	142
Delhi	2	118	24	101	250
Jammu & Kashmir	0	25	2	10	39
Total all India	125	2251	1457	10,012	14,409

5.2 *Two Comparative Case Studies:*

(a). Case Study—1 USA

Floyd, a Hungarian soldier serving in Afghanistan for peacekeeping purposes, sent Christina a friend request, which she accepted. Over time, Floyd and Christina's relationship grew closer. Their relationship grew closer as they each related their own prior stories about their former partners. As time passed, Floyd and Christina agreed to get together sometime during his vacation in Virginia.

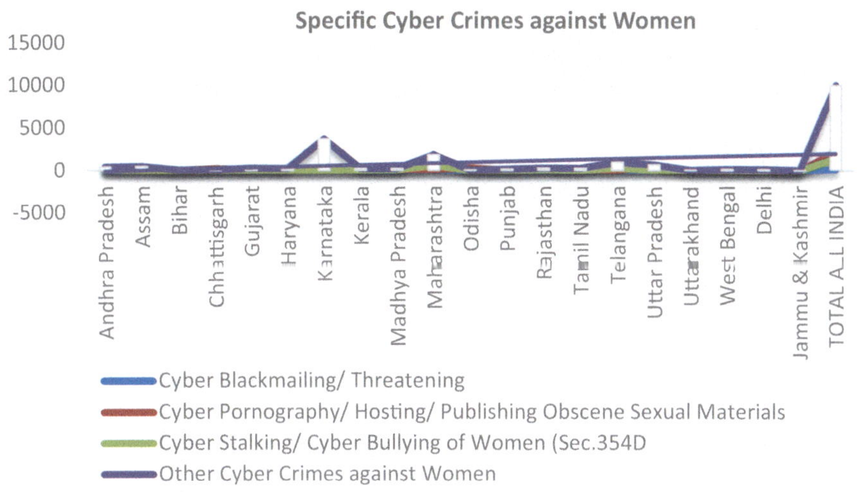

Fig. 4 Specific cybercrimes against women in the Year 2022 (NCRB report)

After they met, they eventually planned to get married and on account of that Floyd told Christina that he would be taking her to meet his parents. His parents on the other hand stayed in California and he said that they cannot come down to Virginia to meet Christina over video call. Soon Christina had a video conference call with Floyd and his parents. After the conversation, the next day she receives an email from an unknown mail Id which has an attachment. When Christina opened that attachment, she found herself in an inappropriate manner conversing and doing gestures to all those three other men in the calls who she thought was Floyd's parents. The email was accompanied with a threat message that if she does not pay 4000$ to the sender's account then this particular video will be all over the net for other people's amusement.

Soon she filed a complaint to track those offenders.

Now let's us analyze what USA has curated for its women's safety in the digital world. The USA's AI State Legislations include the following:

(i). **New York:** Non-Consensual dissemination/publication of intimate images created by digitization, Jail for a year and a $1000 fine.

(ii). **Virginia:** Deepfake Ban (revenge pornography)/ misdemeanors up to $2500 fine.

(iii). **California:** The California Consumer Privacy Act of 2020 enables consumers greater autonomy over the personal data that companies, including social media sites like Facebook and Instagram, collect about them.

(iv). **Right to Publicity:** Class action against digital swap or blend faces exploiting a person's images or voice without permission.

(v). **Indiana** created an AI task force [19].

(vi). **South Dakota** changed its laws to make it clear that having any computer-generated child and women pornography or any visual portrayal of a minor and a woman participating in a forbidden sexual conduct or simulating an unauthorized sexual act constitutes possession of any pornography. A Class 4 crime, punishable by ten years in a state penal facility, is created by the amended statute. Furthermore, a fine of twenty thousand dollars could be applied [20].

(vii). **Tennessee** mandated the adoption of policies by public charter schools, local education boards, and the governing bodies of public universities addressing the use of artificial intelligence (AI) by staff, instructors, and students for educational purposes.

(viii). **Utah** has the Artificial Intelligence Policy Act.

(ix). **The Virgin Islands** established a real time, centralized crime data system within the state police department to combat crime done with AI in real time [19].

(x). **West Virginia** created a special committee on artificial intelligence [19].

Other pending Bills in US State Legislations related to tackling Artificial Intelligence powered crime prevention: [19]

(i). **Alaska:** Relates to the use of artificial intelligence to create or alter a representation of the voice or likeness of an individual.

(ii). **Arizona**: Relates to the use of Artificial Intelligence in the creation of sexual abuse materials.

(iii). **Connecticut:** criminalizes unauthorized dissemination of intimate images that are digitally altered or created using artificial intelligence.

(iv). **Florida:** With AI, the Department of Management Services intends to establish the Government Technology Modernization Council, which would mandate that organizations and individuals develop safety and transparency guidelines for AI-generated content, including photos and videos. It also forbids the use of a real person's image or likeness without permission and mandates disclosures for specific communications, interactions, photos, likenesses, and content [19].

(v). **Georgia:** The government has suggested safeguards against discrimination by automated decision-making and artificial intelligence. It forbids the dissemination of pornographic images of women and children created using computers [19].

(vi). **Hawaii**: It proposes that the Department of Law Enforcement's top officials regularly receive training on offenses involving artificial intelligence [19].

(vii). **Idaho:** It amends the law to make it illegal to distribute digital, computer-generated photos that could be used as evidence of a crime involving the exploitation of another person for sexual purposes [19].

(b). Case Study—2 United Kingdom

An accounting firm owned by Klaus and Rebecca offers small businesses in Central London outsourced bookkeeping and accounting services. The company uses an internet platform to run its operations. Klaus and Rebecca meticulously curated a website portal that allows the client companies to log in and handle various bookkeeping needs for their businesses. This includes processing receipts, tracking expenses, and calculating deductions. Additionally, the website guarantees that no data or privacy will be lost.

Not long before taxes were due, a suspicious email led to the ransomware infection of Klaus and Rebecca's work computers. The malware prevented customers from accessing the company's platform by locking it down. To repair the network, the fraudsters sought $200,000 in cash. Rebecca and Klaus declined to pay. The clients' confidential information was threatened to be made public by the cybercriminals. This type of cyber scam devastated Klaus and Rebecca, and they were at a loss for what to do.

They eventually filed a request to have their rights protected.

In this above case scenario, there are 10 steps to Cybersecurity according to the United Kingdom's Government National Technical Authority for Information Assurance (CESG) that advises organizations on how to protect their information and information systems against today's threats [21].

They are: Secure Configuration; Network Security; Malware Protection; Monitoring; Incident Management; Managing User Privileges; User Education and Awareness; Home and Mobile Working; Removable Media Controls; Establish effective Governance Structure and Risk Appetite.

6 Laws Available in India to Combat the Evolving Threat of Cybercrimes

(A). Information Technology (IT) Act, 2000

India's main piece of legislation pertaining to cybercrimes is the IT Act. It outlines the offenses of hacking, identity theft, data theft, illegal access, and internet fraud. The Act creates the nation's cybersecurity framework and offers legal guidelines and sanctions for cybercrimes.

The IT Act is a specific law specifically concentrating on dealing with the many aspects of the use and misuse of information technology, including the commission of crimes. They are:

- **Section 66A:** This section covers a variety of practices related to delivering offensive communications via communication services, irritating others, including sending emails that intentionally mislead or fool the receiver about their origin (a practice called as IP or email spoofing). These offenses are punishable by up to three years in prison or a fine.
- **Section 66B:** Falsely obtaining a communication device or computer resource that has been stolen is punishable by up to three years in prison, a fine of one lakh rupees, or both.
- **Section 66C:** Other forms of identity theft include utilizing other people's passwords or electronic signatures, among other things.

- **Section 66D:** If someone uses a computer resource or a communication device to cheat, they will be penalized with three years of jail of any kind as well as a fine that might reach one lakh rupees.
- **Section 66E:** Violations of privacy include publishing or sharing someone else's private information without that person's permission, etc. Three years in prison, a fine of two lakh rupees, or both are the possible penalties.
- **Section 67:** The IT Act of 2008 expanded the preceding portion of the ITA to cover child pornography and the retention of records by intermediaries, and it deals with the publication or transmission of obscene material in electronic form.
- **Section 72:** Penalties for violating confidentiality and privacy were all covered.
- Bhartiya Nyaya Sanhita (formerly known as Indian Penal Code, 1860):

There are two key distinctive features of the internet. First of all, it is not limited by space or time, therefore the abuser might be operating from anywhere in the world. Second, users of the internet can remain anonymous. The Information Technology Act of 2000 and the Indian Penal Code, 1860, also known as Bhartiya Nyaya Sanhita (which is effective from July 1st, 2024) are essentially the two main laws in India that deal with cybercrimes against women to a great extent. The BNS is the country's fundamental criminal code, outlining several offenses and the corresponding penalties. They are:

- **Section 356 Defamation—**Anyone who makes or publishes any kind of imputation about another person with the intent to harm that person's reputation—whether by spoken or written words, by signs or visual representations, or both—and knows or has reasonable suspicion that person will suffer—is considered to have defamed that person. (Sec 499 in IPC)
- **Threat and Harassment—**Harassment includes both Sexual and Mental. According to Sec 354A of IPC Sexual Harassment is a crime and Sections 503 and Section 504 deal with criminal intimidation and intentional insult with intent to provoke a breach of peace. Criminal intimidation is a crime which is punishable by Sec 506 of IPC which reference to this statement under Sec 503, i.e. "Criminal intimidation occurs when someone threatens another with harm to their person, reputation, or property, or to the person or reputation of anyone they are interested in. The intention is to create fear in the other person, get them to act in a way they are not legally required to, or not act at all in order to avoid carrying out the threat even digitally." (Clause 76 & 78 of BNS)
- **Sexual Assault—**The Indian Penal Code, Section 354B, defines criminal force against a woman. When a male assaults, uses, or assists a woman in the use of illegal force with the intention of forcing her to undress or be in her underwear, it is considered that he has committed a crime against that woman. This is what sexual assault is defined as. It is always the intention to oppress women as the weaker species through whatever form of offense.
- **Obscenity-** Obscene books, pamphlets, and other materials that are judged to be "lascivious or appeals to the prurient interests" are prohibited from being published or sold under Sections 292 and 293 of the Indian Penal Code. This prohibition extends to obscene advertisements. Obscene acts and songs are prohibited

by Section 294. To the vexation of others, whoever: (1) does an indecent act on any public property; or (2) signs, recite, or utters any vulgar words or song in or near any public area. Since Section 81 of the IT Act expressly indicates that it supersedes the IPC, any violation relating to obscenity in electronic form may be tried under the IT Act rather than the IPC.

- **Section 354D:** The topic of stalking is covered here. A man who stalks a woman is described as one who follows her, makes attempts to get in touch with her, and maintains records on virtually every one of her internet-based activities.
- The Indecent Representation of Women (Prohibition) Act of 1986:

Section 3—It focuses on preventing commercials that depict a woman in an inappropriate manner. It states that no one has the right to publish anything, to plan to publish anything in the future, or to be involved in the publication or display of any advertisement that might be interpreted by society as obscene or that could damage the standing and dignity of women.

Section 4—Specifically targets the creation, dissemination, rental, sale, and circulation of any material, such as books, pamphlets, papers, slides, films, poems, drawings, paintings, photographs, or images, that represent or tend to depict women indecently or in an obscene way.

Section 6—Outlines the legal consequences for individuals who violate the provisions outlined in Sects. 3 and 4. The penalties are as follows:

- Upon the first conviction, the offender may face imprisonment for up to two years and a fine of up to 2000 rupees.
- Upon the second conviction, the offender may face imprisonment for at least six months up to five years, along with a fine ranging from 10,000 to 1 lakh rupees.
- The Criminal Procedure Code 1973 presently known as Bhartiya Nagarik Suraksha Sanhita:

Many major amendments within the Criminal Procedure Code (CrPC), presently known as Bhartiya Nagarik Suraksha Sanhita (BNSS) concerning offenses against women included requirements for First Information Reports to be filed by female official staffs, testimonials to be recorded before a magistrate, delineation of responsibilities for medical practitioners and examination protocols, as well as guidelines for investigations and trials. These procedures address crimes such as acid attacks, rape, and other sexual offenses including those of Online offenses against women.

(E). National Cyber Security Policy 2013:

The security of information, particularly financial including banking data, sovereign data, and personal data, is covered under the National Cyber Security Policy of 2013. The NCSP 2013 policy aims to direct stakeholder about their commitments for the preservation and protection of cyberspace, to sanction the advancement of a safe, secure computing environment, and endow apposite trust and confidence in electronic transactions. The NCSP was constituted by the government in order to strengthen cyberspace's perseverance and safeguard important informative intelligence framework.

A pathway for enumerating a panoramic, collaborative, and indistinguishable retort to acknowledge the problem of cybersecurity at each specific strata of the nation is undermined in the National Cyber Security Policy framework. This policy's main goals are to strengthen cybersecurity capabilities, establishing cybersecurity institutions, substantiating research and development, and promoting international collaboration.

7 Preventing Cybercrime with AI

In the modern digital era, cybercrime is becoming a bigger problem as more and more people and businesses are vulnerable to risks including identity theft, hacking, and online fraud. Artificial intelligence (AI) can effectively prevent and combat cybercrime through the use of machine learning algorithms and predictive analytic. Artificial intelligence (AI)-based solutions for preventing cybercrime are able to examine vast amounts of data, spot trends, and spot abnormalities that can point to possible security risks. Through constant learning from fresh data and adjustment to dynamic threats, these tools can offer proactive defense and improve cybersecurity protocols.

In order to seek help from AI to work on the prevention of Computer-based crime the law enforcement agencies have to identify the SWOT of Cyberspace i.e. Identification of Strength and Weakness and categorically deduce Opportunities and Threats that should be induced.

This SWOT analysis (referred in Fig. 5) is an unending cycle that should be evolving with time and technologies and it has to be maintained by the same. The Cycle starts with Identifying Strengths that can automatically understand the offenders Weakness and with the Strength there is a possibility of utilizing innumerable opportunities for the Government agencies to combat the threat of Cybercrime.

By maximizing the use of AI in cybercrime prevention, organizations, institutions and confederations, and individuals can expand their security protection measures by detecting threats in real time, and mitigate the risks associated with cyber-attacks.

8 Investigation of Cybercrime

Cyberspace is the colloquial term for the computer-generated internet environment, and cyber laws are the set of regulations that govern it. Because this place has a kind of worldwide jurisdiction, all users are governed by these laws. Furthermore, one could consider cyber law to be a subset of law that addresses legal concerns resulting from the usage of networked information technology [22].

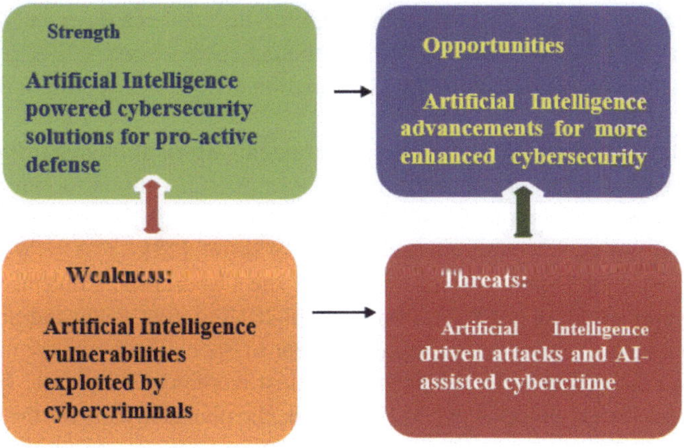

Fig. 5 SWOT analysis windows

Some AI-based Cybercrime Prevention tools that are being used by the Government in mitigating the prevention of Cybercrime against women are Intrusion Detection System, Fraud detection System, Malware Detection through Internet of Things and Blockchain Technology, and also E-mail Filtering.

India's law Executing agencies that are working day and night in combatting cybercrime include the Central Bureau of Investigation (CBI), the Cybercrime Investigation Cell (CCIC), and the National Investigation Agency (NIA). These agencies work together to investigate cybercrimes and bring offenders to justice. Technology is essential to the investigation of cybercrime in India. To capture cybercriminals, law enforcement organizations use a variety of techniques and tools, including network analysis, data mining, and forensic examination of digital devices. However, because cybercriminals use technology, investigators have challenges because they have to stay up to date on the latest tools and techniques used by offenders.

8.1 Procedures Initiated by Government for Cybersecurity of Women in India

- India has implemented various laws and initiatives to protect women from cybercrime under the Ministry of Home Affairs, Indian Cybercrime Coordination Centre(I4C), National Cybercrime Reporting Portal (NCRP), NCIIPC, CERT-IN, Online Women specific Crime Reporting Unit [23].
- A National and Sectoral 24X7 mechanism has been envisaged to deal with cyber threats through National Critical Information Infrastructure Protection Centre (NCIIPC) [23].

- Computer Emergency Response Team (CERT-In) has been designated to act as a nodal agency for coordination of crisis management efforts. CERT-In will also act as an umbrella organization for coordination actions and operationalization of sectoral CERTs [23].
- A method is proposed for obtaining strategic information regarding threats to information and communication technology (ICT) infrastructure, incorporating scenarios of response, resolution, and crisis management through effective predictive, prevention, response, and recovery action [23].
- The Ministry of Home Affairs introduced the Cybercrime Prevention against Women and Children (CCPWC) program in 2018 to give states and union territories financial support for the establishment of specialized cybercrime cells to handle cases pertaining to cybercrimes against women and children. In order to support the investigation of cybercrime cases, the plan also calls for the establishment of a specialized cyber forensic laboratory in each state and union territory. The Indian government has launched other policies to combat cybercrimes against women in addition to these initiatives [23].
- The safeguarding of women's rights in cyberspace is covered by the 2001 National Policy for Empowerment of Women. The policy asks for the creation of gender-sensitive legislation and policies to address gender-based violence, acknowledging that it must be addressed in all of its manifestations, including cyberspace [23].

India is to become a knowledge economy and society enabled by technology through the introduction of the Digital India program since 2015 [24]. The program had provisions for raising citizens' understanding of cybersecurity and digital literacy, especially among women and girls. The initiative still aims to promote women's access to technology and lessen the digital divide by enhancing the quality and availability of digital infrastructure and services. AI can aid in the enforcement of these laws by assisting in the identification, tracking, and prosecution of cyber criminals.

9 Conclusion

John McAfee once stated "The only way to stay away from Cybercrime is to invest in Cybersecurity."

Cybercrime and Cybersecurity are both different sides of the same coin. According to Augustine's privation theory something is evil in virtue of a privation, a lack of something which ought to be present in a particular nature. While it is not evil for a human to lack wings, it is indeed evil for a human to lack rationality according to the end of a rational nature [25]. Similarly having a crime free society is a rational thought but when the thought is not true than modes of combating crime should be rational.

India is experiencing a significant increase in cybercrime, which calls for immediate attention and collaboration from all stakeholders. The negative effects of cybercrime utilizing AI, which can include damage to one's reputation, monetary losses, and risks to the safety of women, emphasize how critical it is to handle this expanding issue in its entirety. India has made significant progress in combating cybercrime, but much more work has to be done. Improving cybersecurity infrastructure, increasing cybersecurity awareness and education, and fostering collaboration between the public and private sectors as well as international organizations are all necessary to fortify a strong defense against cyberthreats aimed at women.

Women should prioritize cybersecurity by utilizing secure online practices, regularly updating software and systems, and being alert to phishing and social engineering schemes.

The Indian laws has come a long way ahead with the new version of IPC, 1860 i.e. the Bhartiya Nyaya Sanhita to be effective from July 1st, 2024 along with the Information Technology Act, 2000 and the National Cyber Security Policy, 2013. In the end, the Law Enforcement Agencies should skill them up to work using Artificial Intelligence to make a concerted, continuous effort in order to be able to curb the social menace like Cybercrime and uproot its existence from its origin. India and its women should be safe and feel safe in and around the society and the digital society by creating a safe digital ecosystem that promotes creativity, trust, and economic advancement by raising awareness, developing a cybersecurity culture, and implementing robust preventive measures.

References

1. Dey A, Dey SK (2023) Cyber security in digitized e-governance: evaluating Indian IT act 2000 (as amended in 2008) in context. Int J Mass Commun 1:1–7
2. Koops BJ (2010) The internet and its opportunities for crime. In: Herzog-Evans M (ed) Transnational criminology manual. Nijmegen
3. Rao KK, Human rights and cyber-space: use and misuse
4. Gupta S, The evolving landscape of cyber crime: challenges and counter-measures. Legal Service India E-journal
5. Saini M, Kaur M (2022) Indian government initiatives on cyberbullying: a case study on cyberbullying in Indian higher education institutions. In: NIH PMC-Covid Collection -2019. Springer Nature. https://doi.org/10.1007/s10639-022-11168-4.
6. Borah U (2020) Cybercrime against women in the digital era: a brief Indian scenario. Int J Res Appl Sci Eng Technol (IJRASET) 8(VII):615
7. Sharma R (2018) Cyber crimes against women in India: an analysis. J Inf Technol Econ Dev 9(2):18–33
8. Rana M, Crimes in cyberspace: right to privacy and other issues.
9. Smith RG (2007) Crime control in the digital age: an exploration of human rights implications. Int J Cyber Criminol 1(2):167
10. Altaf Z, Javed N (2024) AI-generated content poses a looming threat to women in India and Pakistan. The Diplomat. Published on 31.01.2024
11. Cybercrimes cross borders and evolve rapidly. Terrorism Prevention Centre of Excellence. https://tpcoe.gov.pl/
12. Introducing cyberspace (2003) Mapping cyberspace. pp 13–43

13. Sample I (2020) Hat are DeepFakes and how can you spot them? The Guardian. Published on 30.01.2020
14. Table 9A. Crimes in India 2022, vol I. NCRB.GOV.IN
15. Table 9B. Crimes in India 2022, vol I, NCRB.GOV.IN
16. Hathaway OA, Crootof R, Levitz P, Nix H, Nowlan A, Perdue W, Spiegel J (2012) The law of cyber-attack. Calif Law Rev 100(4):817–885
17. Report on the proposal for a regulation of the European Parliament and of the Council on laying down harmonized rules on Artificial Intelligence (Artificial Intelligence Act) and amending certain Union Legislative Acts; Committee on the Internal Market and Consumer Protection Committee on Civil Liberties, Justice and Home Affairs; 22.5.2023—(COM(2021)0206—C9-0146/2021–2021/0106(COD))
18. Texts Adopted (2023) Artificial intelligence act. European Parliament. 14.06.2023
19. Artificial Intelligence 2024 Legislations; National Conference State Legislatures; https://www.ncsl.org/technology-and-communication/artificial-intelligence-2024-legislation (Last visited on 21.05.2024)
20. Felony classes and penalties—Restitution—Habitual criminal sentences, Codified Laws, South Dakota Legislature Legislative Research Council
21. Steps to Cyber security. National Cyber Security Centre. United Kingdom. CESG
22. Bichwa KR (2018) Identify cyber bullying words using clustering for social-media. Int J Comput Sci Eng 6:525–528
23. National Cyber Security policy in a Nutshell (2024) https://www.clearias.com/national-cyber-security-policy-2013/ (Last Visited on 20.05.2024)
24. Economic Times (2021) India's rank drops to 10th in global cybersecurity index: ITU. https://telecom.economictimes.indiatimes.com/news/india-ranks-among-top-10-in-itus-global-cybersecurity-index/83978564 (Last Visited on 20.05.2024)
25. Haratine P, On the privation theory of evil "a reflection on pain and the goodness of God's Creation. Theologica, UClouvain

"Accept": Freedom of Contract While Navigating Apps in the Era of Artificial Intelligence

Purbasha Mukherjee

Abstract In a time when smartphone applications enable convenience like never before, it's critical to be mindful of the rights we provide to the applications. Smartphones and apps have significantly improved our quality of life, but we also can't ignore the possibility of deceptive or occasionally harmful programs, especially when they fail to originate from reliable sources. The phrase App Permissions refers to the ongoing act of selecting the Allow button on the little pop-up windows that appear as soon as we install an app on our smartphone or another device and want to use it. These applications fail to show or provide a clear explanation of why they need certain permission in order to function. By disregarding the reason why this is necessary, we are, in some manner, either handing up control of our device or transferring our personal data to those app developers, oblivious of what the code, bugs, and bots are programmed for. Data Security breaches can result from this, as the data gets permanently stored in those databases. Recently, The Competition Commission of India held Google guilty of abusing its dominant position in search engines and creating appreciable adverse effects on market competition. This article will discuss the online safety and data privacy issues while giving permission to access our location, internal files, camera, GPS, gallery, and contacts, and also the threats it possess on our freedom of contract if dealt unfairly.

Keywords App · Permission · Data · Privacy · Protection · Applications · Contract

P. Mukherjee (✉)
IEM's International Institute of Juridical Sciences (IIIJS), University of Engineering & Management, Kolkata, India
e-mail: purbasha.mukherjee@uem.edu.in

M. Chakraborty et al. (eds.), *Proceedings of 5th International Ethical Hacking Conference*, Lecture Notes in Networks and Systems 1148,
https://doi.org/10.1007/978-981-97-8457-8_21

1 Introduction

Contracts remain the cornerstone of business and, consequently, the entire corporate world. They govern everything from recruiting and hiring of staff to the selling of chocolates to business acquisitions [1]. In a nutshell, without the Law of Contracts to ensure that the agreements we make with others are upheld, business around the world would completely fall apart as would our daily lives. Although it is not a comprehensive regulation, the Indian Contract Act, 1872 controls contracts in India. The Contract Act is extensive on a subject where it addresses a particular concern. However, M.P. Jain, an eminent professor of Delhi University in his article "Law of Contract before its codification" categorically mentions that in circumstances where the Indian Contract Act has been silent, the Judiciary did take the help of English judgments, on principles like—Law of Limitation, Restraint of Trade, Wagering Agreements [2].

One thing which has lately emerged to be crucial to a rapid life is the internet. Information and Communication Technology (ICT) is widely used, and its intrinsic qualities—including speed, instantaneous, and precise information—these qualities have been a valuable tool for economic development [3]. With the recent developments in information technology, software, telecommunications, and computer technology, people's standards of living have changed in ways that were be unthinkable, few years back geographical and temporal limitations are no longer to impede communication. The globe is connected and the concept of a globalized world is created by the widespread and faster transmission and reception of information. Technology has proved to serve as a bridge between people, cultures, and societies. Contracts used to be paper-based, but as the internet has evolved, businesses have shifted from the idea of conventional-based contracts toward online ones. A new type of contract known as "electronic contracts" has emerged as a result of the transition from paper-based to online agreements. It refers to an electronic contract.

This paper will study the fundamentals of E-contracts and their enforceability in digital platform.

2 Understanding E-Contracts

The Indian Contract Act, 1872 has not defined e-contract specifically. The Information Technology Act, as amended in 2008 entails the validity of contracts formed via electronic means [3]. Thus in layman's term, an electronic contract is one that is formed and enforced virtually. Vendors give prospective consumers information about their goods, costs, and conditions [4]. When making an order, buyers consider their choices, try to negotiate terms and rates, place their order, and pay. The sellers then provide the goods that were ordered. This is referred to as electronic commerce, and it involves the exchange of goods and services, as well as information, over computer networks [4]. The Indian Evidence Act, 1872 too recognizes any communication or evidence made electronically as valid evidence and therefore admissible, if found relevant [5]. E-contracts are agreements that adhere to the "Uberrimae fidei" principles. These principles were first propounded by "Lord Mansfield" in the renowned case *Carter vs Boehm,* [6] where the parties are not negotiating at arm's length; rather, a party indicates that he is prepared to sign a contract based only on the data that the other party has supplied. As information being omitted or misrepresented in such transactions, the principle "Uberrimae fidei" acts as a safeguard of e-contracts. The Information Technology Act, 2000 does not specifically address legal ability, but it does assume that both parties to an online contract are competent after it is signed. Stated otherwise, neither party may subsequently protest and claim that the other party lacks competence, rendering the contract invalid. In the event of an electronic contract, the Uberrimae fidei doctrine will be diligently upheld, and any prejudice should not be placed on one party for behaving in a way that is detrimental to the other on the other's declaration that he is competent [6].

2.1 Development of E-Contracts

Businesses realized in the 1960s that a large number of the papers they were exchanging, including bills, delivery, purchase orders, and invoices, had to do with the delivery of products. For nearly every transaction, the identical sets of details were found in these records. Additionally, it was discovered that deals were devoting a sizable portion of their time and resources to entering this data into database, and repeating the process. Businesses were able to save costs by developing a uniform set of formats for electronically conveying such information. Through standardizing a set of formats for electronically transmitting such data, organizations were able to lower errors cut down on printing and mailing expenses, and do away with the necessity for data entry errors.

The various types of e-contracts are: Shrink Wrap Contracts: Under these, parties accept the conditions when they use previously acquired software. When software is licensed, it is packaged with a notice containing the terms and restrictions for using the program affixed to the top of the packaging. Upon opening the product, the user acknowledges that they have read these terms and agrees to abide by them; Click Wrap Contracts: These contracts are typically encountered during the app installation procedure, by using the "I Agree/I Accept" or "Cancel" icon.

2.2 App Permissions in Click Wrap Contracts

The permissions of app use are listed by the app developer in the opened box when downloading the app program online. Terms and conditions of agreements may change based on the type of application. After reading the terms and conditions, the user needs to confirm his consent by using the "I Agree/I Accept" or "Cancel" icon. The final user decision is crucial to the creation of the contract. By clicking the symbol, the user or downloader agrees to the terms stated in the contract. Since the terms and conditions were provided by one party and access to download is only permitted if the user agrees the terms and conditions in their entirety, there is no room for negotiation between the user and app provider throughout the contract [7]. Under Browse Wrap Agreements, the website in question grants users access to and usage of the program by presenting privacy terms as a link at the bottom of the screen that opens on a different page. The terms need to be accepted in order to use the website. If the terms are not given permission, access is completely denied.

The majority of the applications need the user's contacts. This facilitates their marketing touch with them. A person's privacy is violated when their contacts are used without their consent. The users will receive more and more messages regularly. Additionally, the majority of apps demand location. While this could be acceptable in certain situations, it makes it simpler to attack the victim when you know where they are thus posing a security risk. Photo and files of the individual also need to be allowed to download a particular app [7]. This is unsafe as well as problematic leading to privacy breach. Usually, storage space is needed to search through your device and determine whether further programs may be loaded. Since mobile devices have a limited amount of storage, using too much of it might slow down the device and make it difficult to access it speedily or stop it entirely if malware is found to be there. Again the majority of apps also demand having access to the camera [8]. It becomes an issue of confidentiality when someone captures a personal photo and posts it online without proper consent. A vast number of applications also include identities, phone calls, SMS sending, and microphones, all of which might compromise service users privacy and security [8] (Fig. 1).

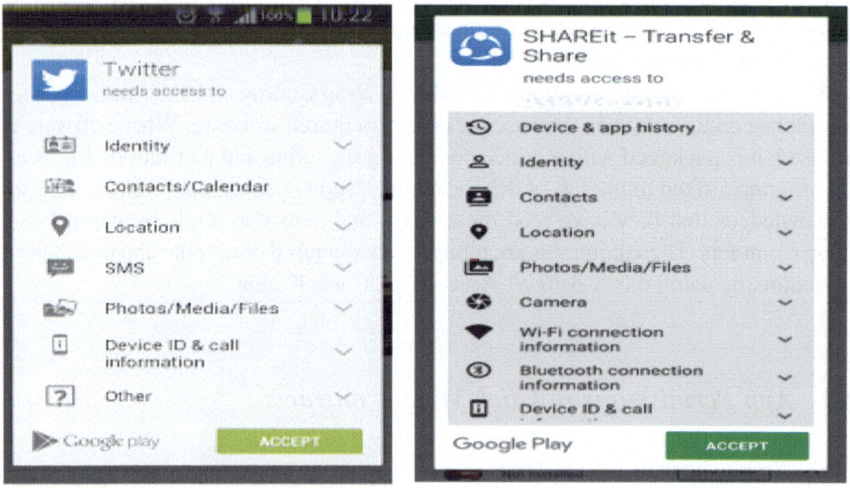

Fig. 1 Sample click wrap contracts [9]

2.3 Contractual Obligation in Click Wrap Contracts

Both offline and electronic transactions must comply with the fundamental principles of contract law. An electronic contract needs the same components as any normal paper contract, including the following ones:

There must be a proposal made. The first requirement for the creation of a legally binding contract is an offer, as without one, there cannot be an acceptance and no contract will be formed. A proposal is all that an offer entails [10]. An offer is a statement giving an intent to do something, giving an impression that as assent to such intention would result in a legally enforceable contract. The statement has to make an effort to act or not act in a way. The one presenting the offer is referred to as the proposer, promisor, or offeror, while the recipient of the offer is referred to as offeree. An offer from the offeror to the offeree is also necessary for the establishment of e-contracts [10].

In the context of electronic contracts, the individual to whom a promise is made by electronic communication is referred to as the addressee, and the promise maker is referred to as the originator. When electronic means are used to transmit an acceptance, in this scenario, the addressee takes on the role of the originator of the acceptance and the promisor becomes the addressee of the electronic offer. Nevertheless, the offeror does not make any direct one-on-one interactions, in online communications. After "browsing" the internet and the reviews, it is the applications that are invited to the users, the same proposes what he wants to install. It is not the application/software that offers the software for installation but only an invitation for an offer (Fig. 2).

Regarding acceptance is equally essential for e-contracts. An offer through the internet cannot deem to be binding legal agreement until it is accepted. Under the Information Technology Act, 2000, originator refers to the person who originates any electronic record and addressee is a person who receives it. When an originator

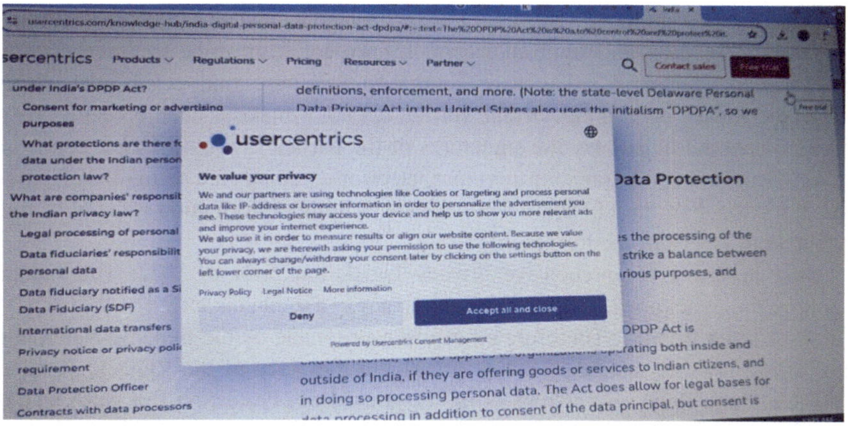

Fig. 2 Invitation to offer the cookies [11]

uses electronic media to convey a proposition to an addressee, the addressee has the option to transmit the originator his acceptance through interest. An electronic offer can be accepted electronically by clicking a button marked "I Accept" or "I Agree," downloading files, or clicking on other buttons [8].

The proposer shall be informed of the acceptance in the event of an electronic contract as well. Regarding e-contracts, there is disagreement on when acceptance should be notified. According to one perspective, the principle of mailbox rule will be infructuous as communication is immediate and ends when the originator receives the acceptance. The second point of view, on the other hand, contends that electronic message reception may not always occur instantly since the recipient must first take action to connect to their mail server or mailbox after receiving the message. In this case, the mailbox rule should be followed [8].

It might be challenging to confirm the identity or ability of the contractual party in an electronic contract. Due to the intricate nature of electronic contracts, it is hard for the party making the agreement to determine the other side's level of competence, as do not engage face-to-face communication and misrepresentation is quite likely in e contracts [10].

3 Legislative Framework Governing E-Contracts

The United Nations Commission on International Trade Law (UNCITRAL) adopted the Model of E-commerce, 19,961 (MLEC) [12] after realizing the significance of this technology and its potential to transform the commercial environment.

Most nations have included provisions of the Model Law in their e-commerce laws. The terms "Electronic Contracts or Online Contracts" are not specified in the Indian Contract Act of 1872 or the Information Technology Act of 2000. Nonetheless, The Information Technology Act, 2000, as amended by the Information Technology (Amendment) Act, 2008, addresses the legitimacy of contracts made electronically vide section 10A. It specifies that a contract is deemed valid if all aspects of its formation, communication, and proposal/acceptance revocation are done via the internet or through electronic records. The mere fact that an electronic contract was made using electronic means or in an electronic format does not make it unenforceable. Any genuine contract must have the signatures of the parties as proof of their approval of the terms and conditions. Furthermore, electronic records are recognized legally under section 4 of the IT Act, where any law requires information accessible and available in an electronic format.

India lacked a comprehensive privacy legislation prior to 2022. The Indian Supreme Court acknowledged the right to privacy as a right guaranteed by the Constitution of India in 2017. Therefore all efforts were made by the Indian government in drafting laws to safeguard Indian privacy in the wake of the Right to Privacy [13] judgment. Previous versions of the Personal Data Protection Bill, such as the Data Protection Bill 2021, [14] were subject to intense examination and eventually proved to be ineffective. The 2023 Act permits the processing and handling the data for lawful uses as stated by law. According to the Act, permission must be freely given, explicit, unconditional, and unmistakably favorable. The amount of data that is collected must be limited to what is necessary for the intended purpose. Consumers must get a notice that clearly explains each of these components, as well as details on the grievance redress forum and the parties' rights. Therefore, it is necessary to determine if the legislation appropriately protects personal data and, as stated in the law's preamble, whether it strikes a reasonable balance between "the need to process such personal data for lawful purposes" and "the right of individuals to protect their personal data" [14].

3.1 Judicial Approach

When app developers began to offer their products in ways like having the program pre-installed on a user's computer or downloaded from the internet, click wrap agreements were used [15]. A pop-up with the licensing terms clicks for the user to read upon in order to download, install, or use the application for the first time. Either "I agree" or "I do not agree" is requested of the user to click. The entire process is ended if the user disagrees [15]. The click wrap agreements frequently prompt one to wonder if the user gave sufficient advance notice of the licensing terms and indicated their approval to them [16].

There is a lot of debate throughout the world about whether these kinds of agreements are legally enforceable. It's still unclear if these agreements can be legally enforced. In response to this question, US courts have decided that if these agreements will be enforceable against users, provided they are given the opportunity to review the privacy terms and are asked to provide their approval before accessing the website or using the service. For such contracts acting to be legally enforceable on a user, the terms must be clearly displayed on the website and give a fair chance to read the terms and conditions before clicking the "I agree" or "proceed" button. Although the conditions don't always show up in the same window or webpage, the license or terms of service are always available before acceptance. An example of this would be a pop-up window that appears before installation or a hyperlink that is included in the application's homepage. The user must be informed for them to be considered to have accepted the terms of service. Courts have determined that the notice requirement is absent in cases, where the terms of service are not readily apparent and/or accessible and consequently, the user may not be legally obligated to adhere to the terms of the e-contract [17].

App provider's attempt to enforce an arbitration clause was rejected on the grounds that the user was not required to accept the terms of the licensing agreement in order to download it; rather, it was published only as a request that the user review the terms, thereby protecting the interests of the customers [18]. A message asking users to review the conditions was located at the bottom of the page, but accepting the terms was not a sine qua non for downloading the app. Click wrap agreements appear to be a viable substitute for paper contracts in the modern digital world. As a result, in order to complete the browsing, people click the "I agree" button as quickly as possible. Maybe in the future, app providers and developers will see the issues with lengthy, complex contracts and take steps to translate the language into plain English. It is best to read the terms before clicking "I agree" at any point. Hence, the next time a box containing pages of conditions printed in a print that is too small to read appears, we should avoid scrolling to the bottom of the page and selecting "accept" right away. Rather, it is usually best to study the provisions to determine what one is actually giving access to. However, one issue brought about by the usage of click wrap agreements is that businesses frequently overuse "legalese" in their service or conditions of use, making users handcuffed. Unfortunately, sometimes people disagree with the conditions and are unhappy with the permissions until after it has already happened. Nevertheless, since they consented, they are obligated to abide by the terms as disagreement will render the app function inappropriately (Fig. 3).

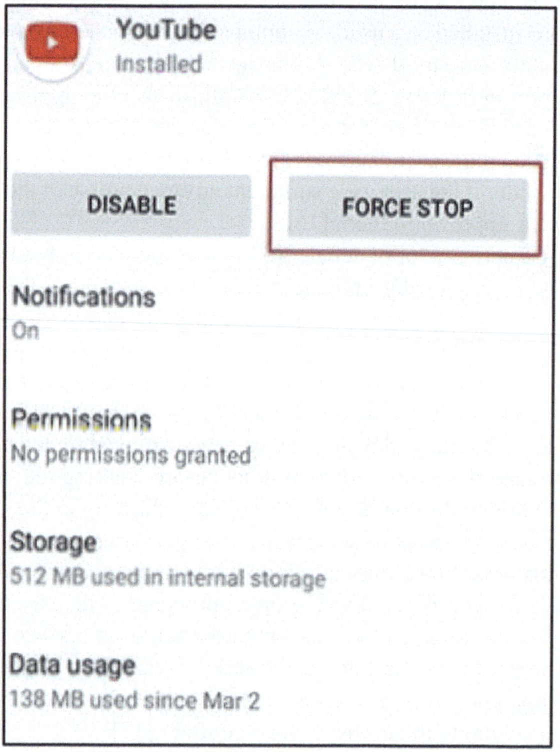

Fig. 3 Denying app permissions disturbs the functioning of the app [19]

4 Role of AI in E-Contract

4.1 In Collecting Data

The collection and storage of data are greatly aided by AI. The online portrayal of marketing involves two parties: person who provides the commercials to be displayed on the viewer's content and a publisher who incorporates the ads into its online content [20]. Online advertising is a means of reaching consumers with promotional marketing messages. Posts videos, photos, and other graphics are some of the several ways that these adverts may be shown in order to visually convey their message. To increase the effectiveness of their advertisements, display marketers frequently target individuals based on particular choices and likings [21]. Cookies are unique computer identifiers that online advertisers commonly use to select which posts to display to a specific consumer. Cookies detect if an end user exits a page without making any deal, which enables an advertiser to retarget that user with ads from the website they previously visited. Advertisers may create a detailed profile of a user's interests by compiling data about the user's online activity from several websites, enabling them to deliver even more targeted advertising [22]. News Feed Ads, typically seen on social media platforms that showcase information and updates in predetermined formats. Any type of content, including items, fan pages, websites, and applications, gets promoted via these advertisements. This approach to gathering data is referred to as behavioral targeting [22]. Typically, this type of data management platform may collect three distinct types of data: This may comprise cross-platform data, data from consumer behaviors, habits, and interests, in addition to website and sponsored media content. The data is directly from the source, which might be ads, hosted social networking platforms, or analytics platforms. Third-party information is often gathered from many websites and other sources. Businesses are allowed to disseminate and sell data obtained from other parties via a range of distribution methods. After receiving this integrated customer data, the supply-side platform may combine information about the end user who will view the advertisement and will act accordingly [23] (Fig. 4).

4.2 Storing Data

AI is also used to enhance a website's exposure in search bars. An internet search engine displays sponsored and nonpaid results based on the query provided. Search engine algorithms typically utilize visual clues to make a website more approachable and appealing [25]. The rise of Google as a primary search platform has gained immense momentum over the last few years. Building campaigns with Google Ad Words was simple for marketers. Consequently, the marketers managed to monitor the campaign by tracking the time spent on a specific advertisement. Artificial Intelligence by storing previous browsing data improves a website's organic search ranks

Fig. 4 Facebook tracking friends based on Instagram connections [24]

on search engine results pages by making the content more relevant to and addictive to other search queries. Social media platforms serve as a great medium to promote products and services socially. Using their social media sites, many businesses use regular updates and exclusive deals to market their products. This business includes sponsored postings, interactive tests, and videos. These advertisements are often found on Twitter, WhatsApp, Snapchat, Instagram, and Facebook [26] (Fig. 5).

Chat advertising, as opposed to static messaging, describes messages that are sent to consumers in real time on certain websites. The operational staff behind the site frequently places advertisements on visitors who are browsing around the sites [22]. This is accomplished via the use of live chat software or monitoring tools put within certain websites. Any form of marketing that includes producing, disseminating, and distributing media together with content in order to draw in and keep consumers is

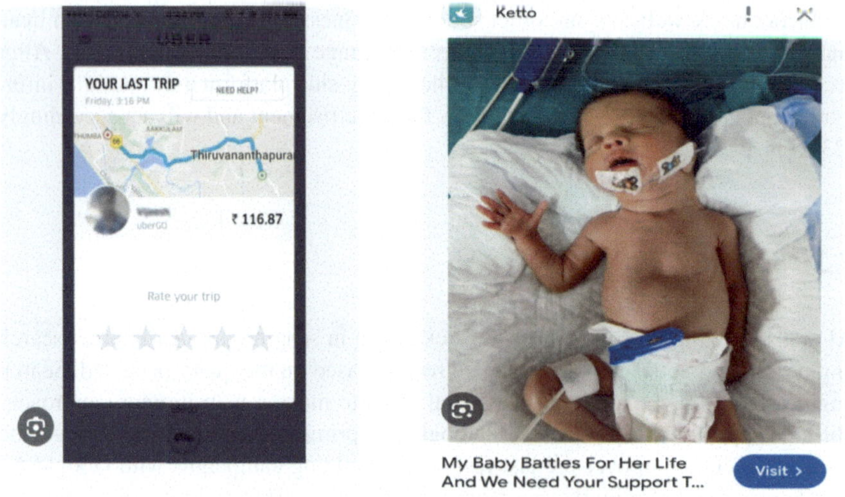

Fig. 5 Data gets stored on the basis of browsing activity [27]

known as content marketing. A multitude of media, including blogs, news, videos, e-books, white papers, infographics, case studies, and how-to manuals, may be used to communicate this information. In today's time marketing outweighs quality and Artificial Intelligence plays a big role in it [21].

5 Conclusion and Recommendation

As of now, no one can deny the importance of E-contracts. The outbreak of Covid-19 has also witnessed the expanding growth of e-contracts. However, there are certain threats which should be addressed [28]. Firstly, data privacy—often unknowingly we provide a lot of personal information on the internet making us vulnerable to online attack. While giving permissions during installation of an App we must be well aware of what access is been permitted. Secondly, when there lays no option to opt out, can the yes be a wholehearted yes. The DPDT Act is also silent regarding this. The consent in such cases cannot be said to be free. Thirdly, another concern with regard to competency of app developers, whether they are equipped enough to develop the infrastructure of the App and maintain it. Another issue regarding jurisdiction. The jurisdiction clause in click wrap and browse wrap contracts, for example, is one party's creation rather than the result of free mutual consent. The originator forms rules applicable to his jurisdiction and beneficial to his interest, not the customer or any other end user [18]. In a click wrap contract, the legitimacy of this type of jurisdiction regulation cannot be disputed.

Some Suggestions in This Regard

- Properly read all the permissions before accepting it.
- Understand if the particular permission is required by that app, and if not, block the permission.
- The nomenclature of the privacy terms must be more layman friendly.
- If app uninstalled, clear all cache and cookies to remove all stored data.
- Be aware of identity theft.
- There should be a strong legal framework to protect data privacy and detect data breach, when necessary.

References

1. Pollock and Mulla. Indian Contract Act
2. Jain MP. Law of Contract before its Codification
3. Sharma SR. Nature of cyber laws
4. Rohas N. Ecommerce - Legal Issues
5. Kapil R. Evidentiary Value of E-Contracts, ES Law College Pune
6. Carter vs Boehm (1766) 97 ER 1162
7. Sharma V. Information Technology law and practice- cyber law and ecommerce

8. Rastogi S. Insights into Electronic Contracts.
9. Fig 1: Sample photos of Click Wrap Agreements taken from https://www.freeprivacypolicy.com/blog/write-clickwrap-agreement/
10. Bakshi PM, Suri RK. Cyber and E-Commerce laws
11. Fig 2: Screenshot taken by the author herself in her smartphone model number Realme UI 4.0 dated 31st May 2024 showing an invitation to offer cookies in order to browse the website
12. UNCITRAL model law on electronic commerce Nanadan Kamath, Law relating to Computers Internet and E-Commerce (2nd, Universal Law Publishing Co.Pvt. Ltd., New Delhi (2006) 557
13. Justice KS. Puttaswamy vs Union of India. (2017). 10 SCC 1
14. Times of India (2023) Digital Personal Data Protection Act, 2023—A game changer. 27th August 2023
15. Sharva S. Online click wrap agreements—an analysis. https://www.legalserviceindia.com/legal/article-369-online-click-wrap-agreements
16. Feldman vs Google. (2007) 513 F. Supp. 2d 229
17. Bragg vs Linden Research Inc. (2007) 487 F. Supp. 2d 593
18. LIC India vs Consumer Education and Research Centre. (1995) AIR 1811
19. Fig 3: Screenshot taken by the author herself in her smartphone model number Realme UI 4.0 dated 31st May 2024 showing consequences of denying access of an App
20. Interactive Advertising Bureau (2016) https://www.iab.com/news/first-quarter-u-s-internet-ad-revenues-hit-record-setting-high-nearly-16-billion-according-iab
21. Zuiderveen Borgesius Frederik J (2015) Improving privacy protection in the area of behavioural targeting. Kluwer Law International
22. Bennett Steven C (2011) Regulating Online Behavioral Advertising John Marshall Law Review
23. Rodgers S, Esther T (2000) The interactive advertising model: how users perceive and process online ads. J Inter Adv 1:41
24. Fig 4: Facebook suggesting friends based on Instagram connections taken from https://www.ubergizmo.com/2015/05/facebook-now-suggesting-friends-based-on-instagram-connections/
25. Baek TH, Mariko M (2012) Stay away from me: examining the determinants of consumer avoidance of personalized advertising. J Adv
26. Bleier A, Maik E (2015) The importance of trust for personalized online advertising. J Retail 91:390
27. Fig 5: Browsing data history gets stored in App database taken from https://www.ketto.org/fundraiser/support-Twin-babies-of-mohini-to-get-a-healthy-life
28. Verma SK, Mittal R. Legal dimensions in cyber space

AI and IPR: Balance Between Innovation and Creativity

Sravya Jammula and Manoggnya Reddy Kanjula

Abstract Are we ready for this new change? With recent developments in technology and innovation, we came across Artificial Intelligence (AI) which is now playing a major role in our everyday tasks and endeavors. It came to that advent that we prefer not to do simple tasks without using AI. From assisting students in academics to facilitating the creation of scholarly articles through tools like ChatGPT, AI has ingrained itself into this modern society. But the pivotal query starts there, how is this AI impacting intellectual property rights and the challenges of granting authorship or ownership rights if there is AI used? This article addresses the ongoing legal issues that include whether any intellectual property made by the technology or AI possesses the right to claim ownership or authorship. Moreover, the article delves into the inquiry of determining the rightful owner in cases of AI-generated works, contemplating whether ownership should vest with the AI entity or another identifiable party. The article concludes by stating the inherent disadvantages and the potential threats that are associated with granting legal rights and authorship to AI.

Keywords Artificial intelligence · Intellectual property rights · Authorship and ownership · Technology · Threats

1 Purpose and Significance

The main purpose of this chapter is to provide an analysis of the impact of Intellectual Property Rights caused by Artificial Intelligence. This chapter will discuss legal perspectives, frameworks, and precedents. Further, the challenges that are occurring due to the development of AI will be explored.

S. Jammula (✉) · M. R. Kanjula
ICFAI Law School Hyderabad, Hyderabad, India
e-mail: jammulasravya29@gmail.com

M. Chakraborty et al. (eds.), *Proceedings of 5th International Ethical Hacking Conference*, Lecture Notes in Networks and Systems 1148,
https://doi.org/10.1007/978-981-97-8457-8_22

The chapter aims to provide insights into the overlap caused by AI and IPR and to provide suggestions or recommendations for professionals in the respective fields. The chapter also addresses the ongoing debate between the IPR and AI based on ownership or authorship rights.

2 Methodology and Scope

This chapter uses a qualitative methodology, preliminary through reviews of existing academic research, legal case laws, and documents. The scope of the chapter is to focus on the impact on Intellectual Property Rights specifically: Copyright law, Trademark law, and Patent law caused by AI. The chapter analyzes the challenges that AI has caused in each of these areas by providing case studies to explain them. Finally, in conclusion, the chapter provides suggestions and recommendations for legal professionals about the way to tackle the situation and adapt to the present increasing technology.

3 Prelude

Artificial Intelligence is a technology that enables problem-solving by simulating human intelligence through computers, phones, or any digital device. Today AI can learn not only human languages but it can also other data types which include deepfakes that are being used more now against many celebrities. Intellectual property means the creation of any work using the mind and person's intellect and the creations include literary and artistic works, designs, symbols, etc. These creations are protected by law which has different provisions for the protection of various IPR. A few types of intellectual property include Patent, Trademarks, Copyright, Industrial Designs, Geographical Indications, and Trade Secrets. Artificial Intelligence has reached a height and now it is playing a major role in the lives of people compared to that of the past decade. It includes various aspects from machine learning, cleaning, doing transactions, etc. While on the other hand intellectual property provides rights to the creators who use their innovation and creativity [1]. The intersection of IP and AI has been evolving hugely these days including a few of the AI tools like ChatGPT and many more. Due to this, many issues are being brought up in the present day. The primary rule of IP is to grant rights of IP to the person who did the work using his own intellectual mind and hard work. With the involvement of IP, the work has become easy and the main question arising here is in the cases of work done by AI who will be having the ownership of such work created? Whether it is the AI or the person?

4 AI and IP: Issues of Ownership

Since AI is evolving and the use of AI has been used in many of the works created, the main question of ownership pertains to whether the ownership rights have to be granted to the AI that created the work or to the person to whom AI generated such work.

AI algorithms can autonomously generate a myriad of creative works, including music, literature, art, and inventions. However, determining the rightful owner of such works poses significant challenges. Unlike human creators, AI lacks legal personhood, raising questions about authorship and ownership. Case studies such as the infamous "Monkey Selfie" case illustrate the ambiguity surrounding ownership when AI is involved in content creation.

The European Patent Office (EPO) has stated that the inventor must be a human being and therefore not an AI system. United States Patent and Trademark Office (USPTO) also stated that the inventor must be human but it did not address the issue of AI-generated works [2]. Many legal scholars argue that current legal provisions are not well equipped with the provisions of AI and the ownership of such AI content and that new legal provisions are needed [3].

These kinds of issues occur in other IPRs as well which include the Copyright. Wherein to generate any of the work of authorship AI can be used to make the work and finish quickly. Copyright protects original works of authorship. A key question is whether AI-generated works meet the originality threshold. Some argue that if an AI is following a set of instructions, the work lacks the requisite human creativity. Others contend that the selection and training of the AI itself demonstrates sufficient human authorship.

However, as per the provisions of the Copyright law, the work must be generated by a human author, and nowhere in the provisions of the Copyright Act does it mention the issues of AI-generated works or the ownership in cases of works that are created by AI. So many legal scholars argue about the provisions stating that they are not up to date to deal with current legal scenarios as it does not address any of the new issue complexities and that new legal frameworks are needed [4].

Due to this many legal questions are being raised regarding the potential owner of such AI-generated works and what would be the impact if AI was given the ownership. The protection of IP holders in such cases becomes a major issue.

5 Legal Perspectives on AI and IPR

Existing IP laws vary in their approach to AI-generated content. Copyright laws typically grant protection to human creators, leaving AI-generated works in a legal gray area. Patent laws require inventions to be attributed to human inventors, raising questions about the patentability of AI-generated innovations. To enhance fairness courts must adopt new legal frameworks, especially for the cases where innovation is involved.

Authors are granted exclusive licenses in cases of copyright. At the same time, AI can produce numerous works which include literary, dramatic, and artistic works along with musical compositions.

The main issue that arose in the debate is the aspect that AI-generated works suffer from lack of creativity: Whatever the work is generated by AI whether it's artistic, dramatic, musical, etc., the main concern is about this creativity and whether the work that is created by such AI systems is holding any creativity, as envisaged in the statute concerning AI.

The second issue is about "human intervention" as there is no human involvement in the creation that is created by a mere technology AI. Only the data training and architectural aspects of AI have human intervention. But the ultimate result doesn't include any such human intervention.

All these issues put together raise one single main issue whether the work generated by AI can be given copyright protection or not, and the debate on this issue continues.

6 Ethical Concerns and Relevant Considerations

The main concern is about the transparency of the work, fairness to everyone, and whom to make accountable in cases of liabilities. Without clear guidelines, there is a risk of exploiting AI-generated works without proper attribution or compensation to the original creators, whether human programmers or AI systems themselves. Ethical frameworks must address issues of agency and responsibility in the context of AI and IPR.

A few challenges may include [5]:

6.1 IP of AI Models: This means who owns the AI model that is developed? These models and data may include any texts, images, content, etc.

6.2 Plagiarism of Content Generated by AI: The identification of the actual owner could be difficult due to the increase in the advancement of AI and due to the use of AI certain plagiarism tools cannot detect AI and that would raise legal and ethical questions on the ownership of the rightful owner.

Considerations:

6.3 Transparency: The first ethical consideration should be transparency about the origin and the use of the algorithms. All the IP holders must be transparent about the sources they are using to get the data for the research and preparation of the IP.

6.4 Justice and Equity: The IP works must be developed with fair and equitable means without any unjust and discriminatory bases.

Proper Regulations: There should be proper frameworks and provisions made with the development of AI there should be provisions made into the legal frameworks as to who is the actual owner of the IP. Such provisions must also contain provisions in cases of AI-generated works regarding the ownership of such works.

Improvement of Plagiarism Tools: With the advancement of AI, plagiarism tools are also advanced to detect the content of AI and to determine the rightful owner. This tool must be implemented in all the plagiarism sites to make it convenient for the authorities to detect AI-generated work.

7 AI and IP Ownership: International Views

The EPO has stated that the inventor must be a human being and therefore not an AI system. USPTO also stated that the inventor must be human but it did not address the issue of AI-generated works [6].

WIPO is currently prioritizing a few of the issues and developing the considerations on such issues raised for IP policy in respect of AI [7].

8 Case Studies—Ownership Issues

A few case studies that can provide a few insights into the ownership issues:

8.1 The Raghav Case: In 2020 Copyright office faced a notable case involving the case of an AI system named Raghav to secure the copyright registration for artwork called "Suryast." Initially, it was rejected for lack of a human author but it was later granted protection when the natural person was named as co-author along with RAGHAV. Later the issue was highlighted on the AI qualification as an artist under the Copyright Act [8].

8.2 The OpenAI GPT-2 Language Model: In 2019, OpenAI unveiled GPT-2, a language model renowned for its ability to produce lifelike text. Its introduction sparked discussions surrounding the ownership and attribution of AI-generated content. OpenAI opted against releasing the full version of the model, citing concerns regarding potential misuse of the technology.

8.3 Dr. Thaler Versus The World: Dr. Stephen Thaler has emerged as a central figure in discussions surrounding the attribution of authorship or inventorship to AI-generated works. Renowned for his endeavors to challenge intellectual property laws, Thaler has notably focused on asserting ownership rights over copyrights and patents produced by AI. His efforts have spanned various legal jurisdictions, including Australia, the European Union, the United Kingdom, the United States of America, New Zealand, and South Africa, all aimed at securing intellectual property protection for his creation, the "DABUS" (Device for Autonomous Bootstrapping of Unified Sentience). However, despite his widespread legal actions, Thaler has seen limited success, with only South Africa ruling in his favor. This article will delve into two of Thaler's prominent cases: the recent decision in the United States concerning AI authorship in copyright and the UK Court of Appeal's ruling regarding AI inventorship in patents [9].

9 Legal and Ethical Issues in AI-Generated Inventions:

Can an AI be considered an inventor? Several possibilities emerge:

9.1 The Programmer: Ownership could be attributed to the programmer who designed and trained the AI, similar to the "human intervention" argument in copyright.
9.2 The Company: Ownership could rest with the company that owns and operates the AI system.
9.3 The AI Itself: A more radical approach could potentially recognize the AI itself as an inventor, raising philosophical questions about artificial consciousness and legal personhood.

The current legal framework for patents is ill-equipped to address these complexities. Legislative action is imperative to establish clear guidelines for AI inventorship, ensuring that genuine innovation receives appropriate protection while preventing unwarranted monopolies on AI-generated inventions.

10 Case Studies on the Patentability of AI-Generated Inventions:

10.1 Dr. Thaler versus The World: "DABUS" (Device for Autonomous Bootstrapping of Unified Sentience). The recent decision in the United States concerning AI authorship in copyright and the UK Court of Appeal's ruling regarding AI inventorship in patents [10].

11 Strategies to Address the Issues and Challenges

The lack of clear rules regarding ownership of AI outputs creates a web of uncertainty for businesses and individuals alike. Several potential solutions can guide the way:

11.1 Legislative Reform: Amending copyright and patent laws to explicitly address authorship and inventorship in the context of AI. New laws must be enacted especially with the concerns about AI.
11.2 Separate Rights: There should be an enactment specifically for AI elaborating all the features, significance, liabilities, and punishments consisting only of AI.

11.3 Contract Agreements: There should be a contract between the AI users and as well as the developers regarding whom the ownership lies on the outcome of the project.

Therefore, with the increase in everyday technology, there should be new legal reforms made.

12 Conclusion

With the advancements in technology, AI keeps growing every day and there will be more issues regarding the AI and IPR intersections. So, the policymakers, and legal framework drafters must ensure that they make necessary amendments to the existing laws and bring in new enactments to clearly state all the legal possibilities and clarifying as to who holds the ownership rights in cases of AI-generated works. The new legal frameworks that will be made must be flexible to change any new changes easily without any complexities.

Thus, AI technology is widely spread and it keeps growing there are certain benefits of these technologies, and are also time efficient. However, the overall overlap between AI and IPR must be considered by the legal policymakers.

References

1. https://www.linkedin.com/pulse/swot-analysis-artificial-intelligenceai-aided-intellectual-5qrpc/
2. Frackiewicz M (2024) The ethics of AI-generated content: navigating the world of deepfakes. https://ts2.space/en/the-ethics-of-ai-generated-content-navigating-the-world-of-deepfakes/. accessed 4 May 2024
3. Ali MA, Kamraju M (2024) Impact of artificial intelligence on intellectual property rights: challenges and opportunities. https://www.researchgate.net/publication/376751087_Impact_of_Artificial_Intelligence_on_Intellectual_Property_Rights_Challenges_and_Opportunities, accessed 4 May 2024
4. Ali MA, Kamraju M (2024) Impact of artificial intelligence on intellectual property rights: challenges and opportunities. https://www.researchgate.net/publication/376751087_Impact_of_Artificial_Intelligence_on_Intellectual_Property_Rights_Challenges_and_Opportunities, accessed 5 May 2024
5. Socratica Learning, Intellectual Property and Ethics of Artificial Intelligence: Challenges and Considerations, https://www.linkedin.com/pulse/intellectual-property-ethics-artificial-intelligence-learning-dxoue/, last accessed 7 May 2024
6. Frackiewicz M (2024) The ethics of ai-generated content: navigating the world of deepfakes. https://ts2.space/en/the-ethics-of-ai-generated-content-navigating-the-world-of-deep fakes/, accessed 7 May 2024
7. WIPO (2024) Artificial intelligence and intellectual property policy. https://www.wipo.int/about-ip/en/artificial_intelligence/policy.html, accessed 9 May 2024
8. Nayantara S, Sheetal M, Nihal S (2024) Intersection of intellectual property rights and AI-generated works – part I, https://www.barandbench.com/law-firms/view-point/intersection-intellectual-property-rights-ai-generated-works-part-i, accessed 8 May 2024

9. AsiaIP (2024) Rise of the machines: ownership of intellectual property created by artificial intelligence, https://asiaiplaw.com/section/in-depth/rise-of-the-machines-ownership-of-intellectual-property-created-by-artificial-intelligence, accessed 4 May 2024
10. AsiaIP (2024) Rise of the machines: Ownership of intellectual property created by artificial intelligence, https://asiaiplaw.com/section/in-depth/rise-of-the-machines-ownership-of-intellectual-property-created-by-artificial-intelligence, accessed 5 May 2024

The Impact of Artificial Intelligence on Legal and Jurisdictional Challenges

Bhupal Bhattacharya, Ripon Bhattacharjee, Sarmistha Bhattacharya, and Md Baharul Islam

Abstract The fundamental challenges identified in this research to making law computable are not addressed by developments in machine learning or even deep learning. Machine learning is primarily concerned with identifying patterns rather than thinking. Predictive models that improve over time as a result of exposure to fresh data are commonly found in machine-learning software. In order to suggest machine learning or deep learning as an option for responding to the issue of making law attainable, one must be able to specify the task that one wishes to have the machine-learning model complete, list the inputs and outputs of the program, and provide the training data that the software will use to create a predictive model. Furthermore, unlike human readers, machine-learning algorithms are incapable of reasoning or comprehending texts written in natural language. In legal AI, classical symbolic AI techniques still play a significant role, even though machine learning may prove beneficial for legal predictive analytics and categorization tasks like e-discovery document analysis. Users who use modern information retrieval systems are able to submit complete documents as search queries and conduct innovative, novel types of information searches. Numerous experts in the legal field will have to reinvent themselves and discover new specializations. In response to that disturbance, the State will need to rearrange itself and attempt to strike a balance. To maintain ethical, responsible, and legally sound procedures that respect human rights and the welfare of society while simultaneously utilizing AI's potential benefits for the legal

B. Bhattacharya
Department of Law, Raiganj University, A State Aided University of Goverment of West Bengal, Raiganj, India

R. Bhattacharjee (✉)
National Law University, Tripura, India
e-mail: ripon.bhattacharjee69@gmail.com

S. Bhattacharya
Alliance School of Law, Alliance University Bengaluru, Bengaluru, India

M. B. Islam
Department of Social Work, Amex Law College, Affiliated to the University of Burdwan, Burdwan, India

M. Chakraborty et al. (eds.), *Proceedings of 5th International Ethical Hacking Conference*, Lecture Notes in Networks and Systems 1148,
https://doi.org/10.1007/978-981-97-8457-8_23

system is the fundamental challenge. This makes it necessary to do research in order to alter legal frameworks so that they can keep up with the rapidly changing field of artificial intelligence.

Keywords Artificial Intelligence · Legal Institutions · AI Algorithms · Impartial Judicial Processes · Jurisdictional Frameworks · Legal Boundaries

1 Introduction and Background

The legal and judicial institutions face a dynamic and dangerous transformation due to uncontrolled development of technologies of artificial intelligence (AI) [1]. AI algorithms are getting developed using pre-existing human data [2], which has the absolute certainty of leading toward unfair results and perpetuating social biases [3], regardless of whether they are utilized in criminal justice or other legal environments [4]. AI decision-making processes maintain algorithmic secrecy generating serious concerns about accountability, transparency, and impartial judicial processes [5].

The fundamental issue of this study is the concerns that arise from the rapid advancement of AI and the limited ability of the legal and judicial institutions to handle the problems that they face today. Legal innovation opens the boundaries of AI in using its algorithms to get applied to legal research, review, and analysis. In some cases, on the basis of its previously fed data and information [6], it attempts to make a most relatable solution which raises concerns for accountability and transparency. AI-driven legal processes have now become a serious unethical issue that the legal system fails to address with all its vast boundaries. The limitations of the law further exaggerate the matter, putting obstacles before the judicial institutions to handle these advances. Innovations in AI are pushing the boundaries in a number of fields, including law, where they can be used to conduct preliminary legal analysis, document review, and legal research. There are instances where it is making suggestions and solutions. Legal systems must contend with a complicated landscape that includes issues such as i. ethical concerns about bias, accountability, and transparency in AI-powered legal processes, and ii. Jurisdictional frameworks that are often inadequate to handle these breakthroughs. iii. Uncertainties in jurisdiction about accountability and liability, particularly when AI systems function internationally; iv. Unforeseen legal concerns about who owns outputs produced by AI, data security and privacy in the context of massive data processing, and the possible influence on fundamental rights like fair trials. This study attempts to explore the topics surrounding the main problem, which is the conflicts between the speed at which AI is developing and the capacity of the legal and jurisdictional systems currently in place to effectively handle the problems it poses.

2 Rise of Virtual Agents

Technological developments have created new avenues for the provision of services in addition to face-to-face interactions between service providers and recipients [7]. One may now enhance data availability and interchange [8], provide customer assistance, and maintain connections, with the help of mobile applications, information platforms, digital diaries, video call platforms [9], electronic health record systems, and many other tools. A growing amount of research has been done recently on technologies like virtual agents and avatars [10]. There is a claim that they can offer clients fresh opportunities to engage with various professions and in their daily lives [11].

Virtual avatars are graphic representations of people (or other characters) created by computers that are typically controlled by people. Conversely, virtual agents are computer programs that mimic human speech while conversing with users. They are frequently built on AI [12]. They can function somewhat "autonomously" when they are AI-based [13], which allows them to supplement or replace professionals in specific roles by use of chatbots [14], conversational AI, and virtual assistants as virtual agents. Interactive Virtual Reality [15] is another new technology that is becoming increasingly popular in the virtual world [16]. VR is defined as a computer-generated 3D world that can be accessed by a Head-Mounted Display [17]. Applications for virtual reality can help professionals with a variety of tasks. Additionally, for instance, they can be used to involve individuals in sporting activities for older adults with dementia among others.

3 Development of AI and Law

AI has historically focused on the creation of rational agents, computer agents, or software programs that strive to attain the greatest expected result based on a performance measure, despite the fact that the term has no widely agreed definition. AI is predicted to revolutionize the legal sector. It can automate tasks that lawyers have historically done, forcing them to reassess their future role in an AI-powered legal system. Determining who is responsible [18] for what actions in cross-border AI operations seems to be difficult to be traced given the increasing complexity of territorial concerns involving AI in legal frameworks. AI's pervasiveness also raises ethical and sociological concerns19 regarding human agency and control [19]. With the advent of predictive cum suggestive AI technologies, it is more crucial than ever to establish comprehensive legal frameworks that will ensure safeguard against moral decay [20], and warrant that international humanitarian law is ethically followed [21] while developing and controlling the AI application [22]. Therefore, in order to reduce the existential threats that the incorporation of AI into legal proceedings poses, stringent ethical guidelines are required.

The most recent advancements in AI technology imply that the history of AI in law needs to be reexamined. Although there have been notable advancements in the use of IT in legal work, these have primarily been brought about by standard IT procedures (data management, processing, storing, retrieval, and communication with rich, quick, and worldwide information via online resources) [23]. Unfortunately, the results of the applications of information technology to legal processes, which include the fundamental essence of law, have not been particularly effective until now. An excellent example of this is the use of IA systems in court proceedings which is happening on a day-to-day basis.

4 Understanding Augmented Reality

There's virtual reality in addition to Augmented Reality (AR). While Virtual Reality (VR) completely replaces the actual world [24], Augmented Reality augments reality with visible electronic material. VR brings users into a virtual environment that they can walk around in and interact with as they would in the real world with the use of goggles and speakers [24]. Virtual reality and augmented reality are competing technologies that may be used for meeting purposes and social interactions with distant parties, depending on which one develops more swiftly. In some contexts, AR and VR might be complementary. For example, users can use AR to enhance in-person interactions while using VR to create fully imagined worlds. Numerous VR applications, which primarily include games, immersive news reporting [26], and social experiments, were launched along with four major VR hardware platforms [27]. Furthermore, the rapid advancement of technology is causing it to become even more realistic [28]. Even though games still make up the majority of VR applications today, however, will eventually interact mostly in virtual rather than actual space [29] (especially if avatars get realistic enough and start to consistently track user facial expressions). VR impacts a wide range of activities [30], including work, sales, training, social life, education, exercise, and even psychoanalysis.

5 Deep-Learning Revolution

The majority of present-day machine learning consists of supervised learning [31]. Software learns from a set of data samples in supervised learning [32], where each sample (input) has a label (output) attached to it [33]. The software picks up a sophisticated mathematical function that illustrates a relationship between the samples and labels via repeated exposure to sample-label Combinations [34]. The machine-learning model is being trained during this procedure [35]. When faced with new, unseen examples for which it does not have human-assigned labels, a trained model can utilize the function it has learnt to predict the correct label. The advent of deep learning, or multilayered neural networks, has been the most significant advancement in AI research during the last ten years. Neural networks use patterns found in input data to translate comparable but unseen data to specific outputs [36]. Every input data sample that goes through a series of layers in a neural network is transformed in a way that facilitates the easier association of the sample with the correct output [37]. The last (output) layer associates the altered sample with a predicted value once it exits a pipeline of internal (hidden) layers [38]. The model gently modifies itself to lessen such errors in the future if the predicted value differs from the sample's true (human-assigned) value. A good model is able to identify unknown data with the correct values when it is trained.

6 AI Applications in Modern Legal Practice

Supervised learning is used by the majority of neural networks, and classification is the most popular supervised learning task [39]. The image is transformed by the units of each layer of the neural network, which then multiplies feature values by randomly assigned weights to either emphasize or minimize the characteristics [36]. The modified data of the image is utilized to forecast the output label at the final stage of the network process. Numerical processing, which does not include verbal processing, is the foundation of computer operation. Ever since AI research began, its creators have aimed to build machines that can speak human language and close the gap between human and computer communication. Natural language processing, or "NLP," is this area of AI [40]. Alan Turing proposed that a computer is intelligent if it can answer questions so well that a human questioner would mistakenly believe that it is a human. This idea is known as the "Turing test," and it exemplifies the basic role that natural language processing plays in AI [41]. The way that traditional symbolic AI approaches to NLP operate is by using predetermined rules to process input language [42]. At its most basic, a regular expression can split up text in human languages into distinct parts [43] that are subsequently utilized as variables in other programs.

7 AI and Its Applications to Law

Calculations are not very easy to apply to law. Legal norms "tend to be incomplete, logically and semantically ambiguous, sometimes inconsistent, and frequently hard to find or even identify," as Kevin Ashley observed [44]. Furthermore, the legal system is adversarial, meaning that there is often no one correct solution to legal issues. Although the technical features and application domains of VAs and VR applications are different, they present comparable opportunities and problems that should be evaluated before implementing them in practice. Virtual Reality applications and VAs can expand the service's geographical scope while providing various advantages. For example, the COVID-19 pandemic has demonstrated that providing distant healthcare services [45] can lower contact and infection rates [46], shielding vulnerable populations from potential exposure [47]. Additionally, it can make a variety of services more accessible to people who are hesitant to interact digitally in person. Moreover, in the case of virtual reality, it might present chances to give safe, regulated experiences that aren't possible in a real-world setting. As a result, by removing parts of the processes from the analog world with all of its limitations (such as time or geographical restrictions), VAs and VR applications can increase the chances for service access. Virtual Reality applications offer venues for therapeutic and care experiences, while Virtual Assistants expand the chances for engagement with a professional [48]. Customers might be able to meet their 3D-generated specialists [49] in a virtual setting that is immersive by means of innovation [50]. On the other hand, VR apps and VAs have lately been researched with the intention of including them in the virtual service repertoire in the future [50]. In addition to the possibilities described above, the use of VR and VAs presents issues with regard to its social, legal, and ethical implications [52]. Since the goal of virtual agents is to replace some face-to-face interactions, such as those between medical professionals and patients, with human–machine interaction, it is important to consider the potential effects on the doctor–patient relationship and the quality of care that may result from this matter. It is reasonable to anticipate ethical and social concerns in the areas of secrecy, privacy, openness, and justice. Intelligent virtual agents with autonomous behavior in particular would have a lot of responsibility when interacting with patients who are in need of assistance. This may also give rise to legal concerns, such as liability in the event of errors. Furthermore, the procedures and environments used in healthcare are altered when it is transferred from the real world into the virtual one. It is necessary to assess the implications of this transferal with regard to safety, care quality, and legal accountability, among other considerations.

8 The Knowledge Explosion in Law and the Limits of Human Capacity

The onerous work of keeping up with the latest judicial decisions, statutes, and regulations that enter into force on a daily basis makes it nearly impossible to comprehend the universe of knowledge that underpins the law. This task entails many hours of study and significant resource expenditures, which are typically billed to legal services clients. Thus, whether AI has an impact on politics or law, it won't drastically alter reality in a single stroke. Since many of these duties and activities will be carried out by intelligent systems, their implementation will likely result in changes to the chores and activities that people conduct. Notwithstanding the underlined challenges, AI methods can help with a variety of specialized regulatory responsibilities. Computer-automatable problems are often those that have predictable, yes-or-no inputs and outputs. It is possible to use this kind of problem structuring with certain legal difficulties. For example, there is no space for judicial discretion because traffic laws are essentially "bright-line rules" that drivers either follow or don't. It should come as no surprise that we have automated law enforcement systems operating as enforcement devices like speed cameras and red-light cameras. These programs are monolithic systems that take over the functions of a judge, court clerk, and police officer inside a specific domain.

9 Due Process Concerns and Automated Legal Decisions

Vagueness is ingrained in the law. It might be challenging or impossible to decide with objectivity whether a term truly includes a given fact scenario because a concept or phrase is ambiguous if its limits are not clearly defined. In law, for instance, the phrase "reasonable" is often used, although it has no set definition that can be computerized. All legal issues are not as applied in case of over speeding through fines. For example, the concept behind red-light cameras cannot be applied to careless driving since determining what is considered recklessness requires a fact-based analysis and the rational application of an ambiguous legal norm. It's hard to say if a defendant's actions in a negligence case were reasonable or not because there are so many unknowns. These include the specific things the defendant did or didn't do and the fact finder's opinion of what a reasonable person would do in the same situation. The same holds true for ideas like good faith. The usefulness of a software to anticipate whether a judge would find that a particular driving behavior is "reckless" would be limited by due process concerns, even in the unlikely event that one could be created. If a driver is found guilty of reckless driving, there are serious repercussions for them. The automated system cannot be trusted to administer such penalties until it is thoroughly understood. There must be good cause to assume that the judgments rendered by the system fairly represent judgments that a human judge would make after considering the identical evidence and the law. That requirement is unreasonably difficult to conceptualize, organize, implement, and test in a feasible system.

Legal reasoning programs have relied on manually composed representations of legal texts instead of the texts themselves [53]. However, due to the dynamic nature of legal concepts and reasoning, it is challenging or unattainable to codify the contents of the majority of legal cases or fact situations in a reproducible manner. Reproducibility would require multiple legal experts to examine the same judicial opinion or fact situation and reliably encode its pertinent information. Due to the prevalence of ambiguity in the law, several legal norms and concepts that can be applied to computerized representations of factual scenarios or legal cases are either extremely difficult or impossible to model robustly [54]. The fact that unwritten exceptions to the law exist and that laws are prone to change over time exacerbates these issues [55].

The process of calcifying an inherently dynamic concept is involved in attempts to encode a legal rule into an algorithm. Stated differently, attempts to render legal reasoning calculable entail distilling vague notions and regulations into manipulable constants, variables, and formulas. Such initiatives may, in theory, aid in elucidating the workings of legal regulations. However, if the goal is to create operational systems that play a role in resolving legal conflicts, we should be mindful of satisfying our desire for false clarity by artificially simplifying the messy complexity of the law. It is hard to imagine a computer program being able to do these kinds of things. Additionally, there are additional basic issues of knowledge representation, which is a branch of AI that looks at how to first communicate facts to a computer so that the machine can make sense of them.

Another issue is that AI systems aren't taught common sense, which is the ability to comprehend fundamental ideas like the idea that a physical object can only be in one place at a time. Such rules typically need to be manually encoded. Encoding such rules has been expensive and has not produced much results over the years.

10 Redefining Reality and Its Legal Implications

Education is at a crossroads because law is still taught in the twenty-first century in a manner that was only modified to meet the demands of the previous century based on the body of knowledge. In light of this, what should be included in law education are abilities that machines do not acquire effectively, as many ordinary legal duties tend to disappear.

Both AR and VR raise legal issues for businesses, users, and courts. Some offer new perspectives on well-known legal issues. With AR and VR, people will threaten, scam, or kill others via technology. Courts will need to comprehend the technology and how it differs from the past in order to determine who is at fault. However, a comprehensive reexamination of legal concepts won't be necessary. Legally speaking, a death threat made by AR or VR is equivalent to a death threat made orally, in writing, by email, or by fax. However, AR and VR will potentially raise fresh legal issues. Networks and equipment that are owned and operated by private companies will be used for virtual interactions. Because of this, those exchanges might be governed by contractual terms and conditions that users won't likely ever notice or take into account, but which severely restrict their right to privacy, property, and liberty. It could be more difficult to adequately govern these interactions because they might not take place in a single physical jurisdiction. This shift in our business practices from primarily conducting transactions in public spaces with public rules, which are mostly found in a single jurisdiction, to private spaces with private rules, where the parties appear to be next to each other but are actually geographically situated in multiple jurisdictions, may lead us to reconsider what exactly qualifies as a legally binding contract and what matters we prefer to have regulated by public rather than private rules.

Furthermore, AR and VR have the potential to highlight deeper issues. The term real in virtual reality isn't appropriate here because what we see and hear in VR are made-up noises and visuals. But until it is experienced, it feels real in a manner that is difficult to comprehend. If Augmented Reality is able to superimpose realistic and vibrant representations of people and objects on top of the actual world we perceive, then the same might hold true. Legal concepts that typically distinguish between bodily touch and physical risk and things that are mere auditory and visual communication may be called into question by this intuitive experience of reality. The distinction between the mental and the intense, between what we see and what we experience, forms the basis of many of our laws. We also distinguish important legal distinctions between conduct and words. It will be more difficult to discern those distinctions between VR and AR, and it might force us to consider carefully why we penalize some behaviors in the real world but not others. In fact, they might even cause us to reconsider what constitutes real in a world where an increasing number of our most meaningful and emotionally authentic experiences don't fit the traditional definition of real in terms of being real. However, they feel real and may have actual physiological effects. Not all technologies that challenge legal doctrine are as new as VR and AR. For example, there are significant insights to be gained from the past 25 years of attempting to apply legal regulations to the Internet [56], as well as from some innovative research on the subject of non- immersive virtual world law conducted ten years ago [57]. Yet rather than being intentionally created, the majority of those laws evolved accidentally [58]. We must now explore new territory in our careful consideration of how the law will relate to VR and AR. The benefit should be a deeper comprehension of concepts we take for granted in the real world in addition to a strong foundation for applying legal doctrine to a few challenging new cases.

11 Legal Challenges

One of the biggest challenges in developing practical expert systems in any field, including law, is the requirement to design suitable data structures to hold facts and the relationships between them. Furthermore, even under the assumption of a vast knowledge base, there exists a particular challenge that no system can surmount: It is impossible to ensure that an additional fact, which has been previously unnoticed and is extraneous to all preceding cases, will not acquire critical importance in the current case. Put differently, there is no assurance that the correct rule of determination for each new case can be found in the body of previous decisions since judicial precedents in the common-law system are made up of court rulings that are intricately linked to the facts that gave birth to the disagreement. Neural networks and deep-learning algorithms typically aren't able to give explanations for their choices. These systems are occasionally labeled black boxes [59] due to the fact that their inputs and outputs are visible, their internal decision-making processes, and the predictive models they generate are not discernible. The phrase explainability problem describes the current state of affairs that even the most sophisticated deep-learning algorithms are unable to provide an explanation for the decisions or predictions they make. A significant barrier to the advancement of AI technologies that supplement or replace the work of judges and attorneys is the comprehensibility problem. A basic feature of the legal system is the requirement to provide rationale for the findings drawn. For instance, it is customary for judges to provide thoughtful opinions in support of each important ruling. Reasoned opinions help people believe that judges make impartial, principled decisions. This crucial task cannot be completed by AI technologies.

All AI software will be essentially specialized for the foreseeable future, created to carry out certain tasks inside a limited domain. Judges, however, have a wide range of responsibilities when it comes to presiding over conflicts. These include determining the admissibility of evidence, making decisions regarding procedural violations like late filings, interpreting ambiguities in statutes and contracts, and more. At the moment, none of these judicial functions can be carried out by computer systems on their own; a computer program requested to act as a judge might have to carry out every one of these functions. If the program is employing a template approach, the answers to these questions may help it determine a sentencing range and create the basic framework to determine the result. By requiring the human judge to support the algorithm at every point where human interpretation is recommended, the program would prevent the algorithm from extending its potential and overcoming its limitations. Software that asks the user for pertinent information and then uses that knowledge to generate documents with templates is one variation of these expert systems. There are certain legal chores that can be made reasonably simple with the use of a template program, such as creating simple contracts, wills, and business rules.

12 Conclusion

The significance of unsupervised learning is increasing. The development of semi-supervised learning further increases its contributions to AI. Unsupervised learning is utilized in semi-supervised learning to produce the labeled data required for a supervised learning problem. However, it's possible that AI won't be able to perform many of these tasks. The fundamental challenges identified in this research to making law computable are not addressed by developments in machine learning or even deep learning. Machine learning is primarily concerned with identifying patterns rather than thinking. Predictive models that improve over time as a result of exposure to fresh data are commonly found in machine learning software. In order to suggest machine learning or deep learning as an option for responding to the issue of making law attainable, one must be able to specify the task that one wishes to have the machine-learning model complete, list the inputs and outputs of the program, and provide the training data that the software will use to create a predictive model. Furthermore, unlike human readers, machine learning algorithms are incapable of reasoning or comprehending texts written in natural language. In legal AI, classical symbolic AI techniques still play a significant role, even though machine learning may prove beneficial for legal predictive analytics and categorization tasks like e-discovery document analysis. Users who use modern information retrieval systems are able to submit complete documents as search queries and conduct innovative, novel types of information searches. Numerous experts in the legal field will have to reinvent themselves and discover new specializations. In response to that disturbance, the State will need to rearrange itself and attempt to strike a balance. To maintain ethical, responsible, and legally sound procedures that respect human rights and the welfare of society while simultaneously utilizing AI's potential benefits for the legal system is the fundamental challenge. This makes it necessary to do research in order to alter legal frameworks so that they can keep up with the rapidly changing field of artificial intelligence.

References

1. Citron DK, Frank P (2014) The scored society: due process for automated predictions. Washington Law Rev 89(1):1–33
2. Barocas S, Selbst AD (2016) Big data's disparate impact. Calif Law Rev 104(3):671–732
3. Eubanks V (2018) Automating inequality: how high-tech tools profile, police, and punish the poor. St. Martin's Press, New York
4. O'Neil C (2016) Weapons of math destruction: how big data increases inequality and threatens democracy. Crown, New York
5. Future of Life Institute (2024) AI policy: global AI policy frameworks and initiatives. Accessed 18 May 2024. https://futureoflife.org/ai-policy/
6. Brynjolfsson E, Andrew M (2014) The second machine age: work, progress, and prosperity in a time of brilliant technologies. W.W. Norton & Company, New York
7. Parker C (2022) The role of mobile applications in enhancing customer assistance. J Digit Innov 14(3):45–60
8. Smith J, Emily B (2023) Digital diaries and their impact on data availability and interchange. Technol Soc 29(2):78–92
9. Young R (2023) Video call platforms and the future of remote interactions. Commun Technol Rev 35(1):30–44
10. Turner A (2023) Virtual agents and avatars: new opportunities for professional engagement. Int J Virtual Reality 18(4):103–120
11. Wang L, David G (2022) The utilization of electronic health record systems in modern healthcare. Healthc Technol Today 22(1):15–27
12. O'Neill P (2018) AI and the law: how AI is shaking up legal research. Harvard J Law Technol. 30 Mar 2018. https://jolt.law.harvard.edu/digest/ai-and-the-law-how-ai-is-shaking-up-legal-research
13. Knight W (2017) The dark secret at the heart of AI. MIT Technol Rev. 11 Apr 2017. https://www.technologyreview.com/s/604087/the-dark-secret-at-the-heart-of-ai/
14. Siau K, Weiyu W (2018) Building trust in artificial intelligence, machine learning, and robotics. Cutter Bus Technol J 31(2):47–53
15. Jerald J (2015) The VR book: human-centered design for virtual reality. Morgan & Claypool Publishers
16. Adamopoulou E, Lefteris M (2020) An overview of chatbot technology. In: Ilias M, Lazaros I, Elias P (eds) Artificial intelligence applications and innovations, IFIP advances in information and communication technology, vol. 584. Springer, Cham, pp 373–383
17. Slater M (1999) A note on presence terminology. Presence Teleoper Virtual Environ 8(5):555–559
18. Freeman D, Ward JA, Rosebrock CC, Antonietta S, Deeley KM, Fineberg MJK, Greenhalgh RL (2017) Virtual reality in the assessment, understanding, and treatment of mental health disorders. Psychol Med 47(14):2393–2400

19. Bostrom N (2014) Superintelligence: paths, dangers, strategies. Oxford University Press, Oxford
20. Bryson JJ (2019) The ethical landscape of robotics. AI Soc 34(1):27–31. https://doi.org/10.1007/s00146-017-0787-6
21. Cath C (2018) Governing artificial intelligence: ethical, legal and technical opportunities and challenges. Philos Trans R Soc A Math Phys Eng Sci 376(2133):20180080. https://doi.org/10.1098/rsta.2018.0080
22. Taddeo M, Luciano F (2018) How AI can be a force for good. Science 361(6404):751–752. https://doi.org/10.1126/science.aat5991
23. Calo R (2015) Artificial intelligence policy: a primer and roadmap. UCLA Law Rev 51(4):399–435
24. Sherman WR, Alan BC (2002) Understanding virtual reality: interface, application, and design. Morgan Kaufmann, San Francisco
25. Azuma RT (1997) A survey of augmented reality. Presence Teleoper Virtual Environ 6(4):355–385. https://doi.org/10.1162/pres.1997.6.4.355
26. BBC News (2022) How virtual reality is changing news coverage. Last modified 15 Mar 2022. https://www.bbc.com/news/technology-56362149
27. Lanier J (2017) Dawn of the new everything: encounters with reality and virtual reality. Henry Holt and Co., New York
28. Slater M, Sylvia W (1997) A framework for immersive virtual environments (FIVE): speculations on the role of presence in virtual environments. Presence Teleoper Virtual Environ 6(6):603–616
29. Sherman WR, Craig AB (2018) Understanding virtual reality: interface, application, and design. In: Paper presented at the IEEE Conference on Virtual Reality and 3D User Interfaces, Atlanta, GA, March 19–23
30. Jones M (2023) Virtual reality's impact on the future of work and social interaction. The Guardian, 30 Apr 2023. https://www.theguardian.com/technology/2023/apr/30/virtual-reality-impact-future-work-social-interaction
31. Surden H (2014) Machine learning and law. Washington Law Rev 89(1):87–115
32. Bishop CM (2006) Pattern recognition and machine learning. Springer, New York
33. Goodfellow I, Yoshua B, Aaron C (2016) Deep learning. MIT Press
34. Patel A (2021) A comprehensive guide to supervised learning. Towards data science. Last modified 20 Feb 2021. https://towardsdatascience.com/a-comprehensive-guide-to-supervised-learning-8f1f08c5b6a
35. LeCun Y, Bengio Y, Hinton G (2015) Deep learning. Nature 521(7553):436–444. https://doi.org/10.1038/nature14539
36. Schmidhuber J (2015) Deep learning in neural networks: an overview. Neural Netw 61:85–117. https://doi.org/10.1016/j.neunet.2014.09.003
37. Deng L, Yu D (2014) Deep learning: methods and applications. Found Trends Signal Process 7(3–4):197–387. https://doi.org/10.1561/2000000039
38. Krizhevsky A, Sutskever I, Hinton GE (2012) ImageNet classification with deep convolutional neural networks. In: Proceedings of the 25th international conference on neural information processing systems (NIPS'12), Curran Associates Inc., Red Hook, NY, pp 1097–1105
39. Jurafsky D, Martin JH (2009) Speech and language processing: an introduction to natural language processing, computational linguistics, and speech recognition. 2nd ed. Prentice Hall, Upper Saddle River, NJ
40. Mitchell M (2019) Artificial intelligence: a guide for thinking humans. Farrar, Straus and Giroux, New York
41. Manning CD, Schütze H (1999) Foundations of statistical natural language processing. MIT Press, Cambridge, MA
42. Friedl JEF (2006) Mastering regular expressions. 3rd ed. O'Reilly Media, Sebastopol, CA
43. Ashley KD (2009) Teaching law and digital age legal practice with an AI and law seminar. Artif Intell Law 17(4):321–341. https://doi.org/10.1007/s10506-009-9072-5

44. Wiederhold BK (2021) Using virtual reality to advance the care of COVID-19 patients. In: Morisson JM (ed) Virtual reality in health and social care. Springer, Cham, pp 145–158. https://doi.org/10.1007/978-3-030-56930-0_9
45. Centers for Disease Control and Prevention (2021) The benefits of telehealth during COVID-19. Reviewed 9 April 2021. https://www.cdc.gov/coronavirus/2019-ncov/hcp/telehealth.html
46. Riva G, Wiederhold BK, Mantovani F (2020) Surviving COVID-19: the neuroscience of smart working and distance learning. Cyberpsychol Behav Soc Netw 23(9):615–620. https://doi.org/10.1089/cyber.2020.29184.gri
47. Schroeder R (2007) The social life of avatars: presence and interaction in shared virtual environments. Convergence Int J Res New Media Technol 13(1):36–48. https://doi.org/10.1177/1354856507072856
48. Lin J, Zhang W (2020) Incorporating virtual assistants into customer service repertoires: current trends and future directions. Serv Sci 12(2):129–144. https://doi.org/10.1287/serv.2020.0321
49. Nobre G (2021) Virtual assistants and customer service: enhancing customer experience with AI. J Serv Res 23(4):501–520. https://doi.org/10.1177/10946705211012345
50. Palomares A, Morales L (2020) Navigating the ethical landscape of virtual reality and virtual assistants. Ethics Inf Technol 22(3):255–272. https://doi.org/10.1007/s10676-020-09544-3
51. Sharma P, Kumar R (2021) Exploring future applications of virtual reality in customer service. J Bus Res 125:384–397. https://doi.org/10.1016/j.jbusres.2020.02.043
52. Posner RA (2007) Legal reasoning and legal theory. Stan Law Rev 59:1055
53. Susskind RE (2017) The future of the professions: how technology will transform the work of human experts. Oxford University Press
54. Sunstein CR (2019) The limits of quantitative legal analysis (and why many social scientists seem unaware of them). Univ Chic Law Rev 35:271
55. Zittrain J (2012) Reflections on internet law: challenging legal doctrines. Harv Law Rev 125(6):1567–1580
56. Weiss MA (2011) The law of virtual worlds: past, present, and future. Int J Legal Inf 39(2):210–235
57. Wu T (2010) The master switch: the rise and fall of information empires. Vintage Books, New York, pp 78–89
58. Pasquale F (2015) The black box society: the secret algorithms that control money and information. Harvard University Press, Cambridge, MA
59. Myers S, Chater N (year) Interactive explainability: black boxes, mutual understanding and what it would really mean for AI systems to be as explainable as people

Implementation of Artificial Intelligence in Criminal Justice Framework: A Comparative Study

Snehatit Sett, Shrabana Chattopadhyay, Avishikta Chatterjee, and Ankan Bhattacharya

Abstract Artificial intelligence (AI) encompasses the development of computer systems capable of performing human-like functions such as sensing, contemplation, and decision-making. Over the past five years, machine-learning-based AI has developed autonomous decision-making systems that imitate human conduct. Where the criminal justice framework is burdened by an abundance of cases, scarce resources, and concerns about equity and transparency, artificial intelligence (AI) offers hope. AI can completely transform the criminal justice framework in India, transforming areas such as tracking, processing evidence, predicting criminal justice outcomes, and legal research. This paper focuses on how AI might improve the criminal justice framework in India. This chapter examines how AI technologies, like SUPACE, SUVAS, and E-Filling, are used in the Indian legal system. It also makes parallels to other legal systems around the world, with a focus on the US and the UK. The chapter demonstrates that, despite the law being rigid, artificial intelligence manipulation has the power to transform and reform the country's legal system. Aspects of AI in criminal justice have advantages and disadvantages for India. To truly realise AI's potential to transform justice, its use must be driven by transparency, accountability, and fairness.

Keywords Artificial intelligence · Impact on criminal justice system · Use of AI in investigation · Justifiability · Use of AI in USA & UK · Potentiality to change the system

---

S. Sett · S. Chattopadhyay (✉) · A. Chatterjee · A. Bhattacharya
Department of Law, IEM's International Institute of Juridical Sciences (IIJS), University of Engineering and Management, Kolkata, India
e-mail: shrabana.uem@gmail.com

M. Chakraborty et al. (eds.), *Proceedings of 5th International Ethical Hacking Conference*, Lecture Notes in Networks and Systems 1148,
https://doi.org/10.1007/978-981-97-8457-8_24

1 Introduction

In recent years, the field of criminal procedural law has been actively engaged in discussions regarding the potential applications of Artificial Intelligence (AI). The legislation does not currently incorporate the idea of AI, but it is characterised by scientific theories, methodologies, and procedures that share the goal of emulating human cognitive capacities using computer systems [1]. The legal and judicial process is set to be revolutionised by advancements in artificial intelligence, machine learning, and big data. In the past five years, advancements in machine learning have enabled the development of autonomous decision-making systems that are based on and imitate human behaviour. This is particularly evident in the advancement of self-driving automobiles, which essentially function as autonomous systems that operate massive vehicles at high speeds. It is the volume of data being gathered from billions of older judgements made by human drivers on which the system is based. These mechanisms are increasingly being used in more and more fields that encompass the process of distilling prior decisions, the legal profession being among them [2].

Artificial Intelligence is the capability of a computer programme to perform functions that equate with human abilities, like voice recognition, decision-making, cognitive thinking, perception of visuals, and experience-based learning, and solving intricate disputes with greater accuracy and speed than humans. A developing sector is used to develop criminal justice frameworks and various investigations. In the investigations relating to criminal offences, numerous barriers as a result of the huge volume of data, small fragments of evidence in a disorderly and difficult setting, conventional laboratory setups, and occasionally insufficient knowledge, all of which can result in failures or miscarriages of justice [3].

The Indian court system is at the present time experiencing a massive challenge, in a sense, by a mountain of cases that are still awaiting their turn to be heard and resolved which affect the fair delivery of justice to the citizens. India has transcended China in terms of population number which has been a record in the recent past. Its citizens are equally reflecting materialism, intolerance, and tendencies towards conflict than what was observed earlier. And so, there has been a lot of increase in both criminal and civil litigation, the upshot being that the number of cases is really very large. As for the effect of digital media on the typical Indian citizen, there is a trend of materialism, intolerance, and a higher likelihood of suing. Consequently, considerable pressure emerges on the Indian judiciary which makes it challenging to efficiently carry out their responsibilities of delivering justice. However, this becomes a challenge, which requires special attention right away in an endeavour of searching for novel approaches that could help in addressing the congestion of cases and at the same time improving the efficiency of the courts. Integrating AI in various departments of the Indian legal framework is a solution to the problem [4].

Presently, we have an excess of unsolved cases in the district and Taluka court of India where there are 4,50,55,792 cases still awaiting trial. The cases include civil as well as criminal that include 1,10,14,992 Civil cases and 3,40,40,800 Criminal cases. In these cases, more than 69.06% have been reported where the complaints have not been addressed for a period of over one year, an issue that raises much concern [5]. There is another area of grave concern that the number of cases listed for hearing in the High Courts and Supreme Court of India keeps on piling up. The High Courts of India maintains a total number of 61,93,054 pending cases all over India, of which 44,34,657 are civil cases and 17,58,397 are criminal cases. Among these, a larger proportion (77.81%) had taken more than one year to await their resolution [6]. By May 22, 2024, the Supreme Court still had undetermined cases, inclusive of 64,106 civil cases and 17,239 cases in its criminal docket [7] These details shed further light on the need for measures to address the build-up of cases in Indian courts. Concerning the aforementioned issue, it is possible to find the following recommendations to solve it: The application of AI is needed in several domains of the justice system. In this context, it plays a pivotal role in addressing the issues related to the pendency of cases and the effectiveness of the Indian justice delivery system. The system incorporates intelligent AI technology to perform such tasks as document scanning and management, case workflow, and research. Thus implementing this measure will greatly reduce the burden on the judges, lawyers, and judicial personnel where they can spend most of their time on other important judicial functions as well as the process of decision-making. Different barriers exist for AI use in software development within India, most primarily because technological advancement is still slow at the lower levels of the judiciary. This present situation emerges when AI integration is applied to the higher courts of law. To precede these recommendations, the following measures ought to be taken to determine AI availment success in the lower courts and driven by the following sound policies and proper implementation strategies [5].

2 Objective

Incorporating intelligence in the criminal justice system can thus provide some beneficial prospects for prediction as well as higher efficiency, accuracy, and fairness in decision-making. However concerns range from bias to privacy and unintended consequences. In this context, this study aims to analyse the stage of AI in criminal justice.

- Examine of AI applications for police, courts, and corrections.
- To Understand the Make of value and challenges when thinking about the use of AI.
- To learn about how AI can speak to ethical issues, in the context of criminal justice [8].

3 Literature Review [9]

A few Recommendations by Dr. Justice V.S. Malimath [10]—These include: separating police law and order from its crime investigative wings and establishing an inquisitorial style of inquiry, curtailing the right to silence as per Article 20(3) of the Constitution, diluting the "innocence until proven guilty" stance allowing courts to operate on the preponderance of probabilities rather than the proof beyond reasonable doubt, informing accused persons and victims of their rights, and appointing more judges to reduce the pendency of cases. Here are a few crucial recommendations.

The 254th report of the Law Commission [11] was submitted by the 20th Law Commission. This report focused on addressing the issues of delay and arrears in the criminal justice system. It also provided clear definitions for terminology such as backlog, pendency, delay, and defaults. This report highlighted the problems of an insufficient proportion of judges to the population and the pace at which cases are resolved.

The committee study focuses on the machine learning capabilities of AI in addressing cyber security, legal, and ethical issues [12]. It explains how AI may use past patterns to prevent unauthorised access to data, detect fraudulent identities, and identify abnormalities. AI implementation could bolster your cyber security framework, an area that is increasingly prone to cyber-crimes. Several experts have warned that, by publicly showcasing the results of AI experiments on open source platforms and domestic firms from sharing parts of their AI experiments around the world, multinational organisations may inadvertently create pathways for rogue actors to leverage AI for malicious intent.

The paper from the Committee of Artificial Intelligence of the Supreme Court examines the criteria for inviting firms to design an appropriate AI system for the court [13]. The report focuses on understanding the necessary background processes and requirements for implementing AI, as well as establishing guidelines for developing the AI infrastructure. The letter states that the SC committee has outlined the necessity for AI to have a natural language reader to comprehend legal terminology, hierarchical classification of case content, automation in e-filing, and other related requirements. The document highlights the importance of providing training to court staff in order to effectively utilise AI software. Additionally, it proposes a five-year period of assistance for addressing any bugs or issues in the AI system. The bidder has been assigned the responsibility of maintaining the AI system, upgrading it, and ensuring its seamless operation. To provide SC additional power, a service termination clause has been inserted.

NITI Aayog [14], in 2018, initiated the National Strategy for Artificial Intelligence (NSAI), which led to the commencement of significant AI projects and implementation across various sectors, including agriculture and education. Therefore, this chapter has focused on the need for AI along with its responsible practices to deliver on the promise of AI. Several barriers to the broader growth of AI have been identified in the research, including a lack of basic understanding of AI among actors in the market and the public, as well as questions surrounding privacy and security risks. The chapter provides crucial guidelines for creating a responsible AI, including security, privacy, equality, accountability, and the preservation of moral standards.

3.1 Foreign Cases

3.1.1 US Case

The evidence presented at trial in State v. Loomis [15] was that Mr. Loomis, a United States citizen, arrived in a motor vehicle at the scene of a drive-by shooting, possessed property taken during the robbery, and fled from police and immigration agents. Using a risk assessment tool called COMPAS, the court predicted the type of jail sentence that appropriately fits the crime. On appeal, the Wisconsin Supreme Court upheld the use of machine learning predictive justice tools, but it also noted that county courts would not be able to rely entirely on these tools.

3.1.2 Indian Cases

(i). In the case of Hussainara Khatoon & Ors against the Home Secretary, State of Bihar [16], the highest court had recognised the right to a fair trial guaranteed by Article 21 of the Constitution. This decision establishes that the trial method holds a crucial role in the country, ensuring the preservation of justice.

(ii). The Supreme Court noted in the case of Imtiyaz Ahmad v. State of Uttar Pradesh and Ors [17] that the ideals of fair justice are undermined by the delay in administering justice, constituting a violation of Article 21. The court additionally requested the Law Commission to present its recommendations for addressing the issue of delayed justice.

(iii). In the case of Parkash Singh vs Union of India [18], the highest court provided recommendations for improving the criminal justice system, particularly the police system. These ideas included the establishment of a State Security Commission to oversee the functioning of the police, as well as the creation of a police complaint authority in every district of each state.

3.2 Journal Articles

The work by Andrew C. Michaels [19] has provided a potential description of AI, but it has imposed some limitations on its scope. The author rejects Eugene Volokh's perspective, as an American legal scholar, that AI has the ability to make compelling judgements in decision-making, ultimately rendering human judges mostly unnecessary. This chapter has identified fundamental principles of AI, but it also emphasises the superiority of human judgement. Human judgement is based on the interpretation of laws within a specific area, which is determined by various advocates, parties, and judges who draw upon their experience, intuition, and legal principles. In contrast, AI is limited to the ability to understand and apply explicit legal rules.

In this work, Cass R. Sunstein [20] explores the area in which AI can be applied in the field of legal thinking. Two assumptions have been made regarding the extent of AI's impact on legal reasoning: either a strong role or a minor role. The user suggests that AI may be integrated into the field of legal reasoning, functioning as an upgraded version of LEXIS & Westlaw. In this scenario, AI would improve the capabilities of the software to differentiate between instances with comparable data by identifying underlying principles.

Li Ma's [21] work examines the benefits of incorporating AI into legal research and education. The paper suggests that utilising AI in the collection and analysis of big data can help convert it into relevant data for research, job preparation, and argumentation. In addition, integrating AI into legal studies can enable the learning of legal reasoning, problem-solving, and the application of logical principles, ultimately leading to more efficient outcomes.

The article highlights the notable capabilities and the potential of AI to develop the criminal judicial framework. By using AI for the delivery of justice, it may perform impartially and rely on logical evidence rather than emotional factors. The hypothesis is that the increasing reliance on AI for decision-making in the future would exacerbate issues related to backlog and delay.

4 Research Methodology

The detection of effective application of AI in the field of criminal justice mechanism for crime detection along with the purpose of prediction and prevention of crime from being committed is the sole objective of carrying out this research. To achieve this goal, the researchers have developed a questionnaire consisting of comprehensive AI-related inquiries targeting the residents of India under various age groups.

The aim is to establish the need and level of understanding in relation to the application of AI in the legal framework. The basic requirement of the questionnaire was to derive insights into the existing criminal judicial framework and establish the necessity of using AI to enhance age-old techniques of crime prevention and detection [22].

5 History of Technological Development in the Indian Legal System [23]

The voyage of digitalisation in the Indian legal system has been met with resistance since its inception, as the courts have refrained from embracing this transformation. The onset of the pandemic, specifically the outbreak of COVID-19, rendered physical court operations impossible. Consequently, the use of digitalisation and AI emerged as a valuable solution. Owing to the situation, the courts were forced to establish the virtual court system, which turned out to be a workable solution for the judiciary. Virtual courts have several benefits over traditional courtrooms, such as being cost-effective and time efficient because they don't require parties to travel and are less hectic.

The government has implemented the e-Courts Integrated Mission Mode Project as part of the national e-governance plan. This project aims to digitise District and subordinate courts across the country, thereby enhancing access to justice through the use of technology. Its goal was to maximise the legal system's efficacy, economy, and transparency while also improving the functioning of the judiciary. Phase 1 of the project took place between 2011 and 2015; it was first implemented in 2007. The goal of this phase was to create databases on the national judicial data grid for courts at the district and talukas levels. It put in place a system to keep an eye on how cases were handled in the Subordinate courts and High Courts. A thorough database of directives, decisions, and case data from High Courts, District Courts, and Subordinate Courts is offered by NJDG. Since the project's second phase began in 2015, 18,735 District and Subordinate courts have been computerised. Furthermore, every court including taluk level courts has a single video conferencing equipment.

In addition, funds have been approved for the purchase of additional video conferencing apparatus for 14,443 courtrooms. 3240 courtrooms and 1272 jails have also set up video conferencing systems. Phase III has been approved by the Indian Supreme Court, while Phase II is nearly complete. The objective of the third phase of the e-Courts project is to create an affordable, easily accessible, dependable, transparent, and efficient legal system in India for everyone who is involved in the administration of justice or is seeking justice. The existing status of the judiciary lacks the necessary infrastructure and capabilities to effectively implement virtual courts. Since the virtual court system is not a substitute for real courtrooms, a thorough overhaul is required to guarantee that all parties to legal processes can access technology. AI can be used to perform different tasks. Such tasks include forecast and prediction, betterment in administrative efficiency, filing automation, scheduling cases in a more intelligent manner, enhancing the case documentation system, and communicating with plaintiffs through chatbots to help with faster disposal of cases.

6 Facilitator of AI [9]

The e-Courts Project, an ingredient under the initiative for e-governance proposed by the e-Committee of the Apex Court of India, is based on the "National Policy and Action Plan for Implementation of Information and Communication Technology (ICT) in the Indian Judiciary 2005". This project was established with the objective of implementing ICT in the Indian justice framework. The objective was to revolutionise the Indian Judicial framework by equipping the courts with essential equipment and information and communication technology to bring the Indian judicial framework at par with international advancements. The Project e-court has effectively met the crucial need for modern justice delivery. Here are several important initiatives that will help integrate AI into the criminal justice system:

1. Digitalisation was implemented in two stages, with the first phase taking place in 2007 and the second in 2015. During the first phase, the focus was on establishing the necessary infrastructure and basic computerisation in all district courts. This included providing LAN, hardware, and Case Information Software (CIS) to facilitate essential case-related services for parties to the suit and advocates. In Phase II, advanced computerisation was implemented to enhance the development of upper-level systems. This includes giving the District Legal Services Authority, Legal Services Committee, National Judicial Academy, and State Judicial Academies access to computerised technology and contemporary hardware. These improvements were made with the intention of making training modules and services more effective and efficient. AI in the judiciary cannot be implemented without computerisation.
2. The Crime and Criminal Tracking Network and Systems (CCTNS) is presently being executed as a Mission Mode Project by the National Crime Records Bureau under the National e-Governance Plan (NeGP) of the Government of India. CCTNS seeks to establish a unified framework to enhance the efficiency of law enforcement. Its improved the capabilities of searching and created a centralised collection of crime and criminal information, including images and fingerprints of criminals. Therefore, it enhanced the ability to evaluate patterns of criminal activity or modus operandi.
3. The Ministry of Home Affairs (MHA) established the Inter-operable Criminal Justice System (ICJS), which is enforced by the National Informatics Centre (NIC) and the National Crime Records Bureau. Improving openness and interconnectedness within the criminal justice framework is its goal. By leveraging a single platform, the Inter-operable Criminal Justice System enables smooth data and information sharing among numerous criminal justice system entities, including courts, police, jails, and forensic science labs. Courts now have access to the charge sheet, case diary, and FIR through the ICJS website. Phase I of the project involved the establishment and stabilisation of individual IT systems. Consequently, record searches may now be carried out on these sites. Phase II, which starts in 2023, adheres to the "one data, one entry" concept. This technique eliminates the need for duplicate data entry since data is entered just once in one

pillar and is accessible in the other pillars. Forensic labs, e-Courts, and e-Prisons are integrated with CCTNs to accomplish this.

7 Definition of Artificial Intelligence

Section 500(3) of the National Artificial Intelligence Initiative (NAII) Act of 2020 provides an exact and clear definition of "Artificial intelligence". According to the provision of this act, AI is defined as a framework that functions using machines and has the ability to make forecasts, suggestions, or choices that can have an effect on real or virtual settings [2]. Crucially, these activities are carried out based on pre-established human goals. The concept of AI is known for being elusive.

The term "artificial intelligence" encompasses the following [4]:

A. Any autonomous framework that can operate in dynamic and uncertain conditions, without substantial human supervision, or that has the ability to acquire knowledge from their interactions and enhance their capabilities. These systems can be created using computer software, real hardware, or in circumstances that have not been considered yet. They are capable of solving tasks that involve human-like thinking, perception, preparation, instruction, interaction, or physical action. AI can be attributed to systems that exhibit a higher degree of human-like behaviour in the context of their tasks.
B. Artificial intelligence (AI) systems that can think like humans, like neural connections and cognitive frameworks.
C. Systems that behave like humans, like those that can pass the Turing test or an equivalent evaluation by using automated reasoning, natural language processing, knowledge representation, and learning.
D. The term "cognitive computing" describes a group of approaches, including machine learning, that try to simulate different cognitive processes.
E. Intelligent software agents and embodied robots are examples of rational systems that use perception, planning, reasoning, learning, communication, decision-making, and action to achieve their objectives.

8 Implementation of AI in India

The Maharashtra police purchased predictive policing software in 2015 as part of their effort to improve crime control through the use of AI technologies. As part of the programme, the department purchased Universal Forensic Extraction Devices (UFED), which are used in digital forensics and investigations, from well-known international manufacturers. These devices can recover data from cell phones, social media, hard drives, and other devices of a similar nature, even if the data has been wiped. Moreover, they are efficient enough to recover audio-visual data from CCTV and drone cameras. As per the crime report published in the year 2016, Uttar Pradesh

(UP) has the highest rate of criminal activity recorded in the entire country. To eradicate such issues, the UP police have worked with ISRO in order to deploy a variety of AI tools, such as crime-tracking analytics and prediction systems.

Artificial intelligence (AI) and other forensic technologies were used by the Police of Delhi to inquire into a sum of 755 cases related to the brutal violence that occurred in the year 2020. Facial recognition and other technologies were used to investigate about 945 cases of surveillance and smartphone videos, which helped in capturing offenders. On 24 February 2021, a 26-year-old girl, Solanki, was shot behind Rajdhani School in Delhi. The suspect's cell phone was confiscated and analysed. The offence was determined through the examination of the Google Map application loaded on his mobile device. The Delhi Police concluded 2020 with the resolution of over 400 cases and the submission of 342 charge sheets [3].

9 Need for AI in Delivery of Justice

A significant obstacle to the court frameworks in India is the large number of unresolved cases. Despite the use of mechanisms like e-court and Alternative Dispute Resolution, the judicial framework was unable to effectively cut down the growing number of pending cases. In addition, the scarcity of court personnel exacerbates the issue. This results in the postponement of justice, which is tantamount to the rejection of justice. It is imperative to provide a non-conventional tool to reinstate the efficacy of the justice delivery framework. An example of such is the implementation of AI in the Judicial system. Despite the Government's efforts in India, to initiate and support the computerisation of police records, paper based conventional records are still widely used instead of AI. This offline strategy hinders the efficiency and expediency of investigating and tracing criminals. In order to conduct a thorough investigation, the police need access to up-to-date information. The application of AI is imminent. AI can be utilised to capture and store biometric information, including facial features, voice patterns, blood type, fingerprints, and records of past criminal activities. The AI-driven web browser allows the Investigating Officer to identify suspects by utilising the recorded digital data [3].

10 Applicability of Artificial Intelligence

Historically, the identification and control of criminal behaviour have primarily been responsive, meaning that they often happen in response to immediate occurrences rather than through a proactive examination of past and developing evidence and data. Usually, a criminal act takes place or is already happening, and law enforcement authorities react to the incident upon receiving a report from a member of the public or the victim. Policing often includes proactive efforts aimed at preventing crime, which may involve randomised actions such as routine police patrols. However, the majority of police are focused on specific objectives and long-term planning. AI has the capability to significantly enhance the efficiency of proactive law enforcement. Advanced algorithms have been developed that can accurately forecast the probability of criminal activity occurring at a specific place and time. These are derived from prior patterns of behaviour. An explicit example of this is the utilisation of speed cameras to identify drivers who exceed the speed limit. Typically, these cameras are positioned in areas with a history of frequent speeding or a heightened likelihood of accidents. Prior driving behaviour is a highly reliable indicator of future behaviour [2].

AI has the potential to greatly alter the decision-making process, enhancing its speed, precision, and efficiency. Judges can derive advantages from utilising AI by swiftly accessing pertinent information and rendering well-informed judgements. The Apex Court of India has implemented various measures to tackle the problem of a large number of matters that are pending in Indian courts, consisting of more than 81,345 cases in the Supreme Court alone [4]. There is an urgent requirement for a remedy to accelerate the judicial process. The Supreme Court established an Artificial Intelligence Committee in 2019 in order to address this concern. The Artificial Intelligence Committee has the responsibility of supervising the application and advancement of AI-based solutions within the judicial realm. The primary objective is to investigate the possible uses of AI, such as pattern recognition, machine learning, and big data analysis, to increase efficiency in several parts of the judicial system [9]. The committee acknowledges the significance of AI in the translation of legal agreements, facilitating legal research, expediting policy-making, and enhancing case monitoring. Through the utilisation of AI, the committee seeks to tackle the difficulties that arise from a large number of cases, complicated geographical factors, and the requirement for effective dispensation of justice.

The SUPACE which stands for Supreme Court Portal for Aid in Courts Efficiency, is an AI tool that has been undertaken under the guidance of the Artificial Intelligence Committee of the Apex Court. It uses machine learning to analyse huge quantities of data relating to cases, providing judges with appropriate data to make well-informed decisions. The SUPACE seeks to combine the capacities of AI and human for the purpose of enhance efficiency.

The Apex Court of India has developed an AI-powered tool known as the Supreme Court Vidhik Anuvaad Software (SUVAS) to overcome the problem of language barrier [3].

11 Implementation of AI-Driven Technologies in Criminal Investigation to Facilitate Modernisation

11.1 Recognition of Patterns

This is an automatic machine recognition process that is classified based on the learning procedure employed to produce the resulting value. Forensic science requires, among other things, the ability to recognise and contrast particular types of patterns within the suspect data. The professionals are tasked with the analysis of an enormous volume of data using rigorous statistical and probabilistic reasoning methods. A pattern may consist of a speech signal, a handwritten cursive word, a fingerprint image or even a human visage. Numerous applications have made use of neural networks, including speech recognition, biometric recognition, face recognition, and iris recognition, among others.

11.2 Identification and Classification of Fingerprint Pattern

Machine learning has achieved remarkable growth in the domains of visual analysis and recognition of patterns due to its elimination of the need for manual feature extraction. Deep learning acquires knowledge of structures and features automatically when provided with an adequate amount of input training data. AI is well-suited for a wide range of duties within automated biometric recognition and categorisation systems due to the aforementioned benefits. Additionally, these AI-based automated systems can significantly reduce the variety of comparisons required for accurate matching.

11.3 Multimedia Analysis

Artificial intelligence enables the mitigation of human error in the evaluation of images, CCTV footage, or video in the course of criminal investigations. Conventional algorithmic approaches that support human experts are constrained to pre-established attributes such as eye colour, eye shape, distance between eyes (for facial recognition), and demographic data (for pattern analysis). In addition to learning complex tasks, the AI algorithms designed for multimedia analysis are also capable of independently developing and determining complicated facial recognition features and parameters required to complete specified tasks. These sophisticated AI algorithms are capable of comparing voices, identifying weapons, matching faces, and detecting complexities such as crimes and incidents. In the realm of hearing evaluation and speaker identification, security workers and experts have been assisted by a multitude of AI-based algorithms. The application of AI extends beyond speech recognition to encompass language instruction and speech fluency assessment.

The legal documents that the Indian judiciary releases are extensive, disorganised, and intricate. Pre-processing and data cleansing are of the utmost importance for any type of analysis, particularly for supervised algorithms that have been trained on said data. Conventional empirical legal research has conventionally tackled this concern through the utilisation of limited-scale datasets, in which legal variables are manually classified and the scope of inference is restricted to a subset of cases that are relevant to a specific issue. The significant obstacle encountered during data pre-processing pertains to the considerable variation in reporting across different states and districts. The variability in data quality is substantial due to the absence of a nationally standardised framework for defining and reporting variables. This complicates the comparison of specific case classifications between states and courts. The data does not contain any standardised identifiers that could be used to track a case's prospective progression of appeals to higher courts. Likewise, the process of monitoring a criminal case from its inception as a First Information Report (FIR) to its conclusion as a judgement is not straightforward.

When discussing the application of AI to criminal cases, two broad categories come to mind [22]:

A. AI for the Detection of Criminality

(i). **Analysis of DNA**: A society that enforces the law on the basis of scientific and forensic evidence can also benefit from AI. This is particularly true of forensic DNA tests, which in recent years have had an unprecedented effect on the justice system. During criminal activities, organisms such as blood, saliva, semen, and epidermis cells can be transmitted via human and object contact. The advancement of DNA technology facilitated the sensitivity of DNA analysis, enabling scientists to detect and process evidence that was previously unusable due to its poor quality, corruption, or inaccessibility. Decades' worth of DNA evidence originating from cases involving cold-blooded murder and sexual assault, for instance, have been submitted to a laboratory for analysis.

(ii). **Gunshot Detection**: The identification of pattern signatures in the analysis of gunshots presents an additional domain where AI proficiency can be implemented. The National Institute of Justice (NIJ) funded Cadre Research Labs, LLC to analyse audio files from smartphones and smart devices as part of one of its projects. The NIJ recognised that the quality and content of firearms could be affected by factors such as the type of firearms and ammunition, incident geometry, and recording device. Cadre scientists are developing algorithms to identify gunshot targets, detect bullet waves in shock waves, ascertain the timing of a shot, ascertain the quantity of available firearms, assign a shotgun, and estimate the likelihood of legal authorities involved in the investigation regarding the class and size of the firearm. These algorithms are being developed using a mathematical model that is well-defined.

(iii). **Public Safety Images and Recordings**: In order to aid criminal investigations, the criminal justice and legal communities utilise videotape and image analysis to acquire information about individuals, objects, and activities. Nevertheless, the examination of video and image data necessitates a substantial workforce, thereby demanding a substantial investment in the story's knowledgeable personnel. Due to the plenty of information, the rapid evolution of technologies such as mobile phones and apps, and the scarcity of specialised personnel with experience processing that information, video, and image analysis is also susceptible to human error ("Assistive AI maintains the human component in safety for everyone | 2021-06-04|Security Magazine", n.d.). Such errors are surmountable with the assistance of AI technology, which enables it to function as an expert. For pattern analysis, conventional software algorithms are constrained by predetermined variables such as eye shape, colour, and the distance between eyes, which hinder the ability to discern human or facial details. Beyond what most people could ever conceive, multimedia algorithm rule and picture AI not only learn complicated duties but also constructs and determines the complexity and constraints of the complicated facial expressions in order to accomplish the tasks. Algorithms possess the capability to discern concealed realities, recognise firearms as well as additional objects, and discern complex occurrences such as ongoing criminal activities, incidents, and the like.

(iv). **Digital Forensic Analysis**: Digital forensics, which is also referred to as computer forensics, is the process of proving something by extracting and analysing digital content devices. There are numerous tools available that scan software, mobile devices, and computers for evidence of potentially incriminating details. In this case, done wisdom is effective because it enhances the strength of computational tools, which expedite the processing of enormous amounts of data that are beyond the comprehension of mere mortals. Software dubbed Magnet AXIOM, which was created by Magnet Forensics of Waterloo, Canada, is a prime example. The instrument is referred to as "a digital examination platform that enables researchers to locate and verify pertinent data using smartphones and computers, with the intention of visualising it to facilitate more effective analysis".

B. AI in the Prevention of Crime

(i). **Predicting Danger Zones**: Imagine a thief who arrives at the location of his next heist only to be met by the authorities. It is indeed possible to accomplish this using AI technology.

(ii). **Forecasting the Identity of the Perpetrator**: Criminals frequently discuss offences they have perpetrated online. AI can aid in the monitoring of the transfer of online content. Cybercrime-related unfamiliar terms can be identified by algorithms. Subsequently, AI programmes have the capability to transmit data to the appropriate law enforcement agencies pertaining to atypical criminal behaviour or correspondence, thereby facilitating the identification of potential offenders.

(iii). **Determining the Pretrial Release Option**: Numerous hazardous objects can be identified by AI algorithms prior to the apprehension of a defendant. Cases that remain pending during the trial, convictions for violent offences, non-appearance at court hearings, and pre-arrest sentencing are some of the potential risks to be aware of. On the basis of a multitude of such factors, AI programmes are capable of precisely ascertaining whether an individual warrants early release.

12 Different Tools for Application of AI

The AI Committee of 2019 is tasked with the establishment of four AI portals, namely E-Court, SUPACE, SUVAS, and LIMBS [4].

A. **SUPACE:** It is an abbreviation for the Supreme Court Portal for Court Efficiency Assistance. It is a mechanism facilitated by AI that gathers crucial data, pertinent facts, appropriate precedent, and legislation. It issues a preparatory report to the court containing a comprehensive summary of all pertinent information regarding the case at hand. Its purpose is not to make decisions; instead, it processes data and provides it to justices who are seeking inputs for a decision, merely in a supporting manner. Initially, it is implemented experimentally by the criminal proceeding's judges of the Bombay and Delhi High Courts. It is crucial to acknowledge that the training process of the AI in SUPACE is to predict patterns and data that have been annotated and extracted by human users. This procedure facilitates the AI's progressive learning and enhancement of its performance. It is advisable that High Court justices adopt SUPACE as a component of the implementation in order to optimise their operational effectiveness and streamline the judicial proceedings.

B. **SUVAS:** The acronym SUVAS represents Supreme Court Vidhik Anuvaad Software. The Apex Court has used this AI tool to increase the easy accessibility of judicial proceedings by translating formulas found in judicial documents. This tool is used to translate judgements, orders, and other judicial documents issued by the Apex Court into nine vernacular languages: Assamese, Hindi, Bengali, Kannada, Odiya, Marathi, Tamil, Urdu, and Telugu. SUVAS presents a multitude of benefits. To begin with, it helps in precise and effective translation, thereby reducing the probability of miscommunication or misinterpretation resulting from language barriers. This is very important in a nation as diverse as India, which is home to a multiple regional language. It provides benefits to individuals lacking English proficiency to access legal information and services by removing the language barriers. This feature allows individuals the ability to engage themselves in the administration of justice and ensure their rights. Subsequently in 2023, this instrument was also employed by the High Court in

Kerala, Delhi High Court, and Karnataka High Court to render judgements in a number of regional languages. The translation of the following cases pertains to the following domains: Criminal law, ordinary civil law, family law, religious and charitable endowments, personal law, simple money and mortgage, eviction under the public premises (eviction) act, land law, agriculture, tenancies, and consumer protection. It serves as a manifestation of the judiciary's dedication to utilising cutting-edge technologies in order to benefit external parties.

C. **LIMBS**: The Legal Information Management & Briefing System (LIMBS) portal is an AI-powered website. It was devised to monitor litigation involving the Union of India. Users representing diverse Ministries/Departments input pertinent information regarding pending cases in the District Courts, High Courts, and the Supreme Court. The case-involved users are subsequently able to view and modify the data's specifics. According to data derived from the LIMBS database, the number of pending cases in the District Courts, High Courts, the Supreme Court, and Tribunals stands at 5.95 lakh at present. Despite this, LIMBS only stores data in a predefined format due to the fact that it is a user-interface programme. The resolution of any inconsistency that may arise in the data provided by the user is the responsibility of the Ministry/Department in question. The case status of matters involving the State Government as a party is not disclosed on the LIMBS Portal, an interface utilised to track legal proceedings conducted on behalf of the Union of India. The portal is intended to facilitate the management of legal affairs by ministries and departments of the central government. However, in accordance with the National Data Sharing and Accessibility Policy (NDSAP), case entry statistics, among other information, are made public via the website data.gov.in.

D. **Project E-Court**: By utilising technology to improve access to justice, the e-Courts project is a trailblazing endeavour that has revolutionised the Indian legal system. The project provides a variety of offerings, which include electronic filing, digitally equipped courts, and case management framework which have increased the judiciary's efficacy and transparency. To enable remote participation in hearing by establishing virtual courts is one of the main aims of the e-Courts initiative. The digital facilities reduce logical difficulties and save time and money. Due to increasing popularity and efficacy of this technology over 1.46 crore cases have been taken up through virtual courts as of January 2023.

The use of these technologies within the judiciary has resulted in enhanced efficiency. Besides, this has also helped in providing justice to citizens residing in various regions of the country. Furthermore, the use of this technology has reduced the administrative duties of magistrates and other court officers, enabling them to allocate their efforts towards more important responsibilities. The Indian judiciary has to adopt new technologies into its systems as technological developments continue to advance in order to make sure that justice is delivered fairly and efficiently [9].

13 Regulations in India and Abroad

The applications of AI can be rooted in every other field. Thus, AI must be continuously regulated to prevent any misuse. One of the many advantages and conveniences that they enjoy also poses a threat to their very existence. The security and preservation of data is the primary concern of AI. It is entirely dependent on big data, and due to the abundance of data, associated hazards must be accounted for. A user's sensitive information may be compromised and rendered hazardous if so. As India lacks a specific legislation pertaining to AI, the government has additional grounds for concern. Intervention is required as this matter has evolved into a national concern. Presently, AI is being embraced and promoted in India at an unprecedented rate. The necessity for regulation emerged as a result of the rapid progression in the implementation of AI. Currently, the government is harmed by the development of new AI-related laws, policies, and guidelines. The Right to Privacy was recognised as a fundamental right by the Constitution in 2017, as part of an effort to safeguard the public. The Justice Srikrishna committee recommends that privacy laws be passed by the government. The Digital Personal Data Protection Act (DPDP Act) of 2023, which was enacted on August 11, 2023, is the first comprehensive legislation for data protection in India. The legislation aims to safeguard individuals' privacy while also establishing a structure for the oversight and control of data. It is applicable to all organisations in India that process the personal information of individuals. Over the course of the next six to twelve months, the DPDP Act is anticipated to be implemented gradually, supplanting the current patchwork of data protection regulations in India.

India has made notable progress as a participant in the Global Partnership on Artificial Intelligence (GPAI) which is an international attempt to encourage ethical and human-centered developments in AI. In the present Indian scenario, there is no specific law addressing the regulation of Al and machine learning (ML). However, in order to formulate a framework for AI, the Ministry of Electronics and Information Technology (MEITY), which controls Al-related strategies as the executive organisation, has recently incorporated four committees. Moreover, seven principles of responsible entrepreneurship have been undertaken by the NITI Aayog, a renowned Indian organisation. These principles encompass safety, equality, privacy, transparency, accountability, and equality [5].

Enterprise technology companies have engaged in collaborative efforts with the Indian government and multiple ministries to execute projects in pivotal sectors including healthcare, education, and agriculture. In addition, an Al standardisation committee has been constituted by the Department of Telecommunications with the aim of formulating interface standards and India's Al stack, thus propelling the nation's Al capabilities forward. Additionally, regulatory agencies in India have acknowledged the operational-improving potential of Al. For example, Securities and Exchange Board of India (SEBI) created the Picture-based Information News Accumulator and Key Information Analyzer as an Al-based tool. SEBI is assisted in its regulatory functions by this system, which makes a database of recommendations after examining a number of stock market shows. Reserve Bank of India (RBI) also plans to implement advanced analytics, Al, and ML significantly to its huge database in order to develop regulatory oversight of banks and non-banking financial corporations (NBFCs). The primary aim of this venture is to secure the RBI's data-driven surveillance ability in order to guarantee efficient regulatory supervision [8].

A significant development in this field is the implementation of the Artificial Intelligence Act by the European Union (EU) in 2021. The preamble of this act is to protect the fundamental rights of individuals whose lives may be affected by Al technology and to ensure their protection. This law has influenced other nations to make regulatory measures regarding AI. As a response to this law, Brazil has enacted a bill with the aim of establishing a legal structure for Al. This shows the nation's dedication to the advancement and implementation of Al. Additionally, Canada presented the Digital Charter Implementation Act to the House of Commons in June 2022. This act is made up of three legislative proposals that are intended to improve Canada's data privacy laws and advance artificial intelligence in a morally responsible manner.

14 Comparative Analysis with Other Countries

The emergence of AI and the advent of digital society have influenced novel approaches to routine tasks across all social sectors. At the current time, AI has evolved into a ubiquitous concept, manifesting itself in various forms throughout the world. The software application responsible for developing AI will direct the system as it generates automated results. Given the current rate of AI development, it is unsurprising that the public and private sectors will be dominated by AI in the near future. As AI is a relatively new technological advancement, it is not regulated by appropriate national or international laws.

The Indian legal industry has experienced a sluggish rate of AI development in terms of new developments. Only 4% of Indian solicitors employ AI in their line of work. Four committees were established by the Ministry of Electronics and Information Technology (MEIT) to examine legislation and regulations pertaining to AI. Based on a proposed data protection statute, the Joint Parliamentary Committee has argued and disputed the Personal Data Protection Bill of 2022.

The British Government published an AI rule book in an effort to increase public confidence in the government's oversight of the nation's AI regulations and to foster innovation. Regarding this manual, AI is characterised as a "general purpose technology" that, similar to the internet or electricity, will profoundly influence numerous facets of human existence and will exhibit substantial variation contingent upon the specific context and implementation. Numerous of these are no longer prognosticable. The AI Act of European Union, which will be regulated by a single body and endeavours to harmonise the law of European Union across all member states, is in contrast to this AI guideline. As regards the rules, the legislative proposals of the EU establish a relatively stable definition of its regulatory approach.

Although individual states in the United States have implemented their own legislation and regulations regarding AI, the nation as a whole approach the matter of artificial intelligence governance in a fragmented fashion. Regarding AI, there exists no national law in the United States. The predominant objective of the majority of state legislation has been to establish a various committee charged with determining how agencies of each state may implement. Most of the state laws have been devoted to the formation of diverse committees tasked with determining how state agencies may utilise AI technology and examining the potential repercussions of AI on consumers and the workforce. This has been the prevailing circumstance regarding the vast majority of enacted legislation.

With respect to international law, in a report of 2014 to the UN Council on Human Rights, the Special Rapporteur on Extrajudicial, Summary, or Arbitrary Executions, Mr. Christof Heyns, debated that the use of lethal autonomous robots (LARs) or lethal autonomous weapons (LAWs) "could potentially be deemed unacceptable due to the absence of a suitable legal framework and the inability to establish a system of accountability that adequately addresses these concerns". The authority to determine an individual's life or mortality ought not to be vested in a robot. A number of non-governmental agencies have advocated that autonomous weapon systems should be banned because they fear that "killer robots" could one day endanger humankind. As stated by the leader of one such organisation, "our concern is that targeting decisions should be made by humans, not sensors [23]".

15 Effects on Legal Systems

The implementation of AI in criminal justice in India will have both positive and negative implications for the interests of numerous legal stakeholders in the country. The principal legal stakeholder shall be impacted as follows [9]:

I. JUDICIARY: Judges perceive AI as a contradiction. On the one hand, AI will assist and streamline intricate duties while reducing human effort and increasing accuracy. As a result, in order to ensure prompt administration of justice, the judge will have accomplished their highest responsibility. Conversely, the judges face imminent replacement, ethical and security concerns, and a current apprehension regarding the veracity of AI-generated reports.

II. LAWYERS: AI provides advocates with a double-edged sword: on the one hand, it facilitates effortless contract formulation and analysis, document review, and legal research argument preparation; on the other, it provides access to pertinent and necessary knowledge. However, it is accompanied by apprehension regarding potential job substitution, technical difficulties, and substandard work.

III. POLICE: AI is regarded as a technical assistant to the police, capable of enhancing the effectiveness of surveillance, crime prevention, and monitoring. Moreover, notwithstanding its merits, the implementation of AI will impose a second set of eyes on police operations, thereby increasing their accountability.

IV. PARTIES IN THE CASE (A): AI is relatively advantageous for the involved parties because, to begin with, it safeguards their constitutional rights to personal liberty and expeditious justice (Article 21 of the Indian Constitution). Secondly, their case proceeding will be more accountable and transparent as a result. It will prevent investigation corruption for the claimant and safeguard the accused's personal freedom by determining their sentences using impartial predictions.

16 Findings and Analysis

In order to assess people's expertise and interest in the intersection of AI and the Criminal Justice framework, researchers have designed a questionnaire of 10 straightforward questions. Those questions were included in a Google form that was disseminated to respondents of various ages and genders. The survey results are succinctly summarised below. Respondents were initially asked to indicate their age range and gender (Figs. 1 and 2) before answering the subsequent questions.

The subsequent question was "Have you ever learnt, studied, or even heard about Artificial intelligence (AI) technology"? Four options were given:

A - Yeah, I have studied AI in school or college.
B - I researched it on my own, so I understand the notion.
C - Although I've heard of the term "artificial intelligence", I don't know much about it.

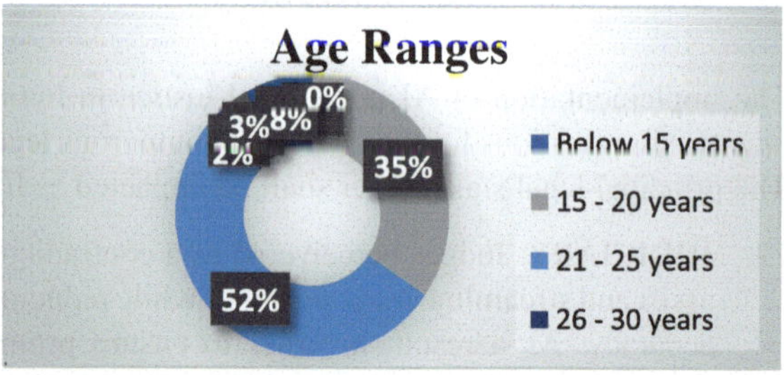

Fig. 1 Result summary of question 1

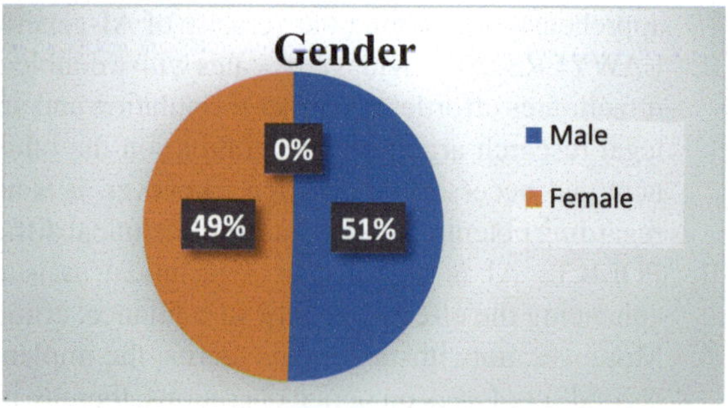

Fig. 2 Result summary of question 2

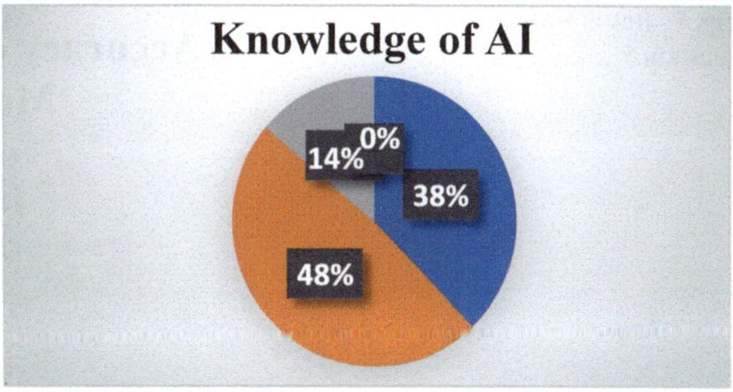

Fig. 3 Result summary of question 3

D - AI is something I've never heard of.

The purpose was to gather information about people's proficiency in AI. The results are summarised as follows. 37.9% of the responders have a decent idea about AI. Most people in the particular audience (48.3%) have self-studied AI and 13.8% of the audience have no thorough idea of the field of AI. The summary result of the question is shown in Fig. 3.

The purpose of the fourth question was to ascertain the responder's level of familiarity with AI applications. The question was "In the first world countries, AI is a blooming field that is used to upgrade the daily life of people. Have you ever heard of such AI Applications"? Of these, 95% are aware of AI applications. Others are ignorant of the uses of AI. Our goal was to gather thoughts regarding the use of AI in criminal justice framework from the target audience, starting with the fifth question (Fig. 4).

The next one was, "Do you believe that the most accurate results can be obtained using the traditional methods of criminal justice"? The pie chart below displays the results (Fig. 5).

The subsequent ask in the survey was "Are you satisfied with the current judicial decisions based on traditional crime detecting methods"? The purpose of the inquiry was to find out if public opinion of the current legal system was favourable. The result as shown in Fig. 6 shows that maximum audience couldn't say that they are completely satisfied.

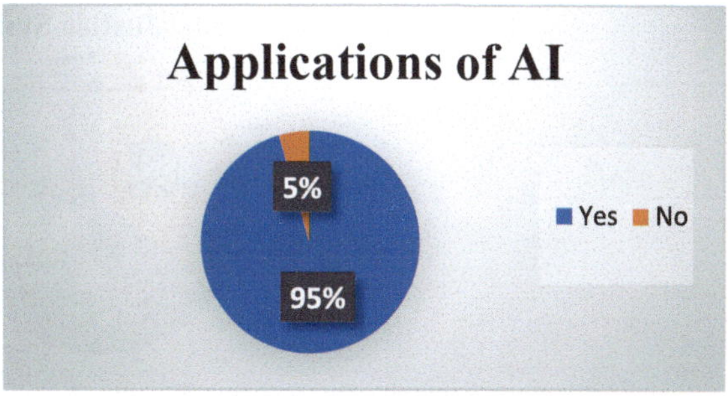

Fig. 4 Result summary of question 4

Fig. 5 Result summary of question 5

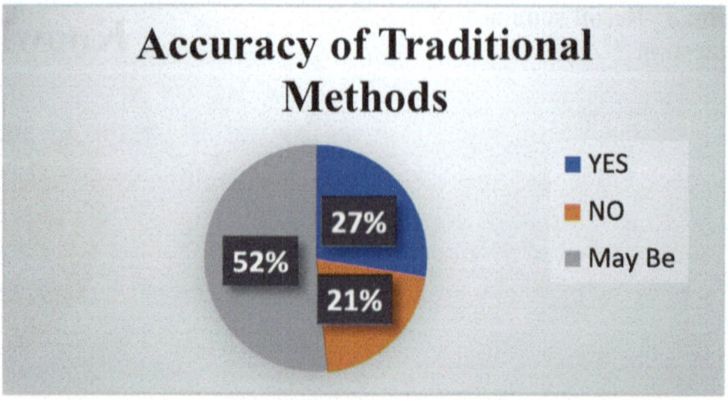

Fig. 6 Result summary of question 6

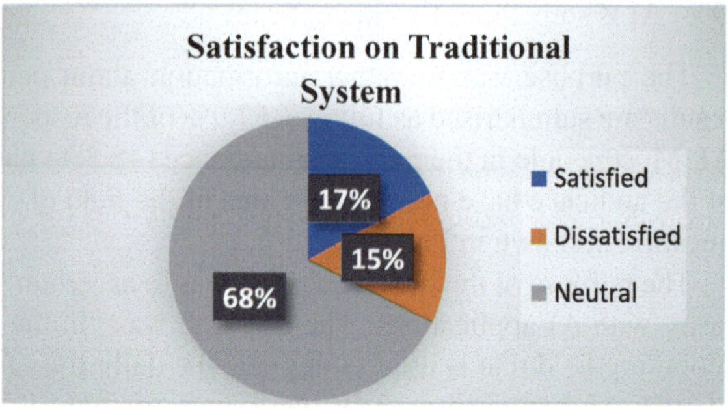

The seventh question was "Do you think that AI should be used by Indian judiciary"?. Most of the audience (64.4%) thinks AI should be implemented in the Indian Justice system to ease the process. 35.6% audience thinks AI shouldn't be implemented now (Fig. 7)

The eighth question of the survey was "Do you think that crime detection method will reduce the crime rate"?. To this question, almost 47.1% audience had a neutral opinion and 41.4% audience gave a positive response. The rest think crime detection methods will not help in the reduction of crime rate in India (Fig. 8).

Fig. 7 Result summary of question 7

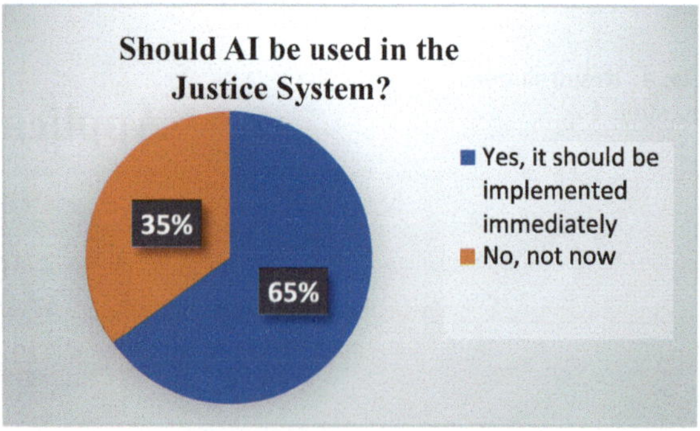

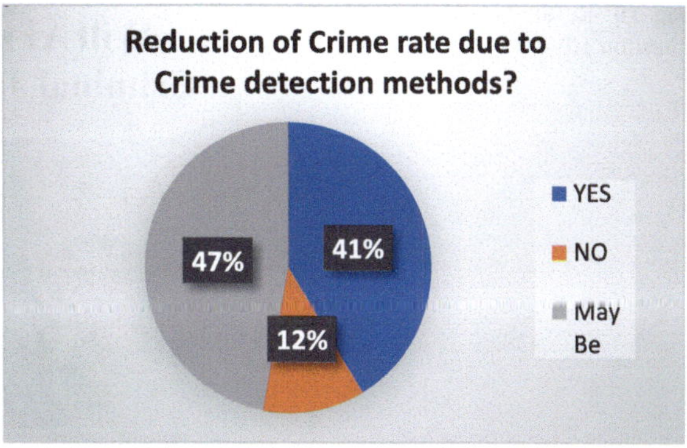

Fig. 8 Result summary of question 8

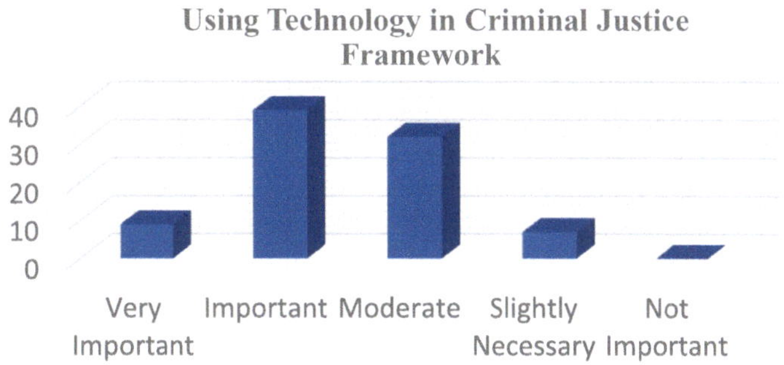

Fig. 9 Result summary of question 9

The next question was "Rate how you feel about using of new technology in criminal justice beyond the existing traditional methods?". Respondents were given a 5-point scale. The responses varied from very important to not important. The chart of results is shown below. The findings indicate that the majority of people are interested in integrating modern technology into the criminal justice system (Fig. 9).

The last question of the survey was "According to you, will the use of AI Applications, speed up the Criminal justice system"? to which almost 60% audience responded positively which implies they firmly believe that using AI in the criminal justice framework will speed up the process of providing justice. Figure 10 provides the complete result of the question.

17 Summary of Findings

From the above findings it can be summarised that, implementing AI tools in the criminal justice framework is a matter of utmost importance. As per the view of the respondents, using traditional methods for resolving justice is not enough. The

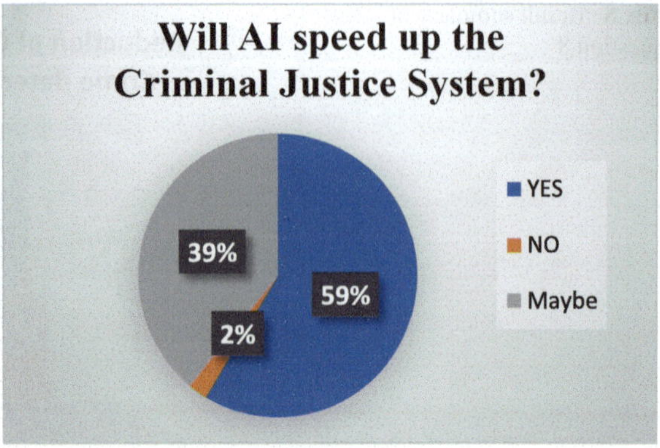

Fig. 10 Result summary of question 10

current judicial system is not capable of resolving disputes in a speedy manner, hence resulting in a backlog of hundreds and thousands of cases. Implementing AI in the judicial framework and also in the criminal investigation sectors will help the system to solve cases efficiently and even help in the prevention of crimes by predictive policing. People are realising the rising need for AI in different aspects. Almost 65% of the respondents wanted AI to be implemented in the criminal justice framework and 59% think that it will speed up the justice system. There's still a fear of replacement of the human mind if AI becomes efficient in the field, but just as the invention and use of computers brought efficiency in human life and rather increased opportunities, we believe implementing AI would do the same. A major issue in the Indian judiciary can be solved with the efficient use of AI.

18 Conclusion

Criminal law addresses the most detrimental behaviours that occur within a community. The incorporation of artificial intelligence into the criminal justice framework in India signifies a paradigmatic change with the objectives of augmenting operational effectiveness, refining decision-making procedures, and mitigating the difficulties linked to backlogged cases [1]. The utilisation of AI in various domains, including case management, forensic analysis, predictive enforcement, and risk assessment, has the potential to enhance investigative capacities and streamline the legal process. SUPACE, SUVAS, E-Courts, and LIMBS [4] are some of the initiatives that have explained the benefits arising out of AI in addressing pending cases, streamlining administrative processes, and increasing access to justice. Technological development has made it possible to create a legal framework that is more transparent, inclusive, and efficient. AI has the capability of significantly effecting and enhancing the operations of every field of the criminal justice framework [2].

However, this AI technology is not devoid of disadvantages and ethical deliberations. Because of the possible algorithmic biases, data privacy issues, and the importance of just and fair technology, diligent usage and constant supervision are crucial. The endless task of reconciling the benefits of AI along with safeguarding personal liberties remains a formidable one, bringing the importance of strong legislations, ethical principles, and public awareness [22].

However, it is ineffectual to debate that AI tools should not be used in the criminal justice framework or that their use should be prohibited. They will be used, and in the upcoming years, an increasing number of legal decisions will be made by algorithms. This research was mainly concerned with the use of AI in the criminal justice system and to prevent, predict, and detect crimes. In order to facilitate the development of a technologically sophisticated criminal justice system in India, a commitment to ethical, transparent, and accountable practices and a nuanced comprehension of the capabilities and limitations of AI are essential.

References

1. Tatyana S, Andrew S (2019) Artificial intelligence in the criminal justice system: leading trends and possibilities. Adv Social Sci Educ Hum Res 441:432–433. https://doi.org/10.2991/assehr.k.200526.062
2. Dan H et al (2020) A framework for the efficient and ethical use of artificial intelligence in the criminal justice system. Florida State Univ Law Rev 47:750–751. https://doi.org/10.31926/but.ssl.2023.16.65.3.5
3. Debakanta M (2023) Artificial intelligence in criminal investigation and criminal justice in Indian perspective. J Legal Stud Res 9(1):1–2
4. Ashutosh M, Ritu S (2023) Artificial intelligence in the indian judiciary: current applications and future prospects. J Namib Stud 33:3479–3482. https://doi.org/10.1109/ICCIKE58312.2023.10131795
5. National Judicial Data Grid (2024) https://njdg.ecourts.gov.in/njdgnew/index.php (last visited 22 May 2024)
6. National Judicial Data Grid (2024) https://njdg.ecourts.gov.in/hcnjdgnew/ (last visited 22 May 2024)
7. National Judicial Data Grid (2024) https://njdg.ecourts.gov.in/scnjdg/?p=home&app_token= (last visited 22 May 2024)
8. Muskan S, Vinit S (2023) Artificial intelligence and criminal justice system in india: a critical study. Int J Law Policy Social Rev 5(4):156
9. Isha S (2023) A legal study on role of AI in criminal justice system in India. J Emerg Technol Innov Res 10(8):680–682
10. Committee on Reforms of Criminal Justice System Government of India, Ministry of Home Affairs, available at: https://www.mha.gov.in/sites/default/files/criminal_justice_system.pdf
11. Arrears and Backlog: Creating Additional Judicial (wo)manpower, The 20th Law Commission was constituted for a period of three years from 1st September, 2012 by the Government of India, Ministry of Law and Justice, Department of Legal Affairs, available at: https://lawcommissionofindia.nic.in/report_twentieth/
12. Ministry of Electronics and IT, Government of India (MeitY), Available at: https://www.meity.gov.in/writereaddata/files/Committes_D-Cyber-n-Legal-and-Ethical.pdf
13. Developing Artificial Intelligence Solution for automation of scrutiny of cases in the Supreme Court of India, Available at: https://main.sci.gov.in/pdf/TN/24122020_044510.pdf

14. The Need for Responsible AI, NITI Aayog, Available at: https://www.niti.gov.in/sites/default/files/2021-02/Responsible-AI22022021.pdf
15. https://harvardlawreview.org/2017/03/state-v-loomis/
16. 1979 AIR 1369, 1979 SCR (3) 532
17. AIR SC 2012 642
18. (2006) 8 SCC
19. Artificial Intelligence (2020) Legal Change, and Separation of Powers, 88 U. CIN. L. REV. 1083, 1083–1084
20. Of Artificial Intelligence and Legal Reasoning (2001) 8 U. CHI. L. Sch. ROUNDTABLE 29:29–31
21. Artificial Intelligence in Legal Education under the Background of Big Data Computation, International Conference on Computation, Big-Data and Engineering (ICCBE) 51, 51–53 (2022), Available at: https://www.researchgate.net/publication/363683401_Artificial_Intelligence_in_Legal_Education_under_the_Background_of_Big_Data_Computation
22. Wickramarathna NA, Edirisuriya EATA (2022) Artificial intelligence & crime prediction: a systematic literature review. Social Sci Hum Open 6(1):334–337
23. Boddu HS, Naveen S (2023) Redefining the paradigm of the indian legal system through artificial intelligence. Nirma Univ Law J 12(2):63–65

Whose Right Is It Anyway? Challenging the Copyright Claim of AI in India

Shambhu Prasad Chakrabarty

Abstract AI has been around for more than 60 years but never so intensive was its claim to its creations. The availability, affordability and efficacy of AI in recent years have challenged humans in almost all domains of creativity. The absence of regulatory mechanisms across jurisdictions has given AI free passage to experiment and engage, in both constructive and destructive ways. AI has challenged the copyright regime as it did to other intellectual property rights. The amazing creativity of generative AI has jaw-dropped millions across the global community establishing its commercial potential. However, on the other, generated multiple questions including its ethical and legal standards including its reliability in many cases. This paper delves into the creative claims of generative AI and its plausibility to be recognized by law. Multiple questions arising thereof are also unravelled with an effort to evaluate the dichotomy, slowly but steadily crawling towards the courts for an answer.

Keywords Artificial intelligence · AI and copyright · Copyright law · IPR · AI and transparency

1 Introduction

Technology has always been the reason for social transformation. Artificial Intelligence (AI) is considered to be the next big thing the world is witnessing after the internet. This is not merely a perception to be debated upon but the ramifications of AI have raised considerable questions pertaining to the absence of and need for regulations. AI as a tool is gaining popularity in almost all sectors of human functioning. Like every technological advancement, AI has the potential to revolutionize a range of industries, albeit at a cost to the rights of stakeholders. This paper limits its exploration to the potential infringement generative AI is posing especially to the creators of art and copyrightable works. There are AI and deep learning enabled

S. P. Chakrabarty (✉)
Department of Law (IIIJS), University of Engineering and Management, Kolkata, India
e-mail: shambhuprasad.chakrabarty@uem.edu.in

M. Chakraborty et al. (eds.), *Proceedings of 5th International Ethical Hacking Conference*, Lecture Notes in Networks and Systems 1148,
https://doi.org/10.1007/978-981-97-8457-8_25

devices capable of recognizing voice commands and creating a playlist suitable to the listener. "Software today can create an artwork based on a model derived from 15,000 portraits over the past six centuries" [1]. Exploring Generative AI's commercial capability, a work generated by this software was sold for $400,000 in an auction [1].

Irrespective of the radical ramifications of AI, a reasonable path needs to be paved by not moving into the radical approach of Spain which promotes a strong regulation against AI nor the liberal line of Japan, which promotes AI uninterruptedly [2].

This paper unravels the debate of furthering innovations by promoting generative AI. The consequential ethical and legal infringements of the original creators whose actual hard work and creations are used to train the AI models to generate output, cannot be overlooked. This mode of functioning of generative AI raises strong ethical and legal concerns regarding the right to access copyrighted work created by human artists without their knowledge, permission and compensation. The current legal regime fails to clearly address these fundamental questions in several such cases. Whether the application of the existing legislations is interpreted strictly to restrict or liberally to accommodate is still not clear in many countries including India. The demand for transparency especially in generative AI has been demanded by many who are suffering and will continue to suffer due to the ramification of AI in this domain and the absence of a clear policy in this regard.

The relevance of the research undertaken is based on the question of the Copyright of AI-generated work, with almost no specific answer. It is important that laws be framed to accommodate AI with adequate compensation to the creators of work used by AI models to generate work.

In order to achieve the desired end of appreciating AI on the one hand and combating ramifications on the other, transparency shall be the key [3]. Once transparency becomes a mandate as it has in some laws promoted in some jurisdictions including that of the European Union, the copyright holders shall get the platform for ascertaining their rights. Adopting this policy may resolve the major issues involved in the ensuing debate.

There have been many instances where AI has outsmarted human intelligence. In 2016 Le Sedol was defeated by Google's AlphaGo [4] continuing the feat achieved by IBM Watson beating human TV quiz Jeopardy champion in 2010 [5]. This legacy can be traced back to 1996 when the world Chess champion was beaten by Deepblue [6].

On the application aspect, AI again has outperformed humans in safety standards in autonomous vehicles equipped with AI, healthcare in performing less invasive procedures, in economy as AI does great in online pricing and stock trading [7].

Anticipating a competing future between humans and machines equipped with AI, where humans may be prohibited from doing these works, is not unrealistic anymore. This hypothesis may be appreciated, but not in every domain. For example, the question about the capability of machines equipped with AI to replicate advocates and judges is a different ballgame altogether [8]. The use of AI in legal research and predicting court rulings is underway to some success. It comes with the criticism of inequality of access, and opportunity to access to same technologies in lawsuits due to economic and many other disparities.

The primary challenges involved in developing an AI-based "computer judge" are twofold. There is no conclusive answer with regard to the setting up of goals (fair ruling). The principles that may be determined to set the goal is important to ensure the justice system but determining the facets involved therein may be complicated and puzzling. To make things more complicated, determining the extent to which the goal was reached (understanding the extent of fair ruling) is even more complex to determine, let alone applying them.

2 Understanding Copyright in the Technological Landscape

As referred to earlier, this paper will focus on the copyright concerns induced by AI, it is important to understand the concept of copyright in the technological landscape. Copyright exists in a tangible form of expression of an original work of authorship [9]. The relevance of copyright multiplied manifold with the broadening of the technological landscape across the planet. So do the innumerable easier modes to infringe. The strength of copyright laws lies in preventing and punishing the wrongdoer and promoting creativity. It ensures the legal and moral rights of the owner the exclusivity to exploit the work and earn a livelihood over their creations [10].

3 AI and Copyright: Who is the Owner?

Ordinary use of technology to create a work is one thing while using generative AI to create a work is a different genre altogether. Work generated by the use of AI is independent and intuitive [11]. Creating works using AI could have very important implications for copyright law.

Using various tools to generate a copyrighted work is a *sine que non,* of the author. A painter using paint and brush on a canvas, and the writer using a pen on the paper. This does not preclude the creator from the creations. With the evolution of digital technology, creators were assisted by computers, photographers using digital cameras, software assisting and supporting the photographer with the choice to illuminate the picture taken by the camera according to their choice, darker or brighter. The decision has always been of the artist. The decision-maker was neither the software nor the machine, but only the artist. Thus, creative works using various tools qualify for copyright protection, subject to it being original and copyrightable. In major jurisdictions law acknowledges human authors as creators entitled to copyright protection.

In contrast, the latest genre of generative AI can actually make a decision (which is required to create an original work) without human intervention by using the algorithm inherent in the machine incorporated by the programmers on the basis of which the machine takes the decision. "Most jurisdictions, including Spain (IP Spain, 1987) and Germany [12] state that only works created by a human can be protected by copyright" [13, 14].

4 Allied and Consequential Challenges

Copyright laws are inherently human-oriented as is the majority of the laws with some exceptions. This objective can clearly be spotted in various determinations including the US copyright office and the US Supreme Court, where the term "human" is categorically included as creator.

"The U.S. Copyright Office has long granted copy-right only to works created with the significant involvement of a human hand or "author," a policy that the courts have routinely reaffirmed, says Tushnet" [14]. "The U.S. courts are generally in agreement that you need a human being sufficiently in the loop to have an author and a lot of AI-generated works are not that" [15]. It is also to be understood that there can be hybrid works as well where improvements or changes (significant or not) are done by human intervention to develop something. Such cases need to be delved into independently by law.

Recent guidelines by the US Copyright Office into this question states that only works of "human authorship is eligible for copyright protection" [16].

5 Copyright Protection of Generative AI: How It All Started and the U-turn!

AI-generated work has never been the subject matter of copyright protection until the US Copyright granted copyright to one such application which opened the Pandora box of debate and discussion in this area. The said case has been illustrated in brief hereunder:

5.1 Case Study 1. Zarya of the Dawn: The U-turn on AI by US Copyright Office

Kris Kashtanova who was granted copyright for "Zarya of the Dawn" last year in September 2022 was considered to be the first AI-generated copyrighted work and the news rocked the digital world by storm. However, a little more than a month later, the

Fig. 1 AI generated picture of Zarya [19]

US Copyright Office required her "to show that there was substantial human involvement in the process of creation of this graphic novel [17]". The office later cancelled the registration process on the ground that "the information in [her] application was incorrect or, at a minimum, substantively incomplete [18]" (Fig. 1).

5.2 *Case Study 2. AI-Generated Selfie of Prominent Personalities*

Very recently there was a set of AI-generated historical figures available on the internet, taking selfies including prominent figures like Mohandas K Gandhi, Arthur Luther King and many others [20]. These pictures have got the much-needed sparkle from the existing pictures, sketches and diagrams of these famous people, available on the web, and in many cases picked up by the AI algorithm from paid archives without any permission, let alone any payment. The pictures thus generated by AI, how attractive it may look, eventually falls short of the basic legal, ethical and moral standards to be considered as "work" under any legal definition (Fig. 2).

6 Contemporary Issues

AI has some inherent dilemmas. The issues of authenticity, reliability and transparency are the three commonly raised fundamental flaws which require specific introspection.

Fig. 2 AI-generated picture of MK Gandhi [21]

6.1 *Bias in AI*

AI has a few grey areas that need serious retrospection from a legal perspective. The potential bias [22] inherent in algorithmic decision-making cannot be ignored in light of various misuse of AI. A significant issue arises when government officials start depending on AI predictions to determine aspects which influence and affect ordinary people. As AI depends on an underlying computer model developed by a set of computer programmers which predicts on the basis of embedded data, a set of biased data inputs can be fatal to the viability of such predictions and be rebutted in a court of law very easily. "Multiple critics have raised the possibility that computer models that learn patterns from data may be subtly biased against certain groups based upon biases embedded in that data" [22].

6.2 *Interpretation and Transparency*

Another interesting issue that the law delves into with its interaction with AI is the basis on which AI systems come to a decision and how transparent the decision-making mechanism is, leading to its interpretability. AI systems have been alleged to be non-transparent due to their inability to interpretability. In many cases, even the programmers who created the AI system fail to identify the underlying mechanism of interpretability. One set of critics demands that the "AI mechanism must be explainable, interpretable or at least transparent" [23]. The other set of critics encourages the AI systems to be developed in such a way that they will automatically produce explanations for the reasons for such decision-making [24]. This, in turn, allows the lawmakers to determine the legality of the "reasons" on which AI has drawn its decision.

6.3 Are All Generative AI Violative of Copyright?

Even when the Large language models (LLM) engages the AI with the existing human works, all generative AI output will not constitute copyright infringement. If the LLM is developed and deployed in such a way that it will only use those data which are outside the ambit of copyright protection, [25] e.g. publications beyond the copyright protection period, e.g. literary works beyond the lifetime and 60 years of the author or a photograph used beyond the protection period of 60 years.

The principal validates the legality of such practices across IP beyond their period of protection.

Interestingly all exceptions referred to in the copyright laws can provide further impetus to the legality of works produced by generative AI.

7 The Turning Point! Will Transparency in Generative AI Be *Lex Loci*?

Transparency may be considered an indispensable aspect of any AI policy. The current generative AI is strong enough to complicate the transparency aspect which will only be stronger with its forthcoming functions. Needless to say, recent AI-related regulations have focused on transparency. The "European Declaration on Digital Rights and Principles for the Digital Decade" in its preamble clearly states that it will "promote a European way for the digital transition, putting people at the centre" [26]. In the chapter "Interactions with algorithms and artificial intelligence systems" there has been a commitment by the European Parliament, the Council and the Commission together to "ensuring transparency about the use of algorithms and artificial intelligence, and that people are empowered and informed when interacting with them" [26]. In other words, the requirement of algorithmic transparency is to ensure transparency to the rights holder. This right may affect developers of AI tools as well as the users of such tools. The former enforces the disclosure of how the generative AI tool functions and the latter (the users) the obligation to disclose to the public the level of AI tools used to develop a work.

7.1 European Directive on Copyright in the Digital Single Market (2019/790)

The directive ensures that the authors and performers who have licensed their work receive information from the licensees. Article 19 provides "*up to date, relevant and comprehensive information on the exploitation of their works and performances from the parties to whom they have licensed or transferred their rights, or their successors in title, in particular as regards modes of exploitation, all revenues generated and remuneration due"*. Article 17 para 8 further clarifies that the "*Member States shall provide that online content-sharing service providers provide right-holders, at their request, with adequate information on the functioning of their practices with regard to the cooperation referred to in paragraph 4 and, where licensing agreements are concluded between service providers and right holders, information on the use of content covered by the agreements*" [27].

7.2 More Transparency: The US Copyright Office Directives

The potential of generative AI to overwhelm the existing legal and ethical framework has prompted the inclusion of algorithmic transparency to distinguish between human and non-human creations, which will be considered as an inherent factor as to the grantworthiness of IP protection. The recent US Copyright Office guidelines on "Copyright Registration Guidance: Works Containing Material Generated by Artificial Intelligence" states that,

> *"[a]pplicants have a duty to disclose the inclusion of AI-generated content in a work submitted for registration and to provide a brief explanation of the human author's contributions to the work."* [28]

8 Are Machines Liable?

One of the cardinal challenges AI has posed to law is the personhood theory. To date, no uniformity could be achieved in this issue. Personhood as a concept is notional in nature. The importance of a legal person is primarily threefold, determination of responsibility, rights and commercial exchanges. Personhood theories have also evolved with time paving the way for artificial legal persons, e.g. corporations. The AI revolution has now engaged in the question of "electronic personality". To achieve clarity, this part of the paper will delve into the personhood principles and machines; the mens sit rea of legal liability and machine liability as a misnomer in the copyright domain.

8.1 The Personhood Principles and Machines

The concept of legal personhood is a jurisprudential question that has its answer in the philosophical understanding of human understanding. De jure and de facto persons differ exactly on the point of liability. A person de facto, may not be a person *de jure* and falls short from the purview of legal liability as in the case of a minor or a person of unsound mind. Again, a person who is not a natural person can be a juristic person, as in the case of a Company under the Companies Act, 1956 (India). A company is considered as a separate legal entity from its shareholders and is capable of having its own property and the capacity to sue or be sued. Its decision is its own and the law treats such entities as juristic persons. In a similar line, European Institutions have been discussing seriously the capability aspect of AI having the potential to be a new specie, a juristic person. The report drawn on Civil Law Rules discussed in 2016 in the EU Parliament proposes legal status as, "electronic person" to robots which are self-learning, autonomous robots making them unpredictable [29].

Irrespective of the debate that ensued and the complicated questions that arose, it has become certain that these entities can make decisions and at times anticipate better than human brains. Moreover, if AI can do so, society would most certainly want to regulate them. It is the interaction of AI with society, which invokes the need for laws to regulate such relationships. As long as the interaction is absent, there is no need for the law to be active. Ethical considerations would have been sufficient in such cases. However, the fact is, AI has engaged in every facet of human life. AI like all other technological marvel is capable of good and bad. AI which was often thought to be a great promise to transform our world for good, has met the road of hell, which was paved with good intentions. As robotic surgery is a boon to human civilization, an autonomous weapon system, is surely a bane. Thus, the law is needed to determine the indulgence and promotion of AI in some domains, with absolute prohibition in others.

8.2 *The* Mens Sit Rea *Principle of Legal Liability*

The increasing sophistication of technology and the availability of affordable generative AI has created a situation where human intervention may not be essential to create a work which can be copyrighted. The element of intelligence required to develop copyrightable work is now available in machines. The idea of owning a copyright comes with the challenge of liability in case of infringement being done by such machine of another copyrightable work. The current legal framework requires either explicit or implicit mental state requirements to divest liability. *Actus reus nisi mens sit rea* has been the long-standing principle for imposing sanctions on the wrongdoer. Thus, to incur liability, the machines will require the capacity to operate in an independent mental state. Can the intelligence in AI fit in the idea of "mental state"?

> *"Although the law has yet to fully grapple with questions such as whether machines are (or can be) sufficiently humanlike to be the subjects of law, philosophers have long contemplated the nature of machines"* [30].

8.3 *Machine Liability! is It a Misnomer in the Copyright Domain?*

Machines have been used for copying without the knowledge of the copyrightability of the subject matter. For example, if someone buys a Spiderman DVD (of Sony corporation) and writes the content on a blank DVD disk (of Sony corporation) in his Sony Vaio laptop (of Sony Corporation) and distributes it amongst his friends for paid viewing, is there an issue of copyright infringement?

Today, with the advancement of technology the situation has become far from simple. Machines are now capable of engaging with the choice of the consumers and use their algorithms to provide contents which may very well be copyrightable works.

In the former situation, the person copying the DVD may be held responsible, but for the latter, the situation gives rise to a set of complicated questions primarily because machines cannot function independently without the programs and algorithms set for them to function. Thus, the DVD copier/writers are not illegal devices (misfeasance and not malfeasance) and the user is liable. For the machines the creators of that software or the owner of the machine which is capable of copyright infringement is answerable.

> *"In the copyright space, we might wonder whether technology creators or owners can be directly liable for copyright infringement when a bot fetches an infringing copy of a song in response to a user's request for that song or when software taught on portraits produces an artwork that is copied from and substantially similar to an existing portrait on which the software was trained"* [30].

9 The Legal and Ethical Landscape Across the Planet

Amidst a variety of confusing and rather conflicting legal positions, the UNESCO ethical guidelines of 2022 [31] have brought some clarity into the formational ethical standards to be followed while dealing with AI. The European Union Artificial Intelligence Act, 2024 has introduced a plethora of legal changes in EU in the next few years, depending on the acceptability of EU members [32]. The position in the US and China is bifurcating while India is observing the global dynamism of AI laws very closely.

10 Conclusion and Recommendation

The legal landscape of Generative-AI as of date is blurry and bifurcative. AI has initiated a paradigm shift in the way people live their lives and articulated a social transformation for the law to evolve and engage with novel challenges not confronted before. The domain of Intellectual Property Rights is also evolving with the fast intervention of AI. WIPO is coming out with policy guidelines to address the volatility induced by AI. The IP policy toolkit, "Getting the Innovation Ecosystem Ready for AI" [33] is one of the latest. In order to achieve the desired end of appreciating AI on the one hand and combating ramifications of generative AI on the other, transparency shall be the key.

Transparency will lead to licensing opportunities and accountability in case of unauthorized and unlicensed use of copyrighted works. This will consequently promote a system of accountability into the system. Once transparency becomes a mandate in some laws (e.g. the EU AI Act, 2024) promoted in some jurisdictions including that of the European Union, the copyright holders shall get the platform for ascertaining their rights. Adopting this policy may resolve the major issues involved in the ensuing debate. India is taking a close look into the legal development in this intersecting domain of AI and law. We may expect a legal intervention in this regard from the Indian Copyright Office very shortly.

References

1. Is Artificial Intelligence Set to Become Art's Next Medium? Christie's (2018) https://www.christies.com/features/A-collaboration-between-two-artists-one-human-one-a-machine-9332-1.aspx [https://perma.cc/R6NB-RU5F]
2. Molina ÓA, Bernal MJ, Wolf DL, Herreros B (2024) What is Spanish regulation on the application of artificial intelligence to medicine like? Hum Social Sci Commun 11(1):1–7
3. de Werra J (2023) AI transparency: an emerging principle in the IP ecosystem? J Intellect Property Law Pract 18(6):407–408
4. Curran NM, Sun J, Hong JW (2020) Anthropomorphizing AlphaGo: a content analysis of the framing of Google DeepMind's AlphaGo in the Chinese and American press. AI Soc 35(3):727–735
5. Thompson C (2010) What is IBM's Watson. New York Times Magazine (June 2011). http://www.nytimes.com/2010/06/20/magazine/20Computer-t.html
6. Higgins O, Short BL, Chalup SK, Wilson RL (2023) Interpretations of innovation: the role of technology in explanation seeking related to psychosis. Perspect Psychiatric Care. https://doi.org/10.1155/2023/4464934
7. Yang H, Liu XY, Zhong S, Walid A (2020) Deep reinforcement learning for automated stock trading: an ensemble strategy. In: Proceedings of the first ACM international conference on AI in finance, pp 1–8
8. Koenig ML, Oseid JA, Vorenberg A (2018) Ok google, will artificial intelligence replace human lawyering. Marq L Rev 102:1269
9. Buccafusco C (2016) A theory of copyright authorship. Va L Rev 102:1229
10. Hansmann H, Santilli M (1997) Authors' and artists' moral rights: a comparative legal and economic analysis. J Leg Stud 26(1):95–143
11. Chen V, Liao QV, Wortman Vaughan J, Bansal G (2023) Understanding the role of human intuition on reliance in human-AI decision-making with explanations. Proc ACM Hum Comput Interact 7(CSCW2):1–32
12. Bibliotheksverband D (2016) Bibliotheken fordern Modernisierung des Urheberrechtsgesetzes. Bibliotheksdienst, 50(7)
13. Soroka L, Danylenko A, Sokiran M (2022) Legal issues and risks of the artificial intelligence use in space activity. Philos Cosmol. https://doi.org/10.29202/phil-cosm/28/10
14. Andres G (2023) Artificial Intelligence and Copyright, available at, https://www.wipo.int/wipo_magazine/en/2017/05/article_0003.html, accessed on 17 July 2023

15. Colleen W (2023) AI: delving into the legal and ethical challenges of a game-changing technology. Harvard Law Bull 72(2), accessed on 7 July 2023, available at https://hls.harvard.edu/wpcontent/uploads/2023/06/HLB_Su23.pdf
16. Karl H (2023) OpenAI CEO briefs house judiciary on automated copyright issues. accessed on 7 July 2023, available at https://communicationsdaily.com/news/2023/05/18/OpenAI-CEO-Briefs-House-Judiciary-on-Automated-Copyright-Issues-2305170045
17. Franklin G (2023) Copyright office pilot public records system mistakenly reflects cancellation of registration for AI graphic novel. IP WATCHDOG (2023), available at, https://ipwatchdog.com/2023/01/24/copyright-office-publishes-retracts-official-cancellation-registration-ai-graphic-novel/id=155686/. [https://perma.cc/3QMV-NY5C]
18. Letter from Van Lindberg (2022) Taylor English Duma, LLP, to Kasunic RJ, U.S. Copyright Office (Nov. 21, 2022), available at, https://drive.google.com/file/d/1Idhn8eb9t883mm_U4CxAQQ_aANTI7UTX/view [https://perma.cc/RD4U-YXUV]
19. Gray J (2023) Copyright eligibility for AI-generated images: the threshold of human creativity. Gonz. L. Rev 59: 507
20. AI-generated images of Mahatma Gandhi, Mother Teresa, Jawaharlal Nehru and others clicking selfies stun the internet, Times of India, March 22, 2023 available at, https://timesofindia.indiatimes.com/etimes/trending/ai-generated-images-of-mahatma-gandhi-mother-teresa-jawaharlal-nehru-and-others-clicking-selfies-stuns-the-internet/articleshow/98907706.cms
21. Mehul D (n.d.) Gandhi and Bose in the time of AI: artist imagines world leaders from the past taking selfies. Firstpost, Available at, https://www.firstpost.com/photos/artist-imagines-world-leaders-from-the-past-taking-selfies-using-ai-12332682.html, accessed on 27 October 2024
22. Tene O, Polonetsky J (2017) Taming the Golem: challenges of ethical algorithmic decision-making. NCJL Technol 19:125
23. Hall P (2016) Predictive modeling: striking a balance between accuracy and interpretability. Retrived from https://www.oreilly.com/ideas/predictive-modeling-striking-a-balance-between-accuracy-and-interpretability
24. Luke J (2018) AI is useless until it learns how to explain itself, Towards Data Sci (4 Jan 2018), https://towardsdatascience.com/ai-is-unless-until-it-learns-how-to-explain-itself-7884cca3ba26 [https://perma.cc/PK3D-FFUW]
25. Samuelson P (2015) Justifications for copyright limitations & exceptions. Copyright law in an age of limitations and exceptions
26. European Declaration on Digital Rights and Principles for the Digital Decade' 2023/C 23/01, PUB/2023/89. Available at https://eur-lex.europa.eu/legal-content/EN/TXT/?uri=OJ:JOC_2023_023_R_0001 (accessed 10 Apr 2023)
27. Dusollier SVC, Benabou VL, Bently L, Derclaye E, Dreier T, Geiger C, Xalabarder R (2020) Comment of the European copyright society. Addressing selected aspects of the implementation of articles 18 to 22 of the directive (EU) 2019/790 on copyright in the digital single market. J Intellect Property Inf Technol Electron Commer Law 11(2):132–148
28. Copyright Registration Guidance: Works Containing Material Generated by Artificial Intelligence, (2023-05321) 88 FR 16190, available at, https://www.federalregister.gov/documents/2023/03/16/2023-05321/copyright-registration-guidance-works-containing-material-generated-by-artificial-intelligence, 16190–16194
29. European Civil Law Rules in Robotics, Directorate-General For Internal Policies, Policy Department C: Citizens' Rights And Constitutional Affairs Legal Affairs, available at, https://www.europarl.europa.eu/RegData/etudes/STUD/2016/571379/IPOL_STU(2016)571379_EN.pdf, accessed on, 10 May 2024
30. Chatterjee M, Fromer JC (2019) Minds, machines, and the law. Columbia Law Rev 119(7):1887–1916
31. Resseguier A, Rodrigues R (2021) Ethics as attention to context: recommendations for the ethics of artificial intelligence. Open Res Eur 1:27
32. Neuwirth RJ (2023) Prohibited artificial intelligence practices in the proposed EU artificial intelligence act (AIA). Comput Law Secur Rev 48:105798

33. Getting the innovation ecosystem ready for AI, An IP policy toolkit (2024) WIPO, available at, https://www.wipo.int/publications/en/details.jsp?id=4711#:~:text=An%20IP%20policy%20toolkit&text=The%20purpose%20of%20this%20IP,as%20AI%20becomes%20increasingly%20autonomous, accessed on 15 May 2024

The Convergence of Artificial Intelligence, Internet of Things, and Healthcare: Legal Considerations for a Connected Future

Jayanta Ghosh, Vijoy Kumar Sinha, and Ardhendu Sekhar Nanda

Abstract AI and IoT may supplant human actions, and results have had a more profound impact on bioethics across the globe. Technology-based healthcare using AI and IoT has exploded in the global health business. Human-to-human care has succeeded in many domains, from industrial efficiency to human life. Gaps formed between the human psyche to allow AI and IoT may be scrutinized, and ethical problems raised. Initial testing revealed loopholes in meta-data collecting, threatening digital privacy, and disrupting healthcare protection and insurance. A minor error can cost a human life. Low risk of physical injury, precise outcomes, and speedier healthcare data governance make IoT services popular. Big data ensures unpredictability in data handling cloud storage solutions, which may be very confidential. Government regulations may struggle to maintain unique health IDs and considerable tracking of health services. The newest applications of mind-based deductions to real life put both parties' mental health at risk. The global market for digital healthcare has surpassed 5.5 billion USD. The healthcare industry is trying hard to replace human intervention with human growth and development. AI and IoT technologies are available to aid healthcare practitioners with administrative activities, clinical documentation, patient outreach, image analysis, medical device automation, and remote patient monitoring. This research will examine AI and IoT in healthcare, both from a practical and international legal standpoint. The international legal scenario will help to make a synthetic analysis for future legal directions.

Keywords Human · Healthcare · IoT · Artificial intelligence · Data · Technology · Law

J. Ghosh (✉)
The West, Bengal National University of Juridical Sciences, Kolkata, India
e-mail: joyantaghoshcool@gmail.com

V. K. Sinha
Centre for Regulatory Studies, Governance and Public Policy, The West Bengal National University of Juridical Sciences, Kolkata, India
e-mail: vks161991@gmail.com

A. S. Nanda
National Institute of Securities Markets and Maharashtra National Law University, Mumbai, India

M. Chakraborty et al. (eds.), *Proceedings of 5th International Ethical Hacking Conference*, Lecture Notes in Networks and Systems 1148,
https://doi.org/10.1007/978-981-97-8457-8_26

1 Introduction

AI and IoT influence nearly every facet of modern life, from entertainment to commerce to healthcare to education. Everyone uses Netflix and Amazon, and they all know what movies and shows their customers enjoy [1]. Google also knows what symptoms and ailments its customers are searching for themselves. This information may be utilized to create a highly personalized profile of an individual, which can be extremely useful for understanding and targeting behavior and anticipating healthcare trends shortly [2]. People are pretty optimistic about the potential benefits of using artificial intelligence (AI) in healthcare, from diagnosis to therapy. In many activities, AI algorithms currently outperform humans, such as analyzing medical imagery or correlating symptoms and biomarkers from electronic medical records (EMRs) with illness definition and prognosis [3].

With an ever-increasing need for healthcare services, many nations face a lack of medical professionals, particularly physicians. Aside from trying to stay up with new technical breakthroughs, healthcare institutions must also contend with patients' increasing demands for higher quality service and results from their providers, as seen by the success of consumer products such as Amazon's and Apple's [4]. A new way to offer healthcare has been enabled by wireless and smartphone developments in health tracking applications and search platforms and a new way to communicate with healthcare providers via remote interactions, which is accessible from anywhere and at any time. Underserved areas and locales without experts can benefit from these services, cutting expenses and protecting patients from contracting infectious diseases. Development nations with increasing healthcare systems and healthcare infrastructures tailored to match current demands might also benefit from telehealth technology [5].

AI-driven solutions are becoming crucial to the healthcare ecosystem in next-generation healthcare technologies. AI has the capacity to enhance almost every aspect of healthcare delivery and operation. AI's capacity to decrease expenses is a substantial incentive for its adoption in the healthcare industry. Current projections suggest that AI applications would lead to a yearly reduction of USD 150 billion in US healthcare spending by 2026 [6]. Reactive to proactive healthcare models, which emphasize health management over illness treatment, have resulted in substantial cost savings. Reduced hospitalizations, medical visits, and therapy are all predicted as a result. AI-based technology will be crucial in continuously monitoring and guiding people, allowing for earlier diagnosis, focused treatment, and more efficient follow-ups, thereby helping to preserve their health. The AI-associated healthcare market is projected to reach USD 6.6 billion by 2021, with a compound annual growth rate of 40% [7]. This research can potentially transform the healthcare system and the role of AI and IoT in human and machine conflict. It will be a path-breaking techno-legal research analysis that leads to a comprehensive legal scenario for the future scope of the technocrat community.

2 Challenges in the Healthcare System

Vulnerabilities in the healthcare system can be discovered via AI & IoT. Even though this technology can be helpful for people looking to remedy a system's vulnerabilities, attackers also utilize it to find and exploit holes in their target's defenses. The amount of vulnerabilities that attackers may exploit has increased, including zero-day vulnerabilities, as technology, especially those with poor security requirements, such as IoT devices, surges in use [8]. Attackers frequently utilize artificial intelligence (AI) to find and exploit vulnerabilities far more quickly than developers can fix them. It should be noted that developers are at a disadvantage when trying to secure a system or device because they must find and fix every possible vulnerability, whereas attackers only need to find one. This makes automatic detection a vital tool for attackers, but developers can also use it [9].

The Healthcare industry has significantly improved its technology in the last couple of years [10]. The use of AI has changed the whole system of hospitals around the globe by helping physicians make intelligent decisions and improving the accuracy of viewing patients' reports. As the healthcare industry increases daily, it is becoming an attractive place for cybercriminals to attack. As other IT industries have faced many data breaches by cybercriminals, the same goes for the healthcare industry. In 2013, Kaiser Permanente, a healthcare provider in the US, announced that its patient health information was compromised after a USB flash drive was stolen [11]. In 2012, the company's forensic investigation section identified more than 47,000 security issues. Additionally, it identified 621 verified breaches [12]. Recent findings indicate that about 94 percent of hospitals in the United States have seen at least one security breach during the last two years [13]. In addition to this, the report also states that most attacks originated from China, the US, and Eastern Europe [14].

Previously, the healthcare industry used to keep records in thick folders, and now the patient can access their medical histories and their test results from online portals, leading to many vulnerabilities [15]. As healthcare evolves with new technologies and legislation, the security threat to personal data is also changing. Here are some significant healthcare data security challenges.

2.1 Transfer of Electronic Health Records

Previously, the health records of patients were kept physically. Nowadays, most healthcare institutions have adopted the Electronic Patient Record System to keep a record electronically so that doctors can share the data of patients [16]. This sharing of data between them in a network that stores extensive medical data creates a tempting opportunity for cybercriminals. Many countries have expanded their patients' privacy rights with the help of data protection and created a new burden for network providers to maintain compliance and healthcare data security.

2.2 User Error

Another challenge is the simple patient-user error where they access their medical records from the provider's portal and store that data in unencrypted folders of the cloud or the patient shares that record with another person via email, which creates a bridge for the hackers to access the patient's data [17]. Usually, the user is not being cautious about this while sharing. The user must be mindful of what you are storing and where you are storing with solid encryption.

2.3 Cloud and Mobile Adoption

The majority of healthcare institutions store the data in the cloud, leaving the patient data prone to vulnerabilities to data thieves.

2.4 Outdated Technology

Running a hospital is not an easy task. It is costly when we prioritize buying the latest MRI technology and increasing the staff to meet the demands and fall short of the budget [18]. Due to a shortage of funds, it is impossible to maintain the new technologies and purchase new servers, leading to a data breach as it becomes outdated.

2.5 Maintaining Compliance

Many countries have the responsibility to maintain compliance. Like in the United States, the HITECH Act offers an incentive for EHR and HIE adoption and creates a responsibility to maintain compliance, and healthcare providers have to notify their patients if there is a data breach in their network [19]. In addition, healthcare institutions have to comply with their laws and other data regulations, whichever applies to them accordingly.

2.6 Privacy-Preserving

The encroachment upon patient confidentiality is an increasing worry in this era of advanced technology. The healthcare sector uses Internet of Things (IoT) devices to transmit vital signs to healthcare cloud platforms. Due to the lack of robust

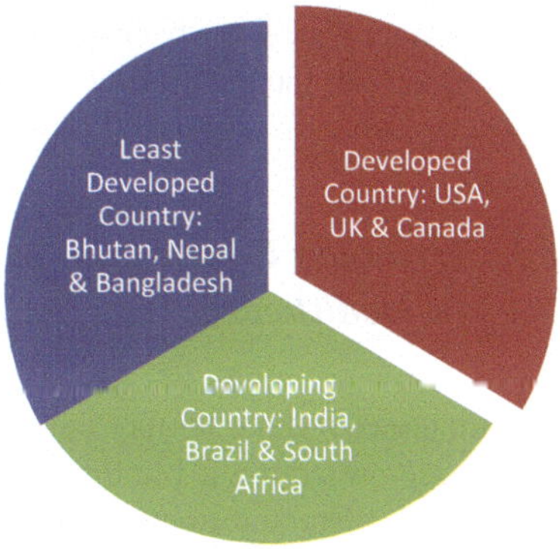

Fig. 1 Sample of developed countries, developing countries, and least developed countries

encryption, the storage of sensitive information on the cloud becomes an enticing target for data theft by attackers [20]. Forbes magazine revealed an event that highlights concerns about patient privacy. Target Corporation discreetly mailed baby care vouchers to an adolescent girl without the knowledge of her parents, as stated in the report. This occurrence underscores the need to carefully evaluate the protection of patient privacy.

3 Methodology

The researcher has taken the doctrinal approach to make this study. For better authentication of the research, a comparative model analysis was applied. This model approach has justification for the nature of the country's status that uses AI and IoT for healthcare. These stages of policy analysis have been taken care of with developed countries, developing countries, and least developed countries. (Fig. 1).

3.1 Sample

See (Fig. 1).

4 Discussion and Results

AI & IoT' smartness, or intelligence, may be found at micro and macro scales. Even though these seem like a future wave of self-driving cabs and talking refrigerators, they are not. Data, devices, and connections are becoming essential considerations for intelligent things. Big Data Analytics can assist in uncovering hidden insights from the data once it has been analyzed [21]. Therefore a dynamic level of discussion helps figure out the position of AI and IoT with human and machine interference.

4.1 Dynamic Connection of Machine and Human in Healthcare

Ever imagined robotic surgery inducing a complicated procedure on the human body? Healthcare protection is the first bay that makes us rely on a symbiotic relationship between patients and doctors, but the multidisciplinary growth in the technology is uncontrollable. AI has attracted varied interests from the medical industry with 'maximum benefits and minimum risks'. Financial boost urged the necessity of over-passing ethical implications via the healthcare passage and miracles in paramedics. Inculcating mobile or remote healthcare options fixed the urge of ailing patients. Even mental health is kept balanced after introducing IoT-based services when probable patients can constantly keep their health under the radar of watch. Health privacy is snooping on the individual's interest to keep the data limited to watchable social spheres for the patient's well-being.

Hard drives come with the commercialization of the medical industry, where bioethics is required. It is being challenged all over the world. One of the worst things related to ransomware attacks is some hospitals in the NHS, UK, and Canada. Surprisingly, the attacker's Cryptocurrency payments are considered the best form of cryptocurrency unlock system. The rise of social media due to vulnerability in health-care IT infrastructure increased the severity of cyber-attacks. The medical technology industry is also facing. Problems with rough handling of plates because subscriptions to medical technology are proportional to income. The whole of human existence is undermined by defining equality as a necessity. Without quick and widespread trust, terrorism is possible or worse. No one can stop. With every second of service delivery, if technology is not tested, medical care. This can have a negative impact on business and can be decentralized in the local language. The judicial system is highly biased toward the rich, and the poor are particularly difficult to strike [22].

4.1.1 Robotic Merits

Three major push factors: rising healthcare demands, exorbitant costs, and a struggling workforce have paced up the medicos' AI interference. Diplomatic relations between industry experts and medical experts got substantial innovations. Taking surgery as a particular medical procedure performed upon the human body, robot assistance can lead to accuracy and effective results in a disclosed manner. Smart bots are a boon to the human dynasty, where every human being receives equal attention, which amounts to 'basic care' in human psychology. Heuristic algorithms call for automated tasks where the AI shakes hands with IoT in superior pricing.

Cognitive AI-based software can automate workload, and it has also helped during the difficult times of the COVID-19 pandemic [23]. Hospitals in the United States are well-equipped with camera-based software that can deduce patients' temperature before entering the hospital, and chatbots can show the way for checkups. Sensor-based devices can screen the fever of patients and manage around 4000 patients in a row [24].

Interconnected 'effects' and robots outside of product lines are generally new, and enactment erecting good prerequisites was for the utmost part intended for effects working in disposition, for the utmost part in ultramodern conditions. Corrections of these enactments, primarily the General Product Safety Directive [25], are just planned formally. Online protection and promising enterprises are also constantly tended to in independent bits of enactment, as though policymakers neglected to perceive the connection between network protection and security on account of digital factual fabrics, including particulars or clinical widgets. Administration robots cooperate with people and, in the medical care area, guests are regularly in a weak position, making these troubles more introductory. For this case, a teleoperated careful robot has been addressed by judges, and actual damage may have been the outgrowth of a nocuous programmer who had finished this [26].

Cyber laws assent the applicable provisions, policies, and preventive remedies in the healthcare sector. Irrespective of spam, machine errors, and mismanaged robotic tasks, the IoT platform controls the allowance of algorithmic commands, which the healthcare service provider allows to comply with the policies guided under the lex terrae. Galloping experiments in the robotic world are in the secondary stage of development, where the innovation costs are pretty high to commercialize machines that can be available at a higher pace. Moreover, every experiment would come out with a new inner framework of algorithms and structure, which may over cross the IP of another or mismatch the laws of the State in a high probability [27]. There will not be a guiding force unless there is a technology transfer agreement between the technologically forward states, regulated corporations into these robotic behaviors, and commerce houses. As humans live peacefully under the rule of law, robots are 'quasi-humans' programmed artificially for support, and there must be a stricter regime to control them because of their 'derived nature'.

4.1.2 Oxygenation of Technological Future

Parallel to robotic-assisted Medicare, the 'attention' element or 'observation' is fulfilled by additional help from nurses and midwives. Skilled workers via consensual working regulations may go undetermined apart from the agreed terms and conditions. We have out-of-scope conditions when extra help becomes late, unsatisfactory, or unskilled to provide proper care. Transforming healthcare promotes efficient and productive resource bots who can look after the children, adults, and elderly in a scheduled format. The machines can be made available round the clock at a signal, gesture, expression, or any other mode comfortable to the patient. Robots are data-driven technologies, and a cyberattack may compromise the adequacy of the robot's operation and the users' safety in any possible manner [28].

Rehabilitative robots who can impart knowledge, care for the patients, and converse cohesively are essential for keeping regular one-on-one notes with mentally challenged patients. These can support a short-term, long-term, palliative, or productive healthcare effort at centers, homes, or hospitals. Care assistive robots are helpful in elderly couples to manage their lifestyle as per their needs, medical interventions, and regularized decisions on healthcare logistics. Aged couples or single living human beings need a constant emotional push to live longer on an optimistic note, free from any risks of social tension, mental harassment, time-based care patterns, and so on. The robotic elements encourage formal commands, setups, and checkups via one common hardware connected to the web and IoT helps decide on required medications, diet, and water requirements [29].

'Emotional AI' is the latest trend required by individuals, retailers, and hospitals. Recognition of one human being by one machine is essential to deduce loyalty and reflect the customized needs or requirements. Medico skills are 'high touch engagement,' which illustrates a symbiotic relationship or 'general trust' between the doctor and the patient. Extremely vital, creating positive notes via the 'jolly mode' of both is essential for medical success, unless the patient is too chronic. Hasbro, an AI cat, has been developed under the Affordable Robotic Intelligence for Elderly Support (ARIES) project under the grant of USD 1 million to give palliative care to the elderly. Supporting aging couples is a noble task, and technology has smartly curbed the minimization of human support [30].

4.1.3 Medicinal AI

IP and technological agreements form the root and base of the robotic assistance or IoT device-based service, including oxygen count, blood pressure measurement, and further. Patents and trade secrets are inherent calls for organizations that may lead to the formation of cartels. Legally accepted, but morally preemptive, that competition would energize the public to seek more information and the government intervenes in reality. Chemical support of drugs is irreplaceable because of gravitational medical knowledge, but an attempt to avoid it is also quite imbalanced. Quantitative reduction of drugs is a fair call, but one-time vaccination is the key to the current requirements. Based on such IP support, if the IoT devices are made with such chemical detecting sensors in the body, it will be pretty practical to analyze the blood component now and then via hand-strapped scammers under the mode of 'watch' [31].

Medicinal malpractice is a criminalized offense, and medical professionals' detection of medical negligence is deeply regretted later. On that note, the trust of human beings floats on the surface if they 'need to visit doctors' or 'remote help and assistance' (without surgery) can ease the work. Medical science does not guarantee utmost care or wellness, but they always try to bring down the side effects and commonality of disease by treating it in various ways possible. For example, reducing the seriousness of Cancer via multiple therapies, drugs, and treatment has made it less complicated than before 2000. IoT can contribute in the form of data storage, analytics, and deductive observations [32]. The medical experts perform the next step on how to proceed with the steps and suggestions to heal the patient. Human progress concerning time is restricted to boundaries because of liabilities, but AI has tripped the time gap and human innovation.

4.2 *Societal Challenges of AI and IoT in Digital Health*

In the AI revolution, we live in a hybrid world of algorithmic language overlapping our natural language. Codes help transmit adaptable functions, and mechanical energy is put into action to complete an assumed utilitarian task. Healthcare has been considered an impact area where innovation knocks the door of patients to diagnose, treat, and nourish in a practical scheme. Health inequality in social, economic, and geographical is running with huge gaps, irrespective of the development status of the countries. WHO has the least focused treatise on 'technologically balanced healthcare devices'.

Spam fear threatens service-related investments and trusted applications shield the spam by detecting the same via AI. Google has developed innovative AI to protect consumers from ads and phishing messages [33]. Malware software should be avoided to protect patients' health interests in all applications without any chance of vulnerability or bad profits. Data mining should be a restricted activity when it comes to medical diagnostics [34]. As per the financial capacities, the recitation of 'basic' medical care varies from socially and economically developed areas, societies, and the public.

4.2.1 Liberally Medical AI vs. Conservative Methods

Justice is as it is when one would look at the sky and place the cases of medical laws side by side. The medical profession begins with Hippocrates' oath that puts them into a medical career's 'fundamental duties'. Oath, being restricted to human beings only, the machines do not understand the terms and conditions directly or at par with the same liabilities. Doctors who pass on responsibilities are themselves liable for the growth in their medical career by 'management techniques,' which are curbed by the machines or devices apart from secondary human help. Liabilities do not end up but have been virtually dealt with by the artificial form of companies linking the goodwill of hospitals, pharmaceutical brands, paramedic products, medical instrument agencies, etc. Transferring the liability has been through the ropeway of innovation in terms of economic significance and technological assertions.

In contrast, the biggest failure that pops up during the eruption of multiple errors, unethical practices, no consent data transfer, biometric migration, and more, which makes medical science a bad image. As machines cannot replace humans similarly, machines cannot be replaced by human beings. Although human beings have ethical and moral restrictions, machines have technical limitations.

Insurance (either medical or product) guarantees the coverage of damage for any substantial failure, but the repercussions must be weighed by the corporate legal departments, appealed in courts, and other adjudicating bodies. Conservative stances show non-human interference to be classified into two terms: exclusive and inclusive medical treatment. Exclusive medical treatment is between the patient and the doctor exclusively in either digital or non-digital mode. Inclusive medical treatment includes using devices and computer-programmed medical equipment to observe the patient by the doctor or by the patient on their own. Beyond insurance, there are 'claims' in the court of law or hampering goodwill via social media, which is quite common to win the claims [35]. Mockery of medical science is not tolerated in exchange for healthcare services for which it is refrained from the more prominent brands to introduce many technologically innovative products from one branch of a legal company.

IoT is worldwide engineering dependent on the World Wide Web which works with the trade of administrations just as merchandise [36]. It assumes a significant part of the stock chain network worldwide. As an expected outcome, IoT impacts the security and security issues of different partners. Critical protections that guarantee the design's strength if there should arise an occurrence of a security break, verification of information, control of the entrance of classified data, and customer security needed to be set up. During the innovative work, the administrative instrument should be set up both worldwide and in neighborhood locales. Independent of the unique idea of the test, it is a lot conceivable to start an instrument to make administrative judgments at the worldwide level with private players adjusting to such necessities. The practicality of this system will generally rely on the worldwide acknowledgment of a comparable norm of the right to security and information protection [37].

4.2.2 Law and Justice in Bioethics: Focus on AI and IoT

They were affirming 'health' to be a fundamental asset of human rights; the World Health Organisation seeks for highest standards for global healthcare for all countries. A state's role in providing 'adequate qualitative healthcare' goes immensely wide under constitutional law. National health policies are introduced under 'Universal Adult Coverage' and 'Agenda for Sustainable Development in the launching phase'. Executions vary as per the government notions, initiatives, and efforts adopted by the citizens. Law and order promote bioethics across the globe through standard medical services and advanced medical services. The trendy functionality of research and exercise has to do with the idea of the activity. Clinical evaluation makes use of check structures to increase the ampleness and safety of the latest interventions or unproved intercessions; several capsules and strategies in like manner use have by no means been offered to randomized managed fundamentals. All matters being equal, diverse methodologies are described that are robust and secured through very successful assessment and found out taking the whole lot into account. Investigators taking a gander at HIV or hepatitis B transmission quotes among intravenous prescription clients now no longer the use of needle alternate undertakings might coordinate detailed evaluation, as proven through identified modern standards, through giving needles to the check assembling and retaining them from getting the benchmark bunch. The clever legitimization of research needs to be made explicit and sensible. Clinically indistinguishable greatness is an essential circumstance of all medical and trendy prosperity evaluation, and a plan of preferred care remedy to govern packs is an unimportant want for maximum research ethics sheets. Confirmation of the preferred, and whether or not or now no longer it must spot, time, and neighborhood, is an area of advancing the conversation [38].

Nuremberg Trials after World War II instilled the role of bioethics in the arena of health research and public health after bloodshed. 'Consent' was the first-hand element that restricts medical scientists' human trials and experiments, and 'humanitarian importance' was given more effect than the achievement of 'medical success'. Smart contracts consent in a digital manner within the scope of terms and conditions of usage of the application. Consent holds a compassionate link with digital privacy and gives quantum importance to the degree of transmission of info. Data analytics predicts a smoother way to ascertain raw data, and significant liability was left ridden with the medical scientist, which may affect the life of the human being. UK Data Protection Act, 1998, EU General Data Protection Regulation, 2018, and US HIPAA Privacy Rules entangle the jurisdictional limitations of data privacy in empowering healthcare support. Location services traced by search engines, cloud-based data storage, and fused data scratching have been the ugly affairs in the medical ecosystem. Bioethics does not agree with AI and IoT to cross paths with human emotions or trust by breaching privacy.

Social engineering began in the late 1940s, and the health industry has boomed up for the better good. The Universal Declaration of Human Rights in 1948 preached for the fundamental right to health and well-being, followed by the UN Covenant on the protection of human rights, predominantly civil and political rights. This special covered protection for women and children is adequately addressed by the 'feminist' approach of technological innovation in medicare. We have exceptionally particular bots for women and children designed according to the stage of life and age groups. Bioethics has asserted the liberal views of the society by accepting the Declaration of Helsinki, 1964. The well-being of patients was given a higher stance over the interests of medical science in all possible manners following the qualifications of research experts, clinical permissions, reasonable goals, and visions. Human liability became tertiary due to gaps in medical research to protect harsh punishments, as research and development are for the public with compromise on the lives of a few wishing to volunteer on their own.

Lifestyles call for policies and laws, which change every day as per the requirements of the public, affordability of the State, and maintenance system. India has announced the inculcation of unique health IDs (UHIDs) under the Ayushman Bharat Digital Mission (ABDM) to simplify hospital procedures and track the actual data on every citizen of India. Global match on health regulations is restricted to the state due to 14-digit health ID, but the motive would inculcate essential well-being. The sandbox integration method is used to store patients' data via electronic health records (EHRs) on blocks. Initiatives may have an application to IoT-based services, as in the case of AarogyaSetu. Recent tracking of COVID-19 is conducted via the same application, along with vaccinations, nearby affected, and more features. Loopholes may be accustomed, but the greater reach is the better way to sensitize the pandemic [39]. The Indian Government has made mandatory application usage and chatbot systems to diagnose the deadly virus with the common motive to eliminate the virus. Government healthcare policies are flowing on a higher norm, and a better impact is expected to reach the Indian Republic.

4.3 International Legal Perspective

See (Table 1).

Table 1 List of countries with legislation

Category	Country	Legislation
Developed Countries	United States of America	*'The Health Insurance Portability and Accountability Act of 1996'*
	United Kingdom	*'Data Protection Act, 2018'*
		'The Health Service (Control of Patient Information Regulations) 2002'
	Canada	*'The Personal Information Protection and Electronic Document Act 2000'*
Developing Countries	India	*'Clinical Establishments (Central Government) Rules, 2012'*
		'Information Technology (Amendment) Act, 2008'
		'Right to Information Act, 2005'
		'Indian Medical Council (Professional conduct, Etiquette and Ethics) Regulations, 2002'
	Brazil	*'Brazilian General Data Protection Law 2020'*
	South Africa	*'National Health Act, 2004'*
Least Developed Countries	Bangladesh	*'The Medical Practice and Private Clinics and Laboratories (Regulation) Ordinance, 1982'*
	Bhutan	*'National Health Policy 2011'*
	Nepal	*'The Public Health Service Act, 2075 (2018)'*

4.3.1 Developed Countries

United States of America

The Health Insurance Portability and Accountability Act of 1996

The 'Health Insurance Portability and Accountability Act of 1996' (HIPAA) [40] mandates the creation of nationwide guidelines to safeguard the confidentiality of patient health data. The rules are referred to as the 'HIPAA Privacy Rule' and the 'HIPAA Security Rule'. The primary objective of the Privacy Rule is to safeguard the confidentiality of people's health information while facilitating the appropriate dissemination of this information to enhance the provision of excellent healthcare. The Privacy Rule pertains to the use and dissemination of data about persons who are subject to this rule. This includes medical facilities, physicians, and other organizations responsible for managing health data. Every healthcare professional is accountable for transmitting health information pertaining to the transactions they carry out.

United Kingdom

Data Protection Act, 2018

The Data Protection Act 1998 [41] was implemented with the purpose of safeguarding the personal information of persons residing in the European Union and the United Kingdom. The change was made in 2018 to adhere to the General Data Protection Regulation. The Data Protection Act of 2018 outlines the procedures that data controllers must adhere to in order to ensure the security of personal information belonging to its users. Additionally, it serves to hinder any unlawful entry or use of this information. The purpose of this legislation is to safeguard individuals' data privacy and facilitate the management of such data by businesses. Additionally, it mandates that companies must guarantee the security of their data and prevent unauthorized use.

The Health Service (Control of Patient Information Regulations) 2002

The 'Health Service (Control of Patient Information Regulations) 2002 (S.I. 2002/1438)' [42], modified by section 117 of the Care Act 2014, outlines the protocols for handling patient information in England and Wales for medical research and treatment. Additionally, there are measures in place to handle patient information without obtaining permission in situations when it is not feasible to do so.

Canada

The Personal Information Protection and Electronic Documents Act 2000

Under The Personal Information Protection and Electronic Documents Act, 2000 (PIPEDA) [43], organizations must obtain the consent of their customers before they collect, use, or disclose their personal information. There are also various rights that individuals have regarding their personal information. This privacy policy applies to the use of your personal information. If you have given permission, then the organization must obtain consent again before doing so. Although PIPEDA does not provide specific safeguards organizations can use, it does stress the importance of keeping sensitive information secure. The penalty for failing to prevent a data breach or keep records related to it is up to $100,000. In order to comply with the PIPEDA, organizations should implement adequate data protection safeguards. This can help prevent unauthorized access and use of their personal information.

4.3.2 Developing Countries

India

Clinical Establishments (Central Government) Rules, 2012

The health industry's nature of health information is susceptible to exploitation according to the 'Clinical Establishments (Central Government) Rules, 2012' [44]. This is shown by the absence of a comprehensive regulatory framework aimed at safeguarding the privacy of this crucial asset. Under the 2012 Clinical Establishments Rules, medical establishments were mandated to preserve and maintain electronic medical records.

Information Technology (Amendment) Act, 2008

The 'Information Technology Act of 2008' [45] (IT Act) has made it possible to impose civil and criminal penalties against individuals who violate the confidentiality of their data. If a corporate entity suddenly ceases keeping records and safety procedures in place, it could be held liable for the actions of the affected individuals under section 43A [46] of the IT Act, 2008. The SPDI Rules apply to all commercial entities and even healthcare facilities that deal with sensitive personal data.

Right to Information Act, 2005

In India, 'Right to Information Act, 2005' [47] comes with the contention that *'the practical regime of right to information for citizens to secure information under the control of public authorities in order to promote transparency and accountability...... for matters connected therewith or incidental thereto'* [48] This is the preamble of the Act 2005, and section(j) speaks about the definition of 'right to information'. The question is whether the data that was collected and maintained by the public authority is safe or not.

The data protection aspect of this Act is being taken care of regarding the individual's right to know. In the case Bennett Coleman v. Union of India,(1994) 6 SCC 632, the court held that the freedom of the press ensured that all citizens have the right to speak and publish their views. In Indian Express Newspaper (Bombay) v. Union of India, (1994) 6 SCC 632, the concept of freedom of speech and expression has been acknowledged by the Supreme Court of India. In order to allow members of the public to communicate their ideas and beliefs freely, the court stated that '*the people*'s *right to know is the fundamental principle involved in this area*'.

Indian Medical Council (Professional Conduct, Etiquette and Ethics) Regulations, 2002

The provisions of the Indian Medical Council (Professional Conduct, Etiquette and Ethics) Regulations, 2002 [49] set professional standards for medical practitioners. They also establish internationally recognized norms for the consent and collection of personal information of patients. This code of conduct establishes a set of internationally recognized norms for collecting, using, and disclosing the personal information of medical patients. It also sets a uniform limit for the number of times medical practitioners can access and disclose this information. These regulations apply to the security of the information that medical practitioners collect about their patients. Doctors are obliged to protect the privacy of their patients during all stages of their procedures and provide complete and accurate information about them. The only exception to this requirement is if the law requires disclosing certain information or severe risk to a specific individual. The current legislation and regulations require updating and amendments.

Brazil

Brazilian General Data Protection Law 2020

Before introducing Law No. 13,709/2018, Brazil did not have comprehensive legislation regarding data protection. On August 14, 2018, the Brazilian General Data Protection Law 2020 (LGPD) was enforced [50]. There have been other rules approved in the past regarding privacy. Although the law does not provide a clear definition of what constitutes health data, it generally refers to any piece of information that can be used to identify or contact a natural person. Sensitive personal data is not defined in the law. Instead, it refers to information that relates to individuals and their health. Health data is protected when it falls within the scope of the LGPD and constitutes personal data. This includes data related to an identified natural person, such as physical or mental health. Health data is prohibited for economic purposes. There are exceptions to this rule, such as providing health services or applying financial transactions. The use of health data by insurance companies for scoring is strictly forbidden under the law. The purpose of the data collected for public health studies is to provide valuable and accurate information about health conditions. However, the data should not be processed in secure and controlled environments. If possible, the data should not be anonymized during the study. However, the data must be anonymized prior to its disclosure. The National Data Protection Authority may also impose higher standards for the security measures used to safeguard sensitive data, such as health data.

South Africa

National Health Act, 2004

The National Health Act, 2004 (NHA) [51], is a law that regulates the management of healthcare in South Africa. Its main objective is to provide a uniform healthcare system in the country. Confidentiality is a vital aspect of a healthcare system's operations. In order to ensure that patient's health records are protected, various sections of the NHA were enacted. The principles for disclosing health records are contained in sections 14 and 15 of the Health Information Act. The patient has given consent to the disclosure. A court order exists to compel the health worker to disclose the information. Section 17 of the Health Records Act protects patients' health records. It also carries severe penalties for the infringement of their privacy.

4.3.3 Least Developed Countries

Bangladesh

The Medical Practice and Private Clinics and Laboratories (Regulation) Ordinance, 1982

The 'Medical Practice and Private Clinics and Laboratories (Regulation) Ordinance, 1982' [52] is a legislation that is applicable to all private clinics and laboratories in the nation. They must save patient data and contact information electronically. Bangladesh uses the Electronic Patient Record System to manage hospital and clinic data. This technology is used in government hospitals since Bangladesh has so much patient data. This is due to a lack of awareness about the appropriate safeguards and security protocols for their data. Organizations and hospitals cannot depend on the patient information they possess if they lack a comprehensive privacy and security strategy. This problem has the potential to impact the caliber of their patient treatment.

Bhutan

National Health Policy 2011

National health policy [53] aims to promote self-reliance and sustainability in healthcare delivery. It aims to achieve this by increasing efficiency and profitability while maintaining strict confidentiality. Section 7 of the policy states that patient health data should be maintained and maintained through the Centralised Health Management Information System (CHIS). Health records and information systems should be interconnected to enable faster decision-making. A healthcare facility should have a modern and accurate information system that enables users to efficiently access and record their medical records. Due to the HIMS, the data is vulnerable and can be easily accessed by hackers. This system is a central component of every healthcare facility in the country. Unfortunately, many countries have weak data protection laws that allow hackers to access healthcare information easily. The authority may also impose higher standards for the storage and processing of sensitive data.

Nepal

The Public Health Service Act, 2075 (2018)

The Public Health Service Act [54] was enacted to provide the entitlement to access cost-free basic healthcare services in Nepal. This legislation is applicable to all health institutions that are neither government nor commercial businesses. Section 14 of the Act pertains to the matter of secrecy. Confidentiality pertains to the maintenance of complete privacy about any information concerning the health state of a service user. This provision stipulates that the information shall not be divulged without the recipient's explicit written authorization. The dissemination of information shall only occur if the recipient has provided written authorization and the court's directives have been adhered to. In today's digital era, electronic equipment is used for every illness diagnosis, resulting in the storage of all information collected for a diagnosis on the cloud. This indicates that it is susceptible to being exploited. The Individual Privacy Act of Nepal is considered the most efficient method of safeguarding the basic right to privacy, as provided by the constitution of the nation. Additional legislation, such as the 'Country Civil Code 2074 (2017) and the National Penal (Code) Act (2017),' also includes rules pertaining to data protection. The Privacy Act and the Privacy Regulation are the only federal rules governing the protection of personal data.

5 Conclusion

The potential of AI and IoT in healthcare might be described as a rapidly growing area of study. These improvements provide healthcare systems with a substantial opportunity to proactively forecast health risks and effectively diagnose, treat, and monitor patients both inside and outside of the hospital. The expansion of technology-supported health services will lead to the supplementation or replacement of traditional health service delivery techniques with AI and IoT, allowing health systems to provide more flexible care models. The implementation of IoT in healthcare will depend on the establishment of a robust and explicit code of conduct for data management, privacy, and cybersecurity, which will govern the provision and use of IoT devices in the healthcare sector. Subsequent research should prioritize the investigation of IoT technology, its impact on the healthcare system, and the individuals who make use of it. AI and IoT permeate every aspect of our daily life. From cell phones to automobiles, and maybe most crucially, our healthcare, everything is connected these days. This new technology will continue to test the limits of what is possible while also enhancing long-established conventions that have been dormant and unchallenged for centuries. Decentralized healthcare management is increasingly gaining attention due to giving patients' healthcare data with openness, trust, security, and privacy. Because of this, the network is not vulnerable to a single point of failure. International laws of healthcare data management are discussed in this research by addressing healthcare. Healthcare laws of the countries are not adequate to the level of protection of patients' information. With smart contracts, a rule-based AI layer is added for spotting rogue nodes and making informed decisions reflected in the developed nation's legal system. However, most countries struggle with the proper legal system, which can curve the gap between humans and machines. It shows that the framework performs admirably from different angles. It may be re-assessed to research trustworthy AI and IoT to improve the system's reliability and build a more efficient techno-legal paradigm.

References

1. Kotler P, Kartajaya H, Setiawan I (2021) Marketing 5.0: technology for humanity. John Wiley & Sons
2. Korda H, Itani Z (2013) Harnessing social media for health promotion and behavior change. Health Promot Pract 14(1):15–23
3. Bohr A, Memarzadeh K (2020) The rise of artificial intelligence in healthcare applications. Artificial intelligence in healthcare. Academic Press, pp 25–60
4. Maspero C, Abate A, Cavagnetto D, El Morsi M, Fama A, Farronato M (2020) Available technologies, applications and benefits of teleorthodontics. A literature review and possible applications during the COVID-19 pandemic. J Clin Med 9(6):1891
5. Sen G (2012) Universal health coverage in India: a long and winding road. In: Economic and Political Weekly, pp 45–52
6. Tumai WT (2021) A critical examination of the legal implications of Artificial Intelligence (AI) based technologies in New Zealand workplaces (Doctoral dissertation, The University of Waikato)
7. Patra R, Bhattacharya M, Mukherjee S (2021) IoT-based computational frameworks in disease prediction and healthcare management: strategies, challenges, and potential. IoT in healthcare and ambient assisted living. Springer, Singapore, pp 17–41
8. Kitchin R, Dodge M (2019) The (in) security of smart cities: vulnerabilities, risks, mitigation, and prevention. J Urban Technol 26(2):47–65
9. Kuzlu M, Fair C, Guler O (2021) Role of artificial intelligence in the Internet of Things (IoT) cybersecurity. Discover Internet Things 1(1):1–14
10. Khezr S, Moniruzzaman M, Yassine A, Benlamri R (2019) Blockchain technology in healthcare: a comprehensive review and directions for future research. Appl Sci 9(9):1736
11. Banu R, Taranum A (2016) Big data collateral and solitude issues in healthcare. Int J Electr Electron Comput Sci https://ijeecse.webs.com/02.pdf
12. Ibid
13. Ibid
14. Patil HK, Seshadri R (2014) Big data security and privacy issues in healthcare. In: 2014 IEEE International Congress on Big Data, available at: https://www.researchgate.net/publication/282280458_Big_Data_Security_and_Privacy_Issues_in_Healthcare Accessed 20 Oct 2021
15. Ayala L (2016) Cybersecurity for hospitals and healthcare facilities. Berkeley
16. Keshta I, Odeh A (2021) Security and privacy of electronic health records: concerns and challenges. Egypt Inf J 22(2):177–183

17. Servos D (2012) A role and attribute based encryption approach to privacy and security in cloud based health services (Doctoral dissertation)
18. Darzi A (2018) Better health and care for all: a 10-point plan for the 2020s
19. Mello MM, Adler-Milstein JULIA, Ding KL, Savage L (2018) Legal barriers to the growth of health information exchange—boulders or pebbles? Milbank Quart 96(1):110–143
20. Rayan RA, Zafar I, Tsagkari C (2021) Blockchain technology for healthcare cloud-based data privacy and security. Integration of WSNs into internet of things. CRC Press, pp 335–349
21. Ghosh J, Sinha VK (2024) Big data analytics in industry 4.0 in legal perspective: past, present and future. In: Kumar A, Sagar S, Thangamuthu P, Balamurugan B (eds) Digital transformation. Disruptive technologies and digital transformations for society. Springer, Singapore
22. Collier R (2017) NHS ransomware attack spreads worldwide. CMAJ: Can Med Assoc J 189(22):E786–E787. https://doi.org/10.1503/cmaj.1095434
23. Antebi L (2021) Challenges in using AI. In: Artificial Intelligence and National Security in Israel. Institute for National Security Studies, pp 97–112. http://www.jstor.org/stable/resrep30590.17
24. Wittbold KA, Carroll C, Iansiti M, Zhang HM, Landman AB (2020) How hospitals are using AI to battle Covid-19. Harvard Business Review: https://hbr.org/2020/04/how-hospitals-are-using-ai-to-battle-covid-19
25. The European Commission seeked advice for Consumer Safety Network (CSN) for revising the General Product Safety Directive under the 'Sub-Group on Artificial Intelligence (AI), connected products and other new challenges in product safety'
26. MIT Technology Review, Security Experts Hack Teleoperated Surgical Robot, April 24, 2015, https://www.technologyreview.com/2015/04/24/168339/security-experts-hack-teleoperated-surgical-robot/
27. Yaacoub JPA, Noura HN, Salman O et al (2021) Robotics cyber security: vulnerabilities, attacks, countermeasures, and recommendations. Int J Inf Secur. https://doi.org/10.1007/s10207-021-00545-8
28. Gerke S, Minssen T, Cohen G (2020) Ethical and legal challenges of artificial intelligence-driven healthcare. Artif Intell Healthc. https://doi.org/10.1016/B978-0-12-818438-7.00012-5
29. Moravec H (1999) Rise of the Robots. Sci Am 281(6):124–135
30. "Researchers Adding AI, Medication Reminders to Companion Robots for Seniors." MobiHealthNews, 6 Nov. 2017, https://www.mobihealthnews.com/content/researchers-adding-ai-medication-reminders-companion-robots-seniors
31. de Oliveira Fornasier M (2019) The applicability of the internet of things (IoT) between fundamental rights to health and to privacy. Revista de InvestigacoesConstitucionais 6(2):297–322
32. Shafqat S, Abbasi A, Amjad T, Ahmad HF (2018) Smart health simulation representing a hybrid architecture over cloud integrated with IoT. Future of information and communication conference. Springer, Cham, pp 445–460
33. For more details, See, Right to Information Act, 2005
34. Ghosh J, Sinha VK (2022) How far data mining is legal!. Legal analytics: the future of analytics in law
35. Mattioli M (2017) The data-pooling problem. Berkeley Technol Law J 32(1):179–236
36. Ganesh Babu R, Karunakaran A, Manikandan G, Kalimuthu Kumar S, Selvameena R (2021) IoT in smart automation and robotics with streaming analytical challenges. Emergence of cyber physical system and IoT in smart automation and robotics. Springer, Cham, pp 103–118
37. Chakrabarty SP, Jayanta G, Souvik M (2021) Privacy issues of smart cities: legal outlook. Data-driven mining, learning and analytics for secured smart cities. Springer, Cham, pp 295–311
38. For more details, See, The Medical Practice and Private Clinics and Laboratories (Regulation) Ordinance, 1982
39. McDonnell A, Van Exan R, Lloyd S, Subramanian L, Chalkidou K, La Porta A, Yadav P (2020) COVID-19 Vaccine predictions: using mathematical modelling and expert opinions to estimate timelines and probabilities of success of COVID-19 vaccines. Center for Global Development Washington (DC)
40. For more details, See, The Health Insurance Portability and Accountability Act of 1996

41. For more details, See, The Data Protection Act 1988
42. For more details, See, The Health Service (Control of Patient Information Regulations) 2002
43. For more details, See, The Personal Information Protection and Electronic Document Act
44. For more details, See, Clinical Establishments (Central Government) Rules, 2012
45. For more details, See, Information Technology Act of 2008
46. Ibid
47. For more details, See, Right to Information Act, 2005
48. Ibid
49. For more details, See, Indian Medical Council (Professional conduct, Etiquette and Ethics) Regulations, 2002
50. For more details, See, Brazilian General Data Protection Law
51. For more details, See, National Health Act, 2004
52. For more details, See, The Medical Practice and Private Clinics and Laboratories (Regulation) Ordinance, 1982
53. For more details, See, National Health Policy
54. For more details, See, The Public Health Service Act, 2075 (2018)

AI and Industry 6.0

Ai-Driven Customer Retention

Pratiksha Agarwal, Arun Gupta, and Mohuya Chakraborty

Abstract Artificial intelligence (AI) has had a significant impact on a wide range of industries, including robots, e-commerce, healthcare, and banking. The wealth of personal data has made it possible to convert human civilization's knowledge into digital form, which in turn has made it possible to use machine learning to create new knowledge and behaviors. AI has demonstrated significant promise for creating customer relationship management (CRM)/Enterprise Resource Planning (ERP) system retention strategies. AI is capable of analyzing consumer data to forecast habits, preferences, and churn risks through the use of sophisticated algorithms and machine learning. By anticipating client wants, personalizing interactions, and providing incentives that are specifically tailored, organizations can cultivate long-term consumer loyalty. AI can also enable real-time adaptation in CRM/ERP systems, improving customer satisfaction and increasing retention rates. This paper presents a Systematic Literature Review using the Preferred Reporting Items for Systematic Reviews and Meta-Analyses model for AI-based customer churn prediction in several industries and proposes a predictive analysis model of churn using a hybrid ensemble supervised machine learning algorithm for an online subscription service. The results are quite satisfactory with accuracy of 82%, precision of 78%, recall of 82%, specificity of 82%, and F1 score of 75%. The paper further discusses AI-driven customer retention policies' opportunities, issues, and challenges.

Keywords Customer relationship management · Customer retention · Artificial intelligence · Customer experience · Customer personalization · Customer churn · Predictive analytics · Customer service automation · Supervised machine learning · Hybrid ensemble · Enterprise resource planning

---

P. Agarwal (✉)
SAP America, Palo Alto, CA, USA
e-mail: pratikshaag86@gmail.com

A. Gupta
Confiance Tech Solutions, Los Angeles, CA, USA

M. Chakraborty
Cybersecurity Centre of Excellence, Institute of Engineering and Management, University of Engineering and Management, Kolkata, India

M. Chakraborty et al. (eds.), *Proceedings of 5th International Ethical Hacking Conference*, Lecture Notes in Networks and Systems 1148,
https://doi.org/10.1007/978-981-97-8457-8_27

1 Introduction

Artificial intelligence (AI) is revolutionizing commercial operations, particularly in customer service. AI-powered chatbots and virtual assistants are being used across industries to address customer inquiries, resolve issues, and provide a positive experience, leading to increased customer satisfaction, loyalty, and advocacy. Marketing and sales are crucial business functions that drive revenue and growth. AI aids in analyzing customer behavior, personalizing campaigns, and predicting sales trends. The healthcare industry relies on AI for innovation and technological advancements, while the finance industry provides capital access, manages risk, and efficiently allocates resources. The manufacturing industry, spanning sectors like automotive, electronics, and consumer goods, is crucial for job creation, innovation, and exports. AI is revolutionizing the manufacturing sector by facilitating automation, supply chain efficiency, quality control, and predictive maintenance. The transportation sector, which includes air, sea, rail, and road travel, is essential to trade, economic development, and ties within communities. AI is useful for traffic control, self-driving cars, route optimization, and predictive maintenance. In response to shifting consumer preferences, the retail sector—which includes physical stores, e-commerce sites, and mobile apps—adapts. Important developments in this regard include omnichannel retailing, personalized shopping experiences, and sustainability initiatives. [1].

Human Resources (HR) oversees workers, making sure labor regulations are followed and promoting a positive workplace environment. HR is adopting machine learning (ML) technology—a subset of artificial intelligence (AI) and data analytics—for creative approaches in the modern business environment [2, 3]. Since customer retention affects income and profitability, it is critical to corporate strategy. Customer relationship management (CRM)/Enterprise Resource Planning (ERP) systems assist businesses in managing client relationships, enhancing customer support, and increasing revenue. A corporation has a competitive edge when its client base remains loyal, revenue rises, and expenses decrease as a result of excellent customer retention.

AI supports firms in making defensible choices regarding client retention strategies. Anticipating customer attrition is essential for developing retention strategies. Churn is defined as the proportion of customers that discontinue using a company's products or services within a set time frame. It offers information about keeping customers around. For example, a company with 560 customers at the beginning and 520 at the end would have a 7.14% churn rate since it has lost 7.14% of its customers. In order to identify at-risk consumers and put proactive retention initiatives in place, churn prediction is essential for businesses. Businesses that analyze consumer data and behavior and offer tailored promotions, discounts, or improved customer service can predict future churn rates. This combo lowers turnover, boosts customer loyalty, and boosts profitability for the business.

This paper analyses various AI-driven approaches toward customer retention policies using a Systematic Literature Review (SLR) [4] using the Preferred Reporting

Items for Systematic Reviews and Meta-Analyses (PRISMA) [5] model and eventually designs and evaluates the performance of a hybrid [6] ensemble ML-based churn prediction model for an electronic online subscription service.

The following are the paper's primary contributions.

1. Conducting SLR of AI-based customer retention policies using the PRISMA model.
2. Finding Research Gaps (RG), formulating Research Questions (RQ), and making RG-RQ analysis.
3. Proposing and analyzing a hybrid ensemble machine learning-based churn prediction model for an electronic online subscription service and explaining the results with Explainable AI.

This is how the paper is structured. After the introduction in Sects. 1, 2 explains the research methodology, followed by deliberations on relevant publications in Sect. 3. A hybrid ensemble ML model for churn prediction of an electronic online membership service is presented in Sect. 4. The simulation results and performance evaluation are shown in Sect. 5, and in Sect. 6 there are discussions on the opportunities and problems associated with AI-driven customer retention mechanisms. Section 7 highlights future scopes as the paper comes to a close.

2 Research Methodology

SLR is a methodical procedure that is used to gather, assess, and compile research findings pertaining to a certain study problem. The research question must be clearly stated. A thorough search strategy must be developed. Relevant databases, keywords, and inclusion and exclusion standards must be found. The studies must then be screened using the inclusion and exclusion criteria, and the best ones must be chosen. Each study's data, which includes participant information, interventions, outcomes, sample size, and other pertinent data, is then analyzed using standardized forms.

There are four basic phases involved in SLR that are enumerated below.

1. **Phase 1 (Planning)**—During this phase, the research question is defined, inclusion and exclusion criteria are specified, and the search method is created. Creating a detailed plan for carrying out the review is the aim.
2. **Phase 2 (Conducting)**—The search approach is used in this step to find pertinent studies. Usually, to do this, databases are searched, titles and abstracts are screened, and finally possibly pertinent papers are fully screened. The various databases that are searched in this SLR are "IEEEXplore", "Semantic Scholar", "Springer", "Science Citation Index", "ScienceDirect", "Taylor & Francis", "Web of Science", Google Scholar" etc. Forms for data extraction might also be

created at this stage. The search strategy was influenced by keywords like "customer retention", "churn", "AI-based churn", "machine learning", customer relationship management, "customer experience", "customer personalization", "customer churn", "predictive analytics", "customer service automation", "supervised machine learning", "hybrid ensemble", etc. Three criteria were utilized to limit the study: (1) peer-reviewed articles (2) published in reputable English-speaking journals and conference proceedings, and (3) studies published between 2008 and 2024.

3. **Phase 3 (Synthesizing)**—Data extraction and synthesis take place after the identification of pertinent studies. This entails reviewing the entire body of evidence, comparing the results of various studies, and summarizing the conclusions of specific investigations. The SLR results, which offer important insights into the current body of research can direct practice, policy, or future research, as illustrated using the PRISMA model in Fig. 1. The figure illustrates that a total of 95 papers were found by utilizing the pertinent keywords. The abstract and title of 35 articles were used to weed them out. Sixty papers were deemed qualified following an initial eligibility search. Fifteen duplicates were removed using software available on the internet, leaving 45 manuscripts prepared for full-text scanning. Twenty articles were deleted because they were unrelated. Nine publications unrelated to the inquiry were removed, leaving twenty-five papers for review and analysis. Sixteen articles were nominated for the final investigation to execute a qualitative and quantitative synthesis. "var" is used to represent these numbers.

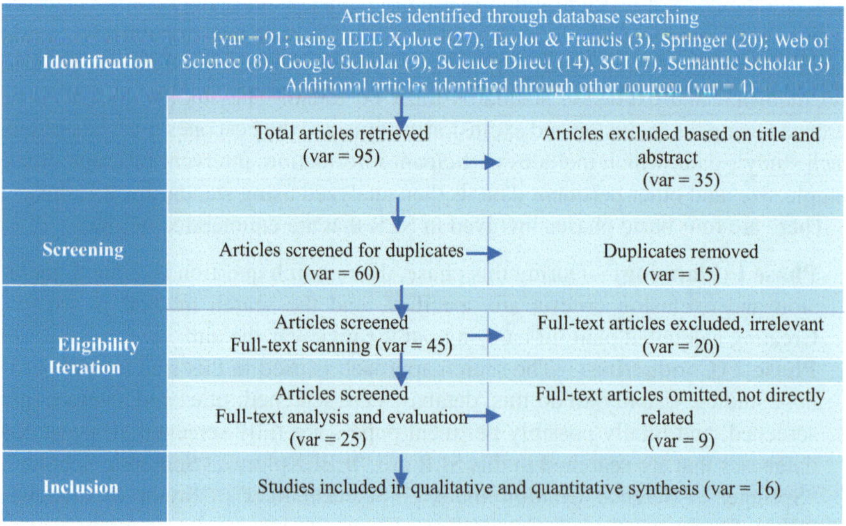

Fig. 1 PRISMA diagram of the current body of research

4. **Phase 4 (Reporting)**—Reporting the SLR's findings is the last phase. This entails outlining the search plan, condensing the results, and talking about the significance of the findings. Giving a thorough and understandable summary of the body of research on the study subject is the aim.

3 Related Papers

Sixteen papers identified in Sect. 2 through PRISMA model are analyzed here to report their findings in the current body of research. With an emphasis on ML predictions [7], suggests a combination of ML models for anticipating client churn. The model can more accurately capture the core characteristics of disgruntled customers and performs better than the single model in terms of accuracy and predictive impact. By efficiently tackling the difficulty of estimating customer turnover, this method is critical for firms looking to reduce customer churn and sustain client loyalty [8]. Employs a stacked autoencoder network to extract characteristics from data and logistic regression to classify customers to forecast customer attrition using deep learning. The greedy layer-wise unsupervised learning approach trains a model, then uses logistic regression layers for accuracy assessment. Tuning the model for customer attrition minimizes client loss and responds promptly to operational shortages [9, 10]. Employ a multilayered convolutional neural network and a deep learning approach to forecast churn on telecommunication datasets to predict aspects of usage, customer, support, and context, thereby aiding businesses in identifying and removing churn factors.

Consumer complaints frequently represent poor service quality, and emotion plays a major impact on consumer satisfaction and loyalty. In an effort to increase customer happiness, [11] examines complaints from information service providers in order to develop an emotional warning model for client attrition. Increased market competition leads to revenue loss and consumer attrition. To mitigate this, companies should analyze potential clients' experiences, employ prediction techniques, and regularly update their strategies. Axiomatic Fuzzy Set (AFS) and parallel DBSCAN clustering are combined in a hybrid approach and examined to notice that parallel DBSCAN clustering runs noticeably slower than parallel K-means clustering [12].

In crowded markets, customer relationship management plays a critical role in keeping clients [13]. Develops a more precise churn prediction model by utilizing support vector machines to analyze newspaper subscriptions. The study compares the predictive performance of random forests and logistic regression, highlighting the importance of parameter optimization and client/company relationship variables in churn prediction. Financial institutions are mitigating client attrition by employing integrated deep-learning network models to forecast customer attrition or retention. In [14], the authors utilize a model divided into two levels: Level 1 uses a logistic regression model, and Level 0 employs three fundamental Deep Neural Network models. To anticipate churn, [15] research presents two hybrid models combining self-organizing maps (SOM) with back-propagation artificial neural networks (ANN). Hybrid models outperform single neural network baseline models in Type I and II errors and prediction accuracy, while ANN + ANN hybrid model significantly outperforms SOM + ANN hybrid and ANN baseline models.

An AI-based model that can forecast client attrition and provide a customized service charge is presented in this study. For regression models and churn prediction, it employs K-nearest neighbors, Decision Trees, Random Forests, and Support Vector Machines. The feature selection approach lowers the cost of data collection [16]. To increase the efficacy of cross-selling in retail, [17] suggests a hybrid model that combines self-attention mechanisms and recurrent neural networks. The model estimates customers' likelihood of making future purchases and can handle both sequential and non-sequential features.

Das and Nayak [18] examines the segmentation of customers based on their tastes and preferences through the use of data mining techniques, both supervised and unsupervised to perform a methodical analysis of various marketing and customer relationship-building strategies, while [19] explores customer clustering using recency, frequency, and monetary analysis for digital start-ups and identifies three customer segments: new, best, and intermittent, highlighting the importance of understanding and addressing these clusters to optimize customer engagement and business growth. In [20] research, an innovative approach to anticipate telecom customer attrition is presented by applying a stratified sampling technique and a binary logistic regression model. XAI strategy for online gaming that uses transparent black-box models and Multiview data sources for player churn prediction and game cheat detection is presented in [21], while [22] compares four classification algorithms and uses a hidden Markov model to predict churn in casual social games, finding neural networks have the best prediction.

4 Proposed Hybrid Ensemble Churn Prediction Model

This section presents a hybrid ensemble ML-based churn prediction model, which combines four different types of supervised machine learning models to solve a churn binary classification problem of whether the company has an at-risk customer (marked by a binary1) or no-risk customer (marked by a binary 0).

4.1 System Model

The system model comprises three blocks that are Data Preprocessing, Classification, and Explanation as depicted in Fig. 2.

To be able to prepare raw data for the creation and training of ML models, data preprocessing is an essential stage in the ML process. It involves getting the required dataset, data cleaning and formatting including handling missing data, encoding categorical data, and feature scaling (viz., normalization which shifts and rescales

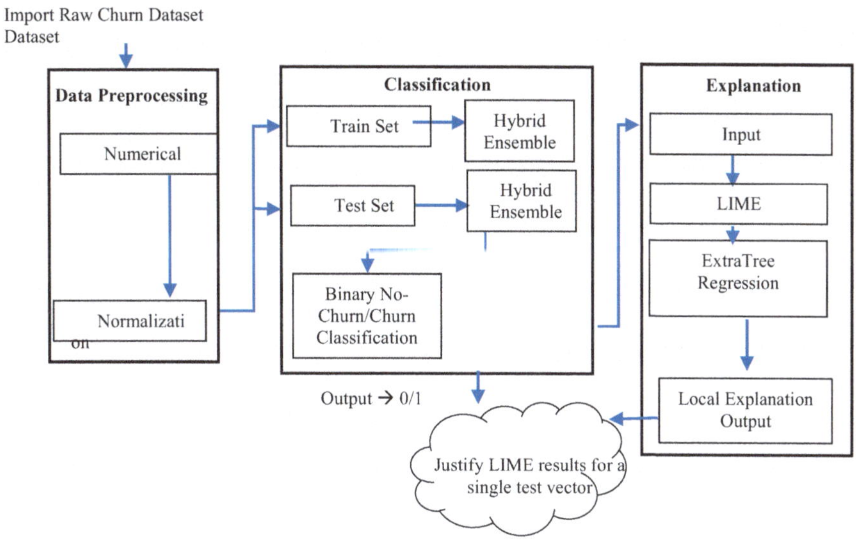

Fig. 2 Architecture of Churn prediction model

data to make it fall between [0,1]). The formula for normalization is:

$$X_{new} = \frac{x - X_{\min}}{X_{\max} - X_{\min}} \tag{1}$$

where x is the original feature vector, X_{min} is the minimum value of that feature vector, and X_{max} is its maximum value.

Classification comprises splitting the data into training and testing data. Normally, 75% of the data is used for training and the remaining 25% is used for testing. The split data is then fed to the classifier for predicting a binary classification of No Churn (0) or Churn (1) output. This is followed by an explanation of the predicted results using Local Interpretable Model-agnostic Explanations (LIME) that performs a local explanation of the ith test sample interpreting the importance and relevance of the important features that led to the predicted output.

4.2 Dataset

The dataset of the online subscription service comprises 19 features and 243,787 data points [23]. The specifications of the dataset are briefly explained in Table 1.

Table 1 Dataset features

Feature name	Feature type	Data type	Description
AccountAge (AA)	Feature	Integer	The age of the user's account in months
MonthlyCharges (MC)	Feature	Float	The amount charged to the user every month
TotalCharges (TC)	Feature	Float	The total charges incurred by the user over the account's lifetime
SubscriptionType (ST)	Feature	Object	The type of subscription chosen by the user (Basic, Standard, or Premium)
PaymentMethod (PM)	Feature	String	The method of payment used by the user
PaperlessBilling (PB)	Feature	String	Indicates whether the user has opted for paperless billing (Yes or No)
ContentType (CT)	Feature	String	The type of content preferred by the user (Movies, TV Shows, or Both)
MultiDeviceAccess (MDA)	Feature	String	Indicates whether the user has access to the service on multiple devices (Yes or No)
DeviceRegistered (DR)	Feature	String	The type of device registered by the user (TV, Mobile, Tablet, or Computer)
ViewingHoursPerWeek (VHPW)	Feature	Float	The number of hours the user spends watching content per week
AverageViewingDuration (AVD)	Feature	Float	The average duration of each viewing session in minutes
ContentDownloadsPerMonth (CDPM)	Feature	Integer	The number of content downloads by the user per month
GenrePreference (GP)	Feature	String	The preferred genre of content is chosen by the user
UserRating (UR)	Feature	Float	The user's rating for the service on a scale of 1 to 5
SupportTicketsPerMonth (STPM)	Feature	Integer	The number of support tickets raised by the user per month
Gender (G)	Feature	String	The gender of the user (Male or Female)
WatchlistSize (WS)	Feature	Float	The number of items in the user's watchlist
ParentalControl (PC)	Feature	String	Indicates whether parental control is enabled for the user (Yes or No)
SubtitlesEnabled (AE)	Feature	String	Indicates whether subtitles are enabled for the user (Yes or No)
CustomerID (CID)	Identifier	String	A unique identifier for each customer
Churn	Target	Integer	The target variable indicates whether a user has churned or not (1 for churned, 0 for not churned)

4.3 Hybrid Ensemble Classifier

Ensemble learning is an ML technique that combines the predictions from multiple models to create a more accurate and stable prediction [6]. It is a method that boosts the learning system's overall effectiveness by utilizing the combined intelligence of several models. The hybrid classifier in this paper comprises five ML algorithms that are briefly described below.

K-Nearest Neighbor (K-NN): The K-NN algorithm finds K-nearest neighbors to a data point using a distance metric like Euclidean distance. The data point's class is determined by the majority vote or average of the K neighbors, allowing the algorithm to adapt to patterns and make predictions based on local data structure. Euclidean Distance calculates the net displacement between two states of an object [24]. The Euclidean distance is given by

$$distance(x, X_i) = \sqrt{\sum_{j=1}^{d} \left(x_j - X_{i_j}\right)^2} \tag{2}$$

The K-NN algorithm's K value is crucial for determining the number of neighbors and should be chosen based on input data. Higher K values are recommended for more outliers or noise. Cross-validation methods can help select the best K value for a dataset. The steps of K-NN are as follows.

Step 1: Select the optimal value of K.

Step 2: Calculate the Euclidean distance.

Step 3: Find Nearest Neighbors.

Step 4: Vote for Classification.

Logistic Regression (LR): When using the LR method to binary classification, the sigmoid function—which accepts input as independent variables and generates a probability value between 0 and 1—is utilized. The logistic regression model transforms the linear regression function continuous value output into categorical value output using a sigmoid function, which maps any real-valued set of independent variables input into a value between 0 and 1 [25]. The mathematical function called the sigmoid function is utilized to convert expected values into probabilities. It converts any real number between 0 and 1 into another value. The LR must have a value between 0 and 1, and as it cannot be more than this, it takes the shape of a "S" curve. The logistic or sigmoid function is another name for the S-form curve. The threshold value, which indicates the likelihood of either 0 or 1, is a notion used in LR. For example, values above the threshold tend toward one, and those below the threshold tend toward zero. The output of the sigmoid function, $\sigma(z)$, represents the probability of the positive class (class 1). The log-odds (logit) are calculated as the natural logarithm of the odds ratio:

$$log - odds = \ln \frac{\sigma(z)}{1 - \sigma(z)} \tag{3}$$

Applying the sigmoid function to the linear combination of input features yields the final prediction given by

$$Prediction = \begin{cases} 1 \; if & \sigma(z) \geq 0.5 \\ 0 \; if & \sigma(z) < 0.5 \end{cases} \tag{4}$$

The steps of LR are as follows.
Step 1: Initialize the coefficients randomly or with zeros.
Step 2: Repeat until convergence or maximum number of iterations.
2.1: Calculate the linear combination of input features and coefficients.
2.2: Apply the sigmoid function to get the probabilities.
2.3: Calculate the log loss.
2.4: Calculate the gradient of the loss function for the coefficients.
2.5: Update the coefficients by subtracting a fraction of the gradient.
Step 3: Make predictions for the test data using the final coefficients.

Decision Tree (DT): Regression and classification tasks are two areas in which a DT algorithm excels. This kind of supervised learning algorithm uses input data to model and predict results. A DT's topology is similar to that of a tree, with inside nodes testing attributes or features, branches corresponding to attribute values, and the final judgment or forecast is represented by the leaf nodes [26]. Three important parameters are used to determine splits and make predictions for classification tasks.

1. Gini Impurity is used to measure the impurity or purity of a node in a decision tree. Mathematically, it is expressed as:

$$Gini(node) = 1 - \sum_{c}^{i=1} p(i|node)^2 \tag{5}$$

where c is the number of classes, and p(i|node) is the proportion of instances of class i in the node.

2. Entropy is used to measure impurity. Mathematically, it is expressed as:

$$Entropy(node) = -\sum_{i=1}^{c} p(i|node)\log_2(p(i|node)) \tag{6}$$

3. Information Gain is used to decide which feature to split. It calculates the energy or Gini impurity reduction that results from dividing the data according to a specific attribute. Mathematically for entropy it is given by:

$$Information\ Gain(feature) = Entropy(parent) - \sum_{i=1}^{m} \frac{N_i}{N}.Entropy(child_i) \tag{7}$$

where m is the number of child nodes after the split, N_i is the number of instances in the ith child node, and N is the total number of instances in the parent node.

The steps of DT are as follows.

Step 1: Begin with the root node, which stands for the complete dataset.

Step 2: Search for the most significant attribute that divides the data into the most distinct groups.

Step 3: Create new branches by dividing the data into smaller subsets based on the response to the previous step's inquiry.

Step 4: Continue from steps 2 through 4 until it reaches the last "leaf nodes", which stand for the anticipated results or classifications.

Random Forest (RF): The RF algorithm combines the predictions of multiple decision trees to reduce overfitting and improve accuracy [27]. A sort of ensemble learning known as "bagging" (sometimes called "Bootstrap aggregating") involves training several base models concurrently and independently on various subsets of the training set. Bootstrap sampling, which selects data points at random with replacement, is used to create each subgroup. For best results, RF's several hyperparameters can be adjusted. A few significant hyperparameters are the number of trees in the forest, the maximum depth of the trees, the minimum number of samples needed to split an internal node, the minimum number of samples needed at the leaf node, the maximum number of features to take into account for the best split, bootstrap samples, and weights associated with classes. The steps of RF are as follows.

Step 1: Choose the number of trees B and the number of features F to use at each node.

Step 2: Create an empty list to store the trees.

Step 3: Repeat B times.

3.1: Draw a bootstrap sample from the training data.

3.2: Grow a decision tree from the bootstrap sample.

3.3: Add the tree to the list.

Step 4: Make predictions for the test data using the trees.

4.1: Initialize an empty list to store the votes.

4.2: Predict and append the label for the test vector to the votes in each tree.

4.3: Aggregate the votes by majority or average.

4.4: Append the prediction to the output list.

Naïve Bayes (NB): A group of classification algorithms based on Bayes' Theorem are known as naive Bayes classifiers. Bayes' theorem is a probabilistic formula that calculates the probability of a hypothesis (class label) based on prior knowledge of conditions that might be related to the event (features) [28]. In the context of Naive Bayes, the theorem is expressed as follows:

$$P(X|Y) = P(X) \cdot P(Y|X)/P(Y) \tag{8}$$

Here,

- $P(Y|X)$ is the class Y posterior probability in light of observation X.
- $P(X|Y)$ is the probability of seeing X considering that the class is Y.

- *P(Y)* is the class Y prior probability.
- *P(X)* is the likelihood of seeing X.

There are a few fundamental Naive Bayes' assumptions. Feature conditional independence, which states that data features are conditionally independent of each other, meaning the presence or absence of one feature does not affect the other.

$$\text{i.e., } P(X|Y) = P(x1 \mid Y) \cdot P(x2 \mid Y) \ldots P(xn \mid Y) \tag{9}$$

where × *1*, × 2..., *xn* are the features of the observation *X*.

Continuous features are presumed to be regularly distributed within each class if they are continuous. The distributions of discrete characteristics are multinomial: Within each class, a discrete feature is assumed to have a multinomial distribution. Attributes hold equal significance: It is believed that each characteristic contributes equally to the class label prediction. There should not be any missing values in the data. The steps of NB are as follows.

Step 1: Calculate the prior probabilities of each class from the training data.

Step 2: Calculate the conditional probabilities of each feature value given in each class.

Step 3: Make predictions for the test data using the Bayes rule.

The Hybrid ensemble classifier uses K-NN, LR, DT, RF, and NB to build a hybrid classifier where each of the five machine learning models has been defined 5 times resulting in a combination of a total of 25 learners. The model is trained and tested on a churn dataset from Kaggle [23]. The final class prediction is determined using the Max Voting Classifier method. To assess the model's performance on unobserved data, cross-validation is employed, whereby each validation step is repeated numerous times with a distinct fold serving as the validation set. To generate a more reliable measure of the model's performance, the average results from each validation step are calculated. The boosting model is used in this task. The hybrid ensemble model is a more robust and efficient approach to churn prediction. The hybrid ensemble classifier model is depicted in Fig. 3.

5 Simulation Results and Performance Analysis

In this section, simulation results including data preprocessing, churn classification, and explanation using the Explainable AI (XAI) approach followed by performance evaluation and discussion of the results are presented.

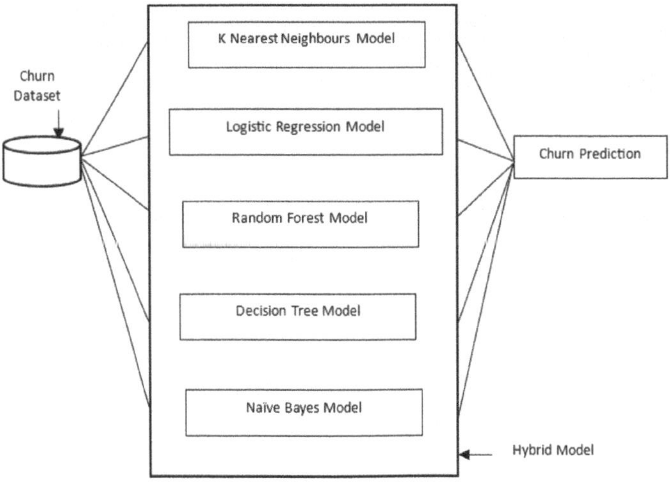

Fig. 3 Hybrid ensemble classifier model

5.1 Simulation Results

Simulation results comprise data preprocessing output in the form of a heat map and correlation matrix between the target and the feature vectors defined in Table 1 followed by classification performance metrics in the form of accuracy, precision, recall, specificity, F1 score, and confusion matrix. After classification LIME results give an explanation to the classification outputs.

5.1.1 Data Preprocessing

The extracted features are those whose correlation $\geq \pm 0.1$ as presented in Table 2 and illustrated in Fig. 4. It may be seen that out of 19 features, only 6 features (viz., AA, MC, TC, VHPW, AVD, and CDPM) are included for further classification.

5.1.2 Classification

The extracted features are fed to the hybrid ensemble classifier and the performance metrics are presented in Table 3. The confusion matrix (Fig. 5) shows the number of true values and predicted values [29]. The number depicting True Label "No Churn" predicted as "No Churn" are called true negatives (TN); that depicting True Label "Churn" predicted as "Churn" are true positives (TP); that depicting True Label "Churn" predicted as "No Churn" are called false negatives (FN); and the number depicting True label "No Churn" predicted as "Churn" are called false positives (FP). A classification model's efficacy is evaluated by plotting the rate of TP against FP,

Table 2 Feature extraction of Churn dataset for online subscription website

Sl. No	Extracted feature	Quality	Status	Sl. No	Extracted feature	Quality	Status
1	AA	– 0.197	**Include**	11	AVD	– 0.146	**Include**
2	MC	0.100	**Include**	12	CDPM	– 0.129	**Include**
3	TC	– 0.120	**Include**	13	GP	0.012	Drop
4	ST	– 0.035	Drop	14	UR	0.022	Drop
5	PM	0.018	Drop	15	STPM	0.084	Drop
6	PB	– 0.000	Drop	16	G	0.007	Drop
7	CT	– 0.011	Drop	17	WS	0.021	Drop
8	MDA	– 0.001	Drop	18	PC	– 0.005	Drop
9	DR	– 0.002	Drop	19	SE	– 0.012	Drop
10	VHPW	– 0.128	**Include**				

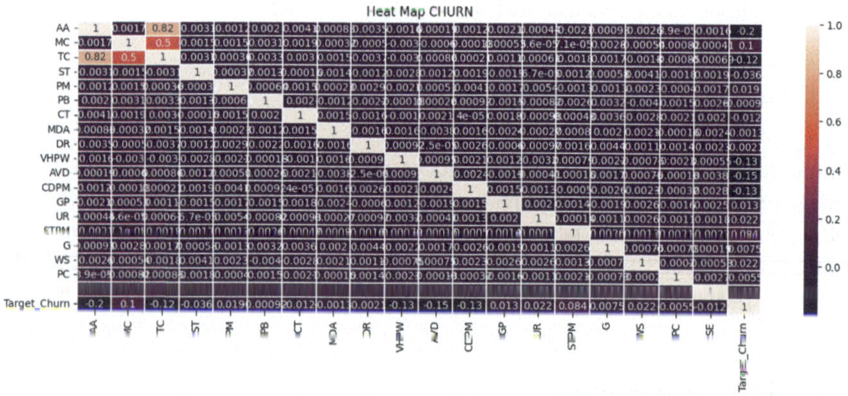

Fig. 4 Heat map of Churn dataset

which produces the Receiver Operating Characteristic (ROC) curve. The probability that a classifier would score a randomly selected positive instance higher than a randomly selected negative instance is known as the area under the ROC curve or AUC [30].

Figure 6 depicts the ROC–AUC curves of all the classifiers for the dataset under consideration. The expressions for the various performance metrics [31] are shown below.

$$\text{Accuracy} = \frac{\text{TP} + \text{TN}}{\text{TP} + \text{TN} + \text{FP} + \text{FN}} \tag{10}$$

$$\text{Precision} = \frac{\text{TP}}{\text{TP} + \text{FP}} \tag{11}$$

Table 3 Performance metrics of Churn prediction hybrid classifier model

lassifier	Training accuracy	Testing accuracy	Cross validation accuracy	Precision	Recall	Specificity	F1_ Score	Confusion matrix
K-NN	0.8401	0.8108	0.7972	0.7492	0.7945	0.7945	0.7637	[38657 1288] [7936 877]
LR	0.8224	0.8219	0.8224	0.7788	0.8219	0.8219	0.7661	[39268 677] [8002 811]
RF	0.8224	0.8192	0.8155	0.7788	0.8219	0.8219	0.7661	[39945 0] [8813 0]
DT	0.8224	0.8220	0.7290	0.7330	0.7250	0.7250	0.7289	[32862 7083] [6355 2458]
NB	0.8213	0.8209	0.8211	0.7788	0.8216	0.8216	0.7544	[38928 1017] [7717 1096]
Hybrid	0.8207	0.8211	0.8224	0.7805	0.8216	0.8216	0.7544	[39630 3015] [8383 430]

True Label	False (No Churn) N	TN	FP
	True (Churn) P	FN	TP
		False (No Churn) N	True (Churn) P
		Predicted Label	

Fig. 5 Confusion matrix

$$\text{Recall} = \frac{\text{TP}}{\text{TP} + \text{FN}} \tag{12}$$

$$\text{Specificity} = \frac{\text{TN}}{\text{TN} + \text{TP}} \tag{13}$$

$$\text{F1 score} = \frac{2*\text{Precision}*\text{Recall}}{\text{Precision} + \text{Recall}} \tag{14}$$

5.1.3 Explanation

The Hybrid classifier uses one of the methods of XAI called Local Interpretable Model-agnostic Explanations (LIME) for model-agnostic and local explanations of results obtained in step 2. LIME treats any supervised learning model as a "black-box"

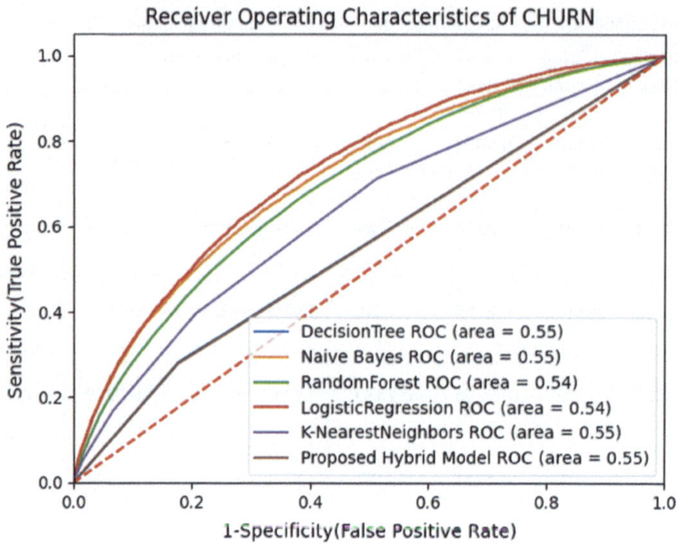

Fig. 6 ROC–AUC curves

distinct from the model once trained using the dataset. Local explanations provide accurate local explanations in the context of the observed or sample. Upon receiving a test sample and prediction model, LIME executes procedures such as sampling and acquiring a surrogate dataset, weighing each row based on its similarity to the original sample or observation, and using a feature selection method similar to Lasso. LIME uses the acquired features to train an Extra Trees Regression on the samples, ensuring that the output forecast and the one produced by the initial prediction model have comparable magnitudes. This emphasizes the significance and relevance of the acquired qualities [32]. Finally, LIME gains justifications for the 6 important features of the 22nd test sample as shown in Fig. 7.

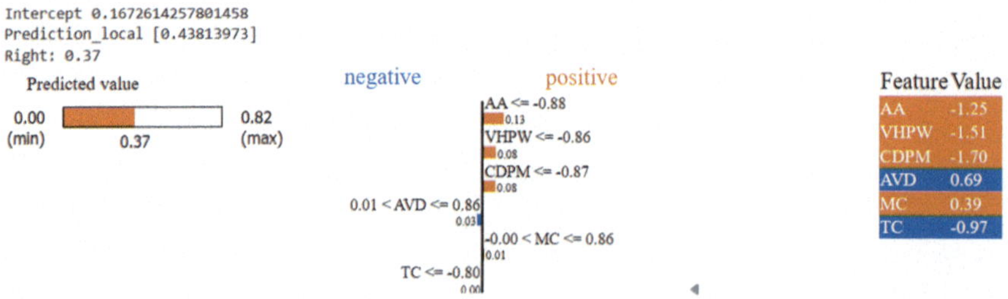

Fig. 7 LIME explanation

5.2 Performance Evaluation and Discussion

Table 3 depicts the binary classification performance of churn prediction including training, testing, and cross-validation accuracy, precision, recall, specificity, F1-score, and confusion matrices. All the individual classifiers and the hybrid model have an accuracy score of ~ (79–84%) for "Churn" and "No Churn" binary classification. The confusion matrices of all the classifiers show a smaller number of "Churn" as compared to "No Churn". Figure 6 shows the ROC–AUC curves. The AUC values for all the classifiers lie between 0.54 and 0.55 indicating that the classifier has some discriminatory power beyond random chance.

The output of LIME may be interpreted from Fig. 7. The first four values just above the visualizations are:

1. Test sample: The 22nd test sample has been explained with 6 important features.
2. Right: This is the forecast made for the provided test vector by our prediction model, which in this case is an extra-trees regressor.
3. Prediction_local: This represents the result of training a linear model with only the top k feature outputs from LIME on the perturbed samples (obtained by sampling around the test vector according to a normal distribution).
4. Intercept: The intercept is a constant component of the linear model's prediction for the supplied test vector.

$$\text{So, Predicted\_local} = \text{w1x1} + \text{w2x2} + \cdots + \text{wkxk} + \text{Intercept} \quad (15)$$

where wi is the ith weight and xi is the ith test vector.

Regarding the visualizations, it can be observed that the hues orange and blue, respectively, represent positive and negative connections between the target and the feature. Positive associations show that as one variable rises, the other usually falls; negative associations show the opposite trend. For example, duration spent on studies and the exam grade are positively associated whereas smoking and life expectancy are negatively associated. The following is an interpretation of the dataset's predictions. Relatively medium predicted value (0.37/0.82max) of the 22th test sample showing 6 important features can be attributed to the following classification reasons:

a. "**AA**", "**VHPW**" and "**CDPM**" has low positive associations and "MC" has medium positive association.
b. "**TC**" has a high negative association and "**AVD**" has a low negative association.

These visualizations help us understand the importance of the features in the predicted result. Features that are positively associated aid in the result whereas the negatively associated features hinder the result.

6 Opportunities, Issues, and Challenges

Opportunities: There are numerous opportunities for AI-driven customer retention policies discussed in the manuscript. Some of these are enumerated below.

Personalized Recommendation: AI algorithms can analyze customer data to provide personalized product recommendations, enhancing the customer experience and increasing the likelihood of repeat purchases.

Predictive Analytics: AI can analyze customer behavior patterns to predict which customers are at risk of leaving. By identifying these customers early, businesses can take proactive measures to retain them, such as offering personalized discounts or targeted marketing campaigns.

Chatbots and Virtual Assistants: These can provide customer support and improve customer satisfaction and reduce the churn rate.

Sentimental Analysis: AI can analyze customer feedback from various sources, such as social media and customer reviews based on which they can suggest improvements in the products or other areas like sales.

Customer Segmentation: AI can segment customers based on their preferences and behavior, which help organizations create marketing strategies to cater to that segment.

Feedback analysis: AI can analyze customer feedback to identify trends and insights that can help businesses improve their products and services.

Predictive customer service: AI can predict customer service issues before they occur, enabling businesses to proactively address customer concerns and improve overall satisfaction. Some manufacturing companies use prediction techniques with AI and ML to forecast issues with the system prior to their occurrence.

Augmenting ERP software with the CRM: To improve customer service and follow-ups, this aids in the creation of a central database involving customer data, transaction history, and interactions according to the CRM architecture.

Issues and Challenges: While AI can significantly enhance customer retention by personalizing experiences and predicting churn, it is imperative for companies to address the limitations mentioned below to protect customer security and privacy and maintain trust.

Data Security and Privacy: Professionals are apprehensive about the possible disclosure of confidential client data, particularly in regulated sectors such as finance and healthcare. Because AI systems use pre-programmed algorithms and rules, they cannot be very adaptive. These dangers can be reduced by putting strong data retention guidelines into place and using data protection techniques like anonymization and pseudonymization.

Compliance: Although AI systems are predicted to have a major impact on regulatory compliance, their widespread adoption is now restricted by their high failure rate. Additionally, AI-powered chatbots in customer support can find it difficult to live up to client expectations, which could lower compliance.

Algorithm Bias:

AI algorithms may inadvertently perpetuate biases present in the training data, leading to unfair treatment of certain customer groups. It is essential for businesses to regularly audit their AI systems to identify and mitigate bias.

Over Reliance on Technology: Relying too heavily on AI for customer retention efforts can lead to a disconnect between businesses and their customers. Human interaction is important. AI cannot replace empathy and human trust which are crucial for customer retention.

7 Conclusion

Using the PRISMA approach, this research examines sixteen publications to assess its conclusions. Literature review shows proposal of hybrid ML approach for client churn prediction by client clustering and segmentation that outperforms individual models and can effectively identify unhappy consumers. Additionally, use of logistic regression and stacked autoencoder network to estimate client attrition using deep learning has been identified. To create an emotional warning model for customer attrition, few researchers look at complaints from information service providers and delve into the significance of examining the past experiences of prospective customers, utilizing forecasting methods, and consistently modifying tactics to analyze newspaper subscriptions, casual social games, telecommunication, digital start-ups, etc., datasets. Furthermore, the paper designs and analyses a hybrid ensemble ML model that combines five supervised ML algorithms to predict customer churn of an online subscription service provider. Performance analysis of churn and no-churn classification results show accuracy of 82%, precision of 78%, recall of 82%, specificity of 82%, and F1 score of 75%. Additionally, XAI analysis using LIME for local explanations shows that a few of the dataset's features like the age of the account holder, number of viewing hours per week, amount of content downloaded per month, and monthly charges positively affect the prediction whereas, other features like average viewing duration and total charges incurred negatively affect the predicted result of a particular instance for the hybrid model. The remaining features showed no correlation with the predicted result.

References

1. Álvarez López JA (2022) Case studies of real AI applications. In: Landeta Echeberria A (ed) Artificial intelligence for business. Palgrave Macmillan, Cham
2. Zhai Y, Zhang L, Yu M (2024) AI in human resource management: literature review and research implications. J Knowl Econ. https://doi.org/10.1007/s13132-023-01631-z
3. Tewari I, Pant M (2020) Artificial intelligence reshaping human resource management: a review. In: IEEE International Conference on Advent Trends in Multidisciplinary Research and Innovation (ICATMRI), Buldhana, India, pp 1–4. https://doi.org/10.1109/ICATMRI51801.2020.9398420

4. Kitchenham B, Charters S (2020) Guidelines for performing systematic literature reviews in software engineering
5. Moher D, Shamseer L, Clarke M, Ghersi D, Liberati A, Petticrew M, Shekelle P, Stewart LA (2015) Preferred reporting items for systematic review and meta-analysis protocols (PRISMA-P) 2015 statement. Syst Rev 4:1–9
6. Ardabili S, Mosavi A, Várkonyi-Kóczy AR (2020) Advances in machine learning modeling reviewing hybrid and ensemble methods. In: Várkonyi-Kóczy A (ed) Engineering for sustainable future. INTER-ACADEMIA. Lecture notes in networks and systems, vol 101. Springer, Cham
7. Upadhyay S, RM (2023) Customer churn prediction using machine learning. In: 14th International Conference on Computing Communication and Networking Technologies (ICCCNT), Delhi, India, pp 1–7. https://doi.org/10.1109/ICCCNT56998.2023.10307748
8. Cao S, Liu W, Chen Y, Zhu X (2019). Deep learning based customer churn analysis. In: 11th International Conference on Wireless Communications and Signal Processing (WCSP), Xi'an, China, pp 1–6. https://doi.org/10.1109/WCSP.2019.8927877
9. Agrawal S, Das A, Gaikwad A, Dhage S (2018) Customer churn prediction modelling based on behavioural patterns analysis using deep learning. In: International Conference on Smart Computing and Electronic Enterprise (ICSCEE), pp 1–6
10. Mishra A, Reddy US (2017) A novel approach for churn prediction using deep learning. In: IEEE International Conference on Computational Intelligence and Computing Research (ICCIC), pp 1–4
11. Zhang X, Gou J (2015) Warning model of customer churn based on emotions. In: International Conference on Logistics, Informatics and Service Sciences (LISS), Barcelona, pp 1–3. https://doi.org/10.1109/LISS.2015.7369683
12. Dulhare UN, Ghori I (2018) An efficient hybrid clustering to predict the risk of customer churn. In: 2nd International Conference on Inventive Systems and Control (ICISC), Coimbatore, India, pp 673–677. https://doi.org/10.1109/ICISC.2018.8398883
13. Coussement K, Van den Poel D (2008) Churn prediction in subscription services: an application of support vector machines while comparing two parameter-selection techniques. Expert Syst Appl 34(1):313–327. https://doi.org/10.1016/j.eswa.2006.09.038
14. Vu VH (2024) Predict customer churn using combination deep learning networks model. Neural Comput Appl 36:4867–4883. https://doi.org/10.1007/s00521-023-09327-w
15. Tsai C-F, Lu Y-H (2009) Customer churn prediction by hybrid neural networks. Expert Syst Appl 36(10):12547–12553. https://doi.org/10.1016/j.eswa.2009.05.032
16. Yasin O, Huseyin S (2024) Optimising customer retention: an AI-driven personalised pricing approach. Comput Ind Eng Sci Direct. https://doi.org/10.1016/j.cie.2024.109920
17. Kalkan IE, Şahin C (2023) Evaluating cross-selling opportunities with recurrent neural networks on retail marketing. Neural Comput Appl 35(8):6247–6263
18. Das S, Nayak J (2021) Customer segmentation via data mining techniques: state-of-the-art review. Comput Intell Data Min: Proc ICCIDM 2022:489–507
19. Kasem MS, Hamada M, Taj-Eddin I (2024) Customer profiling, segmentation, and sales prediction using AI in direct marketing. Neural Comput Appl 36:4995–5005. https://doi.org/10.1007/s00521-023-09339-6
20. Li P, Li S, Bi T, Liu Y (2014) Telecom customer churn prediction method based on cluster stratified sampling logistic regression. In: International Conference on Software Intelligence Technologies and Applications & International Conference on Frontiers of Internet of Things, pp 282–287
21. Tao J et al (2023) Explainable AI for cheating detection and churn prediction in online games. IEEE Trans Games 15(2):242–251. https://doi.org/10.1109/TG.2022.3173399
22. Runge J, Gao P, Garcin F, Faltings B (2014) Churn prediction for high-value players in casual social games. In: Proceedings of the IEEE Conference in Computer Intelligence. Games, pp 1–8
23. https://www.kaggle.com/datasets/synful/churn-data

24. Sreevalsan-Nair J (2022) K-nearest neighbors. In: Daya Sagar B, Cheng Q, McKinley J, Agterberg F (eds) Encyclopedia of mathematical geosciences. Encyclopedia of earth sciences series. Springer, Cham
25. Kleinbaum DG, Klein M (2010) Introduction to logistic regression. In: Logistic regression. Statistics for biology and health. Springer, New York. https://doi.org/10.1007/978-1-4419-1742-3_1
26. Fürnkranz J (2011) Decision tree. In: Sammut C, Webb GI (eds) Encyclopedia of machine learning. Springer, Boston
27. Breiman L (2001) Random forests. Mach Learn 45:5–32. https://doi.org/10.1023/A:1010933404324
28. Webb GI (2011) Naïve Bayes. In: Sammut C, Webb GI (eds) Encyclopedia of machine learning. Springer, Boston
29. Ting KM (2017) Confusion matrix. In: Sammut C, Webb GI (eds) Encyclopedia of machine learning and data mining. Springer, Boston
30. Melo F (2013) Area under the ROC curve. In: Dubitzky W, Wolkenhauer O, Cho KH, Yokota H (eds) Encyclopedia of systems biology. Springer, New York
31. Somogyi Z (2021) Performance evaluation of machine learning models. In: The application of artificial intelligence. Springer, Cham. https://doi.org/10.1007/978-3-030-60032-7_3
32. Shah R, Pawar A, Kumar M (2024). Enhancing Machine Learning Model Using Explainable AI. In: Tiwari S, Trivedi MC, Kolhe ML, Singh BK (eds) Advances in data and information sciences. ICDIS 2023. Lecture Notes in Networks and Systems, vol 796. Springer, Singapore. https://doi.org/10.1007/978-981-99-6906-7_25

The Future of CRM/ERP: Harnessing AI and Blockchain

Arun Gupta, Pratiksha Agarwal, and Mohuya Chakraborty

Abstract The integration of blockchain technology and artificial intelligence (AI) with customer relationship management (CRM) and Enterprise Resource Planning (ERP) is examined in detail in this study. Integrating AI and blockchain technology with ERP and CRM is a game-changing strategy for enhancing ecosystem trust, data security, and customer experiences. Large-scale consumer data is analyzed by AI algorithms to forecast behavior, streamline workflows, and personalize interactions—all of which increase productivity. Blockchain ensures data integrity and secure, transparent transactions. CRM/ERP systems are improved by artificial neural networks (ANNs), a branch of AI, which raise revenue and improve customer satisfaction. Blockchain gives businesses control over critical data by offering an unbreakable ledger. Smart contracts streamline procedures and reduce conflicts by automating and enforcing CRM processes. The transparency of blockchain technology fosters reciprocal trust between businesses and customers. Two case studies are dealt with here using Kaggle datasets: (i) Creation of a CRM/ERP for online retail through data analysis and (ii) Linear regression analysis of an ANN-based sales prediction model. The performance is assessed through analysis of the prediction of the most valued customers based on the profit of customer segmentation for the CRM/ERP and through R-squared value (0.816), accuracy (0.82), precision (0.81), F1-score (0.82), recall (0.81), and Root Mean Square Error (0.792) for the ANN-based model. Finally, some studies on blockchain integration have been presented.

A. Gupta (✉)
Confiance Tech Solutions, Los Angeles, CA, USA
e-mail: arguptaonline@gmail.com

P. Agarwal
SAP America, Palo Alto, CA, USA

M. Chakraborty
Cybersecurity Centre of Excellence, Institute of Engineering and Management, University of Engineering and Management, Kolkata, India

M. Chakraborty et al. (eds.), *Proceedings of 5th International Ethical Hacking Conference*, Lecture Notes in Networks and Systems 1148,
https://doi.org/10.1007/978-981-97-8457-8_28

Keywords Customer relationship management (CRM) · Enterprise resource planning (ERP) · Artificial intelligence (AI) · Blockchain · Artificial neural network (ANN) · RFM · CLTV

1 Introduction

Businesses use a collection of procedures, techniques, and technological tools called Customer Relationship Management (CRM) and Enterprise Resource Planning (ERP) systems to track and analyze consumer interactions and data throughout a customer's lifetime. It seeks to understand the company's client profile and communicates accordingly [1]. Enterprise Resource Planning (ERP) is a software system that aids organizations in automating and managing core business processes for optimal performance. While both CRM and ERP are essential software tools for corporate operations, they have distinct uses: ERP is a full-featured system created to oversee and coordinate a company's essential operations. It concentrates on internal operations to enhance productivity and streamline procedures across several departments, including sales, supply chain, manufacturing, finance, and human resources. CRM, on the other hand, is primarily concerned with handling communications with both present and future clients. With the ultimate goal of strengthening client relationships and boosting sales, it assists organizations in tracking customer information, sales, marketing efforts, and customer support interactions. CRM and ERP both aim to attract new customers beyond existing ones by analyzing customers, identifying audiences with specific behaviors, and creating marketing tailored to that audience. With CRM and ERP systems, the customer gains trust, resulting in dependency on the firm or product. Two techniques that are frequently used to analyze customer value are Customer Lifetime Value (CLTV) and RFM where Recency, Frequency, and Monetary are three words combined to form the abbreviation RFM. Businesses can maximize their marketing efforts by using CLTV, the total amount of money a client will spend on your business throughout their lifetime. RFM is frequently used to divide marketing strategies into distinct groups. Recency is the recentness of the customer's activity, whether it be a purchase or a visit; examples include the time since the last order or the last time the customer used the product; frequency is the number of transactions or visits made by the customer; examples include the total number of transactions or the average interval between transactions or engaged visits; and monetary refers to the customer's intention to spend or their purchasing power, including the average or total value of their purchases. Data analysis of customer information is the basis of the CRM/ERP framework and is useful for sales forecasting and predictive analytics. To forecast future sales patterns, artificial neural networks (ANNs) can be used to evaluate previous sales data and customer contacts [2]. This supports companies in making defensible choices on the distribution of resources, consumer interaction, and sales tactics. Strengthening client relationships, boosting sales performance, and enhancing the efficiency of marketing and sales teams are all possible outcomes of ANN integration with CRM/ERP systems.

Additionally, by incorporating blockchain into the CRM/ERP, the transactions remain private. Improved automation and efficiency in client contacts are also provided, along with data security, transparency, and trust [3]. Claim that while AI can improve scalability and security while resolving the customization and governance challenges for blockchain-based technologies, blockchain can provide explainability, privacy, and trust to AI-based applications. Blockchain and AI differ from one another in a number of technical aspects, but they may also be leveraged to improve upon one another's deficiencies. To put it another way, blockchain and AI are the yin and yang of digital business; while blockchain helps businesses execute, verify, and record, AI helps businesses comprehend, recognize, and make decisions [4].

The major contributions to the paper include:

1. Study and analysis of an AI-based CRM/ERP system of an online retail.
2. Design of a sales prediction model based on ANN.
3. Discussion of integration of blockchain in CRM/ERP systems.

The rest of the paper is presented as follows. In Sect. 2, deliberation of case study 1 dealing with AI-based CRM and ERP system study for online retail is presented. This is followed by case study 2 including the system model and simulation result analysis of the ANN-based sales prediction model of online retail in Sect. 3. Section 4 presents the analysis of the integration of blockchain into CRM/ERP systems. The paper concludes in Sect. 5 with some highlights on prospects and scopes.

2 Case Study 1: Data Analysis of CRM/ERP System for Online Retail

2.1 System Design

In this study, data analysis of CRM/ERP is conducted by using a dataset of Online Retail containing all the transactions occurring for a UK-based and registered, non-store online retail between 1st December 2009 and 9th December 2011 [5–7]. This study and analysis are mainly conducted to identify the most profitable customers, the features liked and appreciated in the products or services of the company by these customers, probable new customers, and to understand the techniques of attracting new customers.

2.2 *Data Analysis*

2.2.1 Data Preprocessing

The dataset [5–7] comprises 8 features or attributes and 525,461 data points. The features are "Invoice No", "Stock Code", "Description", "Quantity", "Invoice Date", "Price", "Customer ID", and "Country". Data is pre-processed to remove outliers and missing values (if any). This is done by checking the descriptive statistics of the numerical values of "Quantity" and "Price", which shows the presence of outliers as depicted by the box plots (graphical representations of the distribution of the dataset explained in Sect. 3.3) in Fig. 1a and b respectively. The outliers are removed without damaging the dataset followed by deletion of the missing values as shown by box plots in Fig. 1c and d respectively. These figures indicate that the data are right-skewed.

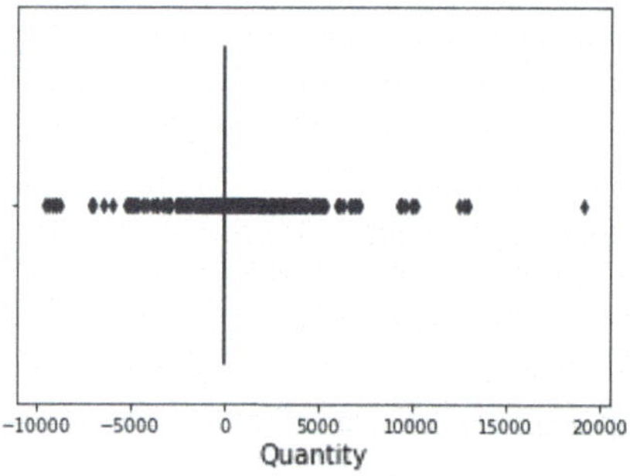

a Box Plot of "Quantity"

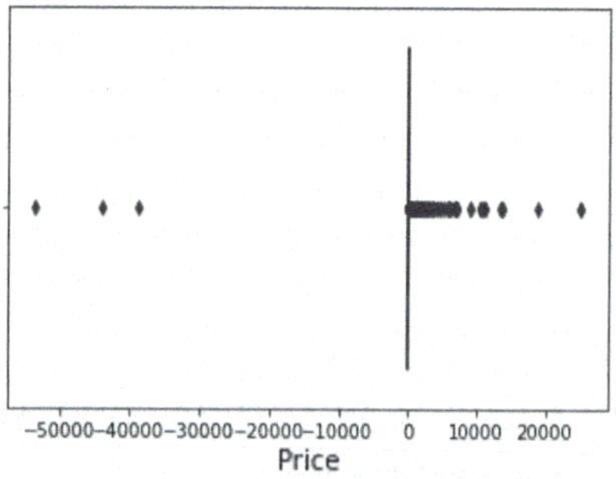

b Box Plot of "Price"

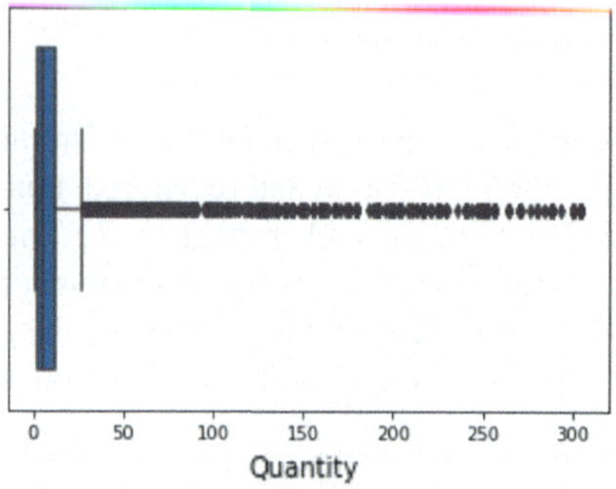

c Box Plot of "Quantity" After Outlier Removal

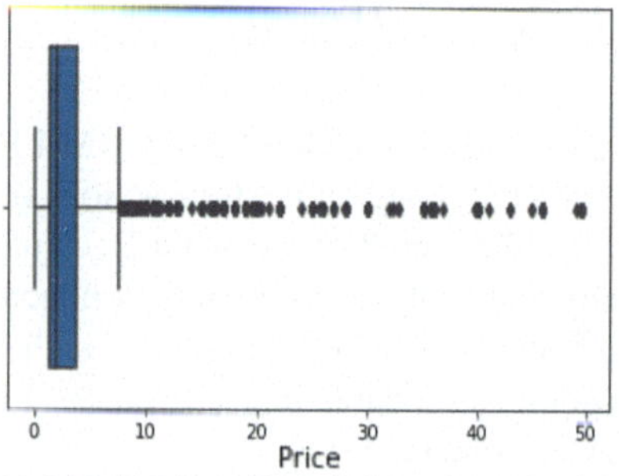

d Box Plot of "Price" After Outlier Removal

Fig. 1 Box plots

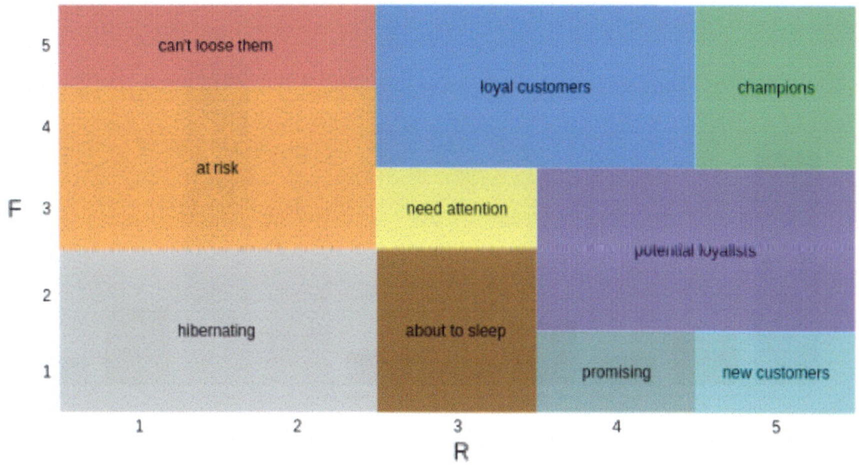

Fig. 2 RFM segmentation

2.2.2 RFM Creation

The next step is to create the RFM, which represents a method used for measuring customer value. An RFM analysis can show the most valuable customers for any business by segregating the ones who buy most frequently, most often, and spend the most [8]. The RFM metrics such as Recency, Frequency, and Monetary are calculated. The RFM segmentation based on the recency score and frequency score with monetary > 0 is done so as to identify various categories of customers, such as: (a) Cannot Lose Them, (b) At Risk, (c) Hibernating, (d) Loyal Customers, (e) Champions, (f) Need Attention, (g) About to Sleep, (h) Promising, (i) Potential Loyalists, (j) New Customers. This is shown in Fig. 2. With better RFM segmentation, companies can better address certain segments in a personalized manner, based on their needs and preferences. Figure 3 shows the bar plots of the frequency of the segmented customers.

2.2.3 Cltv

The amount of money a client will spend with a business over the course of their whole payment relationship is known as CLTV. It is the monetary value that a customer will give to a company during its relationship-communication with a company and can extract the future situation from the current situation of the customer [9].

CLTV is calculated by segmenting people into A, B, and C categories where A > B > C. Figure 4 shows the calculated CLTV of RFM segments as per Eq. (1).

$$CLTV = Customer\ Value(CV) \times Average\ Cuatomer\ Lifespan(ACL) \tag{1}$$

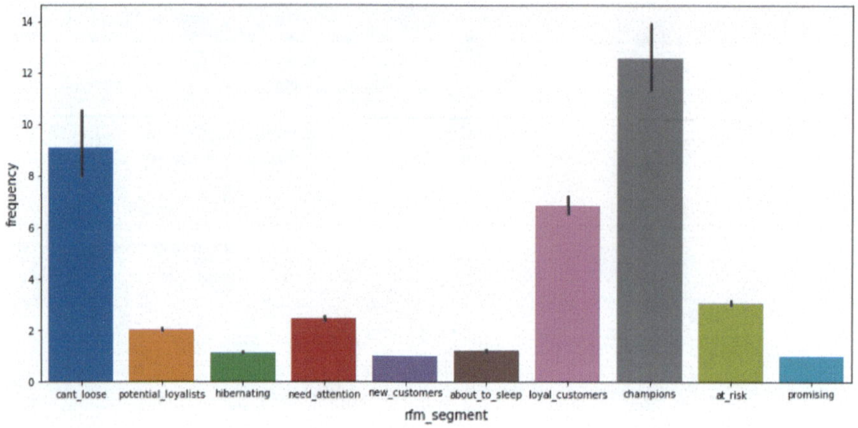

Fig. 3 Frequency of RFM segments

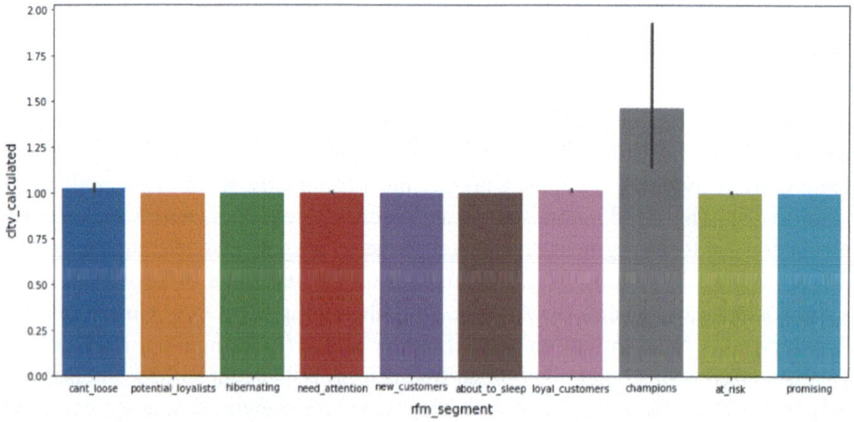

Fig. 4 Calculated CLTV of RFM segments

Here,

CV = Average Purchase Value (APV) × Average Number of Purchases.

ACL = Sum of Customer Lifespans/Number of Customers.

APV = Total Revenue/Number of Orders.

The scatter plot between monetary and frequency is depicted in Fig. 5. There is no correlation between the frequency of transactions and their monetary value. Only repeat customers—that is, customers with a frequency > 0—have been taken into account in the simulation; a frequency of 0 indicates that a customer is a one-time customer and is deemed to have passed away.

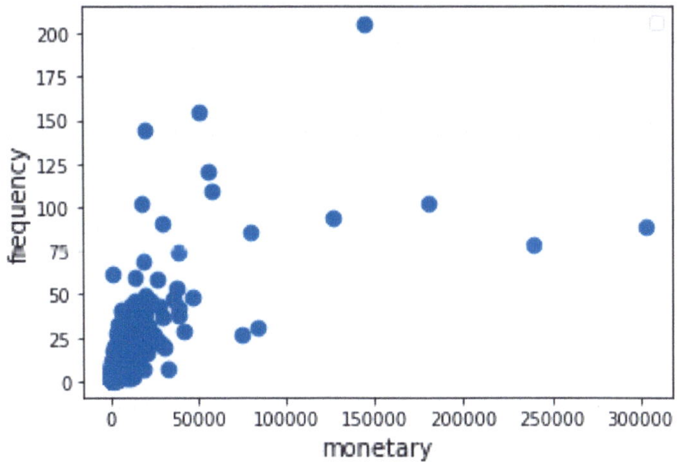

Fig. 5 Scatter plot between monetary and frequency

2.2.4 Prediction

Finally, the CLTV segment is predicted based on the profit as shown in Fig. 6. For clients in various segments, different conclusions are drawn based on the comparative analysis of each indicator. After that, several campaigns based on the choices made are created. Following that, the A/B test can be used to gauge the campaign's productivity by presenting several offers to the general public.

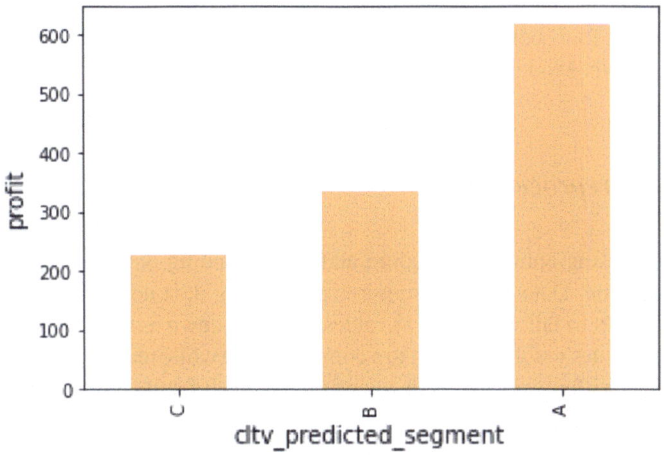

Fig. 6 Predicted CLTV segment with profit

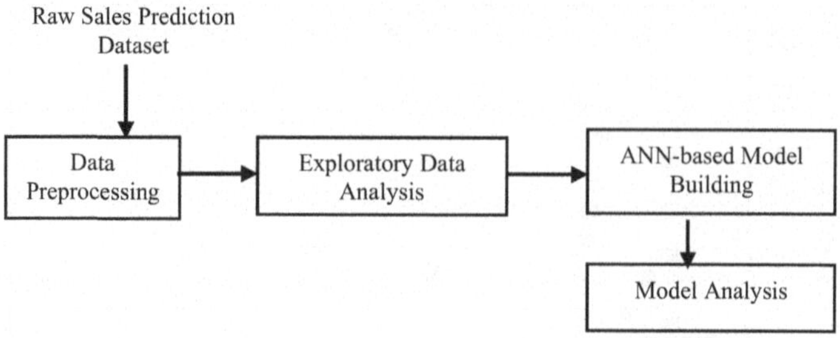

Fig. 7 ANN-based system model

3 Case Study 2: ANN-Based Sales Prediction Model

This study gives an overview of the system model and simulation result analysis of an ANN-based online retail.

3.1 System Model

An ANN-based sales prediction model is designed based on the monetary value spent on different platforms for marketing such as "TV Advertising", "Newspaper Advertising", and "Radio Advertising". Here ANN has been adapted for linear regression tasks by configuring it with just a single neuron as an output of a single linear line (TV Advertising Sales Prediction) with no hidden layer. The proposed system model is shown in Fig. 7.

It consists of four sections: (i) Data Preprocessing, (ii) Exploratory Data Analysis (EDA), (iii) ANN Model Building, and (iv) Model Analysis.

3.2 Data Preprocessing

Data preprocessing comprises reading and understanding data, inspection of data, and data cleaning. Data cleaning requires performing data normalization to rescale the entire dataset to fall within [0, 1] values. This is known as min–max scaling. It is used to align values within a dataset to a common scale while maintaining the integrity of value range differences. In order to ensure that each feature contributes correctly to the model's predictions, normalization is essential for datasets including features that fluctuate in units or magnitude. The mathematical expression for normalization is

$$X_{norm} = \frac{(X - X_l)}{(X_h - X_l)} \tag{2}$$

where X is the original value, X_l is the lowest value for the feature, and X_h is the highest value for the feature.

3.3 Exploratory Data Analysis

EDA was performed on the advertising dataset and the relationship between "TV Advertising" and "Sales" using an appropriate predictor variable was analyzed. The various types of analysis that are performed in this study are briefly presented below [10].

Box Plot: It is a graphic representation of a dataset's distribution, providing clear summary statistics like median, quartiles, and outliers in an easy-to-understand manner.

Scatter Plot: It is a mathematical diagram that shows the correlation between two variables in a set of data using Cartesian coordinates.

Heat Map: It is a two-dimensional data visualization method that uses a color and matching correlation value to show the magnitude of specific values inside a dataset.

Figure 8 depicts the outlier analysis of the features of the dataset such as TV, Radio, and Newspaper using box plots for the present study.

It shows that there are no considerable outliers present in the data. The TV and Radio show normal distribution whereas Newspaper is positively skewed. The boxplot of the Sales is depicted in Fig. 9, which shows no outliers with slightly negative skewed distribution. The heat map (correlation matrix) of the dataset presented in Fig. 10 depicts the highest correlation of TV with sales, followed by Radio and Newspaper. Finally, the scatter plots of sales with the dataset's features are shown in Fig. 11, which indicates higher advertising spend is associated with higher sales. Radio advertising has been shown to have a not a distinct positive correlation whereas Newspaper advertising has no correlation with sales.

3.4 Model Building

The sales prediction dataset comprises four features and 200 datapoints [11]. The features are TV, Radio, Newspaper and Sales. At this stage, 75% of the dataset is used for training and the remaining 25% is used for testing. An ANN-based simple linear regression model is used for sales prediction. In this study, dependent variable is "Sales" and the independent variables are "TV, "Radio", and "Newspaper". The most common form of regression analysis is linear regression, in which one finds the line (or a more complex linear combination) that most closely fits the data according to a specific mathematical criterion [12]. Simple Linear Regression is mathematically described as follows.

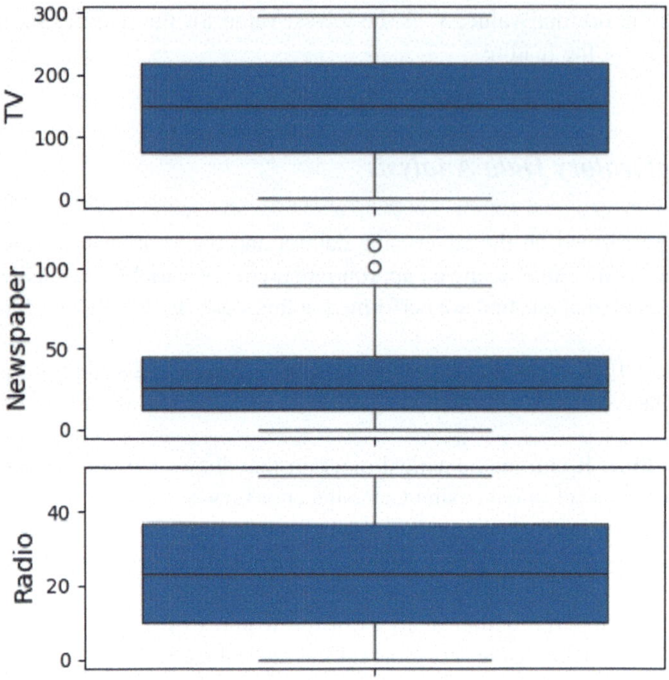

Fig. 8 Outlier Analysis using box plots

Fig. 9 Plot of "Sales" data

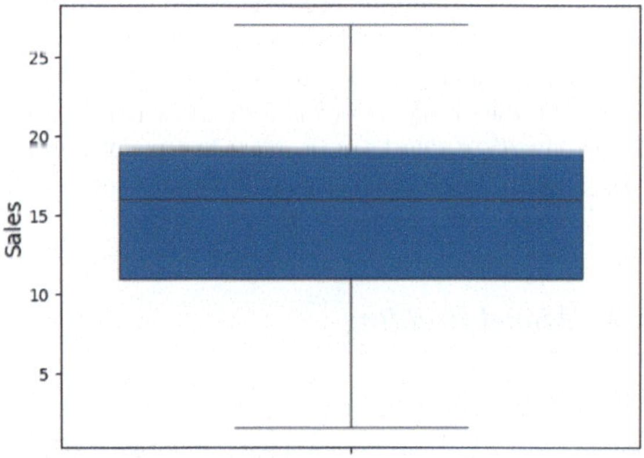

$$Y = c + m_1X_1 + m_2X_2 + \cdots + m_nX_n \tag{3}$$

where m_1 is the coefficient for the first feature, m_n is the coefficient for the nth feature, Y is the output, and c is the bias (intercept). The model weights, or coefficients, are denoted by the m values. In this case:

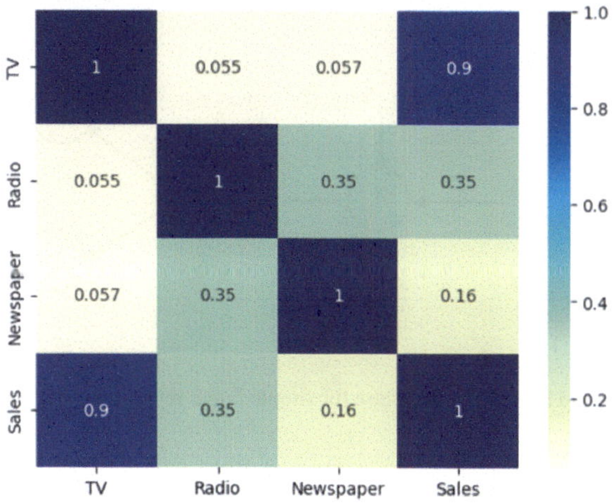

Fig. 10 Heat map of the dataset

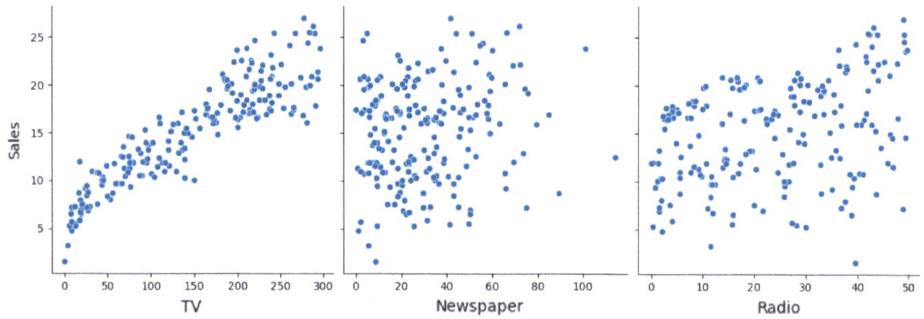

Fig. 11 Scatter plots of "sales" dataset's features

$$Y = c + m_1 TV \tag{4}$$

The Linear Regression learns the parameters by the following mathematical expression.

$$\beta_1 = \frac{\sum (X_i - \overline{X})(Y_i - \overline{Y})}{\sum (X_i - \overline{X})^2} \tag{5}$$

X_i is the feature X, $\overline{X}$ is the mean of X, Y_i is the output Y, $\overline{Y}$ is the mean of Y. The variance of feature X is shown by the denominator, while the numerator represents the covariance of the data. The outcome β_1 is the slope of the linear regression line. The intercept determines the location of the line's y-intercept.

ANNs are chosen for regression to uncover the complex non-linear relationship between the features and the target. ANNs are one of those methods as each layer of an ANN has an activation function, allowing them to learn the intricate link between the

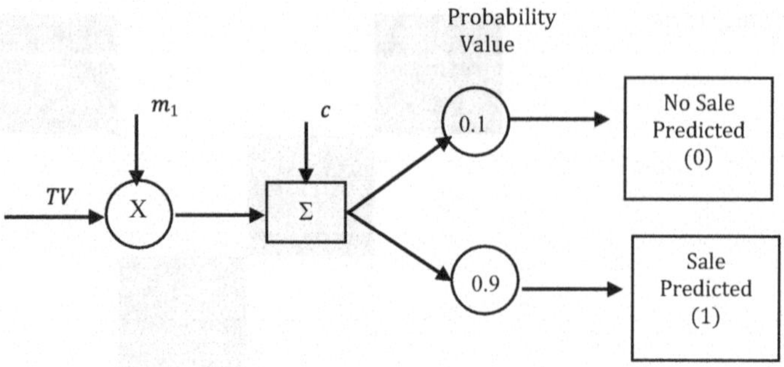

Fig. 12 ANN architecture for simple linear regression

feature characteristics and the target. For simple linear regression, with one feature as in this case, a basic ANN with one neuron for a single input feature (single node) in the input layer, no hidden layers, and one output neuron for the prediction (single neuron) that use a linear activation function in the output layer is used since the prediction of a continuous value of "Sales" target is performed here. This architecture is shown in Fig. 12. The input neuron receives the feature value, and the output neuron produces the prediction. To reduce the error between the anticipated and actual values, the weight and bias term in the connection between them would be modified during the training phase by using any optimization algorithm to find the values of m_1 and c that lead to the best-fit line for the data. The probability value calculated at this point is fed to a comparator that determines whether it is higher or lower than the threshold value. A higher value indicates sales prediction (1) whereas a lower value indicates no sales prediction (0).

3.5 *Model Analysis*

Ordinary Least Squares (OLS) regression analysis, a form of inferential statistics is performed subsequently. In statistics, OLS is a type of linear least squares method for choosing the unknown parameters in a linear regression model (with fixed level-one [13] effects of a linear function of a set of explanatory variables) by the principle of least squares: minimizing the sum of the squares of the differences between the observed dependent variable (values of the variable being observed) in the input dataset and the output of the (linear) function of the independent variable. The summary operations of the regression line fitted are shown in Fig. 13. Table 1 shows the performance metrics of the experimental result needed for data analysis from Fig. 13.

The explanation and visualization of how well the model fits the data are briefly explained here. A respectable R-squared value of ~ 0.816 indicates that TV accounts for 81.6% of the variation in "Sales". The statistical significance of the model fit

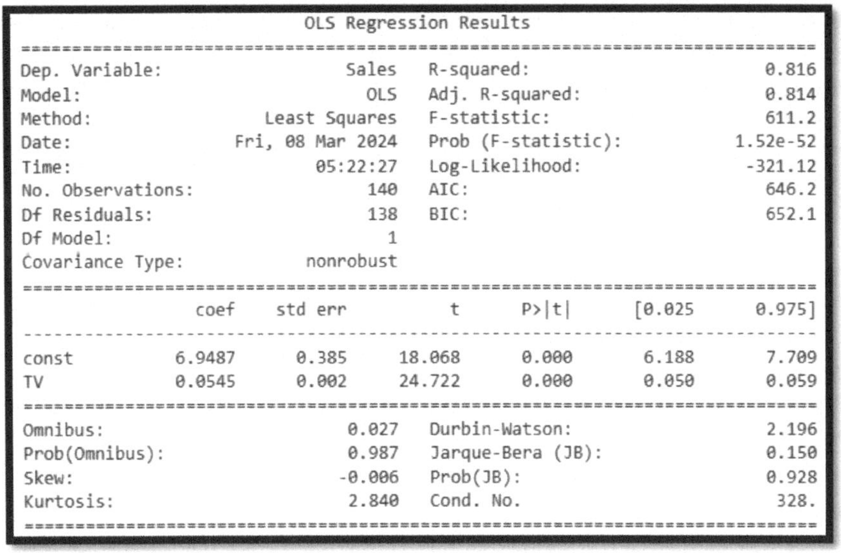

```
                            OLS Regression Results
==============================================================================
Dep. Variable:                  Sales   R-squared:                       0.816
Model:                            OLS   Adj. R-squared:                  0.814
Method:                 Least Squares   F-statistic:                     611.2
Date:                Fri, 08 Mar 2024   Prob (F-statistic):           1.52e-52
Time:                        05:22:27   Log-Likelihood:                -321.12
No. Observations:                 140   AIC:                             646.2
Df Residuals:                     138   BIC:                             652.1
Df Model:                           1
Covariance Type:            nonrobust
==============================================================================
                 coef    std err          t      P>|t|      [0.025      0.975]
------------------------------------------------------------------------------
const          6.9487      0.385     18.068      0.000       6.188       7.709
TV             0.0545      0.002     24.722      0.000       0.050       0.059
==============================================================================
Omnibus:                        0.027   Durbin-Watson:                   2.196
Prob(Omnibus):                  0.987   Jarque-Bera (JB):                0.150
Skew:                          -0.006   Prob(JB):                        0.928
Kurtosis:                       2.840   Cond. No.                         328.
==============================================================================
```

Fig. 13 OLS regression results

Table 1 Performance metrics

Sl. No.	Item	Value	Sl. No.	Item	Value
1	Dependent variable	Sales	10	p-value	0.00
2	Model	OLS	11	Confidence interval of constant term	(6.188, 7.709)
3	Number of observations	140		Confidence interval of coefficient	(0.050, 0.059)
4	Df residuals	138	12	R-squared value	0.816
5	Df model	1	13	F-statistic	611.2
6	Constant term c	6.9487	14	AIC	646.2
7	Coefficient term m_1	0.0545	15	BIC	652.1
8	Standard error of constant term	0.385	16	Log-likelihood	-321.12
	Standard error of coefficient term	0.002	17	Adjusted R-squared value	0.814
9	t-statistics of constant term	18.068	18	Covariance type	Non-Robust
	t-statistics of coefficient term	24.722			

and the fact that the explained variation is not solely due to chance are indicated by the extremely low p-value of the F-statistic. The training dataset's input and output are plotted in a scatter plot in Fig. 14, which shows how well the line of regression covers the data. The histogram of the error terms between output training and output training predicted data depicted in Fig. 15 shows that the residuals follow normal distribution with a mean of 0.

Figure 16 shows the patterns in the residuals. The figure demonstrates that the model fit has some predictive potential and is not accidental. The residual terms' normality permits various deductions about the coefficients. The fact that the variance of the residuals increases with input suggests that there is a substantial amount of variation that this model cannot account for. The test data is then fitted with a regression line by adding a constant to make the test data more comparable to the train data. Next, using the fitted regression line's predict attribute, the output values corresponding to the input test data are predicted.

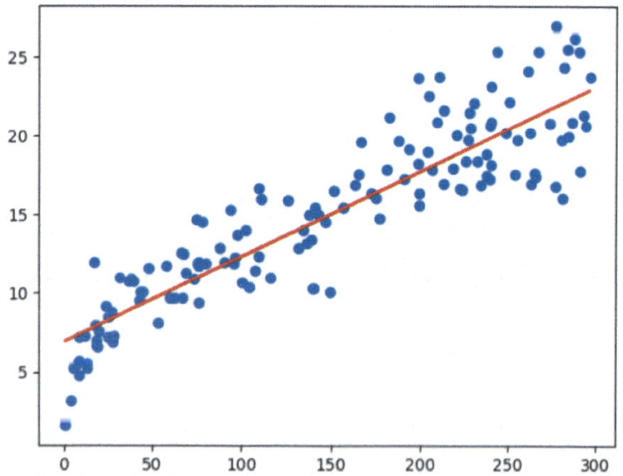

Fig. 14 Scatter plot between input and output of training data

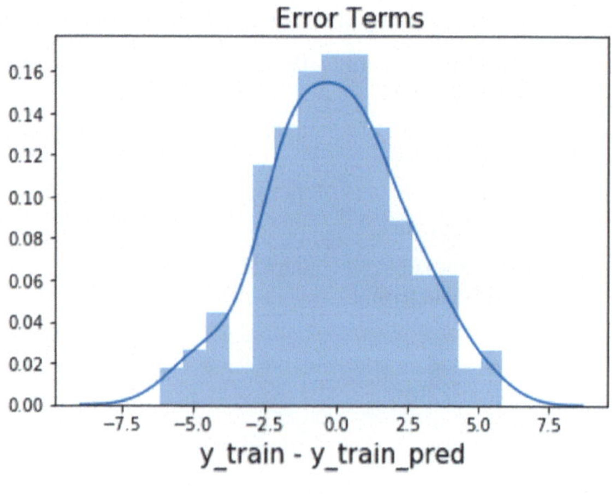

Fig. 15 Histogram plot of error

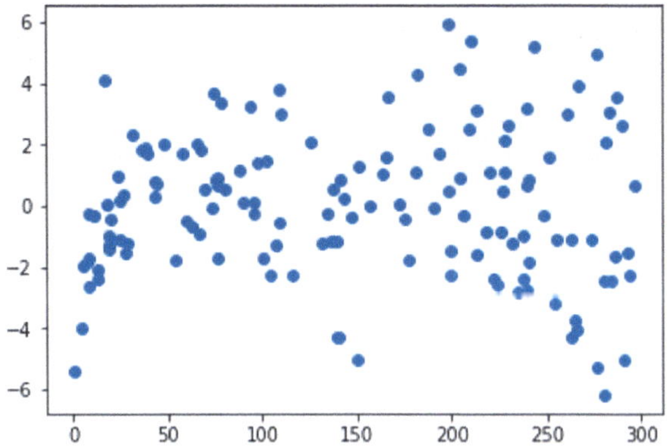

Fig. 16 Scatter plot between input train data and residual

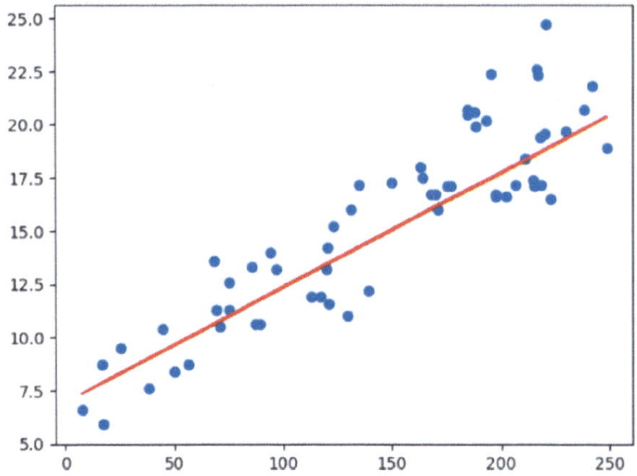

Fig. 17 Scatter Plot between input and output test data

The scatter plot between input and output test data is shown in Fig. 17. It depicts that the regression line fits quite well with the data with $c \sim 7$. The prediction results show a Root Mean Squared Error (RMES) of 0.792, accuracy of 82%, precision of 81.5%, F1-score of 82.6%, and recall of 81.7% which are quite acceptable.

4 Blockchain Integration to CRM/ERP Systems

The remaining part of the study involves understanding integration of blockchain for secured customer transactions using proper authentication mechanisms and smart contract in AI-based CRM/ERP systems. AI-based CRM/ERP systems can benefit from blockchain technology to provide improved privacy, trust, transparency, operational efficiency, and data security. The efficiency of these systems can be increased as blockchain offers an accurate, tamper-proof, and secure ledger for data. Blockchain empowers automatic data entry and validation, thereby reducing human intervention

Table 2 AI vs Blockchain

AI	Blockchain
Centralized infrastructure	Decentralized and distributed infrastructure
Decisions by machine learning unexplainable to human beings	Explainable to humans
Lacks transparency and trust	Trustworthy
Probabilistic	Deterministic
Adapt over time	Immutable

and increasing efficiency. Sensitive data are anonymized before processing and so it enhances customer privacy. Innovative business models based on blockchain smart contracts may be designed that will permit customer interaction and service delivery management by merging blockchain and AI [14]. User verification is one of the most important procedures in establishing trust between the parties before any transactions are performed. Hyperledger Fabric is a permissioned blockchain framework that may be used to design authentication mechanisms for user verification [15]. This could entail setting up a network, maintaining user credentials, authentication, and registration, and developing an authentication smart contract based on chaincodes. The smart contract authenticates users using their credentials, maintains pertinent data, and confirms user registration. Encryption techniques allow for the control of access to resources and services, while identity management systems manage user identities and rights on the network.

Table 2 illustrates the technological differences between blockchain and AI, but they can also be used to strengthen each other's weaknesses [16]. It is possible to build a safe, decentralized, and impenetrable authentication system that offers transparency and confidence in the authentication procedure by utilizing Hyperledger Fabric for authentication in AI-based CRM/ERP systems. Hence the system integration should include AI within the blockchain infrastructure to enhance trust and predictability in a CRM/ERP system.

5 Conclusion

This paper provides studies and design analysis of a CRM/ERP system using AI-based data analysis of customer information in online retail followed by proposing an ANN-based sales prediction model using a simple regression model. The metrics such as RFM and CLTV are calculated and analyzed for predicting the most valued customers based on the profit of customer segmentation. Performance metrics of the sales prediction model show an R-squared value of 81.6%, accuracy of 82%, precision of 81.5%, F1-score of 82.6%, and recall of 81.7%, and RMSE of 0.792, all of which are within the acceptable range proving the efficiency of the model. This study demonstrates how well blockchain integration with CRM/ERP systems works, greatly benefiting all parties involved. Businesses thinking about integrating this technology should assess how these technologies can meet the unique needs and problems of their operations. With a wide range of possible uses, including supply chains, healthcare, banking, and accounting, there are many chances for innovation and advancement. The investigation, analysis, design, and evaluation of an AI-blockchain-integrated CRM/ERP system is within the purview of future scholars.

References

1. Kumar V, Reinartz W (2018) Customer relationship management, concept, strategy, and tools. In: Textbook Springer Texts in Business and Economics (STBE). https://www.springer.com/series/10099
2. Tian Y, Shu M, Jia Q (2021) Artificial neural network. In: Daya Sagar B, Cheng Q, McKinley J, Agterberg F (eds) Encyclopedia of mathematical geosciences. Encyclopedia of earth sciences series. Springer, Cham
3. Salah K, Rehman MHU, Nizamuddin N, Al-Fuqaha A (2019) Blockchain for AI: review and open research challenges. IEEE Access 7:10127–10149
4. Haleem A et al (2022) Artificial intelligence (AI) applications for marketing: a literature-based study. Int J Intell Netw 3(2022):119–132. https://doi.org/10.1016/j.ijin.2022.08.005
5. /kaggle/input/online-retail-ii-data-set-from-ml-repository/Year 2010–2011.csv
6. /kaggle/input/online-retail-ii-data-set-from-ml-repository/Year 2009–2010.csv
7. /kaggle/input/online-retail-ii-data-set-from-ml-repository/online_retail_II.xlsx
8. Taşabat SE, Özçay T, Sertbaş S, Akca E (2023) A new RFM model approach: RFMS. In: Singh G, Goel R, Garg V (eds) Industry 4.0 and the digital transformation of international business. Springer, Singapore
9. Channa HS (2019) Customer lifetime value: an ensemble model approach. In: Balas V, Sharma N, Chakrabarti A (eds) Data management, analytics and innovation. Advances in intelligent systems and computing, vol 808. Springer, Singapore
10. Rao AS, Vardhan BV, Shaik H (2021) Role of exploratory data analysis in data science. In: 6th International Conference on Communication and Electronics Systems (ICCES), Coimbatre, India, pp 1457–1461
11. https://www.kaggle.com/datasets/ashydv/advertising-dataset
12. Huang M (2020). Theory and implementation of linear regression. In: International Conference on Computer Vision, Image and Deep Learning (CVIDL), Chongqing, China, pp 210–217
13. Zdaniuk B (2014) Ordinary least-squares (OLS) model. In: Michalos AC (ed) Encyclopedia of quality of life and well-being research. Springer, Dordrecht
14. Khan SN, Loukil F, Ghedira-Guegan C et al (2021) Blockchain smart contracts: applications, challenges, and future trends. Peer-to-Peer Netw Appl 14:2901–2925. https://doi.org/10.1007/s12083-021-01127-0
15. Cai L, Li Q, Liang X (2022) Application development: fundamentals of hyperledger fabric. In: Advanced blockchain technology. Springer, Singapore. https://doi.org/10.1007/978-981-19-3596-1_5
16. Kumar S, Lim WM, Sivarajah U et al (2023) Artificial intelligence and blockchain integration in business: trends from a bibliometric-content analysis. Inf Syst Front 25:871–896. https://doi.org/10.1007/s10796-022-10279-0

Can AI Help Indian Judiciary to Reduce Its Burden of Cases? Exploring the Potential of AI in Judicial Decision-Making Process

Arin Mukherjee and Shambhu Prasad Chakrabarty

Abstract At this juncture, the thing that is most concerning to the legal fraternity is the number of pending cases in India, so finding an appropriate solution by exploring the potential use of AI, is what the paper aims at. Therefore, the paper considers the option of using AI to assist the courts in the speedy adjudication of cases, which is fundamental in a welfare state because justice delivery mechanism is the primary means of justice delivery. Although in the European Union and in China the concept of Robo-Judge has come up which adjudicates particular types of cases, this paper propose to incorporate independent Generative AI tool for adjudication pending of cases to work as an Augmented Intelligence whose primary purpose would be to assist the judge in the research work of a case and to come up with a probable judgement. It would depend upon the discretion of the judge whether to accept the suggestions generated by AI or not. The judge will be at liberty to use his own reasoning while giving a judgement and may use AI as a mere expert opinion. An important factor to achieve this end is ensuring the accuracy of data accessible by the AI tool, responsible for the suggestions generated.

Keywords AI & law · Aoi and law · AI in judicial decision-making · Unburdening the judiciary

---

A. Mukherjee (✉)
Department of Law, IEM's International Institute of Juridical Sciences (IIJS), University of Engineering and Management, Kolkata, India
e-mail: mukherjeearin23@gmail.com

S. P. Chakrabarty
Department of Law (IIJS), University of Engineering and Management, Kolkata, India
e-mail: shambhuprasad.chakrabarty@uem.edu.in

M. Chakraborty et al. (eds.), *Proceedings of 5th International Ethical Hacking Conference*, Lecture Notes in Networks and Systems 1148,
https://doi.org/10.1007/978-981-97-8457-8_29

1 Introduction

The fundamental element of a democracy is the supremacy of law, which is regarded as key to justice. Although the two key pieces of this puzzle are the access to justice and the independence of judiciary, without which the whole idea of democracy becomes redundant because the idea of democracy is to make the people entitled to govern themselves. That makes it a collective unit, as mentioned by Abraham Lincoln, "Government of the people, by the people, for the people". In light of the same statement, it can be said that as working for each other, working with each other, and complementing the society at large. If it is not ensured in a proper manner, then the supposed base of Justice that is democracy would remain as a gimmick, and the statement of Martin Luther King becomes very relevant, "Injustice anywhere is a threat to justice everywhere".

The discussion surrounding access to justice began a movement that has had a long-lasting, global influence [1]. The Florence Access to Justice Project [2], led by eminent academic Professor Mauro Cappelletti, launched the idea of access to judicial institutions in the late 1970s [3]. Cappelletti identified three "waves" in the evolution of this idea. The first wave [4] focused on providing legal aid to the underprivileged, while the second and third waves expanded representation to diffuse interests including consumers, environmentalists, and ADR respectively. One of the most expensive and time-consuming activities in modern life is getting access to the judicial system. Justice ought to be swift and unbiased. The justice delivery system is still an unsustainable solution to all legal disputes since not everyone can afford the time and money that it needs [5] if we go by the principle of "justice delayed justice is denied" [6]. It has been stated earlier that the court system is overloaded by the huge amount of outstanding cases and this has been the case all over the world, and several nations are considering technology as an alternative to human procedure for the speedy redressal of the enormous backlog of pending cases. China for the first time has introduced the system of automatic adjudication, following that path USA has also worked towards introducing the "Robo-judge" [7].

According to the data published by the World Justice Project, India is placed at 77th [8] out of 144th nations on the Rule of Law index which was created considering various aspects such as the absence of corruption, order, and security, regulatory enforcement, and civil and criminal justice. If we talk only about the regulatory function index which portrays the scenario relating to access to courts and the procedure, India stands at 88th place among 140 participatory nations. The findings are not unexpected because the Indian legal system has long been plagued by a backlog unresolved cases. This stockpile has been caused by a variety of issues, including overly complex, unenforced procedures, automatic appeals, systemic vacancies on the bench, and seriously misaligned incentive structures [9].

According to the Law Commission of India, it is not a lack of clear procedural legislation that causes court delays, but rather their faulty application or outright disregard [10]. The administration of judicial institutions through manual procedures has become incredibly challenging due to the enormous quantity of open cases, leading to a pattern of chronic failure. The government claims that this has had a detrimental impact on judicial productivity in its Report on Strategy Plan for Deployment of ICT in the Indian Judiciary [11]. The Indian Supreme Court made it quite obvious that this situation needs to be taken care of. As it has been observed by the Supreme Court on numerous occasions that the speedy trial is a part of fundamental rights the state should make positive efforts, that the backlog of cases is reduced [12], and if it is not dealt with then it is a gross violation of the fundamental right and which is being violated every day under the sun.

Innumerable efforts have been taken by both state and private actors to unburden the judiciary, including the formal introduction of Alternative Dispute Resolution strategies. The concept of ODR, often, is considered as an option to minimize backlog and improve efficiency to develop and maintain an efficient and just ODR platform [13].

Introduction of technology to facilitate justice delivery processes especially during the later and post Covid-19 pandemic and post-covid period is worth to be celebrated. This inclusion has allowed judges and victims to be virtually present in the courtrooms eliminating challenges consequential to physical presence. This inclusion of technology in justice delivery system cannot, however, assist judges to write judgements or reduce the burden of the judiciary.

This paper engages in this conundrum and proposes a customized Artificial Intelligence (AI) Model to assist the judges in coming to their judgements easily and reasonably. The utilization of AI in the courtroom is shown by robo-judges. The reasons for these benefits hinge on many shaky assumptions, despite the fact that there may be several potential benefits to be gained from a machine's efficiency and economy that is able to lower caseloads and clear dockets [14].

The author believes that to solve a problem of this magnitude only enactment or establishing various ODR platforms will not be sufficient, to succeed a properly designed curriculum needs to be introduced, not in a traditional but it should be in a way of clinical legal education because learning becomes effective when it is coupled with past experiences (practical) and examples. Irrespective of some improvement in the justice delivery system which adopted ADR, it could hardly be a cure-all to the multiple vices that have plagued the justice delivery system.

2 Application of AI in the Judicial System

Presently, two simultaneous opinions are getting popular, approving and disapproving the incorporation of AI and other technologies in the judicial system. Yet, according to legal futurists, technology has the power to radically alter both the structure and substance of the law as well as the way that it is practiced, for good. Although it may be difficult to anticipate the future, projections for the legal industry have ranged from pessimistic to upbeat. The general trajectory of technology disruption in the legal system, however, is predicted to be rather constant. The rising disaggregation of legal activity underlies—and exasperates the disruptive potential of legal technology. With this disaggregation, several sources of legal information and services are made possible, which encourages commercialization and raises competition [15]; although some negative propositions are being circulated in the air regarding the efficacy of AI in the judicial system and the author argues in favour of the inculcation of AI and Machine Learning and proposes that every single thing has a positive and negative side and it completely depends upon us how the thing is being used. In this regard, the author refers to his childhood days when an essay topic was very popular, "technology: boon or bane", and it is pretty much evident to all of us how technology has proved to be a boon to all of us despite having so many drawbacks, and we are in one hand maximizing the boon of technology, and on the other hand, we are making effort to minimize rather eradicate the bane of technology. In light of this philosophy, the author argues how AI can play a key role in increasing access to justice.

Lawyers are often viewed in the context of legal conflicts as highly trained and skilled professionals who analyse pertinent legal problems and acquire pertinent information. In order to analyse a case's merits and choose the best course of action, attorneys use their judgement. They do this by drawing on their expertise and intuition, but to what extent may legal work be automated this is the question that keeps looping in everyone's mind. Traditional ideas of human competence have been called into question by recent advancements in artificial intelligence, particularly in the areas of natural language processing and machine learning. Nowadays, machines outperform humans at ever more complicated jobs [16]. Until recently, it was believed that only humans were capable of doing these activities. It was stated that these jobs call for human judgement [17]. For instance, it was believed that a machine couldn't handle the complexity of driving. Yet now, autonomous vehicles are very much a reality [18].

Undoubtedly, these technological advancements and the effects of data driven analysis have an impact on the practice of law [19]. Retrospective case analysis has been used by lawyers to pinpoint and comprehend the components or variables that are crucial in reaching conclusions. Due to unique characteristics like lengthier manuscript size, varied internal structure, thorough manual indexing, a complex pattern of document relationships, and a strong reliance on citations, the interpretation and implementation of the law, which heavily depends on the legal texts, is a laborious and time-consuming task. It can be said that artificial intelligence will perform better, and AI, with access to the huge repository of data produced by the courts. However, well-trained AI can model the complex interaction of various factors that can specify an erroneous prediction more thoroughly. Predictions are made using information that is easily accessible, such as the text of legal arguments, complaints, and answers from the parties, as well as the language of pertinent statutes. This is consistent with their shared traits, which also include the text's traits. The AI can demonstrate that they share traits based on their findings. In this map, the case of interest may be found, and its forecast outcome can be obtained by reading the outcome of the group of cases that is most similar to the one discovered. The AI will become more predictable as its complexity rises. The type of data it needs is irrelevant to artificial intelligence. Using unusual categories of data, which are frequently omitted from the text of assertions and arguments, can help all performers do even better [20].

As the usage of AI technology increases, judicial systems are dealing with legal issues relating to the consequences of AI for things like liability, surveillance, and human rights, among others. Moreover, court systems are utilizing AI-enabled or automated technologies for judicial decision-making, which has highlighted questions about justice, accountability, and openness in such procedures [21]. With a number of safety measures in mind, AI and Machine Learning for the delivery of justice will need to be modified. Former Chief Justice of India SA Bobde recently launched the Supreme Court Portal for Aid in Court's Efficiency (SUPACE), which is reportedly the first of its type internationally. It will help facilitate access to information while being non-intrusive when it comes to decision-making, building on the need AI should support the judicial. This method of using AI and machine learning to support but not replace human decision-making is the right one [22]. Justice Bobde stated that he was obstinately opposed to the robotic decision-making method in the court system and that it must never be permitted in a report titled "Responsible AI of Indian Judiciary System". Terming the use of AI in the legal system as a crucial and transformational step, he added that AI should only be used to the amount necessary to help judges, attorneys, and litigants in making the legal system efficient and accessible, but how, when, and what to determine would be left to the courts [23].

3 Use of AI in ADR and ODR

According to Gandhi, the purpose of legislation is to reconcile divided parties [24]. Hence, law has a crucial role to play in the settlement of conflicts, which is "an necessary quality for societal peace, amity, comity, and harmony" [25]. India is only one of the several countries that have adopted various forms of ADR due to the necessity for swift and fair conflict settlement [26]. Furthermore, as the backlog in the courts grows, ADR has become a global requirement [27]. In reality, the Indian Constitution's preamble commands the state to "provide to all the inhabitants of India, justice—social, economic, and political—liberty, equality, and fraternity". This is a clear reference to the purpose of ADR [28]. The Constitution goes on to further clarify these objectives by stating, The State should guarantee that the administration of the judicial system promotes justice… and to ensure that no person is denied the opportunity to achieve justice because of economic or other disadvantages [29]. This principle has been interpreted by the Indian Supreme Court to equate "social justice" with "legal justice", which means that the administration of justice system must provide a cheap, expeditious, and effective instrument for realizing justice by all sections of the people, regardless of their economic or social position or their financial means [30].

As was mentioned previously [31], in a wide variety of legal systems across the world, there is a backlog and delay resulting from several issues, including unbalanced incentive structures among the important parties. These issues may be very severe in contemporary India. India now serves as a test case for rule of law development in general and ADR reform in particular. This case study [32] first analyses India's efforts in ADR, criticizing the function and development of Lok Adalat, before moving on to a stakeholder analysis and the potential of ODR to increase access to justice on the subcontinent. The Legal Services Authorities Act 1987 gave underserved groups access to free, elevated legal representation. It also established Lok Adalat to ensure that the legal system functions to advance justice based on equality of opportunity and that no citizen is denied access to the justice system because of their economic circumstances or other disabilities [33].

The continuous enormous rise of the Indian population and the resulting constraints on court facilities, however, are some issues that go beyond the purview of legal or judicial recomposition. India needs to legislate stricter judicial oversight of claim and defence preparation, more judicial oversight of administrative decision-making, less formal, more conciliation-oriented, and calibrated consensual outcomes in order to maintain the adversarial paradigm of civil justice [34]. Note that ODR lacks definition, which is a crucial problem [35]. Although an effort was made to define Online dispute resolution (ODR) under paragraph 24 of The United Nations Commission on International Trade Law (UNCITRAL) Technical Notes on Online Dispute Resolution 2016 as "A mechanism for resolving disputes through the use of electronic communications and other information and communication technology [ICT]. The process may be implemented differently by different administrators of the process and may evolve over time" [36]. In spite of that effort, the subject matter in contention still remains ambiguous. As a result, some observers define ODR as nothing more than the incorporation of technology into an already established judicial system. When not accompanied by a dispute resolution system, online communications as a way for parties to express grievances shouldn't also be excluded from the definition of ODR. Online customer helpline service tools are all that mechanisms that let customers complain to a company are, and they are not, per se, ODR. We contend that a genuine ODR platform must involve the settlement of a dispute and utilize a neutral facilitator or a neutral decision-maker in order to allow the parties to engage in more than just complaining. [37]

It is high time to see ODR as a huge platform for dispute resolution and focus on developing it as a trustworthy and effective means of dispute resolution mechanism and it should be made in such a way that people accept it as an alternative way of formal judicial process. While many court systems are starting to accept ADR, other groups, especially those who oppose contracts of adhesion that contain obligatory arbitration clauses, are starting to doubt the broad adoption of ADR [38]. ODR will probably have the same rising reluctance towards extensive adoption of ADR. Because the ODR system lacks legislative safeguards and was developed by a corporate sector that might not always have the best interests of the customer as its first goal, people's concerns may be well-founded and arise from mistrust of an untested system [39]. As was mentioned above, a tsunami of legislation in India and the European Union is beginning in reaction to consumer mistrust in ADR. In the United States, a comparable movement is also under progress [40], so it shows that every new thing has to go through some kind of resistance and disbelieve, but that doesn't prove the new system ineffective or inefficient.

The recent development of ODR is very positive in this regard as the first official move towards the establishment of ODR in India was the report of the high-level committee on the deepening of digital payments. A number of relevant laws address the ADR and technological aspects of ODR [41]. Additionally, pre-litigation mediation was made legal in India under section 12A of the Commercial Courts Act, 2015. If urgent interim relief is not required, the parties should first explore the possibility of starting a mediation before proceeding with a lawsuit. The "Mediation and Conciliation Panel" is a group of professionals that must be maintained by the Central Government, according to section 442 of the Companies Act, 2013. The law makes it possible for parties to proceed in front of the National Company Law Tribunal (NCLT) or National Company Law Appellate Tribunal (NCLAT), and seldom the central government, to ask for the dispute to be referred to mediation.

Consumers are given access to mediation under sec. 74 of the Consumer Protection Act 2019; thanks to the creation of "Consumer Mediation Cells" in each district. Moreover, the Consumer Protection Act (E-Commerce) Regulations, 2020, which require e-commerce businesses to have internal grievance redressal systems, laid the groundwork for ODR. The Information and Technology Act of 2000 also applies to ODR's technological components. Sections 4 and 5 of the said legislation acknowledge electronic records and signatures. The Act can be utilized as a support mechanism for the digitalization of the justice delivery system, which urgently needs legal acknowledgement.

In the matter of Shakti Bhog Foods Ltd v. Kola Shipping Ltd. [42], the Supreme Court decided, "According to Sections 4 and 5 of the Information Technology Act", "Online arbitration agreements are legal as long as they satisfy the requirements of a complaint". The Indian Evidence Act, 1872, is another noteworthy legal framework. Sections 65-A and 65-B of the Act recognize electronic evidence and allow for the exchange of virtual documents and the holding of virtual hearings. The Supreme Court permitted remote conferencing for the purpose of choosing an arbitrator in the case of Grid Corporation of Orissa Ltd. v. AES Corporation [43]. In a similar vein, in the case of State of Maharashtra v. Praful Desai, the Supreme Court approved video conferencing as an acceptable method for capturing evidence and witness testimony. In Meters and Instruments Private Ltd & Anr. v. Kanchan Mehta [44], the Supreme Court made the following observations; "Not only should paperless courts be explored, but court congestion should be decreased as well. There seems to be a need to take into account case types that can be partially or completely resolved "online" without the actual presence of the parties by streamlining processes when hotly contested issues do not need to be decided".

In spite of having a variety of laws supporting ADR and technology separately and together, we still have a lot to learn in the area of online dispute resolution. However current government initiatives show a favourable approach to a dearth of technology and infrastructure in the past. The government needs to embrace a progressive approach to technologies to maximize the benefits while also taking precautions to reduce the danger involved in integrating technology into dispute resolution. ODR still hasn't attained the pinnacle of accomplishment, but it is unquestionably on the correct track to overcome challenges like cyber-security and widespread technological understanding.

4 AI and Recent Uses in Justice Delivery System

The recent advances in technology are changing the dynamics of human life and the recent advancement of Artificial Intelligence either tends to replace humans or already replaced them, as the example of Robo Judges has already been given where human judges had been replaced by AI. In furtherance of that, there are other areas of the justice delivery system where AI has been incorporated.

4.1 Case Study 1: Traffic Rules Enforcement

In terms of a nation's social or economic standing, road transport is regarded as being crucial for the daily activities of humanity. There is a need to provide an efficient transport system since the demand for better transport services is increasing at an alarming rate. Intelligent transport systems are situations where we use sensing, analysis, control, and communication technologies to monitor road traffic in order to protect people's lives and property, increase people's mobility, and increase the effectiveness of road transport networks [45]. In the past few years, we have witnessed the incorporation of AI-controlled traffic system, and a recent report has come out which shows that the Bengaluru city police in Bengaluru have been able to file more than 1 million traffic violation complaints annually—the most of any city in the world—and collect close to Rs 180 crore in penalties from disobedient drivers thanks to the adoption of artificial intelligence to control the city's chaotic traffic [46].

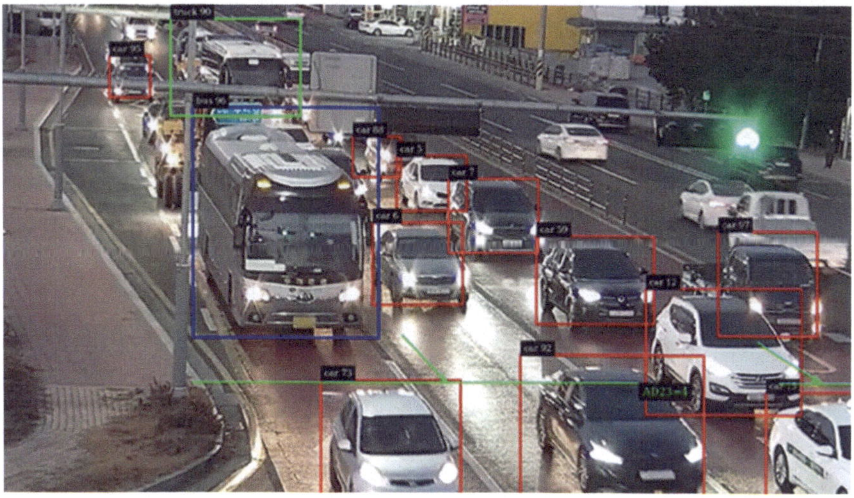

This shows that implementation with proper planning always comes out as a great step towards the mitigation of any problems, and the abovementioned numbers speak for themselves.

4.2 *Case Study 2. AI in the Chinese Judicial System*

The Chinese State Council had implemented an action plan to incorporate AI into their judicial system and planned to make a data platform that integrates trial, personnel, data applications, judicial opening, and dynamic monitoring, and promotes the application of artificial intelligence in evidence collection, case analysis, reading, and analysis of legal documents, to realize the intelligent and experimental ability of the trial system with a view to reform their judicial system [47]. A study focusing on the development and scope of the newly integrated judicial system highlights the factors considered in developing that, such as, firstly, the mode and standardization of collection of evidences, secondly, focusing on the limitations posed by the AI and most importantly the issues relating to the algorithmic outputs, thirdly, as AI completely relies on data so the integrating the data was the top priority because a single trickle of wrong data can completely change the whole output, and lastly but most importantly the thing that is the choice of the judge to accept of proposed judgement by using the reasoning.

The two above-stated case studies rementioned to highlight the various models that have incorporated AI to every aspect of the judicial system whether it is the Robo-Judge system in EU at the ODR system or in the Chinese Judicial system or in the traffic regime in some parts of India to mitigate the traffic-related cases in India, and to chalk out a model for Indian judicial system to make it able to deal with the huge backlog of cases (Fig. 1).

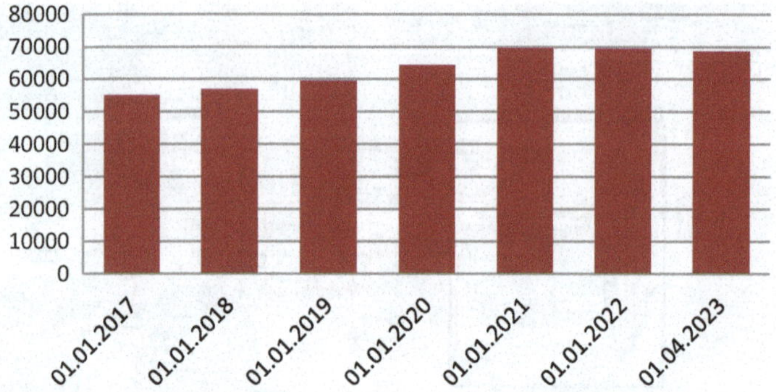

Fig. 1 Number of cases pending before the Supreme Court of India, https://main.sci.gov.in/case-status

This abovementioned chart only sums up the scenario of the Supreme Court of India, then one could probably guess what the scenario in the High Courts and the other lower courts in India is!

5 How AI Can Assist the Judges to Come to a Decision

The model we are proposing to apply to the Indian court system was developed after carefully examining existing models from across the globe. It is based on the observation that many of the talks on AI in the legal context now being had oversimplify the role of judges. In order to comprehend what AI can or cannot accomplish in court chambers, it is crucial to first take into account what the work of judges entails, in the opinion of Judge Aharon Barak of the Israeli Supreme Court, who exposed a complete spectrum of judicial duties of a judge, such as fact determination, law determination, and law application [48].

In light of the abovementioned thoughts, it can be examined to what point AI can provide assistance to the judges. Surely, AI can do for judges a lot of the same tasks AI does for advocates. For instance, much as AI speeds up legal research for attorneys in many situations, it would be prudent to exercise caution when relying entirely on the outcomes of AI legal research until a full-proof database is accessed by AI. AI is largely employed in all of these contexts to support judges in their duties. To put it differently, AI is not doing anything in these circumstances that a team of clerks or court personnel couldn't perform instead. While AI technologies may be beneficial to overburdened and underfunded courts, the job of the judge may not be substantially altered. The concern that now arises is, what happens in instances where AI is utilized to connect directly with the public in place of doing so with the judge? There are various circumstances when this has started to occur [49] (Example China). Examining these all, the thing which can be referred to keeping in mind Indian Circumstances that there are several models that can be used for implementing AI directly into the judicial decision-making process, a judge less model is not we are aiming for at this juncture, maybe we can aim for that later once after enough empirical data being collected after initial incorporation of AI into the judicial system.

6 Conclusion: The Road Ahead

The primary focus of this paper is to highlight about the overburdened judicial system of India and highlighting an untouched area that is speedy trial of pending cases. Although it is fascinating to see that Supreme Court had mentioned in several occasions that speedy trial of the case is a fundamental right under the Indian constitution. A fact that is not so titillating that any positive effort has not yet been made in this regard. This poses several simple questions before society, "whose job it is to protect the people's rights?" "How the use of technology can mitigate it?".

As the paper focuses on the idea of reviving the over-burdened Indian judiciary by aptly using technology like AI, ODR, and ADR and finds from the international experiences, the possibility of succeeding in this endeavour. Earlier in this paper, various models that have been used around the globe got the mention, also it has been highlighted method may prove to be an apt one considering every socio-legal aspects. At this juncture, proper implementation of the plans is the most important one. A thorough inculcation can be done if at least two factors are covered; firstly, training the Judges and the legal professionals, and secondly, training the young professionals who are about to join the profession soon, which cannot be done without proper curriculum planning because it is the young professionals who are going to be in the field for a longer period and who will have the responsibility to train the upcoming generation, so we have to make our young minds ready for the battle of the century because all rest on our shoulder. In this context, the author believes that law can be termed as the sword to fight for revival, the implementation defines the sharpness of the sword, and clinical legal education will train young minds how to use a sword, and with these three things combined we will have a proper mechanism to deal with a problem as gigantic as it is. Putting the situation in the words of Col. Rajeev Bharwan [50], weapon matters the least it has always been the man behind the machine.

Another important thing that needs to be taken care of is machine learning because if AI gets access to false data the result would be influenced by that and getting the desired outcome will be farfetched, so the proposed solution would be to make a separate repository of data of the AI to access which will have all the Indian Laws Cases and Research Articles as well as the foreign laws, and this will only be used as a machinery to aid the judges and not to replace them.

References

1. Julia H, William AB, and Frederick HZ (2005) Access to Justice for a New Century: The Way Forward. Law Society of Upper Canada
2. Mauro C, Bryant GG (1977–1978) Access to justice: the newest wave in the worldwide movement to make rights effective. Buffalo Law Rev 181:181–292
3. Cappelletti M (1981) Access to justice and the welfare state. Cambridge University Press
4. See Ref. [2]
5. Analisa M (2020) Artificial intelligence in the courtroom: increasing or decreasing access to justice?. Int J Online Dispute Resol 6
6. Library of Cong., Respectfully Quoted: A Dictionary of Quotations 183 (2010) (Citing Laurence J. Peter, Peter's Quotations 276 (1977) (Attributing The Maxim's Origin to William E. Gladstone)
7. Professor Daniel Rainey (2020) Lecture in Online Dispute Resolution at the University of the Pacific, McGeorge School of Law, January 2
8. Overall rating which is made after considering all the factors. https://worldjusticeproject.org/rule-of-law-index/global/2022/India/table, last visited on 30th March, 2023, at 17:21
9. Anjanette Raymond and Scott Shackelford (2014) Technology, Ethics and Access to Justice: Should an Algorithm Be Deciding Your Case? SSRN Electron J 35:486
10. The ILC has stated, Law Commission of India, 77th Report, http://lawcommissionofindia.nic.in/51-100/Report77.pdf
11. Report On Strategic Plan For Implementation Of Ict In Indian Judiciary (2005) https://main.sci.gov.in/pdf/ecommittee/action-plan-ecourt.pdf
12. Brij Mohan Lal v. Union of India & Others (2002) SCConline 569
13. Tonio B (2023) European Commissioner for Health & Consumer Policy; https://ec.europa.eu/commission/presscorner/detail/en/IP_12_1381. last visited on 29th March, 2023, at 17: 15
14. See Ref. [9]
15. Emily STP (2017) The future is bright- complicated: AI, Apps & Access To Justice, Vol. 72 Oklahoma Law Review p. 185. See also Richard Susskind (2017). Tomorrow's Lawyers: An Introduction To Your Future 2nd ed.; Stuart A. Forsyth, (2010) Perspectives from a Legal Futurist: Challenges to the Courts and the Legal Community, 51 S. Tex. L. Rev. 913; Symposium, Rise of the Machines: Artificial Intelligence, Robotics, and the Reprogramming of Law, 88 Fordham L. Rev.; Symposium, Artificial Intelligence, Technology, and the Law, 69 U. Toronto L.J. (Supplemental Issue) 1 (2018); Symposium, Artificial Intelligence and the Law, 89 Wash. L.Rev. 1 (2014); Symposium, Legal Reasoning and Artificial Intelligence: How Computers "Think" Like Lawyers, 8 U. Chi. L. Sch. Roundtable 1 (2001); Richard Susskind (2008), The End Of Lawyers? Rethinking The Nature Of Legal Services 273; Albert H. Yoon (2016) The Post-Modern Lawyer: Technology and the Democratization of Legal Representation, 66 U. Toronto L.J. 456, 457 Dana Remus & Frank Levy(2017). Can Robots Be Lawyers? Computers, Lawyers, and the Practice of Law, 30 Geo. J. Legal Ethics 501, p 530
16. Jerry K (2016) Artificial intelligence: what everyone needs to know; Erik B, Andrew M (2014) The second machine age: work, progress, and prosperity in a time of brillliant technologies; Martin F (2016) The rise of the robots (2014); Benjamin A, Anthony N, Albert, Law in the Future, 66 U.T.L.J. p 423
17. Sunstein C (2001) Legal reasoning and artificial intelligence: how computers "Think" Like Lawyers. 8 U Chi L Sch Roundtable 1:19
18. Victor M-S, Kenneth C (2013) Big data: a revolution that will transform how we live, work, and think. pp 6–12; James M et al (2013) Disruptive technologies: advances that will transform life, business, and the global economy; See William MG, Paul EM (1996) Comparative efficiency of informal (subjective, impressionistic) and formal (mechanical, algorithmic) prediction procedures: The Clinical-Statistical Controversy, vol. 2 Psych Public Pol'y Law p 293

19. Richard S (2014) Tomorrow's lawyers: an introduction to your future; Richard S, Daniel S (2016) The future of the professions: how technology will transform the work of human experts; and Raymond HB (2015) Embracing disruption: how technological change in the delivery of legal services can improve access to justice, vol. 78 Albany L Rev p 553
20. Jhanavi G (2021) Artificial intelligence in our legal system. Int J Law Manag Hum 4 <https://heinonline.org/hol-cgi-bin/get_pdf.cgi?handle=hein.journals/judgej59&section=3>
21. AI and the Rule of Law: Capacity Building for Judicial Systems, UNESCO; https://www.unesco.org/en/artificial-intelligence/rule-law/mooc-judges; last visited on 30th March 2023, at 02:31 am
22. See Ref. [20]. See also, Amitabh K, Desh GS (2023) The judiciary's use of AI will be transformative, Financial Express https://www.financialexpress.com/opinion/the-judiciarys-use-of-ai-will-be-transformative/2272714/, last visited on 30th March 2023, at 02:28 am
23. Amit AC (2023) Use of Artificial Intelligence will transform judiciary but technology will not be allowed to decide cases: CJI, Times of India; https://timesofindia.indiatimes.com/india/use-of-artificial-intelligence-will-transform-judiciary-but-technology-will-not-be-allowed-to-decide-cases-cji/articleshow/82183403.cms; last visited on 30th March 2023, at 03:38 am
24. Mahatma Gandhi (1962) The story of my experiments with truth, p 168
25. Jitendra NB (2002) A round table justice through Lok-Adalat (Peoples' Court)—a vibrant-ADR-In India, 1 SCC (JOUR), , available at https://www.kelsa.nic.in/lokadalat.htm last visited on 30th March 2023, at 15:21
26. Owen MF (1984) Against settlement, vol. 93. Yale L J pp 1073–1076; Andrew WM, Thomas LS (1985) For reconciliation, vol 94 Yale L J p 1660 (responding to Fiss); Owen MF (1985) Out of Eden, 94 Yale L J 1669 (responding to McThenia and Schaffer); Carrie M-M (1985) For and against settlement: uses and abuses of the mandatory settlement conference, 33 UCLA L. Rev 485; Judith R (1995) Many doors? Closing doors? Alternative dispute resolution and adjudication, vol. 10 Ohio St J On Disp Resol, p 211
27. Asia Dev. Bank (2006) The role of public administration in building a harmonious society, p 591. See also, Ref. [25]
28. Constitution of India, Preamble
29. See Ref. [9]. See also, art. 39(A) of Indian Constitution; Centre on Pub. Law & Jurisprudence at Jindal Global Law Sch. (2012), Justice Without Delay: Recommendations For Legal And Institutional Reform [Hereinafter Justice Without Delay]
30. Babu vs. Raghunathji, A.I.R. 1976 S.C. 1734.
31. World Justice Project, Rule of Law Index (2024) https://worldjusticeproject.org/rule-of-law-index/downloads/WJPIndex2023.pdf. Last visited on 30th May, 2024, at 03:52 am
32. See Ref. [31]
33. Legal Services Authorities Act, No. 39 of 1987, Preamble
34. Girish Patel, Crippling Lok Adalats, INDIA TOGETHER, India Togather, (2007) https://indiatogether.org/adalat-human-rights, last visited on 30th March, 2023 at 19:00
35. Joseph WG (2003) The pros and cons of online dispute resolution. Duke L Tech Rev, pp 7–9
36. Technical Notes on Online Dispute Resolution 2016, (UNCITRAL), available at https://uncitral.un.org/en/texts/onlinedispute/explanatorytexts/technical_notes; last visited on 30th March, 2023, at 20:35
37. See Ref. [9]
38. Anjanette HR (2013) It is time the law begins to protect consumers from significantly one-sided arbitration clauses within contracts of adhesion, vol 91 Neb L Rev 666. William B (2010) How trust is abused in free markets: Enron's 'Crooked 'E', Social science research network, https://papers.ssrn.com/sol3/papers.cfm?abstract_id=1536532
39. See Ref. [9]
40. Shivasish M, Shambhavi S (2022) India: the ODR policy plan for India—paving the way to a brighter future for dispute resolution in India, Mondaq; https://www.mondaq.com/india/arbitration--dispute-resolution/1161056/the-odr-policy-plan-for-india---paving-the-way-to-a-brighter-future-for-dispute-resolution-in-india; last visited on 30th March, 2023, at 22:14
41. (2009) 2 SCC 134

42. (2002) 7 SCC 736
43. (2018) 1 SCC 560
44. Olayode IO, Tartibu LK, Okwu MO, Uchechi UF, Intelligent transportation systems, un-signalized road intersections and traffic congestion in Johannesburg: a systematic review, Elsevier, Availabe at Science direct Online
45. India Narrative, Bengaluru records highest traffic violations in world as AI catches culprits, 30th September, 2023, 11:01 am. https://www.indianarrative.com/india-news/bengaluru-records-highest-traffic-violations-in-world-as-ai-catches-culprits-93066.html
46. Chen M, Li S (2020) Research on the application of artificial intelligence technology in the field of Justice. J Phys Conf Ser ICAACE
47. Aharon B (2002) The role of a supreme court in a democracy, 53 Hastings L J 1205, 1205; ("The role of the judiciary is to adjudicate disputes according to law. Adjudication involves three functions: fact determination (done mostly by the trial court), law application and law determination); see generally Aharon Barak, A Judge on Judging: The Role of a Supreme Court in a Democracy, 116 Harv L Rev 16:98–99, 110–12 (2002), for an explication of Justice Barak's broader views on how judges mediate between legal text and societal issues
48. Ray WC (2020) Artificial intelligence in the courtroom: the delivery of justice in the age of machine learning. Colo Tech LJ 18:323
49. David RB (1979) The clinical method of legal instruction: its theory and implementation. J Legal Educ Assoc Am Law Schools 30:67
50. Col. Rajeev Bharwan, Interview in an Youtube Channel, quote can be found from 30:00 mins. https://www.youtube.com/watch?v=IpULOwRaayM. Last visited on 31st March, 2023, at 3:54 am

AI-Based Smart Governance

Rahul Vadisetty

Abstract Artificial intelligence (AI) has a broad range of applications in enhancing the efficacy, efficiency, and transparency of government services and procedures. By analyzing large amount of data from various repositories that include data concerning demography, traffic, weather, legal affairs, transport, water supply, education, etc., well-informed policy decisions may be made by the government. The paper employs an AI framework to design an effective traffic management system wherein it evaluates and comprehends a complicated traffic forecast by investigating five machine learning (ML) models, including Multilayer Perceptron (MLP), Gated Recurrent Unit (GRU), Long Short-Term Memory (LSTM), Light Gradient-Boosting Machine (LGBM) Regression, and Random Forest (RF) Regression networks by training and testing a traffic dataset from Kaggle. This framework seeks to significantly take care of traffic congestion at various junction points of a city by analyzing traffic volume at different hours of the day, day of the week, and seasons of the year over a duration of one year and eight months from November 2015 to July 2017. The superiority of the MLP model is demonstrated by multiple data analyses of the traffic dataset, which are followed by simulation and analysis of the prediction model using five ML algorithms. Performance metrics such as root mean square error (RMSE) of 0.235, mean absolute error (MAE) of 0.169, and R2 score of 0.97 are used to demonstrate this. Explainable AI is used to further interpret and explain the outcomes.

Keywords AI · GRU · Infrastructure maintenance · LSTM · LGBM · MLP · RF · Smart governance · Traffic security · Traffic management · XAI

R. Vadisetty (✉)
Wayne State University, Detroit, MI, USA
e-mail: rahulvy91@gmail.com

M. Chakraborty et al. (eds.), *Proceedings of 5th International Ethical Hacking Conference*, Lecture Notes in Networks and Systems 1148,
https://doi.org/10.1007/978-981-97-8457-8_30

1 Introduction

In the current scenario, Artificial Intelligence (AI) has transformed the operations of almost all industries and sectors, including government services. The governmental decision-making process is automated through the use of advanced technologies such as machine learning (ML), computer vision, and data analytics. There are several ways in which AI is being utilized in government services and operations such as making informed choices based on evidence and insights, automating repetitive tasks and streamlining workflows to save time and resources, enhancing public engagement through AI-powered chatbots and virtual assistants to provide faster and personalized assistance to citizens to mention a few [1]. AI has the potential to support governments in predicting trends and demands in vital domains like infrastructure, transportation, and public health. It can also help law enforcement agencies detect fraudulent operations and criminal activity. Ethical considerations and the need for a strong legal framework to foster public acceptance and confidence must be considered when utilizing AI in smart governance. Governments should have robust and intelligent governance structures, such as clear rules, regulatory frameworks, accountability mechanisms, and ongoing review, to safeguard citizens' interests and take advantage of AI's capabilities [2].

One of the main factors contributing to traffic congestion in cities is poor human resource management. Intelligent Transportation Systems can be used to schedule police activities in real-time by integrating machine learning with traffic control forces [3]. In order to promote sustainable ecosystems on roadways, the importance of digitization by using AI, particularly gradient-boosting techniques, is highlighted in [4, 5]. To make well-informed decisions about traffic control strategies and administration, it is important to classify traffic scenarios. This aids in locating traffic lanes, examining the reasons for traffic, and improving efficiency [6]. Deep learning techniques are perfect for anticipating traffic conditions, assisting in the relieving of congestion, and maximizing visitor flow because of their capacity to retain information for extended periods of time [7]. Adopting these AI models in urban control systems represents a paradigm shift closer to more fabulous, intelligent, green, and responsive towns. By leveraging the wealth of statistics from city environments, especially datasets from repositories like Kaggle, researchers and town planners can train these models to apprehend and predict complex styles in city mobility and infrastructure conditions [8]. This background set the stage for our studies, where we aim to harness the abilities of Multilayer Perceptron (MLP), Gradient Recurrent Unit (GRU), Long Short-Term Memory (LSTM), Light Gradient-Boosting Machine (LSBM), and Random Forest (RF) to expand an incorporated technique to enhance visitors' protection, cope with site visitors' waft, and preserve infrastructure, contributing to the broader dreams of modern metropolis tasks [9].

The AI/ML algorithms are black box models that fail to give reasoning to the results. Explainable AI (XAI), as the name suggests, is a procedure and collection of techniques that assist users by providing an explanation for the output and outcomes produced by AI/ML algorithms thus converting the black box models to the white

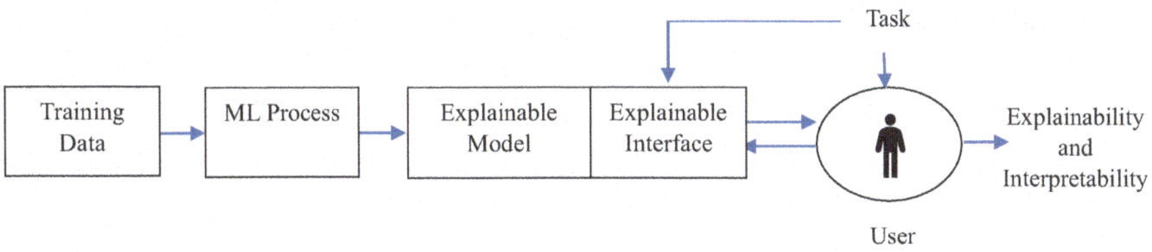

Fig. 1 XAI concept

box models. It refers to a group of steps and methods that allow machine learning algorithms to generate output and outcomes that are trustworthy and comprehensible for human users. Deep learning and XAI are often discussed in relation to each other [10]. The XAI concept depicted in Fig. 1 shows three main components.

1. **ML Process**: The first component is the ML algorithm that draws inferences and forecasts from data.
2. **Explainable Model**: The explanation model gives explanations and interpretations to the predicted results obtained by the ML process.
3. **Explainable Interface**: The explanation model may be produced on multiple platforms, including mobile apps, online applications, and visualizations. The explanation model generates knowledge and insights, which can be accessed and interacted with by people using the user-friendly XAI interface.

The core contributions of the paper comprise.

(i). Design and predictive analysis of different AI-based frameworks for traffic management using datasets from Kaggle repository.
(ii). Explanation and interpretation of the results in step (i) above by using XAI algorithms such as Local Interpretable Model-agnostic Explanations (LIME) for local explanation and SHapley Additive exPlanations (SHAP) for global explanation that provide explainable insights into the factors that are most relevant and influential in the model's predictions.

The remaining part of the paper is organized as follows. Section 2 gives an overview of the five ML algorithms used in the proposed prediction model of traffic management. Section 3 gives the proposed system model including the architecture, the dataset, data preprocessing, and analyses. This is followed by prediction model simulation results, analysis, and explanation of the outcome in Sect. 4. The paper concludes in Sect. 5 with some highlights on future scope.

2 Overview of ML Algorithms Used in Traffic Prediction Analysis

This section presents overview of the five ML algorithms that are used for predictive analysis of traffic dataset.

- **MLP Network:** A feedforward neural network with numerous layers of nodes, modeled after real brain neurons, is called MLP. By using a bias term, weighted input sum, and activation function, it produces output. Weights and biases are determined during training via backpropagation and gradient descent. MLPs can overfit when there are more parameters than in the training set, yet they are useful for pattern recognition, regression, and classification. Dropout and weight decay are regularization measures that assist lessen this problem [11].
- **GRU Network:** By identifying long-range dependencies in sequential input, the GRU is a recurrent neural network architecture that addresses the vanishing gradient issue and other issues with traditional RNNs. GRUs were introduced as an alternative to LSTM units. They use gating techniques to control the flow of information through two gates: reset and update. These gates are well-liked for sequential data tasks including time series prediction, voice recognition, and natural language processing because they improve efficiency in obtaining pertinent information across long sequences, are simpler to train, and are less prone to overfit [12].
- **LSTM Network:** Using a memory cell and a number of gates to control information flow, recurrent neural networks were the focus of development when LSTMs were created to solve their shortcomings. These gates can recognize long-term dependencies in sequential data since they are composed of an input gate, forget gate, and an output gate. These trainable gates ensure effective learning and processing by adaptively adjusting as necessary as they identify patterns in the data [13].
- **LGBM Regression:** LGBM is effective for handling large datasets and multi-dimensional feature spaces. It uses histograms to bin datapoints eliminating the need for evaluating every data point, unlike depth-first tree growth, it grows trees in a leaf-wise manner wherein it selects the leaf with maximum gain during each iteration for faster convergence, has customizable hyperparameters and has higher speed as compared to other boosting algorithms [14].
- **RF Regression:** RF is an ML algorithm that finds effective utilization in regression tasks. It is based on Decision Trees, which starts from the root node and eventually forks into several child nodes until it reaches a leaf node based on variable outcomes. For regression analysis, it predicts a value based on the input features [15].

3 Proposed AI-Based Traffic Management System

Traffic management encompasses the planning, directing, and regulation of both stationary and mobile traffic, encompassing bicycles, pedestrians, and all kinds of vehicles. The application of AI to traffic management has the potential to transform urban transportation by removing the bottlenecks that cause traffic jams in urban areas. Consequently, by cutting down on the amount of time we spend stuck in traffic, we will be able to lower emissions in addition to travel times and traffic

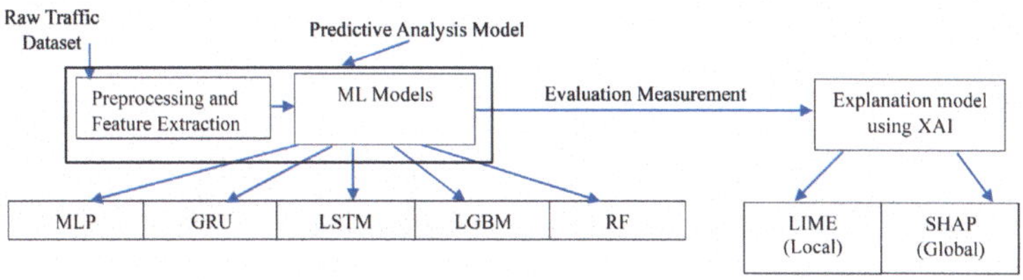

Fig. 2 AI based traffic management system model

congestion. The proposed model uses AI to analyze traffic scenario of a particular dataset from the Kaggle repository and design, simulate, analyze, and explain a traffic management prediction model using five ML algorithms.

3.1 System Model

The architecture of the proposed AI-based traffic management system model is shown in Fig. 2. It comprises two parts: a predictive analysis model and an explanation model using XAI.

3.2 Dataset

This open dataset is taken from Kaggle [16]. It comprises 4 features such as “DateTime”, “Junction”, “Vehicles”, and “ID”, and 48,120 datapoints.

3.3 Data Preprocessing and Analyses

Preprocessing consists of dropping duplicate data in the training data, checking for missing values, and filling them wherever required, feature engineering to extract important features. After this step, the number of features extracted is nine such as “DateTime”, “Junction”, “Vehicles”, “ID”, “Year”, “Month”, “Date_no”, “Hour”, and “Day”. Several analyses are made following the preprocessing steps discussed in the following sections.

3.3.1 Time Series Analysis

A statistical technique called time series analysis is used to examine a collection of data points that were gathered over a period of time. Instead of sporadically or

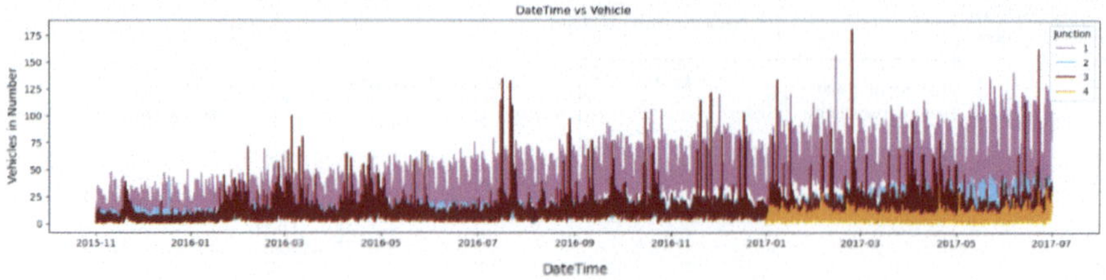

Fig. 3 Time series plot of DateTime with vehicles

arbitrarily, it entails collecting data points at regular intervals over a predetermined length of time. This type of analysis is crucial because it can reveal how variables change over time, providing insights into trends, patterns, and seasonal effects within the data. Time series data points are time-dependent, allowing for the discovery of underlying patterns and trends, essential for understanding changes in variables like sales figures, traffic and weather patterns, and stock prices, despite the potential obscurity of vital information [17]. The time series plot of "DateTime" with "Vehicles" with regard to the four junctions is shown in Fig. 3. It shows an extremely low density of vehicles with a peak of 25 for the duration of January 2017 to July 2017 at junction 4. Junctions 1, 2, and 3 show consistent density throughout the time span from November 2015 to July 2017 with a maximum peak of 175 at junction 3 during March 2017.

With the goal of better comprehending the underlying patterns, Fig. 4 displays the time series analysis of the real traffic for trend, seasonal, and residual studies. The graphic displays variations over time that appear to follow a pattern, indicating that the data may be somewhat seasonal.

- **Trend:** This part presents the observed data in a smoothed format, emphasizing the long-term trends or variations in traffic volume. It is difficult to make firm conclusions without additional context, although the trend line suggests that there may be times of increased or decreased overall traffic volume. Additionally, there are some variations in the trend that might be related to underlying causes influencing traffic volume over an extended length of time.
- **Seasonal:** Since the decomposition was probably completed within a period of seven, the seasonal component in this case is believed to be weekly. It captures the typical pattern within the data that repeats over a specified period of time. The pattern exhibits regular peaks and troughs and is quite consistent. This might be related to regular workweek patterns, as seen by higher traffic volumes on some days of the week and lower on others.
- **Residual:** These are the irregularities that persist in the observed information after removing the trend and seasonal components. The residuals show no visible pattern and seem to be randomly distributed around zero, indicating that the model has successfully captured the trend and seasonality. A small number of outliers, on the other hand, show sporadic variations from the anticipated traffic volume that cannot be attributed to trends or seasonal factors.

Fig. 4 Time series traffic pattern

3.3.2 Traffic Volume Analysis

Figure 5 shows the average traffic volume at different junctions depicting maximum at junction 1 and minimum at junction 4. While junction 4 has the lowest average traffic volume of the four, it may be the least crowded or service an area with lower traffic demand. This shows that junction 1 is a major traffic hub, presumably located in a central or bustling region with substantial vehicle flow. The average traffic volume at Junctions 2 and 3 is comparable to that of Junction 1, being less than half that of Junction 1. These intersections may be feeder roads to the major intersections or they may service less populated areas. Junction 1 may need more attention in terms of traffic signal timing, congestion management techniques, or infrastructure upgrades from the standpoint of city planning and traffic management. Furthermore, a variety of factors, including location, the amenities nearby, the size of the road, and the connection to other areas of the city, may contribute to the variations in traffic volume. It would also be instructive to look at each junction's traffic fluctuation throughout the day to determine if certain are regularly busier than others at particular times of the day. This might be accomplished by making a multi-line plot or heatmap that displays the daily traffic levels at each intersection. The data appears to exhibit a robust seasonal pattern, along with some long-term trends and random noise, according to the decomposition. The analysis's conclusions can be used to build city infrastructure, anticipate traffic levels, and schedule peak traffic periods. While the seasonal component might influence short-term operational decisions, such as traffic signal coordination and congestion management methods, understanding the trend can be especially helpful for long-term planning. The traffic pattern's day of the week analysis is displayed in Fig. 6, with Tuesday, Wednesday, and Thursday seeing peak traffic and Sunday experiencing a decrease. Peak hour analysis is shown in Fig. 7, with peaks at 7 PM and 8 PM and a dip at 5 AM.

In Fig. 8, holiday effects are displayed. According to the data, non-holiday traffic is approximately 58% higher than holiday visitation. It is crucial to examine if traffic volumes have been rising or falling over time, as this may be a sign of changes in population density or urban development. A long-term trend is seen in Fig. 9. The average monthly traffic volume over time is displayed in the chart. Traffic volumes generally show an increasing trend, indicating a rise in traffic over the course of the months. There are some variations, with observable decreases in July 2016 and February 2017, respectively, followed by increases. The real monthly average traffic

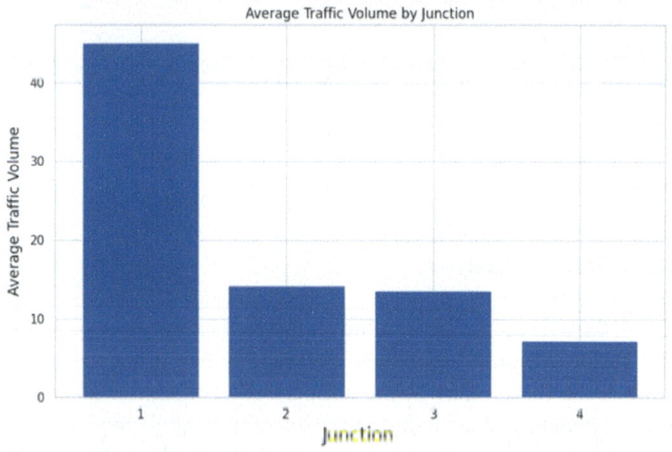

Fig. 5 Average traffic volume at junctions

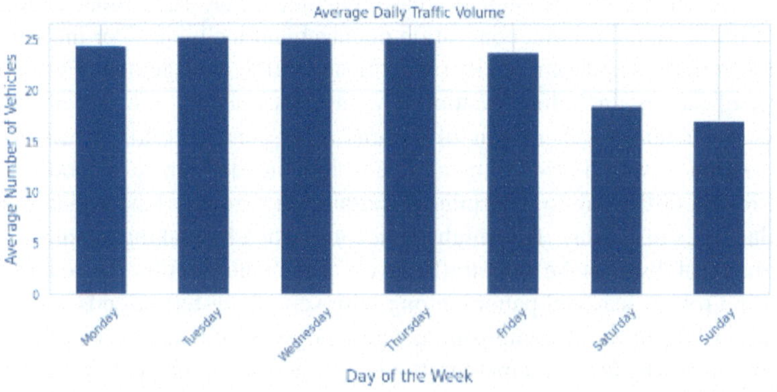

Fig. 6 Day of the week analysis

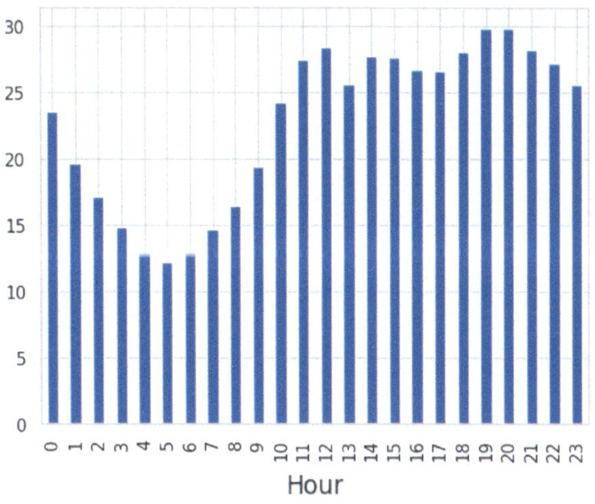

Fig. 7 Peak hour of the day analysis

volume (shown in blue) is superimposed over a trend line (shown in red) in the chart. The linear trend line, which shows a distinct rising trajectory, attests to the long-term rise in traffic levels throughout this time. In fact, this might be a sign of things like increased population density, urbanization, or modifications to traffic patterns. The increased trend implies that techniques for traffic management could need to change to accommodate rising volumes. In addition, this data may be taken into account by lawmakers and city planners when deciding how best to improve public transit and infrastructure in order to reduce traffic.

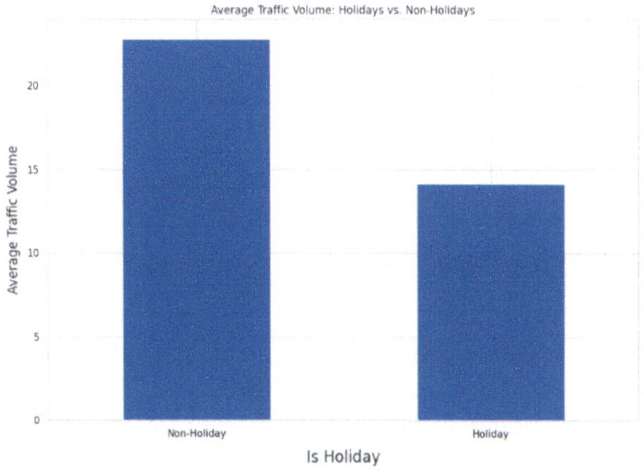

Fig. 8 Average traffic volume: holidays vs non-holidays

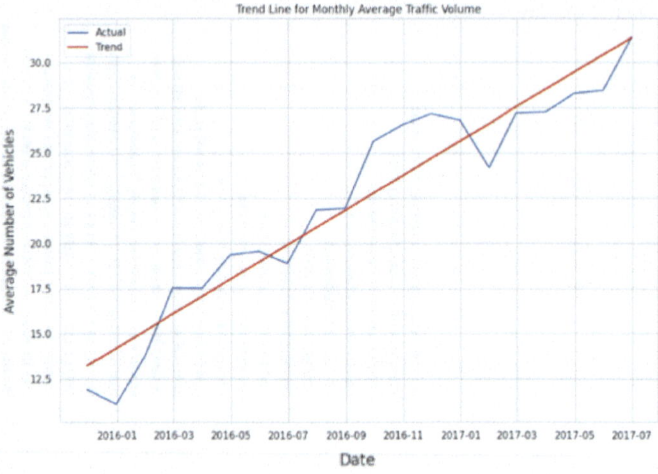

Fig. 9 Long-term trend

3.3.3 Correlation Analysis for Traffic Volume

A statistical technique called correlation analysis can be used to forecast future events, identify patterns and trends in data, and assess the degree of a relationship between two variables. The correlation matrix using heatmap is plotted to understand the correlation among all the variables. Heatmap is a graphical tool that shows the correlation as a color-coded matrix between several variables. It resembles a color chart that illustrates the degree of relationship between several variables. Each variable is represented by a row and a column in a correlation heatmap, and the cells display the correlation between them [18]. Figure 10 shows the heatmap for the traffic volume with regard to "Hour of the Day" and "Day of the Week" both showing correlation values > ± 0.1 and hence have a high correlation to the traffic volume.

3.4 Simulation Results

The simulation environment comprises setting up five ML-based models using Python programming language in Google Colab. The preprocessed traffic dataset [16] is split into two parts: training data (75%) and testing data (25%). The following sub-sections present the prediction results and explanation of the outcome using the two XAI models: LIME and SHAP.

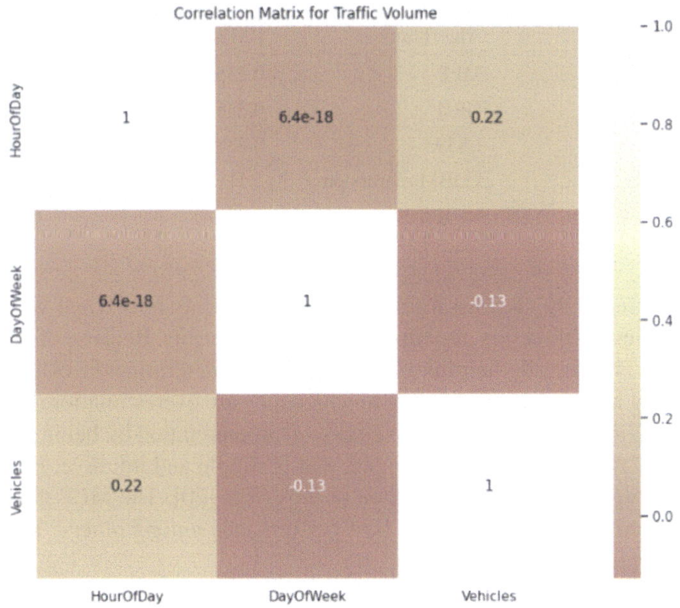

Fig. 10 Heatmap of the traffic dataset

3.4.1 Prediction Results

For prediction analysis, the regression process is utilized. Regression analysis is an approach we can use to comprehend the link between one or more predictor factors and a response variable. Regression analysis is frequently used in forecasting and prediction. For predictive regression analysis of the traffic dataset, five ML models are simulated and the performance metrics such as Root Mean Square Error (RMSE), Mean Absolute Error (MAE), and Coefficient of Determination (R^2 score) are obtained as shown in Table 1. RMSE metric can be used to evaluate how well a regression model fits a dataset. It indicates the average difference between the model's predicted and the dataset's values. A model is better equipped to "fit" a dataset if its RMSE is smaller [19]. MLP model shows its supremacy in the model "fit" as it has the lowest RMSE among others. The following formula can be used to find the RMSE value.

$$RMSE = \sqrt{\sum {(P_i - O_i)^2}/{n}} \tag{1}$$

where, P_i is the predicted value for the ith observation in the dataset; O_i is the observed value for the ith observation in the dataset; n is the sample size

The MAE is a simple yet powerful method for evaluating the accuracy of regression models. Measuring the goal values and the anticipated values yields the average

Table 1 Performance metrics

Algorithm	RMSE	MAE	R^2 score
MLP	0.235	0.169	0.97
GRU	0.236	0.171	0.95
LSTM	0.238	0.172	0.96
LGBM regression	5.11	2.92	0.94
RF regressor	3.97	2.51	0.96

absolute difference. In contrast to other metrics, MAE assigns equal weight to all errors, regardless of direction, as it does not square the errors. Because of this characteristic, MAE is especially helpful when attempting to determine the extent of errors without taking into account the degree to which they are overestimations or underestimates. MAE is robust to outliers, gives ease of interpretation by being of the same unit as that of the original target variable, and is simple and intuitive. Hence MAE provides a more balanced representation of the errors [20]. The MLP model shows the lowest MAE value indicating its best performance among others. The formula for MAE is as follows.

$$MAE = \frac{1}{n}\sum_{i=1}^{n}|y_i - \hat{y}_i| \tag{2}$$

where n is the number of datapoints; y_i represents the actual target value for data point i; $\hat{y}_i$ represents the predicted value for data point i.

The accuracy with which a statistical model forecasts a result is shown by the R^2 score. The dependent variable in the model is a representation of the result. R^2 score has two possible values: 0 is the lowest and 1 is the highest. To put it simply, a model's R^2 will approach 1 the more successful it is at predicting outcomes [21]. MLP shows the highest R^2 score among others.

The performance metrics of other algorithms show the acceptable ranges of GRU and LSTM. While LGBM Regression and RF Regressor show good R^2 score but deficiency in RMSE and MAE values. Hence, MLP is the supreme model that fits extremely well with the traffic dataset under study. The traffic predicted values with true values for the MLP model depicted in Fig. 11 show a close approximation of the predicted values to the true values supporting the values of the performance metrics in Table 1.

3.4.2 Explanation Using XAI Models

In light of the dominance of the MLP model, the XAI models utilized in this work to offer local and global explanations are LIME and SHAP, respectively [22]. ML models can be used to explain predictions for a single instance using a technique called LIME. The approach focuses on explaining individual predictions rather than

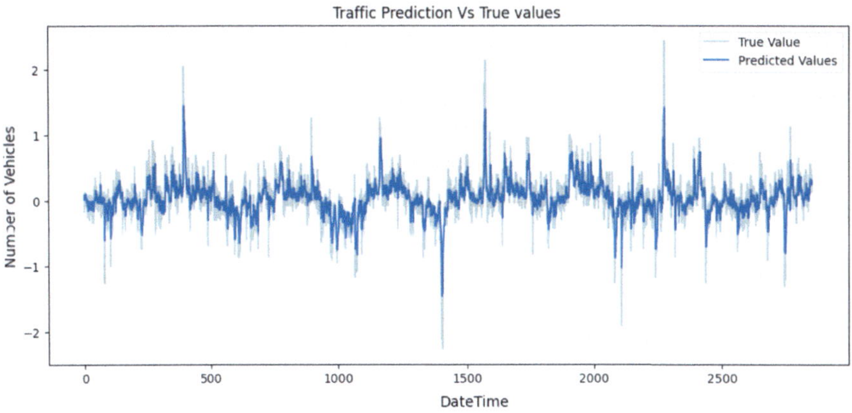

Fig. 11 Traffic prediction vs true values

the model as a whole, and it is model-agnostic, meaning it may be applied to any kind of ML model. Understanding how each input feature affects the model's predictions is made possible via SHAP. Shapley values are used to assign a priority value to each feature for a specific prediction, and it is based on cooperative game theory. This aids in comprehending how each attribute contributes to the output of the model and makes the decision-making process of complex models more transparent and comprehensible by assisting in the comprehension of the contribution of each feature to the model's output. Figure 12 shows the LIME visual explanations to the outcome for an instance of the 12th test sample showing the importance of 9 features. The predicted value is 16.15 (3.12_{min}–106.36_{max}). The blue-colored bars show negative contribution whereas the orange-colored bars show positive contributions to the predicted outcome. The features "Junction", "Hour", and "Date_no" have very low negative correlations, "Day_Sunday", "Day_Saturday", Day_Friday", and Day_Thursday" have zero positive correlation, and "Year", "DateTime" have extremely high positive correlation.

Fig. 12 LIME visual explanation

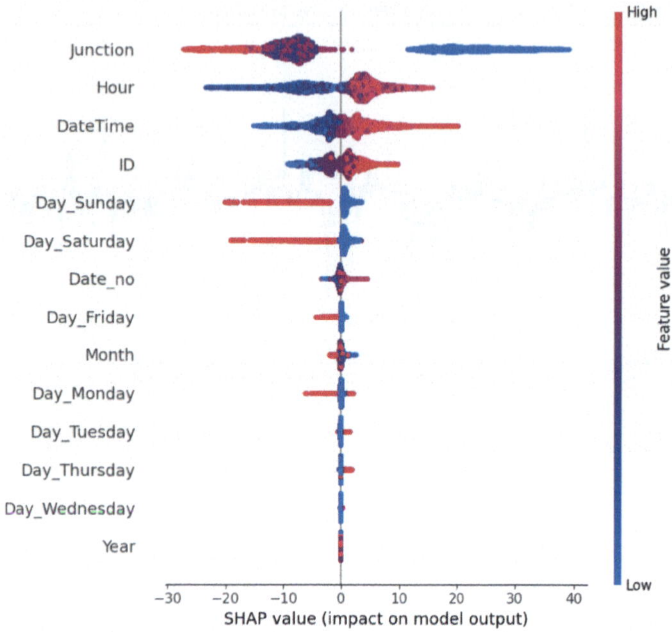

Fig. 13 Mean SHAP summary visualization

Figure 13 shows the global explanation of the predicted result using mean SHAP summary visualization. The plot shows the calculated SHAP values and their impact on the model output. It represents the mean absolute SHAP values for each feature across all predictions. Features that raise the forecast are displayed in red, while those that lower it are displayed in blue. The higher SHAP value of a feature corresponds to the higher prediction.

4 Conclusion

This study looks at successfully addressing full-size demanding situations in traffic safety, traffic management, and infrastructure protection through the revolutionary application of MLP, GRU, LSTM, LSBM regression, and RF regression models trained on a traffic dataset from Kaggle. By harnessing these advanced AI techniques, we have advanced a sturdy framework to interpret complex urban traffic environments. Our research demonstrates the effectiveness of using actual-time object detection, visitor waft prediction, and infrastructure situation assessment to improve city mobility and protection. The models trained on comprehensive datasets from Kaggle ensure that they are prepared to address the nuances of actual global traffic scenarios. MLP model shows its supremacy in traffic predictive analysis among others with the highest R^2 score and lowest values of both RMSE and MAE. Furthermore, the local and global explanations of the model's output using XAI techniques add value to the model's efficiency, effectiveness, and trustworthiness. This painting highlights the feasibility of employing current AI technologies in public infrastructure and unites the groundwork for future advancements in innovative town projects. Our findings highlight the essential role of AI in remodeling urban centers into more steady, potential, and resilient environments, paving the way for similar improvements in urban site visitors and infrastructure control.

References

1. Arora A et al (2024) Towards intelligent governance: the role of AI in policymaking and decision support for E-governance. In: In C, Londhe ND, Bhatt N, Kitsing M (eds) Information systems for intelligent systems. ISBM 2023. Smart innovation, systems and technologies, vol 379. Springer, Singapore
2. Dafoe A (2018) AI governance: a research agenda; governance of AI program. Future of humanity institute, Oxford
3. Nama M, Nath A, Bechra N, Bhatia J, Tanwar S, Chaturvedi M et al (2021) Machine learning-based traffic scheduling techniques for intelligent transportation system: opportunities and challenges. Int J Commun Syst 34:e4814
4. Singh R, Sharma R, Akram SV, Gehlot A, Buddhi D, Malik PK et al (2021) Highway 4.0: digitalization of highways for vulnerable road safety development with intelligent IoT sensors and machine learning. Saf Sci 143:105407
5. Korstanje J (2021) Gradient boosting with XGBoost and LightGBM. In: Advanced Forecasting with Python. Apress, Berkeley
6. Kamaeshwaran M, Jayanthi K, Seshan S (2023) Real-time vehicular traffic flow monitoring and classification of vehicles for indian road scenario. In: Second International Conference on Advances in Computational Intelligence and Communication (ICACIC), Puducherry, pp 1–5. https://doi.org/10.1109/ICACIC59454.2023.10434978
7. Razi A, Chen X, Li H, Wang H, Russo B, Chen Y et al (2023) Deep learning serves traffic safety analysis: a forward-looking review. IET Intel Transport Syst 17:22–71
8. Sanjeewani P, Verma B (2021) Single class detection-based deep learning approach for identification of road safety attributes. Neural Comput Appl 33:9691–9702
9. Hou Y, Li Q, Zhang C, Lu G, Ye Z, Chen Y et al (2021) The state-of-the-art review on applications of intrusive sensing, image processing techniques, and machine learning methods in pavement monitoring and analysis. Engineering 7:845–856
10. Ali S et al (2023) Explainable artificial intelligence (XAI): what we know and what is left to attain trustworthy artificial intelligence. Inf Fusion. https://doi.org/10.1016/j.inffus.2023.101805
11. Mohammadzadeh A, Sabzalian MH, Castillo O, Sakthivel R, El-Sousy FFM, Mobayen S (2022). Multilayer perceptron (MLP) neural networks. In: Neural networks and learning algorithms in MATLAB. Synthesis lectures on intelligent technologies. Springer, Cham
12. Zheng X et al (2023) GRU network-based load allocation for hydro units. In: Hu C, Cao W (eds) Conference Proceedings of 2022 2nd International Joint Conference on Energy, Electrical and Power Engineering. CoEEPE 2022. Lecture Notes in Electrical Engineering, vol 1060. Springer, Singapore. https://doi.org/10.1007/978-981-99-4334-0_28
13. Van Houdt G, Mosquera C, Nápoles G (2020) A review on the long short-term memory model. Artif Intell Rev 53:5929–5955. https://doi.org/10.1007/s10462-020-09838-1
14. Sajid SW, Hasan M, Rabbi MF, Abedin MZ (2023) An ensemble LGBM (light gradient boosting machine) approach for crude oil price prediction. In: Abedin MZ, Hajek P (eds) Novel financial applications of machine learning and deep learning. International Series in Operations

Research & Management Science, vol 336. Springer, Cham. https://doi.org/10.1007/978-3-031-18552-6_9
15. Couronné R, Probst P, Boulesteix AL (2018) Random forest versus logistic regression: a large-scale benchmark experiment. BMC Bioinform 19:270. https://doi.org/10.1186/s12859-018-2264-5
16. https://www.kaggle.com/code/utathya/traffic-patterns
17. Ivanović M, Kurbalija V (2016) Time series analysis and possible applications. In: 39th International Convention on Information and Communication Technology, Electronics and Microelectronics (MIPRO), Opatija, Croatia, pp 473–479. https://doi.org/10.1109/MIPRO.2016.7522190
18. Ma YZ (2019) Correlation analysis. In: Quantitative geosciences: data analytics, geostatistics, reservoir characterization and modeling. Springer, Cham. https://doi.org/10.1007/978-3-030-17860-4_4
19. Rai NK (2022) Chapter ten—RMSE and MAPE analysis for short-term solar irradiance, solar energy, and load forecasting using a Recurrent Artificial Neural Network. Applications of AI and IOT in Renewable Energy, pp 181–192. https://doi.org/10.1016/B978-0-323-91699-8.00010-3
20. Lu WWY (2018) Analysis of the mean absolute error (MAE) and the root mean square error (RMSE) in assessing rounding model. IOP Conf Ser Mater Sci Eng 324(1):012049. https://doi.org/10.1088/1757-899X/324/1/012049
21. Chicco D et al (2021) The coefficient of determination R-squared is more informative than SMAPE, MAE, MAPE, MSE and RMSE in regression analysis evaluation. PeerJ Comput Sci 2021(7):e623. https://doi.org/10.7717/peerj-cs.623
22. Shah R, Pawar A, Kumar M (2024). Enhancing machine learning model using explainable AI. In: Tiwari S, Trivedi MC, Kolhe ML, Singh BK (eds) Advances in data and information sciences. ICDIS 2023. Lecture Notes in networks and systems, vol 796. Springer, Singapore. https://doi.org/10.1007/978-981-99-6906

Artificial Intelligence (AI) in Cybersecurity: A Revolution in Threat Detection and Prevention

Sandeep Sengupta and Subhajit Roy

Abstract The article entitled "Artificial Intelligence (AI) in Cybersecurity: A Revolution in Threat Detection and Prevention" looks at how Artificial Intelligence (AI) has improved measures taken to ensure cyber safety. The methods that have been traditionally used for security cannot match the new sophisticated means of cyber-attacks hence the need for advanced options like AI. This study explores the relationship between cybersecurity and AI as two mutually dependent systems that work together toward achieving their goals. Different approaches within the field of artificial intelligence such as machine learning or deep learning are examined based on their effectiveness when it comes to the detection as well as mitigation of threats posed by cyberspace. It is shown in this paper that artificial intelligence can make threat finding proactive, response time minimal and predictions more accurate thus heightening general security robustness. The writer also assesses how real-life applications of AI have worked in preventing cybercrimes by showcasing some examples which underscore both its benefits and limitations. By doing so he provides readers with hands-on experiences on where they might use this technology while dealing with security issues surrounding information systems within organizations they work for or even personal gadgets like mobile phones. He further goes ahead to consider what could happen if we start using them widely now looking into future challenges plus opportunities associated with applying these machines toward safeguarding data networks against unauthorized access attempts (hackers). Additionally, it gives a balanced view of prospects for success concerning different areas covered under AI and cybersecurity.

Keywords Artificial intelligence · Cybersecurity · Machine learning · Deep learning · Ethical considerations

S. Sengupta (✉) · S. Roy
ISOAH Data Securities Pvt. Ltd., & Researcher NIT Silchar, Kolkata, WB, India
e-mail: sandeep@isoeh.com

M. Chakraborty et al. (eds.), *Proceedings of 5th International Ethical Hacking Conference*, Lecture Notes in Networks and Systems 1148,
https://doi.org/10.1007/978-981-97-8457-8_31

1 Introduction

The experience of the digital age involves a multitude of changes that not only changed the mode of communication, purchase, entertainment, and even society, but also everyday life [1]. Undoubtedly, new technologies are also related to the emergence of novel difficulties, and one of the most critical issues here is cybersecurity. With the world turning into a more digital space, so does the level of technology dependence, and therefore, the hacking and other kinds of cyber-attacks also get more complex and frequent. Besides individual hackers and country-sponsored cyber campaigners, the scheme of digital security has been changing incessantly, being a significant problem both for individuals or companies and governments worldwide [2].

Though conventional security measures can to a certain extent block the threat, they indeed generally turn out to be quite inefficient in the late phases of the attack. It is just all that firewall system, antivirus software, and intrusion detection systems can do in such an environment where attacker are innovating and adapting in every part of their work by the day [3]. The conventional ways have encountered their limitation, and there has been a growth of the fan on searching the new efforts for the solution to the cybersecurity problems. Artificial intelligence, often labeled as AI purpose of which is to create machines that can perform several human tasks, has attracted attention alongside with other meaningful solutions in recent years [4]. Artificial Intelligence, which is prone to learning things from experience and making decisions in accordance with continuous algorithm applications, holds promise to be widely applicable in cyberspace. Applications of Machine Learning (ML) and Deep Learning (DL)—two important AI technologies—have demonstrated impressive results in solving these problems. Unprecedented capabilities, nowadays AI algorithms making it possible to analyze a huge volume of transmitted data, are able to detect regularities and deviations that signal bad intention. AI further improves cybersecurity swiftness and efficiency as it is used by professionals to identify and remove threats. In Fig. 1, it has been indicated the novel finding of the recent research which is combining the application of judicious Artificial Intelligence (AI) to bring a solution for Cybersecurity problem by tracing a number of the world Scopus indexed articles published recently all over the world [5].

Meanwhile, AI in cybersecurity puts a strong emphasis on proactive threat detection, that is to say it enables early identification of potential threats. Rather than the conventional approaches that usually target known signatures or known malicious patterns, AIbased systems using real-small datasets to find out abnormality that may be a sign of a new or completely unfair threats. This preventive technique enables organizations to keep a step forward against hackers or cyber criminals and minimize cyber-attack damages that could occur. Moreover, the AI is able to envisage revolutionizing the way incident response is performed in a cybersecurity setting. Around a cyber-attack, each moment counts and the capability to respond rapidly and adequately through this momentous problem is the same thing as a glitchy operation or a catastrophe. AI systems can automate a lot of tasks in the incident response from

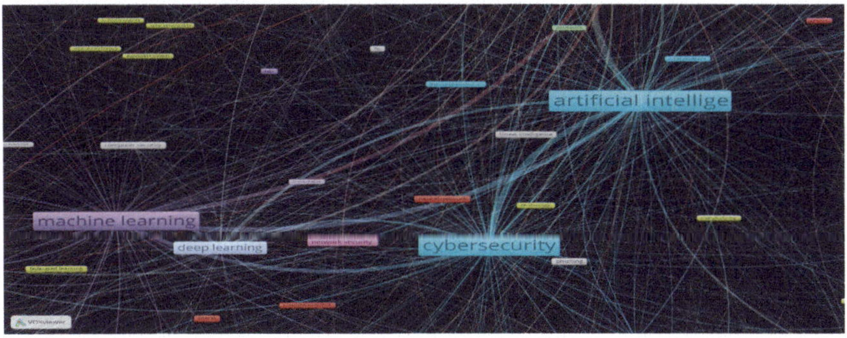

Fig. 1 Showing the recent research articles combining artificial intelligence and cybersecurity

identifying the threat and containing its spread to invoking the right personnel and initiating recovery process. Such action endpoint puts lesser work on human security group by helping them to identify and respond to the incidents throughout the organization faster and in a more coordinated manner [6]. Regardless, AI deployment in the cybersecurity field is facing obstacles. Along with this comes the ethical dilemma as to whether it is ethical and fair to use AI in ways that impact an individual's right to confidentiality and protection of their data. AI systems empowered with analytical capability allowing to amass, refine, and store great quantities of the sensitive information, which magnifies the issue of how such data is being utilized and kept safe. Furthermore, there is also a risk that AI systems could themselves become objects of cyber-attacks which can either be to steal sensitive data or to mess with the system, including to do the same harmfully. Even though these issues are in place, AI technology in cybersecurity, irrespective of the outcome, is a feasible option. The real time application of AI-Powered cybersecurity solutions has already shown their efficacy in identifying possible threats and eliminating them, which results in the lowering of risks of malicious acts including data breaches and other cyber incidents. However, it goes without saying that you should be prepared for these issues while keeping a look ahead and the role of AI in cybersecurity is absolutely expected to increase as organizations constantly come up with ways of ensuring their digital assets are safe from the latest and complex threats [7].

The relationship between AI and cybersecurity seems to be a winning platform to fight cyber threats that will be with us for more years to come. The problems persist, but the opportunity that AI brings to cybersecurity is immense. AI can be effectively used by organizations to increase the cybersecurity power as compared to humans of detection, response, and mitigation which is the essential feature to make the online environment safer and secure for all users.

2 Literature Review

Many cybersecurity researchers are seeking ways of incorporating Artificial Intelligence (AI) as a way of enhancing cybersecurity and reducing cyber-attacks and as discussed under, some of these research works are discussed below in relation to this topic.

(a). The paper of Shah in [8] concentrated on machine learning algorithms applicability for early identification and elimination of insider threats in organizational networks to achieve great effect focusing on the real-world scenarios.

(b). Saxe and Berlin et al. in [9] proposed about Deep learning a malware detection using the example of convolutional neural networks (CNNs) to get 100% accuracy and high precision and recall rates.

(c). Rashid et al. [10] recently conducted a meta-analysis on the efficacy of AI-based intrusion detection systems (CIDS) across heterogeneous networks. It concluded these systems to be effective in detecting threats and curbing false positive rate.

(d). Humayun et al. [11] brought the focus on the ethics of AI in cybersecurity closely in line with the issue of transparency, accountability, and user consent setting the necessary conditions for the proper use of AI-powered security tools.

(e). By means of assessing the various kinds of AI-based anomaly detection algorithms applied for network anomalies detection by Huong et al. [12] found that actually ensemble methods are better than the old-fashioned statistical techniques.

(f). Wu et al. [13] research work used reinforcement learning algorithm in response to the security policy management optimization application, which brought up the opportunity of the AI to dynamically adapt security measures and reflecting the ongoing threats.

(g). Combining a Literature review, Sindiramutty [14] shows, how AI-based threat intelligence platforms have done this in recent years. It is about the drawing of huge amounts of security data to provide actionable information for cybersecurity professionals.

(h). Rich [15] performed a longitudinal study with the aim to examine the specificity of AI-driven deception in the cyberspace as a means for adversaries' detection and deterring, emphasizing the importance of proactive cyber defense to thwart attacks.

(i). Based on Fadi et al. [16] review of existing trends, the authors addressed the utility of AI and block-chain technologies for cybersecurity improvement, among other things, explaining the idea of using the abovementioned technologies to secure decentralized systems against various cyber threats.

3 Challenges of Today's Cybersecurity

As the awareness of cyber threats rises and the edge in cybersecurity increases, the ability of organizations to appropriately deal with their assets seems to be a matter of question. Our modern-day connected world could not have been more venerable; cyber-attacks despite the gravity of the resulting outcomes, have the direction of their sights against businesses, governments, and individuals on the internet. Although cybercrime fight is in progress, overcoming all challenges to the digital world is a very impossible job as it is very complex and dynamic nature than the conventional world. The cyber arena is immersed within itself, extending cyber capacity to individuals and groups from all levels of society, from lone individuals to highly prepared state-sponsored groups, with differing goals and opportunities respectively. The situation or scenario whereby non-state actors, private organizations, and government agencies are perpetually engrossed in an ever-lasting hunt for cyber assets, even further seeking to find vulnerabilities in other systems, makes the struggle to protect their assets from cyber threats even more challenging [17]. This could lead to a myriad of serious sources pulling the strings. The sources might be looking for political aims, financial gains or might even be seeking military implementation. The invention of the digital age has melted the divisions between physical space and virtual space as cyber-space threatens the traditional protection of people from such measures. Nevertheless, there exist undetermined facets within this entropy that enable some order within cyber threats and [which can be counteracted]. The cyber kill chain is the method of looking at cyber-attacks which usually consists of such stages: information collection through surveillance, borne of the enemy, and allowing command and control [18]. It has a basis where each type of intrusion obeys an identical pattern, which allows competitors to address detection and targeted intervention points during the phases. Eavesdropping analysis helps in gratifying the attack phases of the attacker. By knowing their strategies, organizations can understand better how to defend themselves. They should focus on identifying the vulnerabilities and disrupting the attacker's tactics while the attack is at various phases of the kill chain. It translates to recognizing the critical role of early warning systems, and rapid response measures in cybersecurity, stressing the pressing need for continuous monitoring and threat intelligence that will enable us to be in front of the evolving possibilities. Confronting the discrepancies and finding a way out is the interdependent security approach (ISA), where both strategic and technical domains are merged to provide an advanced ground for cyber defense. To begin with, the prime task of the ISA is to provide quick response upon such alerts or alarms with hopes to initiate the early stage (in case there is an ongoing attack) and let the ISA members deter the attack. Through gathering and classifying the real-time threat data, a firm has proactive capabilities of spotting the nascent threats and introducing the right mitigation measures to avoid the instances of attacks. Thus, the modern approach of being proactive in cybersecurity is crucial in the existing fast-progressing threat area, where static ways of counteractions are not often enough to cope with the speed and complexity of modern cyber threats (Fig. 2).

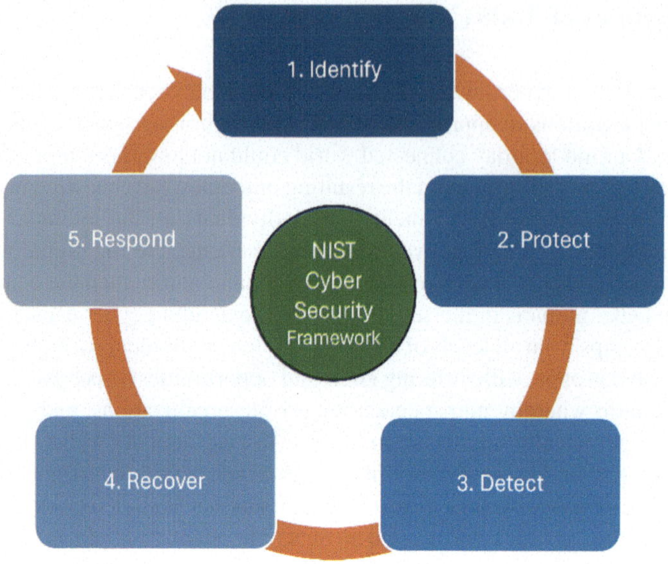

Fig. 2 NIST cybersecurity framework [19]

The ISA framework encompasses four key phases: recognizing, reacting, responding, and recovering which are all associated with the cyber kill chain and referred to as D3R stages. Discovering means to find and inspect threats and determine the best appropriate plan, while reaction/response, means implementing antivirus/countermeasures to lessen the effect of an attack and restore the system functionality. Restoration is just as relevant, being expressed through the process of system/data revival or at least they are brought back to what they were before the attack occurred, with the intention of minimizing negative effects and company downtime. Combined, these phases structure a reliable and iterative cybersecurity approach, which emasculates this art of continuous flashpoint monitoring, analysis, and adjustment. Nevertheless, these organizations and cybersecurity professionals make strenuous efforts of protecting the cyber domain, and in spite of the best of these attempts, the subject of cyber defense still remains a strenuous issue. Cyberspace entail complex and pronounced strategic and technological problems, such as data acquisition, technological issues and the inability of management to regulate processes ethically. Finding the ultimate cure for cybersecurity calls for the adoption of a comprehensive strategy that handles all the issues coherently, a strategy which may include tech innovations, strong policy-making, training programs, as well as sophisticated procedures. It is through this flow of strategies and the integration of a progressive way to cybersecurity that organizations are afforded the chance of safe-keeping their digital properties as well as dealing with risks found across their interconnected world.

4 Challenges in Gathering Cyber Intelligence

Dissecting cybersecurity will show generally a huge level of complexity, as it involves processes like data acquisition, processing, analysis, and exploitation. A vital role in a cybersecurity system is played by the Integrated Security Approach (ISA) which is a cornerstone of the whole system. This method implies aggregation and analysis of various information that contributes to the sophistication of cyber intelligence. Nevertheless, the availability of data tracks and discrepancies in data processing phases are two major stumbling blocks regarding the obtaining and analyzing of relevant data. The digital era brings in whopping volumes of data which organizations have to deal with increasing significantly in value with numbers multiplied at almost all resource levels of a company. Beyond, the diversity of data sources and the incremental of decentralization that takes place on organizational and national dimensions impede the revelation and collection of data at the required level. Although the existence of data heterogeneity necessary for confirmation is recognized, the cyber environment that is in a state of eternal change poses a continuous adaptation to the ever-changing network topologies and behaviors as a priority issue.

On the one hand, a space is created for these big data; however, many challenges arise in the storage and processing context, which are critical first requirements for actual data analysis. However, technical application was faced with a number of obstacles and IDPS system implementation became the key to success in cybersecurity protection. IDPSs regardless of being either software or hardware-based perform a very important task of guarding the systems, being it an individual one or an entire network [20]. These systems are part of misuse detection mechanisms that work on principles of pattern misidentification, which seeks to detect malicious activities by deriving patterns of abnormal behavior, and anomaly detection methods that are based on pattern normal identification. Whereas the dynamic nature of cyber-space implies constant cybercrime revolutions this in turn affects the application of IDPS. New threats are always present and are continuously shaped by the attackers that adopt such behaviors to weaken existing safety protocols.

Not only do flaws in traditional security systems add up to worsen existing problems, but they also contribute to them. IDPS lacks enough detection rates, slow throughput, scaling issue, and high availability/building for automation. The inadequacy of defining the deviant or normal behavior patterns can be a demerit in the system and so, there is a necessity for the fast and accurate identification of the threats. Hence, in determining the threats, the rates of false negative and positive rates can be extremely high and low. On the other hand, the processor and analyzer are almost incapable of processing and analyzing large volumes of data per second, and hence, when any changes in the environment are made, it becomes more complex and resourceful.

To realize their solutions and safety schemes, organizations should opt for multiple security tools or seek intelligence of security from third parties. Despite all this, there are other concerns that are not related to technological measures alone. Other procedures that are inclined to be implemented such as creating and entrenching international as well as intraorganizational protocols are essential to promote and develop an organizational integrated Security Approach for cybersecurity maturity. Besides, by making collaboration and partnerships to exit cybersecurity ecosystem they can boost information exchange and resource sharing among different cybersecurity actors in the cybersecurity landscape [21].

Moreover, the private sector entities applying cybersecurity know-how across different sectors adopt different approaches to the cyber threats in general. The sectors of critical infrastructures put emphasis on safety and security enhancements as a higher priority than the private organizations who consider their own financial position as the priority matter. These complications confirm that this constitutes the main problem of formulating a complete security strategy which would allow technological solutions to go hand in hand with organizational procedures and partnerships.

With an understanding of these issues, organizations need to develop a holistic cybersecurity strategy—one that is built on the technology together with processes, collaborations, and so on. This strategy implies spending money on both technology and personnel, along with a readiness to promote cybersecurity culture and assignment of tasks accordingly all through an organization. Alternately, information security organizations ought to keep on their toes and ensure that they are up to date with the modern-day threats. This entails assessing the current cybersecurity measures ensuring they are not outdated.

At the end of the day, cybersecurity that works entails a stricter preventive approach implemented together with putting early Defense team's response in place. Through the grappling with the difficulties that are associated both with intelligence gathering and cybersecurity solutions implementation, enterprises will be able to improve their resilience to cyber risks and make their digital assets safer in the world that is interconnected more and more.

5 AI in Depth

AI a part of soft computing techniques is one of the oldest in computer science and thus wants to create systems that compete with humans in terms of intellectual capabilities [22]. AI, which started off as a vision, has continued to evolve, having surpassed the expectations of most people and exhibiting extraordinary skills in different domains [23]. An example of this is competitions between human grandmasters in the game of chess; once considered as something that tests how humans are smarter than machines, machines are taking over the reign and become unbeatable with the power of improved computing and advanced search algorithms. This success of AI in addressing complex problems in the short run can be attributed to well-built datasets that have all the relevant domain data pupils need. That is also why, discussions about the future of NLP have already begun in the 1960s when computational linguistics was a direction of research. Among specific NLP programs include those of Google's AI linguistic, where the understanding of human language and activities has just been an initial step to the ongoing evolution of AI.

AI can be considered in this context as a vehicle for enhancing human intellect, providing the alternative of sophisticated problem-solving which requires high levels of intelligence or wisdom to accomplish. In this regard, AI algorithms can be introduced into shaping cybersecurity directions. AI leverage makes cybersecurity processes in the business environment more intelligent and abide by emerging threats more effectively. In this article, we will walk through the role AI plays in combating cybersecurity threats, as we pay special attention to the most exciting innovation within the sector and discuss possible methods of utilizing AI in cybersecurity.

Artificial intelligence (AI) is the most recent, promising, and probably the most powerful technology that can make a significant contribution to cyber defense capabilities of our society, and eliminate or at least lessen the consequences of cyber-attacks. It is not true that this unique undertaking does not have its difficulties as well. Among the major difficulties is the issue of collecting cyber intelligence regarding the target, which includes gathering the data, assessing, synthesizing, and seizing the opportunity to employ the intelligence productively. A considerable challenge known as the scale of data that is being created in the digital environments will be treated by organizations applying techniques of collection and providing the information with the quality and storage. Moreover, the data source's heterogeneity and the dynamics of cyberspace make the task not an easy one as the system needs to collect the needed intelligence.

AI-based solutions are also rather technology-dependent, so they can either stumble upon various technological limitations or provide a poor performance in general. Traditional comfort systems may start to be left out of the race for dealing with the quickly changing danger scene which results in low detection rates, slow throughput, and a lack of scalability and resilience. Making one of these challenges work out needs to create advanced AI algorithms and technologies that can be able to change correspondingly to the environments and ensure the real-time information as well as fast reaction to the threats. In an attempt to overcome these hurdles, there emerge several steps organizations should take with regard to cybersecurity that encompass a set of technologies, processes, and cooperation within the cybersecurity ecosystem. The utilization of AI implies increasing the speed in detecting, preventing, and responding to real-time cyber threats—that enables most cyber-attacks before they can cause great harm. Besides the building of partnerships and information sharing within the cybersecurity community all the sharing of expertise and flow of knowledge will create stronger defense zones from cyber threats [24].

In response to these challenges, organizations must adopt a holistic approach to cybersecurity that encompasses technological innovations, process improvements, and collaboration within the cybersecurity ecosystem. By leveraging AI techniques, organizations can enhance their ability to detect, prevent, and respond to cyber threats in real time, thereby reducing the likelihood and impact of cyber-attacks. Additionally, fostering partnerships and information sharing within the cybersecurity community can facilitate the exchange of knowledge and best practices, further strengthening defenses against cyber threats.

In a nutshell, the joining of AI and cybersecurity affords entirely new prospects for intensifying performance in security systems and protecting information holdings. Through the use of AI, organizations can move a great step forward; they will obtain cyber-threat information, advanced antivirus systems, and overall cyber resilience. Nonetheless, realizing the brim-full potential of AI in cybersecurity brings to the fore the crucial issues of data acquisition, technological concerns, and the necessity of working in collaboration among the cybersecurity community. As AI implementation can be thought of jointly and employed in a clever way by organizations, security risks that are cyber-related can be effectively tackled and the digital infrastructure one is using can be protected in line with the fast-developing inter-connected world.

6 The Integral Role of Artificial Intelligence in Fortifying Cybersecurity Measures

6.1 Is Artificial Intelligence the Definitive Future of Cybersecurity?

Virtual predators who are not restricted by geography and time are prevalent in all parts of cyberspace and strike without any warning. A major attraction of AI is its ability to automate tasks, thereby eliminating unnecessary delays and erroneous entries. Integrating advanced technologies for the extraction of formatted data, along with ultra-deep learning capabilities, from all types of data (regardless of structure, including numbers, text, and speech) makes AI a mighty partner in the war against cyber-crimes. The amazing capabilities of AI in recognizing behavioral changes, like odd login activities or misuse of passwords, is a powerful weapon in regrettable events. To start with, it has to be noted that despite the fact AI does show tendencies toward innovation, it also revolts against humans whose main goal is cyberspace seizure. The art of defending the cyber systems from the ever-growing sophistication of the cyber adversaries requires developers to continually think competitive on the battlefield of AI to devise new strategies which will make them stay ahead in the cybersecurity chess game.

Google, the most recent and one that shows visible progress in AI-based cybersecurity, is its Neural Structured Learning (NSL) which was released in its Tensor Flow machine learning platform. NSL offers training frameworks based on the innovative Neural Graph Learning technique that can be applied to typical datasets and data structures. The training systems are equally efficient for both the professionals with experience in the field of machine learning and the beginners in the subject. The ease-by-which and its broad-range-applicability in improving comprehensive cybersecurity-measures are showcased further by its applications in machine vision-models and NLP [8].

AI components can function in all areas of the traditional cybersecurity domain, making these capacities visible in voice assistants, search queries, and face recognition apps. Nonetheless, AI constitutes an enormous revolution and helps keep fraud detection and prevention in the financial sectors. On the other hand, the application of AI in information security creates a scenario where benefits and hazards present themselves simultaneously. This, in essence, enhances AI technology in digital management strategies and helps businesses and customers to protect; at the same time, AI implementation becomes the new source of concern. It could potentially serve as a double-edged sword, empowering cybercriminals to execute more sophisticated attacks while equipping security practitioners with advanced tools to counter such threats.

6.2 Perspective of AI Executives on AI Integration in Information Security

A survey performed by Capgemini Research Institute shows that Artificial Intelligence (AI) is fast transforming cyber defense strategies from a defensive approach to active adversary hunting. Panelists from both the data security leaders and IT professional demographics across various nations affirm that the application of AI-based solutions is a critical element for addressing cyber threats. The AI effectiveness consensus that comes out of the results conveys the familiarity of AI in boosting the speed during the threat response, growing the precision and competence of the cyber analysts, and the positives that AI can bring to the already established cybersecurity frameworks. Networks are becoming complex, and AI stands out as the primary aspect that will guarantee the enhancement of the security shield in the enterprise against increasingly sophisticated cyber warfare [25].

6.3 Strategic Integration of AI Technology into Cyber Defense Frameworks

AI technology is fully compatible with the pre-existing information defense network, but this integration process is made easier if executed through careful planning, training, and preparation. AI provides from biometric intrusion to better analyzing of risks and predictive analytics and to more intelligent threat interpretation via natural language processing. Nevertheless, the AI-related solutions should be proficient of the information security professionals in using the technologies very gracefully. Recently big cybersecurity players like Palo Alto Networks and CrowdStrike are among the major agents who have guessed the potentiality of sectors and now they are coming up with the state of art capabilities in their cybersecurity offerings. Here also some challenges do exist such as initial investment and permanent training needs [26].

6.4 Addressing Vulnerabilities and Risks Associated with AI Cybersecurity Tools

The implementation of AI for information defense will bring new threats especially about the security of AI-based technologies, because the toughness of malefactors will increase by this new trend. While AI tools are effortlessly capable of detecting and handling malware threats, cyber attackers can exploit and plot their greater and more sophisticated attacks in similar AI technologies. Accelerated by AI and machine learning technology, the number of threats involving cybercriminals multiply by magnitudes, strengthening capabilities to create sophisticated weaponized services and applications. These conditions of confluence, in fact, increase the infrastructure vulnerability to aggressive AI-attacks, so AI-attacks affecting fragile cybersecurity are inevitable. Prevention of emerging threats becomes the main task [27].

6.5 The Menace of Adversarial AI and Its Ramifications for Cybersecurity

The adversarial AI has become a real challenge to computer security, as it is actively used to cheat or control AI algorithms in nefarious ways. The intention of these alterations may be to teal the real data inputs or unscrupulously mischaracterize them, which brings about a lack of dependability and hence inadequate security performance of AI-empowered security technologies. The cybersecurity community including researchers and major shareholders face imminent dangers of adversarial AI so they work tirelessly to evolve their systems and develop countermeasures to deal with these threats. Efforts such as hold trim auto by IBM will shoulder to the fore, meaning that research and interaction must go hand in glove in the fight against adversary attacks [28].

To sum up, with AI comes an enviable capacity to supercharge cybersecurity capability, which makes this new aspect of defense technology should be carefully managed around the potential risks and vulnerabilities that may emerge. Implementing durable security mechanisms, adopting a collaborative model to include the input of all key players in the industry, and taking into account AI resilience parameters allow businesses to benefit from the AI revolution without succumbing to emerging cyber challenges.

7 Expert Systems to Provide Decision Support for Security Professionals

AI-based expert systems are the computer firewall that provides solutions and decision support in the problems, which may be complex and within a specific domain. They consist of two main components: an intelligent system, with the ability to store new knowledge, retrieve the information from the learned expert knowledge, and use the information for reasoning and problem-solving. We may distinguish several kinds of expert systems using different thinking concepts as the basis, including differentiation of several kinds of expert systems based on differences in reasonable thinking methodologies. Each method has its own unique strategy; for instance, case-based reasoning (CBR) method entails solving problems through remembering replays of former cases and modifying their resolution to fit the new situation. The CBR systems can do this in a way that progress is through constant evaluation and revision of previously proposed manifestations in order to enhance their accuracy and capacity to solve unknown problems. Unlike the expert systems (XS), which utilize predefined rules made by experts, and each rule is comprised of the condition and the action, the RBS approaches (RBS) use only facts. RBS is an algorithmic approach that evaluates the problem stepwise and then applies an appropriate solution, but unlike the CBR systems, RBS cannot store and modify rules or generate them as they could be outdated and pose a challenge in adaptation to a dynamic environment [29].

Security experts in cybersecurity domains specifically employ expert systems to help them make sound decisions in auditing data which can help emerge attack activities in networks and systems. The team of experts being the ones in charge, unable to read the data manually, most often perform the analysis using statistical reports that process and retrieve the audited information immensely fast. IA expert systems have pledged to manage cyber environments in real time, even for a large or multiformat system architecture. These types of systems generate alert messages that contain necessary information. Subsequently, security specialists take prompt action based on the detection of malicious activity such that they can select appropriate remedial measures.

It has to be underlined that despite the fact that the expert system relieves decision-makers, it does deprive them of a full authority. Continuing to rely on human expertise and judgment will always be required in the concept of making sense of data and reaching correct conclusions in the difficult and changing cybersecurity environment.

Other reasoning models apart from CBR and RBS such as neural networks and fuzzy logic enable expanding the scope of enriching expert systems through the complexity of the problems the programs are handling. Neural networks which are structure like the human brain and operate on the logic of pattern recognition and data learning are utilized in the processes of machine intelligence. While on the other hand fuzzy logic makes it possible to reason with data that are not exact or precise therefore the expert systems that utilize them are more versatile than those which operate on exact rules that may not be easy to define (Fig. 3).

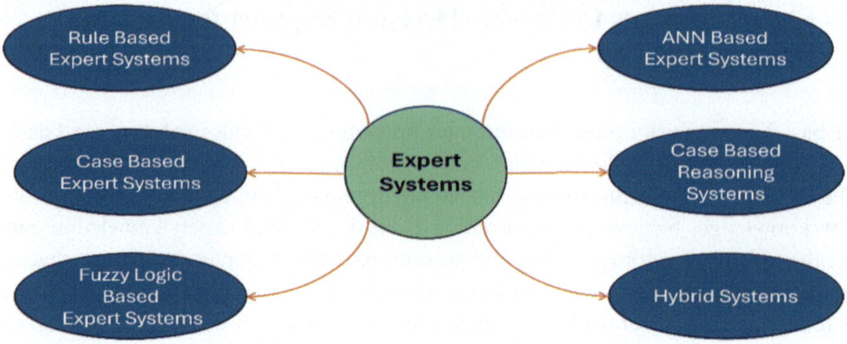

Fig. 3 AI-based expert system for detecting cybersecurity breaches [30]

Coming with the expert systems functionalities which involve machine learning and natural language processing as well portends another promising way of advancing the decision-making capacity in cybersecurity. These integrated systems can procure larger swaths of data and run it through their complex algorithms to provide predictions that are more precise and faster in detecting potential security threats. Hence, the organizations are able to anticipate attacks and defend themselves proactively against cyber-attacks. Finally, to sum all up, expert systems hold a very important place in cybersecurity by offering assistance in the process of forensic audit data processing, malicious activity detection, or appropriate remedial security decisions. Though they do improve the accuracy and completeness of the security tasks, there is plenty of place for human monitoring, especially in providing results interpretation and decision-making. The advanced AI nowadays has not only integrated with the expert systems but they can also work with other AI technologies to fortify cybersecurity defenses and treat emerging security threats.

8 Drawbacks of Artificial Intelligence Within Cybersecurity

Adoption of AI within cybersecurity is highly desirable with it implemented to address these significant technological problems. However, several drawbacks and risks accompany the use of AI in this domain. However, several drawbacks and risks accompany the use of AI in this domain:

According to one of the first main points, the AI-enabled security systems are able to work without the presence of a human being is on the current stage of their development. Regardless of all the progress, some decisions to be made and oversight for humans attributes also to the need to stay highly involved in cybersecurity operations.

Specifically, besides the tracker algorithms like Artificial Neuron Networks (ANNs) and Deep Neural Networks (DNNs), with the more data privacy issues for especially keeping data highly confidential [31]. The confidentiality of personal data could be at risk when immense amounts of it are being investigated and enquiry of both private and public sectors arises. The areas that really need to be clarified are the provisions meant for the use of personal data, and the reason why and to whom the data is divulged. This only goes to add to the citizens' impression of AI-driven solutions to be somehow un-trustworthy.

In addition, poor regulation of AI as well as the all-embracing regulation is one of the most important challenges that stands in the way of AI cybersecurity deployment. Current legal systems might have loopholes that fail to cover AI-driven systems which are unconventional and everchanging. Here arises the debate about the power of AI over mankind and the unwanted results.

Moreover, ethics become key in systems based on AI, when their critical choices are made in place of someone or when they serve as aids in making individual decisions. With moral codes being absent in these formal systems, we are prone to raise the questions about the ability of the decisions to fit into human values and preferences, which underlines the necessity of ethical guidelines and principles in the development of AI algorithms.

In the context of future AI and cybersecurity progress, we need to draw a contrast between the short-term problems we should address and the long-term priorities we should focus on. Although current AI practices are efficient in tackling the urgent cybersecurity challenges nowadays, it is likely that future advancements can lead to a breakthrough of human cognition where new models arise in data processing and decision-making. An example of such a task is network-centric warfare where the technology area is very specialized; income management systems necessary for a quick decision-making process should be automated.

Nevertheless, it does not follow that, in the foreseeable future, Narrow AI may be sufficient for every kind of endeavor, since research works to create General Artificial Intelligence (AGI) are still ongoing. Everyday life will be redefined, as the Singularity defines the age of intelligence exceeding human performances. Although we can predict both positive and negative impacts, the question is how they will be distributed among humans and machines and what society should do to handle those changes. It is safe to say that, apart from the AI route, other technological ways will contribute to the realization of an intelligent mind.

AI provides extensive support in the developments of cybersecurity, but in the similar way, addressing the issues and risk related is also crucial. Establishing the ethical principles, concerns about privacy and regulatory frameworks requires to be balanced with the technological advancements that will expand the opportunities that AI brings to cybersecurity and at the same time minimize the pitfalls that might be created.

9 Conclusion

In short, the application of AI to cybersecurity generates a mixture of opportunities and challenges for the purpose of its defense supervision systems and data. It is clear to see that AI-based security solutions are superior in many ways, like rapid threat detection, monitoring in real time and making decisions on-demand. This, in turn, helps in fighting the ever-evolving nature of cyber threats. On the other side, along with the wonderful effects, we should be tightly observant of other disadvantages and risks at least paying the utmost attention to them. The other significant problem is that AI-powered data security systems are not capable of running by themselves. They require active human oversight as well as from top managers. Artificial intelligence has made tremendous progress but its facility for supplanting human judgment generally is still far from being fully available. This testifies to the necessity of a human touch in the cybersecurity management, which is crucial for operations to be effective and aligned with the dynamic nature of new threats. Additionally, the use of AI techniques in her acquisition of data is privacy-sensitive such as Artificial Neural Networks (ANNs) or Deep Neural Networks (DNNs) increase the risk of privacy issues. By analyzing a tremendous data, an AI-based approach may put at risk the right of the private information, hence there is a need for accountability and transparency in the research. Wrongdoings of these individuals should be investigated and penalized through legal mechanisms to earn the trust of citizens as well as organizations.

Moreover, the inadequate formulation of regulations has also become a substantial obstacle in implementing AI into cybersecurity. Now, the existing legal approaches cannot foresee what is unnatural and predictable for AI, so it is necessary for regulation authorities to set new rules and guidelines that can control this capability ethically. Additionally, the ethical issues are not the sole preserver of these AI system-based security systems increasingly deciding on behalf of people or are useful to decision-making processes. The relative absence of a multicultural ethics within these systems will give rise to certain doubts regarding the issue about the agreement between our values and decisions regarding the development and deployment of AI. This is why it is important to have ethical rules and guidance about AI development.

Considering the future ahead, it is important to differentiate between the tasks that must be done now and tasks that can be dealt with more confidently when time becomes available. At present, AI technologies raise a fast response to urgent cybersecurity problems. Nevertheless, as future AI approaches are bulking up it is quite possible that they will just evolve into whole new concepts of information processing and decision-makings. Balancing AI technical development with the ethical concerns, privacy issues, and regulatory frameworks that provides us the chance to fully explore AI potential for cyber world security while limiting the possible negative sides.

References

1. Brescia R (2020) The future of change: How technology shapes social revolutions. Cornell University Press
2. Shad MR (2019) Cyber threat landscape and readiness challenge of Pakistan. Strat Stud 39(1):1–19
3. George AS, George AH, Baskar T (2023) Digitally immune systems: building robust defences in the age of cyber threats. Partners Univ Int Innov J 1(4):155–172
4. Haleem A, Javaid M, Qadri MA et al (2022) Artificial intelligence (AI) applications for marketing: a literature-based study. Int J Intell Netw 3:119–132
5. Salami Pargoo N, Ilbeigi M (2023) A scoping review for cybersecurity in the construction industry. J Manag Eng 39(2):03122003
6. Van der Kleij R, Kleinhuis G, Young H (2017) Computer security incident response team effectiveness: a needs assessment. Front Psychol 8:315125
7. Jaffe AM (2021) Energy's digital future: harnessing innovation for American resilience and national security. Columbia University Press
8. Shah V (2021) Machine learning algorithms for cybersecurity: detecting and preventing threats. Revista Espanola de Documentacion Cientifica 15(4):42–66
9. Saxe J, Berlin K (2017) eXpose: a character-level convolutional neural network with embeddings for detecting malicious URLs, file paths and registry keys. arXiv preprint arXiv:1702.08568
10. Rashid AB, Kausik AK, Al Hassan Sunny A, Bappy MH (2023) Artificial intelligence in the military: an overview of the capabilities, applications, and challenges. Int J Intell Syst
11. Humayun M, Tariq N, Alfayad M, Zakwan M, Alwakid G, Assiri M (2024) Securing the internet of things in artificial intelligence era: a comprehensive survey. IEEE Access
12. Huong TT, Bac TP, Ha KN, Hoang NV, Hoang NX, Hung NT, Tran KP (2022) Federated learning-based explainable anomaly detection for industrial control systems. IEEE Access 10:53854–53872
13. Wu Y, Ge J, Li T (2022) AI and machine learning for network and security management. John Wiley & Sons
14. Sindiramutty SR (2023) Autonomous threat hunting: a future paradigm for AI-driven threat intelligence. arXiv preprint arXiv:2401.00286
15. Rich MS (2023) Cyberpsychology: a longitudinal analysis of cyber adversarial tactics and techniques. Analytics 2(3):618–655
16. Fadi O, Karim Z, Mohammed B (2022) A survey on blockchain and artificial intelligence technologies for enhancing security and privacy in smart environments. IEEE Access 10:93168–93186
17. Rose J, Spencer C (2020) Immaterial labour in spaces of leisure: producing biopolitical subjectivities through Facebook. In: Re-thinking leisure in a digital age. Routledge, pp 101–118
18. Bahrami PN, Dehghantanha A, Dargahi T, Parizi RM, Choo KKR, Javadi HH (2019) Cyber kill chain-based taxonomy of advanced persistent threat actors: analogy of tactics, techniques, and procedures. J Inf Process Syst 15(4):865–889
19. Staves A, Gouglidis A, Hutchison D (2023) An analysis of adversary-centric security testing within information and operational technology environments. Digit Threats Res Pract 4(1):1–29
20. Wahyu AP, Fauziah K, Nahrowi AS, Faiz M N, Muhammad AW (2023) Strengthening network security: evaluation of intrusion detection and prevention systems tools in networking systems. Int J Adv Comput Sci Appl 14(9)
21. Weersink A, Fraser E, Pannell D, Duncan E, Rotz S (2018) Opportunities and challenges for big data in agricultural and environmental analysis. Ann Rev Resour Econ 10:19–37
22. Roy S, Bhattacharjee K, Bhattacharya A (2017) A modern approach to solve of economic load dispatch using group leader optimization technique. Int J Energy Optim Eng (IJEOE) 6(1):66–85

23. Kumhar R, Roy S, Kumar U, Sarkar A, Mahato TK (2023) Enhancing fault identification in modern day transmission lines through neural network approach. In: 2023 7th International Conference on Electronics, Materials Engineering & Nano-Technology (IEMENTech). IEEE, pp 1–6
24. Manoharan A, Sarker M (2023) Revolutionizing cybersecurity: unleashing the power of artificial intelligence and machine learning for next-generation threat detection. https://doi.org/10.56726/IRJMETS32644
25. Neupane S, Fernandez IA, Mittal S, Rahimi S (2023) Impacts and risk of generative AI technology on cyber defense. arXiv preprint arXiv:2306.13033
26. Dulloo R, Suryawanshi V, Mokal DK, Hagare SR, AI in cyber security: decision support and smart methods
27. Sontan AD, Samuel SV (2024) The intersection of artificial intelligence and cybersecurity: challenges and opportunities. World J Adv Res Rev 21(2):1720–1736
28. Rauniyar A, Hagos DH, Jha D, Håkegård JE, Bagci U, Rawat DB, Vlassov V (2023) Federated learning for medical applications: a taxonomy, current trends, challenges, and future research directions. IEEE Internet Things J
29. Soravito G (2023). Artificial intelligence for risk management: a systematic review (Doctoral dissertation, Politecnico di Torino)
30. Leo Kumar SP (2019) Knowledge-based expert system in manufacturing planning: state-of-the-art review. Int J Prod Res 57(15–16):4766–4790
31. Abiodun OI, Jantan A, Omolara AE, Dada KV, Umar AM, Linus OU, Kiru MU (2019) Comprehensive review of artificial neural network applications to pattern recognition. IEEE access 7:158820–158846